ELEMENTARY
Statistics

ELEMENTARY
Statistics

Henry R. Gibson

State University of New York
Fashion Institute of Technology

WCB **Wm. C. Brown Publishers**
Dubuque, Iowa · Melbourne, Australia · Oxford, England

Book Team

Editor *Paula-Christy Heighton*
Developmental Editor *Jane Parrigin*
Production Editor *Carol M. Besler*
Designer *Jeff Storm*
Visuals/Design Developmental Consultant *Donna Slade*

Wm. C. Brown Publishers
A Division of Wm. C. Brown Communications, Inc.

Vice President and General Manager *Beverly Kolz*
Vice President, Publisher *Earl McPeek*
Vice President, Director of Sales and Marketing *Virginia S. Moffat*
National Sales Manager *Douglas J. DiNardo*
Marketing Manager *Julie Joyce Keck*
Advertising Manager *Janelle Keeffer*
Director of Production *Colleen A. Yonda*
Publishing Services Manager *Karen J. Slaght*
Permissions/Records Manager *Connie Allendorf*

Wm. C. Brown Communications, Inc.

President and Chief Executive Officer *G. Franklin Lewis*
Corporate Senior Vice President, President of WCB Manufacturing *Roger Meyer*
Corporate Senior Vice President and Chief Financial Officer *Robert Chesterman*

The photo—''Boys play marbles''—© FPG International.
The background photo—marbles—© Superstock, Inc.

Copyedited by Pat Steele

A Times Mirror Company

Library of Congress Catalog Card Number: 93–70851

ISBN 0–697–10251–3

Printed in the United States of America by Wm. C. Brown Communications, Inc.,
2460 Kerper Boulevard, Dubuque, IA 52001

10 9 8 7 6 5 4 3 2 1

Contents

Preface

The first purpose of this book is to provide a *students' text* in elementary statistics (clear, concise, and interesting) for students with minimal preparation in mathematics; a fundamental knowledge of fractions, decimals, and percentages would be sufficient. The second purpose is to provide an *instructors' text* in elementary statistics, a text that is both informative (providing step-by-step instruction for students while offering insight into the historical origin of key concepts) and yet firmly rooted in real-world application, addressing important contemporary issues in sociology, psychology, marketing, finance, medicine, health, factory production, education, biology, anthropology, and so on.

Guidelines in Chapter Construction

1. A *historical framework,* along with the text's realistic examples and friendly writing style, will distinguish this book from most others. The origins of the key concepts are probed and chapters are sequenced in much the way statistics developed, with the discovery of one concept leading naturally to the discovery of another. Because of this historical sequencing, topics build logically on prior topics and students appreciate how and why one topic is necessary for an understanding of the next. Many instructors find that this historical framework creates a text that is more interesting to teach (for instance, one of the early applications of the normal curve was to the heights of Napoleon's soldiers; can you guess the average height of a soldier in Napoleon's army? Answer: $\mu \approx 5'0''$) and makes concepts more understandable to the students (for instance, why we use the standard and not the average deviation as a measure of spread or how the t score adjusts for the tendency of small-sample s's to underestimate σ).

2. *Concrete examples* are used to introduce theory, with complicating details footnoted, endnoted, or relegated to later sections. Paragraphs are written clearly, with symbolism and formulas minimized. All chapters were tested by several instructors. Chapters 1–8, for the most part, should be error free since they have already been class tested by ten instructors in more than 26 classes. Chapters 9–11 were also class tested in 6 classes prior to publication.

3. *Exercises:* Students are given answers and fully worked-out solutions to the first five exercises of each chapter and abbreviated answers to odd-numbered exercises thereafter. This gives the instructor the option to assign homework with (1) fully worked-out solutions, (2) abbreviated

answers, or (3) no answers, that is, giving students even-numbered exercises (from question 6 on). In effect, the instructor can adjust the assignments to meet the level of any class.

4. *Real-world applications* are stressed in both demonstration examples and homework exercises, with relevant source material referenced.

    Many topics for the exercises were actually chosen by students. For several semesters, my classes voted on the most interesting newspaper and magazine articles brought in by students, and the winning articles were used as a basis for many of the exercises. Three such winning articles (which are now exercises in chapters 9 and 10) concern the link between creativity and mental illness, the dangers of artificial insemination, and the marketplace theory of dating as it affects later marriage satisfaction.

    Real-world applications are not only presented in the exercises but are also explored in depth in the demonstration examples. For instance, the demonstration examples in chapters 7–11 include use of Elavil, a powerful sedating drug prescribed by psychiatrists for the treatment of depression; social IQ in children as a predictor of life achievement; high-school SAT models for predicting college freshmen GPAs; delusions found to be commonplace in college students; use of nonverbal clues by professionals in assessing character in job applicants; and personality traits, are they born in us or bred from the environment.

    Generally, earlier chapters present simpler situations, and as techniques grow more sophisticated, so do the real-world applications and decisions. In fact, many of the demonstration examples and exercises in the later chapters (especially 7 and 9–11) are based on actual research currently in progress at universities and other institutions across the nation.

5. *Ease for new adjunct instructors:* One of the biggest benefits of the text is its simplicity and ease of use for new adjunct instructors. Often an adjunct instructor teaching statistics for the first time has a difficult time pacing the course and picking suitable topics. Having encountered this problem on several occasions at my college, I wrote the text with this in mind. Chapters flow logically from one to the other with minimal excess.

    For a three-hour course, chapters 1–9 can be used as an overall objective, perhaps with some sections deleted (depending on specific course requirements) and other sections added from chapter 11 (Additional Topics). A four-hour course might include chapters 1–10.

Additional Features

Multiple-Choice Exams   Grading exams *fairly* is one of the most important aspects of a professor's job; unfortunately it is also one of the most time consuming. Another very important aspect is *broadening students' understanding* of the subject matter. To meet these goals, I have developed in my classes over several years a series of exams that allow both *objective* grading and testing of a *broader range* of knowledge.

To accomplish this, more questions per exam were necessary; however, this offers a student who knows the material an opportunity to achieve substantial partial credit even though for some reason their final answer may indeed be incorrect. Essentially, a mathematical setup is first posed, followed by four to seven questions concerning this setup, with each question testing a deeper knowledge. Questions are as independent as possible to avoid follow-through errors. Although these exams are multiple-choice, they can also be given as fill-ins or simply as mathematical setups to be solved.

One exam consists of about six mathematical setups for a total of 33 to 34 multiple-choice questions. My grading system for the semester is as follows:

Exam I:	33 points
Exam II:	33 points
Exam III:	34 points
	100 points

The final examination plus written research reports (explained in section 2.8) add an additional 100 points, for a total course accumulation of 200 points.

The exams are graded by machine (for instance, Scan Tron) in seconds, however, exams are reviewed for additional partial credit, especially with students who do not achieve an acceptable grade.

I do not allow students to keep exams (this avoids hand-down from semester to semester). A tally of the grades are posted on the chalkboard at the beginning of the class review. I quickly answer questions of A and B students (usually about 10 minutes), then exempt these A and B students from the remainder of the session, thus concentrating the bulk of the review where it is most needed, on C, D, and F students.

A properly constructed multiple-choice exam discourages rote memorization (since one must interpret and extract information, not merely do calculations) and rewards students who understand concepts (by allowing them to answer many of the questions even though their final answer may for some reason be incorrect). To sum up, I have found these exams to be objective in grading, allow insertion of questions that test a deeper knowledge, thus rewarding students who understand concepts rather than those who memorize computations. The exams are also relatively easy to implement and grade.

Written Assignments

Written assignments take on two primary forms in this text: first, as a natural part of exercises (students are frequently requested to give brief written explanations concerning the validity of the experiment or interpretation of the results) and, second, in the form of formal written research reports, which can be assigned as overall class projects or computer laboratory projects.

The first form of written assignment, as a natural part of the exercises, is somewhat self-explanatory in the exercises themselves. How to write a formal written research report, the second form, is explained in section 2.8. These reports are often used by corporations and research institutions and, although formats differ slightly, generally include the following information:

Background statement: Essentially this presents the problem, why the study was commissioned. Usually a brief history is presented leading to a statement of the need, purpose, and expectations of the study.

Design and procedures: This states the details of *how* the study was conducted; it also addresses the question of validity and calls attention to possible weaknesses in this area.

Results: The results of the study are summarized in this section in clear, concise, easy-to-read charts and graphs.

Analysis and discussion: Interesting facts and significant findings are pointed out and discussed, especially those findings related to the central purpose of the study that was initially discussed in the background statement.

Conclusions and recommendations: Based on these findings, conclusions are drawn and future action suggested; serious questions concerning validity are again mentioned.

At the end of chapter 2, several research projects are presented, most using situations and data from the demonstration examples presented in the chapter. Likewise, an instructor may use any of the examples or exercises throughout the text (perhaps changing the data slightly) as setups for additional research projects. Again, section 2.8 fully describes how to write a research report.

Grading of research reports may be tempered to encourage students. I use four grades (Bonus [$\sqrt{}\sqrt{}$], Acceptable [$\sqrt{}$], Incomplete [$\frac{1}{2}\sqrt{}$], and Unacceptable [x]), but allow unlimited resubmissions until an acceptable grade is achieved. Essentially only those students who do not hand in the assignment or do not resubmit for an acceptable grade are penalized.

Detailed Step-by-Step Instruction

Students minimally prepared in mathematics require and demand precise detailed step-by-step guidance, especially when introduced to new concepts.

Concrete Examples Introduce Theory; Symbolism and Formulas Are Minimized

Concrete examples (and *not* theoretical examples) introduce new concepts with interesting and relevant real-world situations that pull in the reader with realistic and easily understood problems to be solved.

Fully Worked-Out Solutions to First Five Exercises In Each Chapter

Students are given the answers and fully worked-out solutions to the first five exercises in each chapter and partial solutions or answers to odd-numbered exercises thereafter. This gives the instructor the option to assign homework with (1) fully worked-out solutions, (2) abbreviated answers, or (3) no answers. In effect, the instructor can adjust the assignments to meet the level of any class.

Extensive Chapter Summaries

Extensive reviews of all chapters are included, with key concepts and formulas clearly laid out.

Acknowledgments

Special thanks and appreciation to those who reviewed all or part of the manuscript:

Clifford Adams
Wingate College

Glen A. Just
Mount St. Clare College

Gregory M. Kich
Horry-Georgetown Technical College

Angela Hernandez
University of Montevallo

Gael Mericle
Mankato State University

Lloyd G. Roeling
University of Southwestern Louisiana

J. Larry Martin
Missouri Southern State College

Ernest East
Northwestern Michigan College

James R. Fryxell
College of Lake County

Raymond Patrick Guzman
Pasadena City College

Annalisa L. Ebanks
Jefferson Community College

I am also indebted to the following individuals who have class-tested the text (some continually over several semesters) and have provided valuable input along the way.

Roberta Aaronoff
SUNY/Fashion Institute of Technology

Teresa Perlis
SUNY/Fashion Institute of Technology

Robert Ewen
SUNY/Morrisville

Nancy Portelli
Molloy College

Clare Johnson
SUNY/Fashion Institute of Technology

Lorraine Tawfik
Molloy College

Shantha Krishnamachari
Borough Manhattan Community College

C. K. Wang
SUNY/Fashion Institute of Technology

Rene Mathez
SUNY/Fashion Institute of Technology

I also wish to thank John Catalinotto, Mary V. Di Gangi, M.D., Norma Iorio, Joan Mole, Margarita Martinez, Ted Muzio, Michelle St. André, John Vargas, and Patrick Writt for their excellent suggestions. And special thanks goes to Jane Monroe at Columbia's Teachers College for her invaluable technical input and to editors Earl McPeek, Jane Parrigin, and Carol Besler for their courage and willingness to fight for this unique project. I would also like to thank the Wm. C. Brown production staff, my close friends, and my colleagues for their support and patience. Thanks also to Fashion Institute of Technology for granting me the sabbatical when I needed it the most.

Introduction

S tatistical research touches us all: the clothes we wear, the TV we watch, the cars we drive, the education we receive, the medicines we take, the movies we admire, the warnings against cigarette smoking and other products, and perhaps even the salary we earn. Statistical research touches almost every aspect of our life.

To be more specific, the techniques discussed in this text are used extensively in the fields of psychology, education, health, marketing, agriculture, criminology, factory production, biology, medicine, advertising, economics, and a host of other disciplines in the soft sciences, hard sciences, and nonsciences. In other words, the techniques are used in almost any discipline that requires the analysis of data. Examples will be drawn arbitrarily from all fields since the basic tenets of statistical analysis are much the same whatever your field of interest. ▼

We shall define **statistics** as follows:

Sta-tist'ics
The collection, organization, and interpretation of numerical data for the purposes of

a. summarization, and
b. drawing conclusions about populations based on taking samples from that population.

When statistical techniques are used primarily for the purpose of summarizing data, this is called **descriptive statistics.** Descriptive techniques are discussed mostly in chapters 2 through 4. However when statistical techniques are used for the purpose of drawing conclusions about populations based on samples drawn from that population, this is called **inferential statistics.** Inferential techniques are discussed in chapters 4 through 11. As you can see, the primary focus of this text is on inference.

In'fer-ence
Decision making based on samples drawn from populations.

1.1 Overview of Course (Basic Concepts)

Although many of the following concepts are discussed in depth in subsequent chapters, certain terms are so crucial for a full understanding of the material that a brief overview is presented here.

Population

The term **population** has a unique meaning in statistics. It not only refers to a precisely defined entity or group of entities we wish to study (people, land, insects, or whatever) but also to the specific attribute we wish to measure.

For instance, suppose we wish to study Chicago males divorced in 1988. Although in the real world, this might be considered a precisely defined population of individuals, it is not as yet a population in a statistical sense, because we have no attribute to measure. However, if we had said, we wish to study the *ages* of Chicago males divorced in 1988, then this would be a population in a statistical sense. Notice that a population in statistics must not only consist of the precisely defined entity or group of entities, in this case, Chicago males divorced in 1988, but also to a specific attribute we wish to measure, in this case, *age*. Suppose this population was presented as follows:

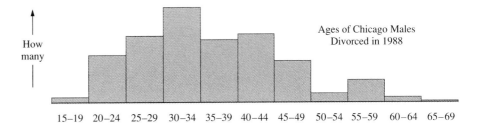

This is called a **histogram.*** The height of the bars represents *how many* in each category. For instance, the category 30–34 years old is represented by the highest bar. This means more men were divorced in this 30–34-year-old age category than in any other age category. Histograms are of enormous value in statistics. A mere glance gives you substantial amounts of information. Notice very few men were divorced in the 50–54-year-old age category. Now, let's look at a different population.

*The histogram is discussed at length in chapter 2.

Suppose we wish to study the *income* of Chicago males divorced in 1988. Although in the real world, the population is still Chicago males divorced in 1988, when we talk about a different attribute, in this case, income, in statistics this constitutes a *different* population. Here the population is incomes of Chicago males divorced in 1988. Suppose we were to actually measure this population and represent the results as follows:

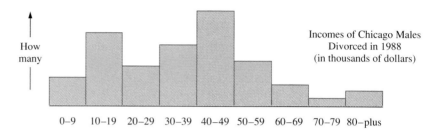

Again, a histogram is used to represent this data. Notice, more men were divorced in the 40–49 ($40,000–$49,000) income category than in any other income category, although the 10–19 income category was a close second. In statistics, the histogram is the workhorse of data presentation. We will use it extensively in this text.

To sum up, when we talk about a population, we not only refer to a precisely defined entity or group of entities (in the above case, Chicago males divorced in 1988) but we also refer to a specific attribute we wish to measure, either age, income, emotional stability, number of offspring, educational level, or whatever.

> **Population**
> Any precisely defined collection of values we wish to study

Sample

Measuring an entire population can be exceedingly time-consuming, costly, and in many cases physically impossible. **Samples,** on the other hand, can often be taken with relative ease.

> **Sample**
> Part of a population

So, where we can, we usually take a sample in the hope the sample will give us accurate information about a population without having to actually measure the population in its entirety.

Samples, if properly taken, can tell us much about a population. However, samples, if improperly taken can mislead. Let us look at some examples of how samples can mislead.

For instance, 5 young men sampled walking out of the student cafeteria at Washington State University yielded an average height of $6'3\frac{1}{2}''$. Can we conclude the average height of *all* Washington State men is $6'3\frac{1}{2}''$? I doubt it. A sample of French women at a local convention showed that 80% were severely underweight. Can we conclude 80% of *all* French women are severely underweight? A poll of 25 joggers in Virginia Beach showed an average annual salary of $183,000. Would this describe the *true* average annual salary of Virginia Beach joggers—or did we, perhaps, run into a pack of Wall Street tycoons on vacation?

What's wrong with the above samples? Chances are, they are not truly representative of the population from which they were drawn.

Of course, this leads us to the question: how do we obtain samples that *are* truly representative of the population from which they are drawn? To start off, we should have a **random sample.**

Random Sample

We have seen in the above examples that it is not enough to take just any sample from a population. Samples can easily give false impressions. Even if we try to be fair when taking a sample, more often than not, the information we obtain will give us misleading information about a population.

To better guarantee a true representation of a population, the sample should be a random sample, which leads to the following definition.

Random Sample

A sample drawn from a population such that each and every member of the population has an equal chance of being included in the sample. In addition, every sample of the same size has an equal chance of being selected.

Many decades of experience have shown that results from *random samples* can be trusted to give reliable information about a population. However, random samples are not easy to come by.

One old-fashioned method used to obtain a random sample and, perhaps, still the best method, is similar to a way lottery or raffle winners are selected. We will call it the **rotating basket** method, and it proceeds as follows:

Rotating Basket Method

First, we assign a number to each member of the population. That is, every Chicago male divorced in 1988 must be assigned an identifying number, just as a number is assigned to someone purchasing a raffle ticket.

Second, each number is tossed into a large basket. The basket is rotated often to ensure proper mixing.

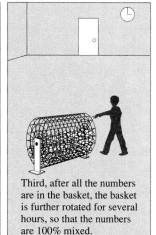

Third, after all the numbers are in the basket, the basket is further rotated for several hours, so that the numbers are 100% mixed.

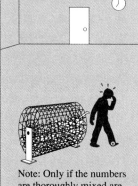

Note: Only if the numbers are thoroughly mixed are we assured of a true random sample. So, it is very important the numbers are mixed completely.

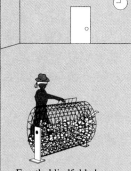

Fourth, blindfolded, we reach into the basket and select out as many numbers as we want in our sample. If we want five in our sample, then we pull out five slips of paper.

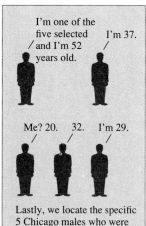

Lastly, we locate the specific 5 Chicago males who were picked and determine their ages.

In this case, since we sampled 5 from the basket, we call this a random sample of size $n = 5$, yielding ages 52, 37, 20, 32, and 29.

In actuality, writing thousands of numbers (sometimes millions) on slips of paper and rotating large baskets can be tedious, so in practice, we very often use a master list, say of all the Chicago males divorced in 1988, assign each a number, then use something called a **random digit table.**

Let's see how it works. Say we have a total population of 5000 Chicago males divorced in 1988. We would proceed as follows:

Random Digit Method

Sam Benton	0001
Mel Sims	0002
.	.
.	.
.	.
Ferris Callo	0299
Al Gee	0300
John Deitski	0301
Juan Eckstin	0302
.	.
.	.
.	.
Chang Hoo	4997
Joe Doet	4998
Harry Fudim	4999
Bill Leeder	5000

Random Digit Table

08141	31926	30566	63607
71798	99058	06766	20575
81679	51276	84580	15001
72925	57040	57012	99058
72844	80198	08598	21410
45929	94739	65371	48423
36663	64130	17730	75755
10610	05794	04717	95862
16688	23757	25018	50541
33967	24160	99725	85113
86980	09066	19347	60203
.	.	.	.
.	.	.	.

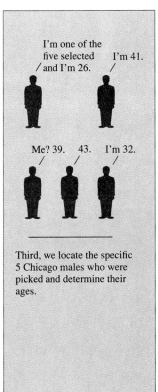

I'm one of the five selected and I'm 26.   I'm 41.

Me? 39.   43.   I'm 32.

First, we write up a master list of the 5000 Chicago males divorced in 1988, assigning each a four-digit identifying number, similar to what we did in the Rotating Basket Method.

Second, we start anywhere in the Random Digit Table, penciling off a section of four-digit numbers. Say for instance, we arbitrarily choose column 3, fifth number down to start. Then, we start circling numbers of 5000 or less in this section.* If we wish 5 in our sample, then we circle only 5 numbers.

Third, we locate the specific 5 Chicago males who were picked and determine their ages.

So, this is another random sample of $n = 5$. This time the sample yielded ages 26, 41, 39, 43, and 32.

The random digit table, in effect, takes the place of the rotating basket. Numbers are written down for us, mixed thoroughly, and presented in table

---

*Since 5000 is a 4-digit number, we circle only groups of 4 digits. Note that in a random digit table, any digit sequence is possible. For instance, the first four-digit sequence we had chosen, 0859, was just as likely to occur as any other four-digit sequence, say 1234 or 0000 or 4444.

form. Our job is simply to reach into the random digit table and pull out any desired number of candidates.

The point is whether we use the rotating basket or the random digit table, everyone has an equal or near equal chance of being selected on each and every pick, and every sample of the same size has an equal chance of being selected. Experience has shown that samples selected in this way generally give reliable information about a population (provided certain assumptions are met, which will be discussed as we proceed).

Let me add: in many industry and research applications, obtaining a random sample is often one of the most difficult aspects of performing a statistical investigation. As a result, a number of other methods have been developed in an attempt to obtain samples representative of a population (stratified, cluster, systematic sampling, random intact groups, random assignment, haphazard sampling, and others), however many of these methods come with risks that are difficult to assess. Therefore these methods will not be used or discussed in this textbook. All sampling in this text assumes use of the rotating basket or random digit table methods.

Internal and External Validity

The ability to achieve a true representation of the population from a sample is often referred to as the ability of the experiment to achieve **external validity.** Although a random sample is an important factor in achieving external validity, a random sample by no means guarantees it. Other factors must be considered before we can safely use our random sample as truly representative of the population, and generally these factors are categorized under threats to either internal or external validity.

> ### Internal Validity
> The certainty that the observations in our sample group are *accurate* measures of the characteristics we set out to measure.

In other words, from our sample group, did we extract honest, accurate, and reliable information? If yes, then we have achieved internal validity.

In the case of Chicago males divorced in 1988, internal validity simply means: were the five ages we obtained from our sample *accurate* reflections of the true ages of the five males? In other words, did we obtain the real ages of these five males? Did we ask them in a telephone conversation and possibly risk one or more of them not telling the truth for any number of reasons? Did we ask to see their driver's licenses or birth certificates, then verify this information with hospital birth records? Put another way: how well-grounded, sound, and supportable are the five measurements of age we obtained?

> **External Validity**
> The certainty that our *methods and presence* in no way jeopardize our ability to use the random sample as a true representative of the population.

For instance, with a caged animal certain feeding habits can be accurately measured and may be repeatable from experiment to experiment (thus achieving internal validity). However, if we measure the feeding habits of those same animals in nature, the results could be quite different. In other words, does the caging of the animals or our presence introduce an influence (and thus threaten external validity, our ability to use the sample results as representative of the feeding habits of these animals in nature)?

To achieve external validity, then, we must ask ourselves not only whether we randomly sampled and achieved internal validity but also whether our test methods or presence in any way influenced our results. This is often a difficult question to answer, but the question must be addressed. Poor methodology has been known to drastically influence results and is often the reason why statistical studies measuring the same phenomenon vary so widely.

Internal and external validity in the case of the Chicago males divorced in 1988 hardly seems a question. Accurate ages are relatively easy to obtain and verify in this particular instance. And we can control threats to external validity rather easily by merely keeping the study secret. However, in most experiments in psychology, education, and a number of other fields, the question of validity turns into a veritable nightmare and, along with the difficulties of obtaining a random sample, often becomes a second or third follow-up course in statistics dealing mostly with these issues.

So, to sum up, not only must we extract honest, *accurate,* and reliable information about that which we are measuring (internal validity), but we must make sure our *methods and presence* in no way influence our ability to use the results as truly representative of the population (external validity).

> All samples used in this text are assumed to be *internally and externally valid random samples.*

1.2 **Why We Sample**

We sample to determine certain characteristics of a population without having to measure the entire population. What are these characteristics, you might ask. Actually, there are many characteristics we might look for, but in broad terms we most often wish to identify either μ or *p,* defined as:

μ (mū)
The arithmetic mean or average* value of a population.

p
The proportion (or percentage) of a population that possesses a certain attribute.

Greek letters are often used in mathematics to represent specific quantities. μ (pronounced mū) is the twelfth letter of the Greek alphabet and is used to represent the average value of the population.

Sampling to Determine μ

One of the most frequent uses of sampling is to gather information about μ, the average value of some population, which is discussed at length in chapters 5 through 8. Let's look at an example.

Suppose we wish to determine μ, the average age of all Chicago males divorced in 1988. To measure the thousands of males in this population would be time-consuming and tedious. Instead, we might simply take a random sample as follows.†

Example ——————— A random sample of $n = 5$ was drawn from a population of Chicago males divorced in 1988 and yielded the following ages: 52, 37, 20, 32, and 29. Calculate the average age of these five men.

Solution To take an average, we add up the ages and divide by 5.

$$\bar{x} = \frac{52 + 37 + 20 + 32 + 29}{5} = 34 \text{ years old}$$

---

*Technically, the word *average* refers to a broad number of measures, however, we shall employ the word *average* in its common usage as the sum of a collection of numbers divided by the number of values in that collection.

†Keep in mind, all samples used in this text are assumed to be both internally and externally valid random samples.

Notice that the symbol $\bar{x}$ (x bar) was used to represent the average age of this sample. This allows us to differentiate between a sample average age ($\bar{x}$) and a population average age (μ).

Because we took a *random* sample and obtained an average age of 34 years old, we can now state μ should be "approximately equal to" 34 years old. In other words,

> $\bar{x} \approx \mu$
> When we randomly sample, the sample average ($\bar{x}$) is approximately equal to the population average (provided certain restrictions are met regarding sample size, discussed below).

Does this allow us to state: the average age of all Chicago males divorced in 1988 is exactly 34 years old? The answer is, no. A *random* sample merely gives an approximation. However, it does allow us to state: the average age of all Chicago males divorced in 1988 is *approximately* 34 years old. ■

One word about sample size: In the above example, we used as our sample size $n = 5$, that is, we used 5 males selected from our population. It is generally preferable to keep your sample size as large as possible. In fact, as a general rule, random samples should be kept above 30.

> It is preferable to keep your sample size at 30 or more.

In the above case, it would have been preferable to have selected 30 or more males. First, larger samples tend to give closer approximations of μ and, second, certain restrictions apply when your sample size is smaller than 30. Perhaps the most formidable restriction when using samples under 30 is that your population must be at least somewhat bell-shaped in appearance, as follows.

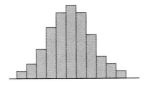

For samples under 30, populations must be at least somewhat bell-shaped* to ensure $\bar{x} \approx \mu$.

If you keep your sample size at 30 or above, this restriction does *not* apply. Samples above 30 tend to give sample averages ($\bar{x}$'s) that are close approximations of μ for almost *any shaped* population.

*This restriction grows more critical the smaller the sample size.

Sampling to Determine p

Another frequent use of sampling is to gather information about p, the proportion (or percentage) of some population that possesses a certain attribute, which is discussed at length in chapters 3, 4, 10, and 11.

Let's look at an example.

Suppose we wish to determine p, the proportion (or percentage) of Chicago males divorced in 1988, say for instance, who are fifty or more years old. Again, to measure the thousands of males in this population would be time-consuming and tedious. Instead, we might take a random sample as follows.

Example ——————— A random sample of $n = 100$ males were drawn from a population of Chicago males divorced in 1988 and yielded 12 males who were fifty or more years old. Calculate the proportion of males who were fifty or more years old. Express this proportion as a percentage.

Solution The proportion of our sample who are fifty or more years old is simply 12 out of 100 or

$$p_s = \frac{12}{100}$$

To convert this to a percentage, we multiply by 100, as follows:

$$\text{Percentage} = \text{Fraction} \times 100 = \frac{12}{100} \times 100 = 12$$

$$p_s = 12\%$$

In other words, because 12 out of 100 males were fifty or more years old, this is $\frac{12}{100}$ or 12% of the sample. Notice that the symbol p_s (p sub s) was used to represent the proportion of males fifty or more. This allows us to differentiate between a sample proportion (p_s) and a population proportion (p).

Since we took a *random* sample and obtained $p_s = 12\%$, the true proportion of Chicago males fifty or more years old must be *approximately* equal to 12%. In other words,

> **$p_s \approx p$**
> When we randomly sample, the sample proportion (p_s) is approximately equal to the population proportion (p).

Does this allow us to state that exactly 12% of *all* Chicago males divorced in 1988 were fifty or more years old? The answer is, no. Remember, random samples merely give us approximations. However, it does allow us to state: *approximately* 12% of all Chicago males divorced in 1988 were fifty or more years old. ∎

Again, it is best to keep your sample size as large as possible. Generally, larger samples give closer approximations of p and there are other benefits, which will be discussed later in the text.

Summary

Definitions:

Statistics: The collection, organization, and interpretation of numerical data for the purposes of summarization (descriptive statistics) and drawing conclusions about populations based on taking samples from that population (inferential statistics).

Inference: Decision making based on samples drawn from populations.

Population: Any precisely defined collection of values we wish to study.

Sample: Part of a population.

Random sample: A sample drawn from a population such that each and every member of the population has an equal chance of being included in the sample. In addition, every sample of the same size has an equal chance of being selected.

Internal validity: The certainty that the observations in our sample group are accurate measures of the characteristics we set out to measure.

External validity: The certainty that our methods and presence in no way jeopardize our ability to use the random sample as a true representative of the population.

Valid random sample: A sample that is randomly selected and possesses both internal and external validity. In this text, we use only valid random samples, unless specified.

We can sample for a number of reasons, but in broad terms we most often wish to identify either μ (mu) or p, defined as follows:

μ: the arithmetic mean or average value of a population.

p: the proportion (or percentage) of a population that possesses a certain attribute or characteristic.

$\bar{x} \approx \mu$: when we randomly sample, the sample average ($\bar{x}$) is approximately equal to the population average (μ). In this case, it is preferable to keep your sample size at 30 or more.

$p_s \approx p$: when we randomly sample, the sample proportion (p_s) is approximately equal to the population proportion (p). As a general rule, it is best to keep your sample size as large as possible.

Sampling techniques, as we will learn, normally come with restrictions and conditions, however when small sample sizes are used, additional conditions and restrictions usually apply. All this is discussed in the following chapters.

Exercises

Note that full answers for exercises 1–5 and abbreviated answers for odd-numbered exercises thereafter are provided in the Answer Key.

1.1 If we define statistics as the collection, organization, and interpretation of numerical data for the purposes of (a) summarization and (b) drawing conclusions about populations based on taking samples from that population, find a newspaper or magazine article or TV advertisement that uses statistics (the *New York Times* and *Newsweek* magazine are good sources). Bring in the article to discuss in class and state whether the data represents a population or sample.

1.2 Suppose we wished to estimate the average age of Chicago males divorced in 1988 and took a random sample of $n = 100$ yielding $\bar{x} = 35.2$ years old.

a. What is the population?
b. What symbol do we use to represent the average age of the population?
c. How many were in your sample?
d. What symbol is used to represent the average age of the sample?
e. Do we know the average age of the population?

1.3   Suppose we wish to take a random sample of $n = 7$ from a population of 5000 Chicago males divorced in 1988. First we assign each male a number from 0001 to 5000. Explain how we might use the following random digit table to select our sample:

```
94620  27963  96478  21559  19246  88097  44926
60947  60775  73181  43264  56895  04232  59604
27499  53523  63110  57106  20865  91683  80688
01603  23156  89223  43429  95353  44662  59433
00815  01552  06392  31437  70385  45863  75971

83844  90942  74857  52419  68723  47830  63010
06626  10042  93629  37609  57215  08409  81906
56760  63348  24949  11859  29793  37457  59377
64416  29934  00755  09418  14230  62887  92683
63569  17906  38076  32135  19096  96970  75917
```

1.4   Suppose we wished to estimate the average age of Chicago males divorced in 1988 and took a random sample of $n = 7$ yielding the ages: 26, 61, 39, 43, 31, 22, and 37.

a. Calculate $\bar{x}$, the average age of the sample.
b. What can we say about μ, the average age of all Chicago males divorced in 1988?

1.5   Suppose we wished to estimate the proportion of Chicago males divorced in 1988 that possess the attribute of *blue eyes* and took a random sample of $n = 100$, discovering that 18 males from this sample had blue eyes.

a. Calculate p_s, the proportion of your sample that had blue eyes.
b. Express p_s as a percentage.
c. What can we say about p, the proportion of all Chicago males divorced in 1988 who have blue eyes?

1.6   Briefly state why we study statistics.

1.7   In a medical study, a researcher wished to estimate the average length of time needed for a particular nurse-in-training to draw a series of blood specimens. A sample of the nurse's work over several months yielded the following times: 10, 6, 5, 14, 6, and 13 (in minutes).

a. What is the population?
b. What is it about the population we wish to determine? And what symbol is used to represent this?
c. What is the sample? Calculate $\bar{x}$.
d. Is this sample representative of the population? Discuss:
   i. randomness.
   ii. internal validity.
   iii. external validity.
   iv. sample size.

1.8   It is often said prior to an election that only a small number of people are needed to predict the outcome of a major election. The statement refers to use of a valid random sample. If we wish to predict the results of an upcoming election, discuss selecting a sample by:

a. polling the class.
b. having your classmates randomly poll friends.
c. selecting 400 names from a telephone book and calling.

1.9   To determine the average height of Washington State University males, a campus newsletter sampled 50 males yielding an average height of $5'9\frac{1}{2}''$.

a. What is the population?
b. What is the sample?
c. How might you obtain a valid random sample?

Organizing and Analyzing Data

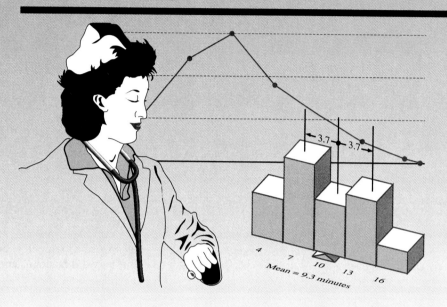

This chapter is concerned with the techniques used to summarize data. Specifically, we will study

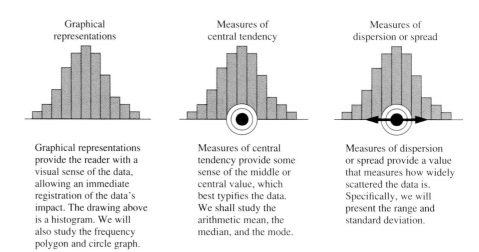

Graphical
representations

Graphical representations provide the reader with a visual sense of the data, allowing an immediate registration of the data's impact. The drawing above is a histogram. We will also study the frequency polygon and circle graph.

Measures of
central tendency

Measures of central tendency provide some sense of the middle or central value, which best typifies the data. We shall study the arithmetic mean, the median, and the mode.

Measures of
dispersion or spread

Measures of dispersion or spread provide a value that measures how widely scattered the data is. Specifically, we will present the range and standard deviation.

We will also explore the usefulness of these graphical representations and measures, especially in relation to the study of inferential statistics, that is, drawing conclusions about populations based on samples taken from those populations. Additional descriptive techniques are presented in section 2.7 and the writing of research reports is presented in section 2.8. ▼

2.1 Graphical Representations

Whether it be population or sample data, the importance of representing data in graphical or picture form cannot be overemphasized. These graphs or picture representations provide the immediate impact of the data in an easy-to-interpret format. Specifically, in this section, we introduce the histogram, frequency polygon, and circle graph shown as follows.

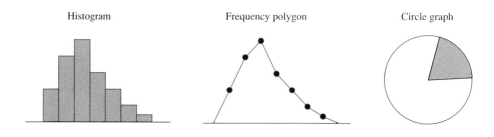

Histogram   Frequency polygon   Circle graph

These graphical representations are used often throughout the text, so understanding them is essential.

Histogram

One of the most frequently used graphical representations is the histogram. Let's look at the following example.

Countrygirl Makeup is a line of facial products marketed to young women, ages sixteen to twenty-four. Although successful when introduced in 1980, in recent years, Countrygirl sales have eroded. An independent research firm was commissioned to gather information about the ages of current users. A nationwide random sample of 100 current users yielded the following ages:

Ages of 100 randomly selected users of Countrygirl Makeup

21.4	18.1	17.6	26.6	47.1
24.5	22.2	22.9	32.7	15.8
31.1	23.1	27.7	29.0	21.6
17.7	32.6	21.6	26.4	31.4
22.1	22.3	31.9	25.7	35.2
21.9	31.3	22.7	27.6	27.9
30.1	23.1	26.4	32.1	22.5
28.2	25.7	33.8	28.9	18.6
26.8	30.5	34.0	21.6	28.2
32.2	27.3	17.5	23.0	32.8
36.0	29.1	42.7	30.5	39.0
26.2	33.2	36.3	22.7	43.1
28.7	26.3	38.6	24.1	21.3
32.1	28.7	25.8	26.0	18.7
18.2	23.9	28.2	20.2	33.1
40.7	40.7	16.6	18.1	42.7
31.1	16.0	38.9	26.7	36.6
21.7	26.7	36.0	37.3	27.1
23.1	28.2	20.6	25.7	26.7
26.9	35.8	23.7	38.2	20.9

When data is presented in this form, we call it **ungrouped.** Each value is recorded exactly as measured. Unfortunately, ungrouped data is difficult to represent graphically. Because graphical representations are often essential for a complete understanding, we may choose to tally such data into groups or categories, in this case, age categories. We would then use the results of these groupings to construct a histogram.

To construct a histogram from the above set of 100 ages, let us proceed as follows.

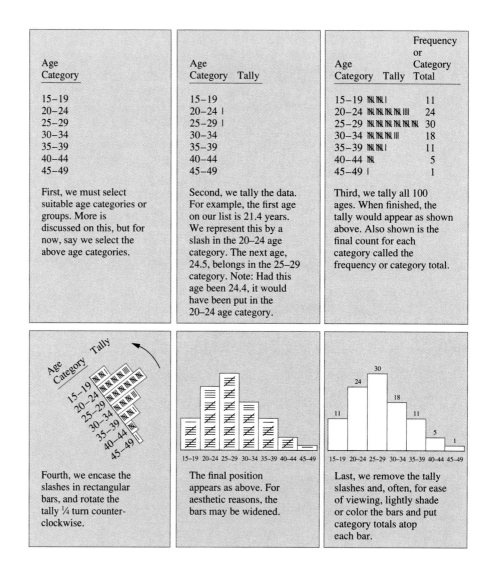

Age Category
15–19
20–24
25–29
30–34
35–39
40–44
45–49

First, we must select suitable age categories or groups. More is discussed on this, but for now, say we select the above age categories.

Age Category	Tally
15–19	
20–24	I
25–29	I
30–34	
35–39	
40–44	
45–49	

Second, we tally the data. For example, the first age on our list is 21.4 years. We represent this by a slash in the 20–24 age category. The next age, 24.5, belongs in the 25–29 category. Note: Had this age been 24.4, it would have been put in the 20–24 age category.

Age Category	Tally	Frequency or Category Total
15–19	ℍℍ I	11
20–24	ℍℍ ℍℍ ℍℍ IIII	24
25–29	ℍℍ ℍℍ ℍℍ ℍℍ ℍℍ ℍℍ	30
30–34	ℍℍ ℍℍ ℍℍ III	18
35–39	ℍℍ ℍℍ I	11
40–44	ℍℍ	5
45–49	I	1

Third, we tally all 100 ages. When finished, the tally would appear as shown above. Also shown is the final count for each category called the frequency or category total.

Fourth, we encase the slashes in rectangular bars, and rotate the tally ¼ turn counter-clockwise.

The final position appears as above. For aesthetic reasons, the bars may be widened.

Last, we remove the tally slashes and, often, for ease of viewing, lightly shade or color the bars and put category totals atop each bar.

To complete the histogram, we add the following components.

1. An overall identifying title, explaining what the histogram represents.

2. A horizontal scale, identifying the attribute we are measuring.

3. A vertical scale, representing *how many in each category.*

The final presentation of the histogram might then be as follows:

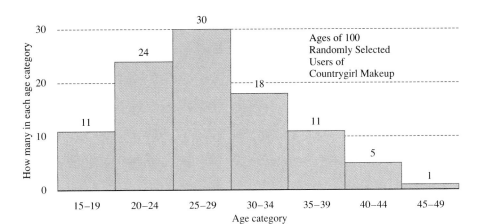

Notice that the histogram bar of the 25–29 age category is almost three times as high as the histogram bar for the 15–19 age category. This means that there were almost three times as many people in the 25–29 age category as were in the 15–19 age category.

Also did you notice that 48 out of a total of 100 in our sample (30 plus 18) were between the ages of 25 and 34? This represents almost half or 48% of our sample. Also did you notice very few users were ages 15–19?

Histograms are quite valuable in helping us unlock the secrets of a population. If a sample is large enough, the shape of the sample histogram will give a good approximation of the shape of the population histogram. In other words, if we were to measure *all* current users of our product, we might find the histogram above to be quite similar in shape to the histogram of our entire population of users.

With this in mind, do you feel Countrygirl executives might reevaluate their strategy of marketing their products to young women ages 16–24? Actually, this is a complex question and executives at companies like Revlon and L'Oreal, along with their advertising agency counterparts, grapple with questions like this on a routine basis.

Basic guidelines in the construction of histograms are as follows:

1. Each category should be the same numerical width. In other words, in our histogram of 100 users of Countrygirl, the first age category, 15–19, contains five ages, 15, 16, 17, 18, and 19, therefore the next category, 20–24, should also contain five ages, which it does. And each subsequent category should contain five ages in sequence, which they do.

2. There should be no overlap. For instance, in our Countrygirl histogram, someone 24.4 years old fits into the 20–24 age category, whereas someone 24.5 years old fits into the 25–29 age category. Each reading fits into *one* and only one category.

Aesthetic guidelines in the construction of histograms would include

3. The histogram should be made to fill as much of the available space as possible by widening or proportionally elongating the histogram bars. Good examples would be the histograms presented in this chapter.

4. Histograms with 5 to 12 categories tend to be the most visually appealing.

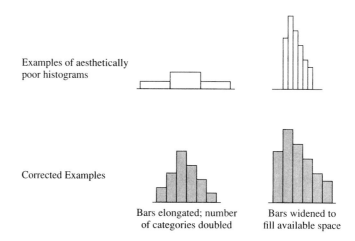

Examples of aesthetically poor histograms

Corrected Examples

Bars elongated; number of categories doubled

Bars widened to fill available space

Population Histograms

Experience has shown that many *population* histograms take on repeating shapes. In other words, there are certain common shapes that seem to reoccur.

Common Shapes of Population Histograms

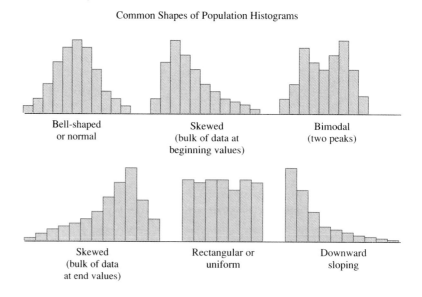

Bell-shaped or normal

Skewed (bulk of data at beginning values)

Bimodal (two peaks)

Skewed (bulk of data at end values)

Rectangular or uniform

Downward sloping

Can you think of populations that might take on the shapes represented in the above figures? For instance, if we measured the heights of all the men or all the women in your town or city, chances are the histograms would be bell-shaped, whereas, if we measured the salaries of all employees in a company, the histogram might be skewed with the bulk of data at beginning values. Can you think of other populations that might fit these shapes?

Frequency Polygon

A frequency polygon is merely a line representation of a histogram. Let's see how it works.

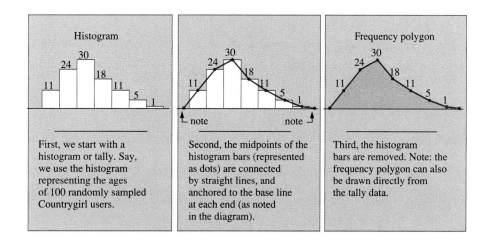

First, we start with a histogram or tally. Say, we use the histogram representing the ages of 100 randomly sampled Countrygirl users.

Second, the midpoints of the histogram bars (represented as dots) are connected by straight lines, and anchored to the base line at each end (as noted in the diagram).

Third, the histogram bars are removed. Note: the frequency polygon can also be drawn directly from the tally data.

The final frequency polygon, titled, might appear as follows.

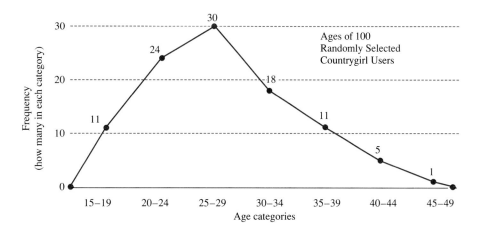

Notice that the word **frequency** was used to title the vertical scale. This is a commonly used word in statistics, meaning *how many in each category.*

Frequency polygons are clear, concise, and easy to sketch and are especially useful when the number of categories in a histogram grows too large to be easily represented. When the number of categories in a histogram grows into the hundreds or more, often the frequency polygon takes on the appearance of a smooth sloping line, which is often referred to as a **curve.**

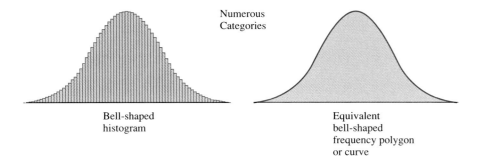

Numerous Categories

Bell-shaped histogram

Equivalent bell-shaped frequency polygon or curve

Circle Graph

The circle graph (or pie chart) is a common pictorial representation for proportion data, that is, the fraction (or percentage) of some sample or population that possesses a certain characteristic or attribute.

In our Countrygirl sample, we might wish to represent the proportion of users who are, for instance, thirty-five or more years old. We would first count how many were thirty-five or more years old, change that number to a percentage, then represent that percentage as a slice in a circle graph, as follows:

Age category	Tally	Frequency or category total
15–19	ʼʼʼʼ ʼ	11
20–24	ʼʼʼʼ ʼʼʼʼ ʼʼʼʼ ʼʼʼʼ	24
25–29	ʼʼʼʼ ʼʼʼʼ ʼʼʼʼ ʼʼʼʼ ʼʼʼʼ ʼʼʼʼ	30
30–34	ʼʼʼʼ ʼʼʼʼ ʼʼʼʼ ʼʼʼ	18
35–39	ʼʼʼʼ ʼʼʼʼ ʼ	11
40–44	ʼʼʼʼ	5
45–49	ʼ	1

First, we start with our raw data, or in this case, we can use our tally, and count the number of users who are thirty-five or more years old. The answer is 11 + 5 + 1 = 17

$$\begin{bmatrix} \text{Percentage of} \\ \text{sample who} \\ \text{are thirty-five} \\ \text{or more} \end{bmatrix} = \begin{bmatrix} \text{Fraction} \\ \text{who are} \\ \text{thirty-five} \\ \text{or more} \end{bmatrix} \times 100$$

$$P = \frac{17}{100} \times 100$$

$$P = 17\%$$

Second, since 17 users were thirty-five or over, we use our formula, given above, to calculate what percentage this is of our total sample. Answer: 17%.

Circle graph

17% (thirty-five or more)

Third, since a circle has 360°, 17% of 360° = 61.2°, so we shade a slice of the circle representing 61.2°. A protractor can be used to draw the angle.

The final circle graph would be titled and might appear as follows:

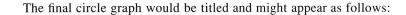

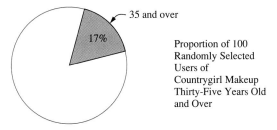

35 and over

17%

Proportion of 100
Randomly Selected
Users of
Countrygirl Makeup
Thirty-Five Years Old
and Over

If we represent the sample proportion with the symbol p_s we can state,

$$p_s = 17\% \text{ (or in decimal form, } p_s = .17)$$

Instead of calculating 17% of 360° = 61.2° and then measuring 61.2° with a protractor, the slice in the circle graph can be estimated as follows:

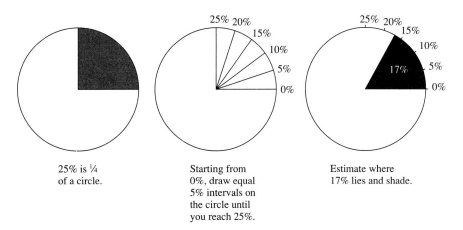

25% is ¼
of a circle.

Starting from
0%, draw equal
5% intervals on
the circle until
you reach 25%.

Estimate where
17% lies and shade.

The circle graph is quite versatile. We may wish to represent multiple categories, such as the percentage of users 35 or more, the percentage of users 25 to 34, and the percentage of users 15 to 24. This would be represented as follows:

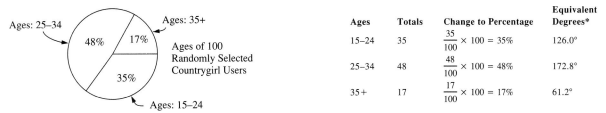

Ages: 25–34

48%   17%

35%

Ages: 35+

Ages of 100
Randomly Selected
Countrygirl Users

Ages: 15–24

Ages	Totals	Change to Percentage	Equivalent Degrees*
15–24	35	$\frac{35}{100} \times 100 = 35\%$	126.0°
25–34	48	$\frac{48}{100} \times 100 = 48\%$	172.8°
35+	17	$\frac{17}{100} \times 100 = 17\%$	61.2°

The following example is presented for practice.

*To change to degrees, take 35% of 360°: .35 × 360 = 126.0°.

Example —————— Out of 50 randomly selected Countrygirl users, 12 possessed the attribute of red hair. Draw a circle graph representing this sample proportion.

Solution   We first convert the quantity 12 out of 50 into a percentage.

$$\text{Percentage} = \text{Fraction} \times 100$$
$$P = \frac{12}{50} \times 100$$
$$P = 24\%$$

Now we know 24% of the sample had red hair, 24% of 360° = 86.4°. We can measure this angle with a protractor or we can estimate 24% as *almost* 25% or almost $\frac{1}{4}$ of a circle.

Answer

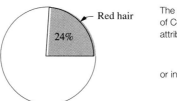

Red hair

The proportion of 50 randomly selected users of Countrygirl Makeup who possess the attribute of red hair is 24%, expressed as

$$p_s = 24\%$$

or in decimal form, as

$$p_s = .24$$

2.2 Measures of Central Tendency (Ungrouped Data)

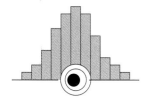

In addition to graphical representations, we often wish to get some sense of the *middle* or *central value* of our data. This is called **central tendency** and the most used measures of central tendency are the arithmetic mean, the median, and the mode.

Arithmetic Mean

Perhaps the most widely used measure of central tendency is the arithmetic mean, what most people refer to as the *average*.

> The **arithmetic mean** is the result obtained when a collection of values are added, then divided by *n*, the number of values. The formula for the arithmetic mean of a *sample* is
>
> $$\bar{x} = \frac{\Sigma X}{n}$$
>
> Sum of the values (Note: the symbol Σ means "sum of.")
> Number of values

Notice that the symbol $\bar{x}$ (*x* bar) was used to represent the arithmetic mean of a sample. If we were to calculate the arithmetic mean of a population, we would use the symbol μ.

Because we shall employ the arithmetic mean throughout the remainder of the text, we shall simply refer to it as the *mean* or *average*. Technically, the words mean and average each refer to a broad number of measures, however we shall use these words only to represent the arithmetic mean.

Example ———————   Suppose in a medical study, a researcher wished to estimate the average length of time needed for a particular nurse-in-training to draw a series of blood specimens. A sample of the nurse's work over several months yielded the following times: 10, 6, 5, 14, 6, and 13 (in minutes).
Calculate the mean.

Solution   Since the sample average, $\bar{x}$, is equal to the sum of the values divided by n, we proceed as follows.

$$\bar{x} = \frac{\Sigma x}{n} = \frac{10 + 6 + 5 + 14 + 6 + 13}{6} = 9 \text{ minutes}$$

The results would be summarized by stating,

$$n = 6 \text{ readings}$$
$$\bar{x} = 9 \text{ minutes}$$

This simply means that for six observations, the average was calculated to be 9 minutes.

Now, the question arises, can we use the sample average to evaluate the nurse-in-training? In other words, because our sample produced an average of 9 minutes, can we assume that if we had measured all the hundreds of times this nurse drew blood specimens, the average of all these hundreds of readings would also be approximately 9 minutes?

In order to use sample values as representative of population values, we must be careful to ensure that a valid random sample was taken. So, the answer to the question would be, yes, provided the sample times were randomly selected over several months, accurately recorded, and the researcher's presence and methods in no way interfered with the normal functioning of the nurse's procedures. In other words, provided we were successful in achieving a valid random sample (actually, for small samples, $n < 30$, other considerations apply, which are addressed in section 2.4; note the expression $n < 30$ means n *less than* 30). ∎

The arithmetic mean or average is often the preferred measure of central tendency for a number of reasons.

Advantages of Using the Arithmetic Mean

1. It is unique and always exists. In other words, in any set of data, there is one and only *one* arithmetic mean and that value always exists.

2. It takes into consideration all the data. No value is left out and every value influences the calculation of the arithmetic mean in some way.

3. It readily lends itself to inferential analysis, that is, in using sample data to estimate population characteristics. For example, with bell-shaped populations, one of the most frequent population shapes we encounter, the sample means cluster much closer to the population mean than, say, the sample medians cluster to the population median. This is an important advantage when we begin to use samples to estimate population characteristics.

4. It is algebraically tractable. In other words, the mean is easily manipulated in complex equations, more so than other measures. This is a vital consideration in advanced work and probably in large part what has accounted for the mean's widespread appeal throughout the centuries.

5. There are even aesthetic advantages to using the mean. It strongly appeals to our sense of balance. In other words, if we represent each value as a block on a number line, the blocks would perfectly balance at one point, and that point always turns out to be the arithmetic mean.

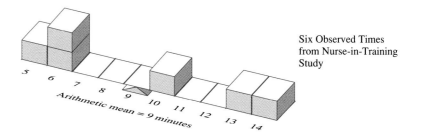

Six Observed Times from Nurse-in-Training Study

Other measures of central tendency do not offer such a powerful array of advantages. Mostly for these reasons, has the arithmetic mean evolved into the preferred measure of central tendency.

However, we must consider other measures, because in certain instances, these other measures offer us information about our data that substantially adds to our understanding. Two such other measures are the median and the mode.

Median

Another commonly used measure of central tendency is the **median.**

> **Median**
> The middle value when data is arranged from lowest to highest value.

Example ———————— For the six values recorded in our nurse-in-training study, 10, 6, 5, 14, 6, and 13, find the median.

Solution

First, line up the data according to size.

$$5, 6, 6, 10, 13, 14$$

Because there is no one middle value, we take the average of the two middle values.

$$\text{Median} = \frac{6 + 10}{2} = 8 \text{ minutes}$$

As a general rule, when the number of values is even, you will have to average the two middle values. When the number of values is odd, you will have *one* middle value and that will be used as the median.   ■

The median is especially useful as a measure of central tendency when data is highly skewed, such as in the following example.

Example ——————

The Technic Company has five employees, which includes the president, yielding the following annual salaries: $30,000; $30,000; $30,000; $30,000; and $200,000. Calculate the median.

Solution

Arrange the salaries in order, then select the middle value.

$$\$30,000 \quad \$30,000 \quad \$30,000 \quad \$30,000 \quad \$200,000$$
$$\uparrow$$

Median = $30,000

Had we calculated the mean for the above salaries, the mean would have been $64,000 [(30,000 + 30,000 + 30,000 + 30,000 + 200,000)/5 = $64,000]. Picture yourself answering an employment ad: Join Technic Company, average salary of employees, $64,000. Do you think the arithmetic mean in this case gives a fair representation of central tendency? Many would consider the median, $30,000, to be a more realistic value here in defining central tendency.   ■

The advantage of the median is that it is unaffected by extreme values and, in certain instances, gives a more realistic measure of central tendency. This is especially true for highly skewed data, such as in the case above. Unfortunately, the median does not lend itself quite as well as the mean to inferential analysis, that is, in using sample data to estimate population characteristics.

Mode

A third measure of central tendency is the **mode.**

Mode
The most occurring value.

Example ——————— For the six values recorded in the nurse-in-training study, 10, 6, 5, 14, 6, and 13, find the mode.

Solution   Mode = 6 minutes since this is the most frequently occurring value.

Unfortunately, not all data sets have modes. On the other hand, some sets have more than one. If two modes occur, we refer to the data as **bimodal.** If more than two modes occur, we refer to the data as **multimodal.**   ■

The mode, as you can see from the above example, can sometimes be misleading as to the true central tendency of data. Although useful when used in addition to the median and mean, the mode should be viewed with caution when used alone.

Comparison of the Mean, Median, and Mode

The following provides a visual look at the mean, median, and mode using the nurse-in-training data. If you recall, we observed the nurse six times over a period of several months and recorded how long the nurse took to draw a specific series of blood specimens.

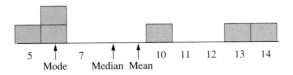

Which do you feel gives a better measure of central tendency for this data? Although the mean is the preferred measure, each adds a little more information.

Example ——————— Estimate the position of the mean, median, and mode for a

a. bell-shaped population.
b. skewed population, with the bulk of data at beginning values.

Solution

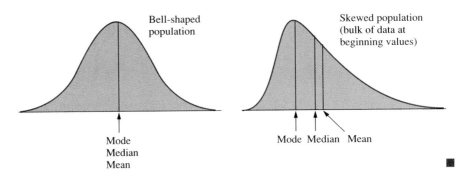

2.3 Measures of Dispersion or Spread (Ungrouped Data)

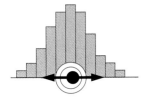

Whereas measures of central tendency attempt to locate the center or middle of the data, measures of dispersion are designed to measure how widely scattered or spread out the data is. We will study two such measures, the range and standard deviation.

Range

The **range** is the difference between the high and low value in your data set.

> **Range**
> High value minus low value.

Example ——————— For the six values recorded in the nurse-in-training study, 10, 6, 5, 14, 6, and 13 minutes, find the range.

Solution Range = high value minus low value = 14 − 5 = 9

$$Range = 9 \text{ minutes}$$

This can also be expressed as: The data ranged from 5 to 14 minutes.   ■

Although easy to compute, the range offers no information about the distribution of data between these two extremes of high and low value and, thus, the range is mostly used as a rough gauge in determining dispersion or spread.

Standard Deviation

> **Standard Deviation**
> A form of average distance from the mean.

The **standard deviation** is a more complex measure and perhaps best explained through the following example.

Suppose we were to represent the six observations in our nurse-in-training study, 5, 6, 6, 10, 13, and 14, as blocks on a number line, as follows:

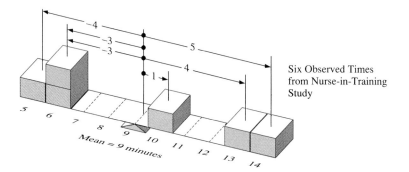

Six Observed Times
from Nurse-in-Training
Study

Notice in the above diagram, we are measuring the distance (in minutes) each value is away from the mean. Don't read on. Please look at the diagram until you understand it.

For instance, the value recorded as 13 is how many minutes away from the mean? The answer is 4 ($13 - 9 = 4$). Symbolically, we would represent this as follows:

$$x - \bar{x} = \text{distance from mean}$$
$$13 - 9 = 4 \text{ minutes}$$

The symbol x represents one value in our data, and $\bar{x}$ is the sample mean or average. When you subtract the two, $x - \bar{x}$, you get the distance a value is away from the mean. To calculate all the distances we use the following chart:

x	$\bar{x}$	$x - \bar{x}$
5	9	−4
6	9	−3
6	9	−3
10	9	1
13	9	4
14	9	5

Now, if we wish to calculate the *average distance from the mean,* we simply add up all the distances and divide by n, the number of readings, which is six.

$$\frac{\text{Average Distance}}{\text{from the Mean}} = \frac{-4 - 3 - 3 + 1 + 4 + 5}{6} = \frac{0}{6} = 0 \text{ minutes}$$

Unfortunately, if we leave in the negative signs, we always get 0. Notice that the negative values cancel out the positive values. This always occurs. However, since we are interested in the absolute distance each value is away from the mean and not whether it's plus or minus, a simple solution would be to use the distances without the negative signs. To do this, we take the **absolute value** of the distances, as follows:

x	$\bar{x}$	$x - \bar{x}$	$\left| x - \bar{x} \right|$
5	9	−4	4
6	9	−3	3
6	9	−3	3
10	9	1	1
13	9	4	4
14	9	5	5

Note that the symbols | |, called *absolute value lines,* allow us to record the distances as positive values.*

Pictorially, we can represent the distances as follows:

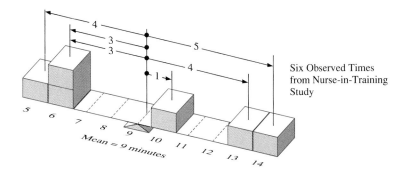

Six Observed Times from Nurse-in-Training Study

Now, to calculate the *average distance from the mean,* we would add up the distances and divide by n, the number of readings.

$$\begin{aligned}\text{Average Distance} \atop \text{from the Mean} &= \frac{\Sigma \left| x - \bar{x} \right|}{n} \\ &= \frac{4 + 3 + 3 + 1 + 4 + 5}{6} = \frac{20}{6} = 3\frac{2}{6} \\ &= 3\frac{1}{3} \text{ minutes}\end{aligned}$$

We now have a measure of how scattered or spread out the data is. We can now say, the average distance a value is from the mean is $3\frac{1}{3}$ minutes.

This measure of spread (along with others) was widely used in the 1800s and referred to by a number of names. Today, the value is most often called the

*Technical note: The absolute value of a number n, represented by $\left| n \right|$, is formally defined as its distance from 0 on the number line without considering direction. For example, $\left| -3 \right| = 3$, $\left| 5 \right| = 5$.

mean deviation or mean absolute deviation (m.a.d.), however we shall simply refer to it as the *average distance from the mean* because, in effect, that is what it is.

Pictorially, this measure of spread might be represented as follows:

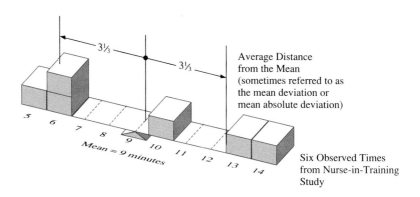

Despite its simplicity, this method for measuring spread is not in wide use today. The measure has a number of disadvantages, not the least of which are the computational difficulties in advanced work. As statisticians tried to apply this measure of spread to inferential statistics (using samples to estimate population characteristics), the problems grew unmanageable, mostly stemming from the use of the absolute value lines to remove the negative signs. For instance, the process of using absolute distances, $\left| x - \bar{x} \right|$'s, to remove the negative signs in complex calculations necessitates breaking up each problem into three cases: (a) when $x < \bar{x}$, (b) when $x = \bar{x}$ and (c) when $x > \bar{x}$. If we pool the absolute distances of many samples, as we do in later work, this mushrooms into a laborious effort. Other difficulties also arose involving discontinuous derivatives when calculus was applied. In other words, it was a statistician's nightmare.

In addition to this practical problem, it has been shown that this measure of spread was not as efficient in estimating the population spread as were other measures. That is, when we calculate the average distance from the mean for many samples drawn from the same population, these values would scatter more loosely around the *true* population value than if we had used other types of measures. If our goal is to use sample data to estimate population characteristics, we want our sample value to be the most efficient estimator of its equivalent population value* and the average distance from the mean was just not as efficient an estimator of its equivalent population value as other available measures.

---

*The term *most efficient* means sample values cluster closer to the population value.

Fortunately, the work of two great mathematicians of the early 1800s, Legendre (1805) and Gauss (1809, 1823), led to the development and ultimate adoption of another form of average distance from the mean, called the **standard deviation.***

The process of calculating the standard deviation involves squaring each distance, which removes the negative sign, for example, $(-3)^2 = +9$, and thus eliminates the need for the absolute value lines, which greatly reduces the computational difficulties in advanced work. In addition, using this new measure, the sample spread value becomes a more efficient estimator of the population spread value. In other words, sample values now cluster more tightly around the population value. Let's see how it works.

Calculation of Standard Deviation

x	$\bar{x}$	$x - \bar{x}$	$(x - \bar{x})^2$
5	9	-4	16
6	9	-3	9
6	9	-3	9
10	9	1	1
13	9	4	16
14	9	5	25

First, we square each distance. For example, the first distance, -4, is squared as follows: $(-4)(-4) = +16$.

Average squared distance (variance)
$$= \frac{\Sigma (x - \bar{x})^2}{n - 1}$$
$$= \frac{16 + 9 + 9 + 1 + 16 + 25}{6 - 1}$$
$$s^2 = \frac{76}{5}$$
$$= 15.2 \text{ squared minutes}$$

Second, we take the average squared distance by summing the squared distances, then dividing† by $n - 1$. This averaged squared distance is referred to as s^2, or the *variance*. Notice that this value, 15.2, is in the units of squared minutes

Standard deviation of sample
$$= \sqrt{\frac{\Sigma (x - \bar{x})^2}{n - 1}}$$
$$s = \sqrt{15.2}$$
$$= 3.899$$
$$= 3.9 \text{ minutes (rounded)}$$

Third, to convert squared minutes back to minutes, we take the square root.‡

*For further historical discussion, refer to chapter 9, section 9.0, under the subheading "Least-Squares Analysis."

†Note that we divided by $n - 1$ (and not n) in our formula for *sample* standard deviation. Experience has shown that dividing by $n - 1$ slightly raises the value of the sample standard deviation and provides, on average, a more accurate estimator of the population standard deviation than if we had divided by n. This has been proven by both experience and theory. However, if we were to calculate the standard deviation of a *population,* we would simply divide by n.

‡Technical note: Actually s^2 (and not s) is the preferred measure of spread, since s^2 is an unbiased estimator of (meaning: on average, equal to) the equivalent population value σ^2. Unfortunately, this is not the case with s. On average, $s \neq \sigma$, however for large samples, the bias is quite small and usually ignored. We will ignore this consideration until later in the text.

The process can be summarized by the following formula:

For Ungrouped Data

$$\text{Sample Standard Deviation} = \sqrt{\frac{\text{sum of the squared distances}}{\text{number of readings minus one}}}$$

$$s = \sqrt{\frac{\Sigma(x - \bar{x})^{2*}}{n - 1}}$$

Let's see how all this might work in an example.

Example

In a medical study, a researcher wished to estimate the average length of time needed for a particular nurse-in-training to draw a series of blood specimens. A sample of the nurse's work over several months yielded the following times: 10, 6, 5, 14, 6, and 13 (in minutes). Calculate the standard deviation.

Solution

We would organize our data in chart form and calculate the standard deviation as follows.

x	$\bar{x}$	$x - \bar{x}$	$(x - \bar{x})^2$
5	9	-4	16
6	9	-3	9
6	9	-3	9
10	9	1	1
13	9	4	16
14	9	5	25

$$\Sigma(x - \bar{x})^2 = 76$$

$$s = \sqrt{\frac{\Sigma(x - \bar{x})^2}{n - 1}} = \sqrt{\frac{76}{6 - 1}}$$
$$= \sqrt{15.2} = 3.899$$
$$= 3.9 \text{ minutes}$$

Pictorially, we might represent the results as follows.

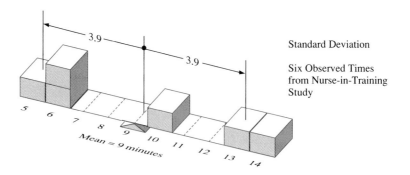

Standard Deviation

Six Observed Times from Nurse-in-Training Study

*The standard deviation may also be calculated with the formula

$$s = \sqrt{\frac{\Sigma x^2 - \dfrac{(\Sigma x)^2}{n}}{n - 1}}$$

known as *the counting formula*. Note that this formula requires only the sum of the x and x^2 columns, however it offers little in the way of understanding the process.

Notice that the standard deviation is still a form of average distance a value is away from the mean, even though we squared the distances, divided by $n - 1$, and took the square root.

Advantages of Using the Standard Deviation as our Measure of Spread

1. The negative signs are removed without use of the absolute value lines, which greatly reduces the computational difficulties in advanced work.

2. The standard deviation is a more efficient estimator of spread than the average distance from the mean.

3. The standard deviation is a *considerably* more efficient estimator of spread than many other measures considered, such as the median deviation.

   Note: The term *more efficient* means sample values cluster more tightly around the population value—that is, on average, sample values give closer approximations to the population value. For further reading on the standard deviation, refer to section 9.0 under the subheading ''Least-Squares Analysis''; also refer to endnote 16 in chapter 9.

2.4 Estimating Population Characteristics

The purpose of a sample is usually to gather information about a population. Two of the characteristics of a population we most frequently wish to know are the

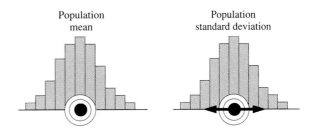

Unfortunately, small samples (under 30 observations) will sometimes give unreliable approximations of population characteristics, depending on the *shape* of your population histogram. (This will be discussed in greater detail in chapters 7 and 8).

One way to avoid this problem is to keep your sample size at 30 or more observations. If our sample size, n, is kept at 30 or more observations, we do not have to worry about the shape of our population. Results from sample sizes of 30 or more give reliable information about population characteristics for almost any shaped population. So, with this in mind, we can state, *it is preferable to keep your sample size at 30 or more.* More specifically, we can state the following.

> For a valid random sample of 30 or more observations, drawn from almost any population
>
> $\bar{x} \approx \mu$
> The sample average, $\bar{x}$, will be approximately equal to the population average,
> μ (mū).
>
> $s \approx \sigma$
> The sample standard deviation, s, will be approximately equal to the population standard deviation, σ (sigma).

We can also use samples of *smaller than 30* observations, but in that case, we must be assured our population is at least somewhat bell-shaped. In bell-shaped populations, small samples give reliable approximations of population characteristics, or at least reliable enough that after certain adjustments, reasonable conclusions can be drawn. So,

> For a valid random sample of *under 30 observations*, we must be assured our population is at least somewhat bell-shaped.

In the preceding examples, we sometimes used samples as small as five or six observations. If our population was somewhat bell-shaped, this would be perfectly okay. However, if our population was far from bell-shaped (let's say, for instance, extremely skewed), then we can *not* depend on these samples to give reliable estimates of population characteristics.

2.5 Measures of Central Tendency and Dispersion/Spread (Grouped Data)

When we work with large bodies of data, it is sometimes more efficient to group the data into categories. In the following example, we will calculate the mean and standard deviation for such data.

Mean

Example ——————— Say in our nurse-in-training study, the researcher took 36 observations of the nurse. Only now, instead of recording individual times, the researcher chose to record the times as part of a category or group, as follows:

Time Category (in minutes)	Number of Observations (tally)
3–5	ʍ I
6–8	ʍ ʍ I
9–11	ʍ II
12–14	ʍ IIII
15–17	III

Calculate the mean.

Rationale

Notice when the data is grouped, the identity of the individual values is lost. For instance, in the first tally, we know the nurse was observed on six occasions where it took 3 to 5 minutes to draw the blood, but we do not know the precise time it took on each of these six occasions.

Calculating the mean from this data proceeds as follows. Because it is quite difficult to use a range of time such as 3 to 5 minutes in a calculation, we merely average the two values $(3 + 5)/2 = 4$ and represent the group as 4 minutes. In reality, some measurements in this category are greater than 4 minutes and some less, but with many readings, experience has shown such measurements tend to average out to near 4 minutes.

Notice in the chart below, we used the letter x to represent this category average of 4 minutes and the letter f to represent on how many occasions this occurred, called frequency.

Time Category	x	Tally	f (frequency)
3–5	4	ᴺᴸ I	6
6–8	7	ᴺᴸ ᴺᴸ I	11
9–11	10	ᴺᴸ II	7
12–14	13	ᴺᴸ IIII	9
15–17	16	III	3

$$\Sigma f = 36 \text{ (or } n = 36)$$

Also note that the sum of the frequency column, Σf, is equal to the sample size, n. In other words, we say $\Sigma f = 36$ or $n = 36$. In either case, it tells us that we observed the nurse on a total of 36 occasions. (Note: we also get 36 by adding up the tally slashes.)

To calculate the mean, we simply add up the recorded times which we had observed and divide by 36.

$$\bar{x} = \frac{\Sigma x}{n} = \frac{\overbrace{4 + 4 + 4 + 4 + 4 + 4}^{\text{six readings}} + \overbrace{7 + 7 + 7 + 7 + 7 + 7 + 7 + 7 + 7 + 7 + 7}^{\text{eleven readings}} + \overbrace{10 + 10 +}^{\text{etc.}} \cdots}{36}$$

$$= \frac{336}{36} = 9.333$$

$$= 9.3 \text{ minutes}$$

This is quite tedious however. Another way is to simply multiply 4×6, 7×11, 10×7, 13×9, and 16×3, then add the results and divide by 36 as follows:

$$\bar{x} = \frac{\Sigma xf}{n} = \frac{4(6) + 7(11) + 10(7) + 13(9) + 16(3)}{36}$$

$$= \frac{24 + 77 + 70 + 117 + 48}{36} = \frac{336}{36} = 9.333$$

$$= 9.3 \text{ minutes}$$

This process is presented more efficiently in chart form in the following solution.

Solution   To calculate the mean, we sum the *xf* values and divide by 36.

Time

Category	x	f	xf
3–5	4	6	24
6–8	7	11	77
9–11	10	7	70
12–14	13	9	117
15–17	16	3	48

$\Sigma xf = 336$

$$\bar{x} = \frac{\Sigma xf}{n}$$

$$= \frac{336}{36}$$

$$= 9.333$$

$$= 9.3 \text{ minutes}$$

Based on these 36 observations, the average time the nurse-in-training took to draw the blood specimens was calculated to be 9.3 minutes. This can be summarized as,

$$n = 36 \text{ observations}$$
$$\bar{x} = 9.3 \text{ minutes}$$

The formula used for grouped data, then, is

Sample average for grouped data

$$\bar{x} = \frac{\Sigma xf}{n}$$

■

Standard Deviation

To calculate the **standard deviation,** we also use a slightly altered formula as follows:

Sample standard deviation for grouped data

$$s = \sqrt{\frac{n(\Sigma x^2 f) - (\Sigma xf)^2}{n(n-1)}}$$

The formula has been algebraically rearranged to avoid having to calculate each individual distance from the mean. It also avoids having to calculate the mean itself. This greatly accelerates the calculations.

Example ——————— For our 36 observations of the nurse-in-training, calculate the standard deviation.

Solution   First, we arrange the data in chart form, just as we did in the preceding example. Only this time, we include an additional $x^2 f$ column. The $x^2 f$ values are obtained by multiplying the *x* value by the *xf* value. For example, in the first row, we obtained an $x^2 f$ reading of 96 by multiplying 4 by 24, $4 \times 24 = 96$ (*x* times *xf* $= x^2 f$).

Time Category	x	f	xf	x²f
3–5	4	6	24	96
6–8	7	11	77	539
9–11	10	7	70	700
12–14	13	9	117	1521
15–17	16	3	48	768
		36	336	3624
		Σf	Σxf	$\Sigma x^2 f$
		$(n = 36)$		

$$s = \sqrt{\frac{n(\Sigma x^2 f) - (\Sigma xf)^2}{n(n-1)}}$$

$$= \sqrt{\frac{36(3624) - (336)^2}{36(36-1)}}$$

$$= \sqrt{\frac{130{,}464 - 112{,}896}{36(35)}}$$

$$= \sqrt{13.942} = 3.734$$

$$= 3.7 \text{ minutes}$$

Let's see what the data looks like using a three-dimensional histogram model.

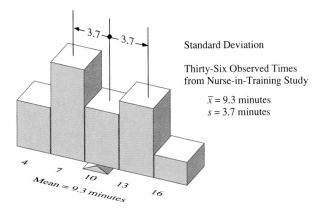

3.7 — 3.7

Standard Deviation

Thirty-Six Observed Times from Nurse-in-Training Study

$\bar{x}$ = 9.3 minutes
s = 3.7 minutes

4   7   10   13   16
Mean = 9.3 minutes

Notice that the sample average and sample standard deviation above ($\bar{x}$ and s) were calculated to be 9.3 and 3.7 minutes, respectively, based on $n = 36$ observations. In a prior section, based on $n = 6$ observations, we had calculated the $\bar{x}$ and s to be 9.0 and 3.9 minutes. Why are the sample results different?

Actually, the two most probable reasons are: first, sample values *approximate* population values—that is, $\bar{x} \approx \mu$ and $s \approx \sigma$, so sample results often differ slightly each time we sample from a population (and the smaller the sample size, the more variation we can expect in sample results); second, the process of grouping itself may have caused us to lose some precision.   ■

Advantages of Grouping Data

1. Ease in collection and organization: slash marks on a tally are much easier to record than individual values, and if we rotate the tally $\frac{1}{4}$ turn counterclockwise, we already have, in effect, a histogram.

2. Ease of interpretation: the central tendency and spread of a grouped-data tally can be *estimated* at a glance. This is not true of ungrouped data.

3. Speed: it is usually much faster to calculate the mean and standard deviation of a large body of data when the data is grouped than if we had recorded individual readings. Although with the hand calculator and personal computer this advantage is not as important as it once was.

Disadvantages of Grouping Data

1. Individual values in a sense lose their identity. As a result, we cannot reconstruct the data into different groupings. For instance, what if we wanted to reconstruct the data into the categories, 2–4, 5–7, 8–10 minutes, etc.? Without the individual readings, we lose our ability to do this.

2. Some precision *may* be lost. Remember, each value is recorded as part of a group. If this precision is vital for some subsequent decision-making process, then grouped data should be used with caution. For instance, in our nurse-in-training example, where we calculated the average to be $\bar{x} = 9.3$ minutes, it is theoretically possible the sample average is actually 8.3 or 10.3 minutes, although it is rare that this would happen. Generally, for large groups of data (hundreds of readings or more), we find the results from grouping data to be quite close to the results if we had measured each individual value. Nevertheless, why introduce this uncertainty when it can be avoided by just recording the individual values.* So, in the case where precision of results is critical, as in much of the work in inferential statistics, grouped data for these purposes should be avoided. The exception would be if the data is of a sufficient size (many hundreds or thousands of readings) in which case we would be less likely to find major differences when we calculated the sample average and standard deviation using grouped versus ungrouped techniques.

Modal Class

> **Modal Class**
> The category that contains the most values.

---

*Actually, certain techniques are available that allow us to record individual values yet maintain some efficiency of grouping. One such technique is called the stem-and-leaf display, which is presented in section 2.7.

The equivalent of the mode for ungrouped data is the **modal class** for grouped data, which is the category that contains the most values.

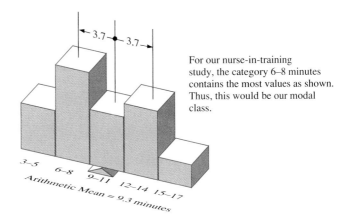

For our nurse-in-training study, the category 6–8 minutes contains the most values as shown. Thus, this would be our modal class.

2.6 *z* **Scores and the Use of the Standard Deviation**

In inferential statistics, we often refer to a particular value in terms of its position relative to the mean. In many instances, we use something called a *z* **score.**

> **z score**
> The number of standard deviations a value is away from the mean.

Although we will eventually introduce a formula, it is best to first learn *z* scores by estimation.

Example ——————— Suppose we took a sample from a grove of recently planted Indiana poplar trees and measured their heights, with the following results.

$$\bar{x} = 14 \text{ feet (average height)}$$
$$s = 4 \text{ feet (standard deviation)}$$

Estimate the *z* score for a poplar tree of the following heights.

a. 8 feet
b. 16 feet

Rationale

First, we would visually represent the mean and standard deviation of the data on a number line.

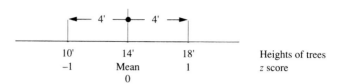

Notice that a tree of 18 feet would be four feet or exactly one standard deviation above the mean. Thus, we say a tree of 18 feet has a z score of 1. Remember, a z score is defined as the number of standard deviations a particular value is from the mean. Because a tree of 18 feet is one standard deviation above the mean, it has a z score of 1. Also notice that a tree of 10 feet is 4 feet below the mean. It has a z score of -1.

Solution

a. To estimate the z score of a tree with a height of 8 feet, we first must expand our scale to include ± 2 standard deviations, then locate the position of 8 feet.

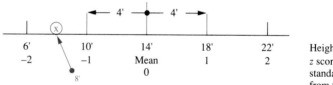

Notice that a tree of 8 feet would be $-1\frac{1}{2}$ standard deviations from the mean, therefore it has a z score of $-1\frac{1}{2}$ (or in decimal form, $z = -1.50$).

b. To estimate the z score of a tree of 16 feet, we locate on the scale where a tree of 16 feet would be.

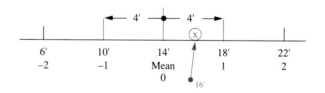

A tree of 16 feet would be $\frac{1}{2}$ of a standard deviation above the mean, therefore it has a z score of $\frac{1}{2}$ (or in decimal form, $z = .50$).

Estimating *z* scores is quite important in the study of inferential statistics, so let's look at a second example.

Example —————— Suppose the diameter of a fiber-optic strand (a glass fiber capable of transmitting hundreds of thousands of times more information than a copper wire) is continuously measured on an electronic assembly line, such that

Fiber-optic strand

Diameter

$\mu = 3.112$ mm* (average diameter)
$\sigma = .008$ mm (standard deviation)

Estimate the *z* score of a fiber-optic strand with the following diameters.

a. 3.120 mm
b. 3.110 mm

Solution Notice that μ and σ were used to represent the mean and standard deviation. This implies every segment of fiber in the entire population had been measured.

a. To estimate the *z* score of a piece of fiber with a diameter of 3.120 mm, we would visually represent the mean and standard deviation on a number line and locate the position of 3.120 mm.

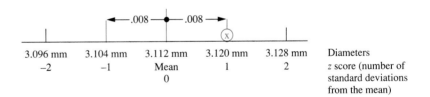

A fiber with a diameter of 3.120 mm is .008 mm or exactly one standard deviation above the mean, therefore it has a *z* score of 1 (that is, $z = 1.00$).

b. To estimate the *z* score of a piece of fiber with a diameter of 3.110 mm, again we represent the mean and standard deviation on a number line, only this time we locate the position at 3.110 mm.

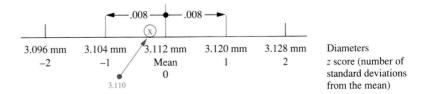

*Note: 3.112 mm (millimeters) is approximately $\frac{1}{8}$ inch.

Because 3.110 mm is .002 mm below the mean and .002 is exactly $\frac{1}{4}$ of .008, we state that 3.110 mm is $\frac{1}{4}$ of a standard deviation below the mean. Therefore, 3.110 mm has a z score of $-\frac{1}{4}$ (that is, $z = -.25$). ■

In inferential statistics, z score notation is commonly used. From chapter 4 on, we will be using it on a routine basis.

Two Important Findings

Along with the mean, the standard deviation is one of the most important measures we have in inferential statistics and much time has been devoted to its study.
One important finding was made by P. L. Chebyshev (1821–1894).*

For any set of data, whether it be population or sample data, no matter what its shape,

at least 75% of your data will lie within two standard deviations of the mean, and

at least 89% of your data will lie within three standard deviations of the mean.

Pictorially, this might be represented as follows.

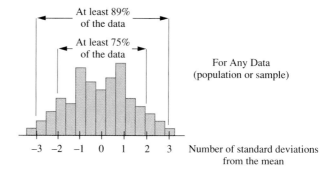

A second important finding was discovered seemingly under different circumstances by De Moivre (1733), Laplace (1781), and Gauss (1809), and this is called the normal population distribution. This distribution (a specific form of bell-shaped data) was encountered so frequently in experiments that the name *normal* was adopted sometime in the mid to late 1800s (from unknown origins).

*Chebyshev's Theorem states: at least $1 - \frac{1}{k^2}$ of any collection of data lies within k standard deviations of the mean. For $k = 2$ at least $1 - \frac{1}{2^2}$ of the data lies within 2 standard deviations of the mean. Since $1 - \frac{1}{2^2} = 1 - \frac{1}{4} = \frac{3}{4}$, we can say: at least $\frac{3}{4}$ of the data (or 75%) lies within two standard deviations of the mean. For $k = 3$, at least $1 - \frac{1}{9} = \frac{8}{9}$ of the data (or 89%) lies within three standard deviations of the mean.

> **For Normal Population Distributions**
> Approximately 68% of the data lies within one standard deviation of the mean.
> Approximately 95% of the data lies within two standard deviations of the mean.
> Approximately 99.7% of the data lies within three standard deviations of the mean.

Pictorially, this might be represented as follows.

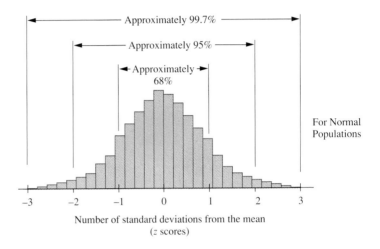

Number of standard deviations from the mean
(z scores)

The normal distribution is one of the most important distributions we study in inferential statistics and it is wise to take a moment to commit the above approximations to memory.

2.7 Additional Descriptive Topics

Although the material presented thus far provides the main thrust of descriptive techniques needed for future chapters, several variations of these techniques are in popular use and will be presented here. Specifically we shall introduce pictograms, stem-and-leaf displays, and box-and-whisker plots. Quartiles and percentiles are also briefly discussed.

Pictogram

Tallies and histograms are often presented in more dramatic form. One of the most common is a **pictogram** where illustrations or pictures are used to represent data, demonstrated as follows.

Suppose a random sample of small-business owners were asked at what age they first became self-employed. A tally of the results is presented below, left. A pictogram for this data is presented at right.

Ages	Tally	Ages	Pictogram
20–24	IIII	20–24	♦♦♦♦
25–29	NH IIII	25–29	♦♦♦♦♦ ♦♦♦♦
30–34	NH II	30–34	♦♦♦♦♦ ♦♦
35–39	II	35–39	♦♦
40+	I	40+	♦

In essence, pictures are used in lieu of slash marks. In the tally (at left), if each slash mark represents one person in the survey, then in the pictogram, each symbol ♦ would represent one person in the survey (note, if each slash mark represented 100 people, then each symbol would represent 100 people).

Essentially the pictogram attempts to offer a more visually appealing display of the data and is often used in business charts and in newspaper and magazine articles for dramatic effect.

Stem-and-Leaf Display

As useful as the tally (or pictogram) is in presenting data, often information is lost when we use such processes. For instance, in the above example, the exact ages of the individuals are lost. One slash mark in the category 25–29 could represent a person who is 26 years old or a person 29 years old. We have no way of knowing.

The **stem-and-leaf display** makes an effort to combine the ease and clarity of a tally while maintaining the original information. Suppose the slash marks in the above tally actually represented the following ages.

24 21 23 22 25 29 28 27 27 26 28 25 34 33 30 31 31 32 32 37 35 29 41

A stem-and-leaf display separates each value into a stem (usually the leftmost digit or digits) and a leaf (consisting of a digit or digits to the right of the stem). For instance, the first two ages 24 and 21 could be represented as 2 | 4 1. Each has the same stem, 2 (the leftmost digit) with leaves 4 and 1. Recording all the data in this fashion, we get

Stem	Leaves		
2	4 1 3 2		(last digit 0 to 4)
2	5 9 8 7 7	6 8 5 9	(last digit 5 to 9)
3	4 3 0 1 1	2 2	(last digit 0 to 4)
3	7 5		(last digit 5 to 9)
4	1		

Notice we laid out the stem-and-leaf display in a manner similar to the tally and pictogram, however, now, if needed, the original ages can easily be retrieved, say for instance, if we wish to calculate a precise sample average.

Box-and-Whisker Plot

Another method of presenting data is the **box-and-whisker plot** (sometimes referred to as a **boxplot**). Essentially the technique attempts to divide the data into four equal groupings using the following technique.

1. List the values in increasing order and locate the median (the middle value)

21 22 23 24 25 25 26 27 27 28 28 $\boxed{29}$ 29 30 31 31 32 32 33 34 35 37 41
↑
Median

2. Mark off the lower half of the data (including the median) and locate the *left hinge,* defined as the median of this lower half. Repeat for higher half of data, locating the *right hinge*.

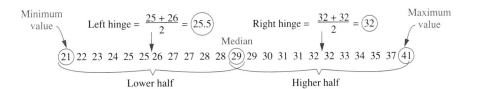

Note: If we have an even number of values, as above, the average of the two middle values is used as the median.

3. Use the 5 values (minimum, left hinge, median, right hinge, and maximum), circled in the above diagram, to construct the box-and-whisker plot as follows:

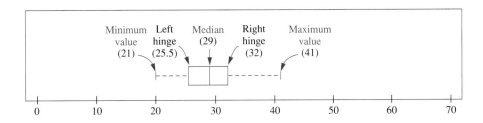

The box between the left and right hinges contains the central core of the data (the middle 50% of all the values), with the lower 25% of values represented by the dashed lines to the left of the box and the upper 25% of values represented by the dashed line to the right of the box. The dashed lines are referred to as **whiskers.**

Essentially, the box-and-whisker plot allows us to quickly identify the data's central core (middle 50%) while displaying the span of those values outside the central core.

Quartiles

Like the box-and-whisker plot, quartiles attempt to divide the data into four equal groups (quarters). The procedure is almost identical, only now the left and right hinges (called Q_1 and Q_3 in quartiles) are established as follows: Q_1 is the median of the values to the left of the overall median value of 29 (however, this procedure does *not* include the median value, 29, in the computation). Q_2 is the overall median. Q_3 is the median of the values to the right of the overall median.

21 22 23 24 25 $\left(25\right)$ 26 27 27 28 28 $\left(29\right)$ 29 30 31 31 32 $\left(32\right)$ 33 34 35 37 41

                        Q_1 Q_2 Q_3

Percentiles

Percentiles is a method used to divide data into 100 approximately equal groupings.

2.8 Writing Research Reports

Research reports are common as a means of formally presenting the results of a research study. Although different formats exist, most include the following information.

Background Statement

Essentially, this is a statement of the problem. Why was the study performed? What questions do you expect to answer? Usually, a brief history of events leading to the commissioning of the study is presented, outlining the need, purpose, and expectations of the study.

For instance, in the Countrygirl Makeup example in section 2.1, it was stated that, "although successful when introduced in 1980, in recent years, sales have eroded," suggesting that a decline in sales has prompted the company to commission the study. Further stated was, "an independent research firm was commissioned to gather information about the ages of current users,'' implying a need to explore the question of whether the ages of the current target population, young women ages 16 to 24, have shifted.

If prior studies have been performed, the results should be included. If there has been much work done in this area, then these results should be fully presented in a separate section.

Design and Procedures of the Study

Essentially, this section answers the question of *how* the study was conducted. If sampling is to be used, which is often the case, discuss your target population and how you intend to sample from this target population. Also discuss how your sampling techniques provide for (1) internal validity and (2) external validity (a major factor in determining external validity is the component of whether a *random* sample was achieved).

Although questions concerning validity can grow quite complex, you might consider the following:

Internal validity: Were honest, accurate and reliable measurements achieved under the given test conditions?

- Was the measuring scale objective?
- How was the accuracy of the measurements verified?
- Did you lose sample results, say with people refusing to cooperate?
- In more complex experiments, requiring measurements over the passage of time as in educational or medical studies, did outside events affect the measurement? Did the natural maturation of subjects affect the measurement? Did some initial test affect the final test?

External validity: Were you successful in achieving a true *random* sample? Did the test methods influence results?

- Did the presence of the experimenter exert an influence?
- Did the test environment create an unnatural setting and thus affect results? In the case where the sampled group is put in the same room and questioned, did one response affect another?

Suggested reading in this area is D. T. Campbell and J. C. Stanley, *Experimental and Quasi-Experimental Designs for Research* (Boston: Houghton Mifflin Co., 1963).

Results

Data is summarized visually in the form of histograms, frequency polygons, circle graphs, and charts and tables of vital information, such as means, medians, standard deviations, and so on.

For instance, in the Countrygirl study, a histogram might be presented with a summary chart of vital information as follows:

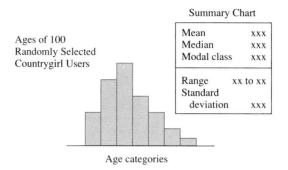

The goal is to provide a clear, easy-to-read *summary* of results. The reader should be able to look at this summary and within seconds understand the complete test results.

Note: Do *not* include raw data or computer printouts in the body of the report. These should be attached at the end as appendixes.

Analysis and Discussion

The visuals (histograms, etc.) and summary charts and tables are analyzed in regard to the central purpose of the study, which was presented at the beginning of the report under background. Interesting facts and significant findings should be pointed out and discussed.

For instance, in the Countrygirl study, one might give the percentage of users in the sample outside the current age range of 16 to 24 year olds, because a large percentage of users outside our current target age range might have relevance to future advertising and promotion decisions. Compare this to what percentage of the sample was *inside* the target population age range of 16 to 24. Also compare the average and median age from the sample with the average age of the current target population.

In addition, suggest reasons why the data turned out the way it did. Were the results surprising? If so, point out why this might happen.

Conclusions and Recommendations

Draw conclusions and suggest future action as if a valid random sample were achieved and the results can be used as representative of the population.

If serious questions exist as to the internal or external validity of the sample, provide a cautionary note and suggest how these problems might be overcome in a future study.

Summary

Graphical representations provide a visual sense of the data, allowing an immediate registration of the data's impact. The following were introduced.

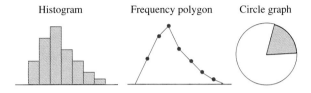

Histogram   Frequency polygon   Circle graph

Experience has shown that many *population* histograms take on repeating shapes, that is, there are certain common shapes that seem to reoccur.

Common Shapes of Population Histograms

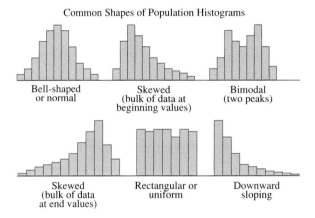

Bell-shaped or normal   Skewed (bulk of data at beginning values)   Bimodal (two peaks)

Skewed (bulk of data at end values)   Rectangular or uniform   Downward sloping

In addition to graphical representations, we often wish to get some sense of the middle or central value of our data. This is called central tendency and the most frequently used measures of central tendency are given below.

Arithmetic mean:

Commonly called the average or mean, this value is calculated by adding up all the values in your data collection and dividing by *n,* the

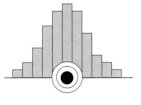

number of values. To calculate the arithmetic mean of a sample, we use the following formulas.

Ungrouped data: $\bar{x} = \dfrac{\Sigma x}{n}$ (sum of the values) (number of values)

Grouped data: $\bar{x} = \dfrac{\Sigma xf}{n}$ (sum of the xf values) (number of values)

Median: The middle value when data is arranged from lowest to highest value. If there are two middle values, we average the two.

Mode: The most frequently occurring value. Some sets of data may have no mode. If two modes occur, we call this bimodal. If three or more modes occur, we call this multimodal.

Modal class: The equivalent of the mode for grouped data is the modal class, which is the category that contains the most values.

Whereas measures of central tendency attempt to locate the center or middle of the data, measures of dispersion are designed to measure how widely scattered or spread out the data is.

Range: The range is the difference between the high and low value in your data set.

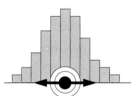

Standard deviation: A form of average distance from the mean. Along with the mean, the standard deviation is one of the most important measures we have in inferential statistics and much time has been devoted to its study. To calculate the standard deviation of a sample, we use the following formulas.

Ungrouped data: $s = \sqrt{\dfrac{\Sigma(x - \bar{x})^2}{n - 1}}$

Grouped data: $s = \sqrt{\dfrac{n(\Sigma x^2 f) - (\Sigma xf)^2}{n(n - 1)}}$

The purpose of a sample is usually to gather information about a population. Two of the characteristics of a population we most frequently wish to know are μ, the population mean, and σ, the population standard deviation. We know from well over a century of practical experience that valid random samples of 30 or more will give reliable information about μ and σ.

More specifically we can state, *for a valid random sample of 30 or more observations, drawn from almost any shaped population.*

$\bar{x} \approx \mu$: the sample average, $\bar{x}$, will be approximately equal to the population average, μ (m$\bar{u}$).

$s \approx \sigma$: the sample standard deviation, s, will be approximately equal to the population standard deviation, σ (sigma).

Samples of under 30 observations can also be used, however *for a valid random sample of under 30 observations, we must be further assured our population is at least somewhat bell-shaped.*

In inferential statistics, we often refer to a particular value in terms of its relative position to the mean. In many instances we use something called a *z* score.

z score: The number of standard deviations a value is away from the mean. For instance, if $\bar{x} = 14$ and $s = 4$, then a value of 18 would be one standard deviation above the mean, expressed as $z = 1.00$.

Exercises

Note that full answers for exercises 1–5 and abbreviated answers for odd-numbered exercises are provided in the Answer Key.

2.1 Countrygirl Makeup is a line of facial products marketed to young women, ages sixteen to twenty-four. Although successful when introduced in 1980, in recent years Countrygirl sales have eroded. An independent research firm was commissioned to gather information about the ages of current users. A nationwide random sample of 50 users yielded the following ages:

					Age Category	Number of Observations (tally)
19.4	31.3	22.7	27.6	27.9		
30.1	23.1	26.4	32.1	22.5		
28.2	25.7	33.8	28.9	18.6	15–19	
26.8	30.5	34.0	21.6	28.2	20–24	
32.2	27.3	17.5	23.0	32.8	25–29	
36.0	29.1	42.7	30.5	39.0	30–34	
26.2	33.2	36.3	22.7	43.1	35–39	
28.7	26.3	38.6	24.1	21.3	40–44	
32.1	28.7	25.8	26.0	18.7		
18.2	23.9	28.2	20.2	33.1		

a. Tally the data into the age categories given above.
b. Construct separately a histogram and frequency polygon.
c. What is the population?
d. What is the sample? How large is *n,* the sample size?

2.2 In exercise 2.1, 21 out of 50 possess the attribute of brown hair.

a. What proportion (or fraction) of the sample, p_s, possesses the attribute of brown hair?
b. Convert this fraction to a percentage.
c. Represent this proportion in a circle graph.

2.3 In a medical study, a researcher wished to estimate the average length of time needed for a particular nurse-in-training to draw a series of blood specimens. A sample of the nurse's work over several months yielded the following times: 11, 7, 13, 7, 5, 8, 10, and 15 (in minutes).

Calculate:

a. Mean
b. Median
c. Mode
d. Range
e. Standard deviation

Discuss:

f. What conditions would be necessary for us to use the mean and standard deviation of the sample as representative of the mean and standard deviation of the population?

2.4 Suppose in our nurse-in-training study, the researcher took 49 observations of the nurse. Only now, instead of recording individual times, the researcher chose to record the times as part of a category or group, as follows.

Calculate:

a. Mean
b. Standard deviation
c. Modal class

Construct:

d. Histogram
e. Frequency polygon

Time category (in minutes)	Number of observations (tally)
3–5	THL II
6–8	THL THL THL
9–11	THL THL THL I
12–14	THL III
15–17	III

2.5 Suppose we sampled from a grove of recently planted Indiana poplar trees and measured their heights, obtaining the following:

$$\bar{x} = 14 \text{ feet (average height of trees sampled)}$$
$$s = 4 \text{ feet (standard deviation of trees sampled)}$$

Estimate the z score for a poplar tree of the following height.

a. 20 feet
b. 10 feet
c. 15 feet

2.6 In a study on shyness at the University of Iowa, a team of psychologists asked 7 participants to rank the anxiety created by being at a party with strangers on a scale from 0 (no anxiety) to 20 (maximum anxiety), yielding the following scores: 17, 19, 16, 14, 19, 15, and 12.

Calculate:

a. Mean
b. Median
c. Mode
d. Range
e. Standard deviation

Discuss:

f. What conditions would be necessary for us to use the mean and standard deviation of the sample as representative of the mean and standard deviation of the population?

2.7 In an educational study of second-graders in Westchester County, N.Y., it took 5 students the following times (in seconds) to put together a simple puzzle: 12, 14, 7, 13, and 9.

Calculate:

a. Mean
b. Median
c. Mode
d. Range
e. Standard deviation

Discuss:

f. What conditions would be necessary for us to use the mean and standard deviation of the sample as representative of the mean and standard deviation of the population?

2.8 A sample of $n = 100$ Countrygirl Makeup users were randomly sampled nationwide, yielding the following ages (taken from the demonstration problem at the beginning of this chapter).

Calculate:

a. Mean
b. Standard deviation
c. Modal class
d. If Countrygirl is currently marketed to 16–24 year olds, and if we managed to achieve a valid random sample above, what effect might these results have on future advertising?

Age category	Number of observations (tally)
15–19	THL THL I
20–24	THL THL THL THL IIII
25–29	THL THL THL THL THL THL
30–34	THL THL THL III
35–39	THL THL I
40–44	THL
45–49	I
	Total, 100 users

2.9 At $n = 70$ boutiques randomly selected throughout the New England sales district, the following number of Rolf Laurie designer bed comforters sold last year were

Calculate:

a. Mean
b. Standard deviation
c. Modal class

Construct:

d. Histogram
e. Frequency polygon

Rolf Laurie comforters sold (last year)	Number of New England boutiques (tally)
0–14	THL
15–29	THL THL THL IIII
30–44	THL THL THL THL III
45–59	THL THL
60–74	THL III
75–89	THL
	Total, 70 boutiques

2.10 If the average amount wagered per person in state lotteries in a particular year was $\mu = \$100$ with standard deviation $\sigma = \$12$, find the z score for a person who wagered

a. $130.
b. $96.

2.11 A vending machine is known to fill cups to an average of $\mu = 7.0$ ounces with standard deviation $\sigma = .4$ ounces. Find the z score for a person that gets a cup with

a. 7.6 ounces.
b. 7.1 ounces.
c. 6.7 ounces.

2.12 Say you took five quizzes in Economics and your average was 76. Four of the five quizzes were graded, 80, 72, 86, and 70; however, one quiz grade was lost. Use the formula for calculating an average to determine the missing grade.

2.13 A young couple, Jason and Rebecca, weighed 170 lbs and 138 lbs, respectively. For their age and body structure, a medical association published the following guidelines:

Male	Female
$\mu = 155$	$\mu = 118$
$\sigma = 12$	$\sigma = 10$

a. Calculate the appropriate z scores for each.
b. According to these guidelines, who would be considered more seriously overweight?

2.14 The Jontnas Company has seven employees with the following annual salaries: $20,000, $20,000, $20,000, $40,000, $40,000, $40,000, and $240,000.

a. Calculate the mean and median salary.
b. Which do you feel would be a more realistic measure of central tendency, the mean or median salary?

2.15 Experience has shown that many *population* histograms take on a similar appearance. In other words, there are certain common or popular shapes that seem to reoccur.

List population values that might take on each of the following shapes.

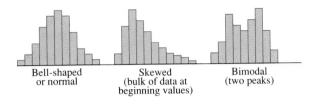

| Bell-shaped or normal | Skewed (bulk of data at beginning values) | Bimodal (two peaks) |

2.16 For $n = 200$ randomly selected New England boutiques, 36 carried Rolf Laurie designer blouses.

a. What proportion (or fraction) of the sample, p_s, carries Rolf Laurie designer blouses?
b. Convert this fraction to a percentage.
c. Represent this proportion in a circle graph.

2.17 A West Coast professor's definition of good character includes the following two qualities, "empathy, meaning regards for the needs, rights, and feelings of others, and self-control, meaning the ability to act with reference to the more distant consequences of current behavior." Suppose a test evaluating good character was administered to twenty-one local politicians, resulting in the following scores (10–49 scale):

18 26 23 21 26 29 30 33 32 34
38 38 36 37 41 40 47 41 42 43

a. Construct both a tally and a pictogram (invent your own symbol) using the following categories: 10–19, 20–29, 30–39, and 40–49.
b. Construct a stem-and-leaf display.
c. Construct a box-and-whisker plot.
d. Locate the quartile points, Q_1, Q_2, and Q_3.

(Reference article is *Newsweek*, "A Sterner Kind of Caring," January 13, 1992, p. 68.)

2.18 "While IQ tests remain excellent predictors of how well one will do in school, they have little or nothing to do with who will earn the most money or prestige, or have the most satisfying social life or relationships," according to *New York Times* article, "New Scales of Intelligence Rank Talent for Living" (April 5, 1988, p. C1). "One factor emerging as crucial for life success is what might be called emotional intelligence. How well people manage their emotions determines how effectively they can use their intellectual ability."

Suppose a test evaluating emotional IQ was administered to nineteen corporate executives, resulting in the following scores (20.0–23.9 scale):

20.7 20.6 21.3 21.8 21.9 21.8 22.4 22.4 22.5 22.4
22.8 22.6 22.9 23.1 23.4 23.5 23.2 23.9 23.7

a. Construct both a tally and a pictogram (invent your own symbol) using the following categories: 20.0–20.9, 21.0–21.9, 22.0–22.9, and 23.0–23.9.
b. Construct a stem-and-leaf display.
c. Construct a box-and-whisker plot.
d. Locate the quartile points, Q_1, Q_2, and Q_3.

Research Projects

The following research projects are offered for class or computer laboratory assignment (note: computer usage is optional). Each requires some preliminary class discussion and the student will issue a two to five page research report for each project.

Class Project A

Use the information and data from the Countrygirl Makeup example in section 2.1.

a. Discuss in class how one might design such a study to provide for (i) internal validity and (ii) external validity.
b. If students have access to a statistical computer package, feed in the raw (ungrouped) data and use the computer to analyze this input—that is, to calculate the mean, median, standard deviation, and other vital information.
c. If students do not have access to a statistical computer package, analyze the data using the grouping techniques as demonstrated in homework exercise 2.8.
d. Issue a research report, as described in section 2.8.

Class Project B

In a medical study, a researcher wished to estimate the average length of time needed for a particular nurse-in-training to draw a series of blood specimens. A sample of 49 observations of the nurse's work over several months yielded the following times (in minutes):

7.3	6.9	15.9	5.4	13.1
9.9	8.4	9.6	4.1	11.1
12.1	14.1	11.3	10.5	4.7
5.7	5.1	10.7	7.3	9.9
12.5	7.3	6.7	7.1	8.1
5.7	16.5	10.8	11.0	8.9
13.7	10.6	7.4	8.5	4.2
15.4	8.1	11.8	9.5	6.3
5.4	3.9	12.0	13.7	10.6
6.6	9.5	8.0	9.4	

a. Discuss in class how one might design such a study to provide for (i) internal validity and (ii) external validity.
b. If students have access to a statistical computer package, feed in the raw (ungrouped) data above and use the computer to analyze this input—that is, to calculate the mean, median, standard deviation, and other vital information.
c. If students do not have access to a statistical computer package, analyze the data using the grouping techniques as demonstrated in homework exercise 2.4.
d. Issue a research report, as described in section 2.8.

Class Project C

Needed for project: Magazine clippings of two different male models.

Background: A chain of exclusive men's shops in your region, A. L. Lewton, is in need of an on-going model to represent their high-priced line of clothes. Executives of the chain have narrowed the selection to two finalists. Although a professional study was considered, the executives felt the cost was prohibitive and asked your professor to sample the class, feeling that students, young men and women, would be more at the forefront of current taste in clothes. Your professor warned the executives that there might be some questions as to the validity of the results, however the executives insisted.

To gather this data, any plan can be devised. One possible plan might be as follows:

1. Put the following scale on the chalkboard.

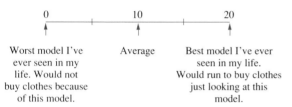

2. Students are informed that magazine clippings of the two male models, labeled Model A and Model B, are shown to them and *each* model is to be rated on the 0 to 20 scale, using the following criteria:

Suitability for modeling clothes

- manly structure
- pleasing looks

Appropriateness for A. L. Lewton image

Overall appeal

3. To ensure honest unbiased responses, the vote is to be secret and no one is to discuss the models.

4. One student will be in charge of distributing and collecting the voting slips while the professor walks around the room showing the clippings of the two models, side by side, to various parts of the class.

5. A student then collects and reads the votes aloud as the professor lists the votes on the chalkboard, as follows:

A	B
x	*x*
x	*x*
x	*x*
.	.
.	.
.	.

In a computer environment, the students will input the data as the votes are read aloud, first for Model A, then for Model B.

a. For this study, discuss (i) internal validity and (ii) external validity.

b. Feed in raw (ungrouped) data into the computer and calculate the mean, median, standard deviation, and other vital information.

c. If students do not have access to a computer package, analyze the data using the grouping techniques discussed in this chapter.

d. Issue a research report, as described in section 2.8.

Hints: Discussion of validity may include:

Was the scale objective? Does a vote from one person of, say, 12 reflect the same meaning as a vote of 12 from another person? Did student comments influence other students? Would different photographs of the models influence the ratings?

Was a random sample from the population achieved? Did the professor influence the vote in any way? Did the environment of the class or laboratory present a proper atmosphere for the voting or affect the vote in any way?

Results might include two histograms with the scales precisely lined up, making it easy for the reader to visually compare the outcome.

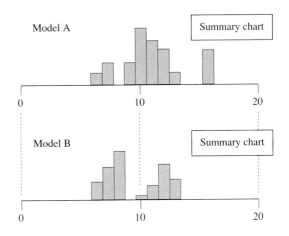

Note: these histograms are for demonstration purposes only.

Probability

T he study of probability grew out of the gambling dens of Europe over three centuries ago, when a few known mathematicians felt challenged to calculate optimum strategies for winning certain bets and, later, when patrons, often the aristocracy, further encouraged mathematicians with sums of money for providing such valued information. Much of what we study today is still based on this early work, especially in our analysis of proportion data and broadly in the use of basic terminology and underlying principles throughout the text.

In this chapter, we will study the concepts of probability mostly to help us understand and define what is meant by p, the proportion or percentage of a population that possesses a certain attribute, such as, the proportion or percentage of a population that possesses the attribute of red hair or diabetes. Later, we will discuss what we can expect when we sample from such a population.

In broad terms, we define **probability** as follows.

Probability
The proportion (or percentage) of times an event will occur in the long run, under similar circumstances. This probability is expressed as a number between 0 and 1 or as a percentage between 0 and 100.

Two methods we can use to obtain this probability are the empirical approximation and classical approach. We shall start with the empirical approximation. ▼

3.1 Probability Defined: Empirically

Empirical Approximation to Probability
The fraction of times an event *has actually occurred* over a great many experiments conducted over a long period of time under similar circumstances, expressed as

$$P(\text{an event}) \approx \frac{\text{Number of Times Event Has Actually Occurred}}{\text{Total Number of Experiments (or Attempts)}}$$

As the number of experiments increases, this empirical fraction gets closer and closer to the true probability.

It is important to note that use of the empirical fraction requires an experiment be performed a great many times over a long period and under similar circumstances.

Let's see how it works.

Example ——————— Out of 100 consecutive tosses of a dart at a Thursday night tournament, a player strikes the bull's-eye 30 times. Calculate the probability of a bull's-eye for this player.

Solution   Although we might be tempted to say,

$$P(\text{an event}) \approx \frac{\text{Number of Times the Event Has Actually Occurred}}{\text{Total Number of Attempts}}$$

$P(\text{bull's-eye}) \approx \dfrac{30}{100}$, this is incorrect.

This does not meet our full definition of empirical probability because we have not performed the experiment a great many times over a long period, and therefore this may or may not be the true probability. So the answer to this question is: from this limited information, *we cannot determine the probability* of a bull's-eye for this player. ■

Well, then, you might ask, how *do* we obtain this probability? One way is to merely continue to perform the experiment (tossing the dart) a large number of times. As the total number of dart throws increase, this fraction (which we will call the **cumulative fraction**) gets closer and closer to the true probability, expressed in the following law.

▼ *Law of Large Numbers*

If we continue to repeat an experiment a great many times under similar circumstances, the cumulative fraction of successes will tend to draw closer and closer to the *true* probability.

Let's see how it works.

Say, for instance, for the first 100 tosses, we get 30 bull's-eyes. Then for the next 100 tosses, we get 12 bull's-eyes. Although the two fractions for each experiment would be 30/100 and 12/100, the *cumulative* fraction is 42 bull's-eyes (30 + 12) out of 200 total tosses, which would equal in percentage:

$$\frac{42 \text{ bull's-eyes}}{200 \text{ tosses}} = 21\%$$

Note: 42/200 can be converted to a percentage as follows:

Percent = Fraction × 100

$= \dfrac{42}{200_2} \times \cancel{100}^{1} = 21\%$

We can present this cumulative fraction in chart form as follows:

Number of Tosses	Bull's-eyes	Cumulative Fraction	Cumulative Percentage
100	30	30/100	30%
100	12	42/200	21%

If we plotted these results on a graph, it would look as follows:

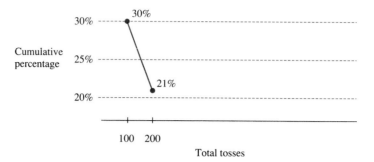

Now, what if we recorded a third set of 100 tosses where 9 bull's-eyes were achieved. The cumulative fraction would be 51 bull's-eyes (30 + 12 + 9) out of 300 total tosses, which equals 51/300 or 17%. And if we recorded a fourth set of 100 tosses, and so on for many sets of 100 tosses, we could represent this as follows:

Number of Tosses	Bull's-eyes	Cumulative Fraction	Cumulative Percentage
100	30	30/100	30%
100	12	42/200	21%
100	9	51/300	17%
100	21	72/400	18%
100	33	105/500	21%
100	27	132/600	22%
100	8	140/700	20%
100	12	152/800	19%
100	37	189/900	21%
100	21	210/1000	21%
100	10	220/1100	20%
100	14	234/1200	$19\frac{1}{2}\%$

Note that the cumulative fraction adds up all the prior bull's-eyes and all the prior tosses and presents this as one fraction. If we plot each cumulative percentage, we get

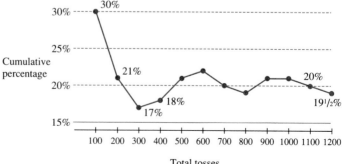

Notice that even though the number of bull's-eyes for each 100 tosses drastically fluctuated in the chart, ranging from 8 to 37, creating a rather ragged pattern at the beginning, the long-term line begins to smooth out as the number of tosses increases.

Experience has shown, in the long run, provided no change in dart throwing ability of the player or other factors that might affect this ability, the line will eventually grow flat (horizontal), sticking very close to one particular value. When this happens over many tosses, we call this percentage the *true* probability of a bull's-eye for this player.

If we examine the preceding graph, we see the probability seems to be leveling at about 20%. If indeed 20% is the true probability, we might represent this probability on a circle graph as follows.

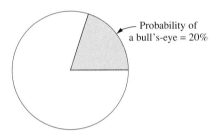

Probability, in essence, is a *population* value, the true percentage of times the player will hit the bull's-eye in the long run under similar circumstances. In effect, it represents what will happen in millions and millions of tosses. This probability may be presented as a percentage or decimal as follows:

$$P(\text{bull's-eye}) = 20\%$$
$$P(\text{bull's-eye}) = .20$$

Later, when we discuss sampling, this population value, 20%, will be referred to as the population proportion, p, expressed as

$$p = 20\% \quad \text{or}$$
$$p = .20$$

This empirical method for assigning a probability to an event is often used in the fields of psychology, education, biology, business, and medicine. For instance, if a surgeon says you have a 95% chance of surviving an operation, the surgeon is usually referring to an empirical probability. That is, in the long run, over many similar operations in the past, about 95 out of 100 people have survived.

Subjective Probability

One word of caution: too often, people will offer you probabilities off the top of their heads, which are no more than personal judgments. It is wise to request and examine the source of all probabilities offered when these probabilities are to be used in any subsequent decision-making process. Although some people are surprisingly astute in their ability to assess the probability of a situation, which is called **subjective probability,** others are not.

Subjective Probability
Assigning a probability based on personal judgment.

In this text, we use only those probabilities derived from the empirical method, as described in the last few pages, or from the classical method, which is discussed in the next section.

3.2 Probability Defined: Classically

The empirical definition of probability demands that we repeat an experiment a great many times before we can estimate the true probability. However, in practical situations this is not always possible. In a situation where we can be assured that every member of a set has an *equal chance of being selected,* then we have a second method for obtaining a probability. This method is called the classical approach.

Classical Probability
In an experiment of n different possibilities, each having an *equal chance of occurring,* the probability a particular event will occur is equal to

$$\frac{\text{Number of chances for success}}{\text{Total number of equally likely possibilities}}$$

This can be expressed as

$$P(\text{event}) = \frac{s}{n} \quad \begin{array}{l}\text{(number of chances for success)}\\ \text{(total equally likely possibilities)}\end{array}$$

Often the word, event, is replaced with the word, success, as follows:

$$P(\text{success}) = \frac{s}{n}$$

Example ——————— Suppose you attend a party of 20 people, of which 3 are famous TV celebrities. Now let's pretend a huge Green Giant were to walk up, lift the roof, reach down into the party and *randomly* pluck up 1 person by the collar. What is the probability the person selected is a famous TV celebrity?

Solution

This is a typical probability experiment. Out of $n = 20$ different possibilities (in this case, 20 different people), each having an *equal chance* of being selected, there are 3 chances for success. Since we have 3 chances for success out of 20 different equally likely possibilities, the probability of selecting a famous TV celebrity is given by the following formula:

$$P(\text{success}) = \frac{s \text{ (number of chances for success)}}{n \text{ (total equally likely possibilities)}}$$

$$P(\text{of selecting a famous TV celebrity}) = \frac{3}{20} \qquad \blacksquare$$

Often these fractions are expressed as percentages or decimals. For instance, the fraction $\frac{3}{20}$ may be expressed as 15% or .15. To convert the fraction $\frac{3}{20}$ to its equivalent percentage or decimal, we perform the following operation:

$$\% = \text{Fraction} \times 100$$
$$= \frac{3}{20} \times 100$$
$$= \frac{3}{20_1} \times \overset{5}{\cancel{100}}$$
$$= 15\%$$

Decimal = Numerator of fraction ÷ Denominator
$$= 3 \div 20 = .15$$

or

$$20\overline{)3.00} = .15$$
$$\underline{-2\,0}$$
$$1\,00$$
$$\underline{-1\,00}$$

So, instead of saying that the probability of selecting a famous TV celebrity is $\frac{3}{20}$, we might state it as 15% or .15. Whether we use $\frac{3}{20}$ or 15% or .15, these all indicate the same probability, which is 3 chances for success out of 20 possibilities.

Two Fundamental Properties

In dealing with probability fractions, there are two fundamental properties.

▼ Property 1

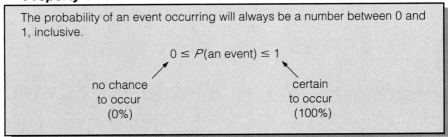

The probability of an event occurring will always be a number between 0 and 1, inclusive.

$$0 \leq P(\text{an event}) \leq 1$$

no chance
to occur
(0%)

certain
to occur
(100%)

A probability of $P = 0$ means the event has no chance of occurring. If at our party of 20 people we had *no* famous TV celebrities in attendance, then if the Giant reached in and selected one person, the probability of selecting a famous TV celebrity is 0 chances for success out of 20 possibilities or

$$P(\text{success}) = \frac{0}{20} = 0$$

$P = 0$ is the minimum probability. This means the event has no chance of occurring and can be expressed as $P = 0\%$.

A probability of $P = 1$ means the event is certain to occur. If at our party of 20 people we had 20 famous TV celebrities (in other words, all were famous TV celebrities), then if the Giant reached in and selected one person, the probability of selecting a famous TV celebrity is 20 chances for success out of 20 possibilities or

$$P(\text{success}) = \frac{20}{20} = 1$$

$P = 1$ is the maximum probability. This means the event is certain to occur and can be expressed as $P = 100\%$.

▼ Property 2

> The probability of an event occurring *plus* the probability of the event *not* occurring = 1.
>
> $$P(E) + P(\text{not } E) = 1$$

Basically this property says there is a 100% probability that the event will either occur or not occur.

Example ——————— What if at our party of 20 people, of which 3 are famous TV celebrities, a Green Giant *randomly* plucks up one person by the collar, what is the probability the person selected is *not* a famous TV celebrity?

Solution Since we now have 17 chances for success (people who are *not* TV celebrities) out of 20 equally likely possibilities,

$$P(\text{success}) = \frac{s \text{ (number of chances for success)}}{n \text{ (total equally likely possibilities)}}$$

$$P(\text{of } not \text{ selecting a famous TV celebrity}) = \frac{17}{20}$$

This can also be expressed as 85% or .85.   ■

Note that in our party example, the probability of *not* selecting a TV celebrity ($\frac{17}{20}$) plus the probability of selecting a TV celebrity ($\frac{3}{20}$) is equal to 1. This is a direct illustration of Property 2, as follows:

$$\text{Property 2: } P(E) + P(\text{not } E) = 1$$

$$\text{Expressed in fractions: } \frac{3}{20} + \frac{17}{20} = \frac{20}{20} = 1$$

Expressed in percentages: 15% + 85% = 100% = 1

Expressed in decimals: .15 + .85 = 1.00 = 1

Example

For a 52-card deck, if we randomly select one card, what is the probability that the card will be

a. a club?

b. not a club?

c. a king or a queen?

d. a jack and a spade?

Solution

To apply our probability formula, we must be assured every card has an *equal chance* of being selected. *Random* selection guarantees this, so now we can proceed.

a. Each deck is divided into four suits: 13 clubs, 13 diamonds, 13 hearts, and 13 spades. Because we have 13 chances for success (13 clubs) out of $n = 52$ equally likely possibilities,

$$P(\text{success}) = \frac{s}{n} \quad \begin{array}{l}\text{(number of chances for success)}\\\text{(total equally likely possibilities)}\end{array}$$

$$P(\text{club}) = \frac{13}{52} \quad (\tfrac{13}{52} \text{ can also be expressed as 25\% or .25})$$

b. Since we have 13 clubs in a deck, we must have 39 (52 minus 13) *non*clubs, thus 39 chances for success.

$$P(\text{success}) = \frac{s}{n} \quad \begin{array}{l}\text{(number of chances for success)}\\\text{(total equally likely possibilities)}\end{array}$$

$$P(\text{not club}) = \frac{39}{52} \quad (\tfrac{39}{52} \text{ can also be expressed as 75\% or .75})$$

c. Each deck has 4 kings and 4 queens. Since we have 8 chances for success (4 + 4 = 8) out of 52 equally likely possibilities,

$$P(\text{success}) = \frac{s}{n} \quad \begin{array}{l}\text{(number of chances for success)}\\\text{(total equally likely possibilities)}\end{array}$$

$$P(\text{king or queen}) = \frac{8}{52} \quad \begin{array}{l}(\tfrac{8}{52} \text{ can also be expressed as}\\\text{approximately 15\% or .15)}\end{array}$$

d. Note that the conditions for success in this problem require that we have a jack *and* a spade. The word, *and*, in statistics means that *both* conditions must be met for success. Since in a deck of cards we have only one card that satisfies both conditions, namely the jack of spades, we have only 1 chance for success out of 52 equally likely possibilities.

$$P(\text{success}) = \frac{s}{n} \quad \begin{array}{l}\text{(number of chances for success)} \\ \text{(total equally likely possibilities)}\end{array}$$

$$P(\text{jack } and \text{ spade}) = \frac{1}{52} \quad \begin{array}{l}(\frac{1}{52} \text{ can also be expressed as} \\ \text{approximately 2\% or approximately .02)}\end{array}$$

AND and OR Statements

Notice that the words **and** and **or** have very specific meanings in statistics.

And   The word *and* between two or more conditions or events implies *all* must be met for success.

Or   The word *or* between two or more conditions or events implies that *either* one or more may be met and that will give success.

Although many shortcut formulas are available to solve probability problems, each comes with restrictions that limit their use to a well-defined set of circumstances that can be quite confusing. It is best to first learn to solve these simple experiments (where we select *one* from a set of possibilities) by the methods described above. The following are offered for practice.

Practice Exercises

Practice 1 ───────── From a 52-card deck, if we *randomly* select one card, what is the probability the card will be

a. a heart *or* an ace?
b. a king *and* an ace?

Solution   **a.** Because we have 16 chances for success (13 hearts plus the aces of clubs, diamonds, and spades) out of 52 equally likely possibilities,

$$P(\text{heart } or \text{ ace}) = \frac{16}{52} \quad \begin{array}{l}\text{Note that the ace of} \\ \text{hearts was already} \\ \text{counted in the 13 hearts.}\end{array}$$

b. An *and* statement means both conditions must be met for success. Since there are 0 chances for success,

$$P(\text{king } and \text{ ace}) = \frac{0}{52} = 0 \quad \begin{array}{l}\text{Note that in selecting } one \\ \text{card, it is impossible to} \\ \text{get both a king } and \text{ an ace.}\end{array}$$

Practice 2 ———————— A chain of family video stores has their movies rated G, PG, R, X, or XX with the following probabilities:

$$P(G) = .31 \quad P(R) = .30 \quad P(XX) = .04$$
$$P(PG) = .25 \quad P(X) = .10$$

If you were to randomly select a video, what would be the probability the video would be rated

a. not G?
b. R or X or XX?

Solution **a.** $P(G) = .31$ means 31 out of 100 were rated G. Therefore, 69 (100 − 31 = 69) must have been rated something other than G. So,

$$P(\text{not G}) = \frac{69}{100} \quad (\text{or } .69)$$

b. Out of 100, 30 were rated R, 10 were rated X, and 4 were rated XX. Thus, 44 (30 + 10 + 4) were rated either R or X or XX out of 100. So,

$$P(\text{R or X or XX}) = \frac{44}{100} \quad (\text{or } .44) \qquad ■$$

Practice 3 ———————— In a regional survey of 1000 customers, a particular cable TV show was rated either favorable or unfavorable. Out of 760 favorable responses, 400 were female. And out of the 240 unfavorable responses, 80 were female.

If you were to randomly select one respondent from the survey, what would be the probability the respondent would be

a. female?
b. favorable and female?
c. favorable and male?
d. unfavorable *or* female?
e. male, given we already know the respondent voted favorable?   (Note: this is referred to as a **conditional probability**.)

Solution The information above can be summarized as follows:

1000 Customers

Favorable	Unfavorable
400 F	80 F
360 M	160 M

a. Because 480 were female (400 + 80),

$$P(\text{female}) = \frac{480}{1000} \quad (\text{or } .48)$$

b. Since it is stated that 400 of the favorable responses were female,

$$P(\text{favorable and female}) = \frac{400}{1000} \quad (\text{or } .40)$$

c. 760 rated cable TV favorably, of which 400 were female. This implies 360 were male. Thus,

$$P(\text{favorable and male}) = \frac{360}{1000} \quad (\text{or } .36)$$

d. An *or* statement implies *either* condition will give success. However, we must be careful not to count the same person twice. Because we have 240 unfavorable responses plus 400 *additional* females (from the favorable responses), we have 640 chances for success (240 + 400) out of 1000. Thus,

$$P(\text{unfavorable } or \text{ female}) = \frac{640}{1000} \quad (\text{or } .64)$$

Note: if we had reasoned there were 480 *total* females plus *240* total unfavorable responses, we would have counted 80 females *twice.*

e. The clause "given we already know the respondent voted favorable," is referred to as a *conditional* and the question referred to as a *conditional probability.* In effect, this conditional "given . . . favorable," limits the total set of possibilities to 760 favorable responses, out of which 360 males give us success. Thus, P(male, given we know the vote was favorable) = 360 chances for success out of 760 total possibilities.

$$P(\text{male, given favorable}) = \frac{360}{760} \quad (\text{or } .47).$$

Note: there were 400 females and 360 males in the 760 favorable responses.

Use of Mathematical Formulas in Simple Experiments

Although it is usually easier to solve simple experiments (where we select *one* from a set of possibilities) without using a formula, formulas are available and are most often preferred. We will demonstrate two common formulas with practice problem 3, parts d and e.

Practice 3(d) ————   Referring to practice problem 3, part d, if you were to randomly select one respondent from this survey, what would be the probability the respondent would be unfavorable *or* female?

Solution   The solution to practice problem 3, part d, can be solved by something known as the **addition rule:** Let E_1 = the first event and E_2 = the second event defined in a sample space, then

$$P(E_1 \text{ or } E_2) = P(E_1) + P(E_2) - P(E_1 \text{ and } E_2)$$

This subtracts out the elements counted twice.

Defining E_1 as unfavorable and E_2 as female,

$$P(\text{unfavorable or female}) = P(\text{unfavorable}) + P(\text{female}) - P(\text{unfavorable and female})$$

$$= \frac{240}{1000} + \frac{480}{1000} - \frac{80}{1000} = \frac{640}{1000}$$

Note: 80 females had to be subtracted out, expressed as the probability 80/1000, since they were counted twice, once in the unfavorable group and a second time in the female group.

If we have the situation where $P(E_1 \text{ and } E_2) = 0$, meaning the two events cannot occur together, then the events are referred to as **mutually exclusive** and the above formula reduces to $P(E_1 \text{ or } E_2) = P(E_1) + P(E_2)$.

Practice 3(e) ———— Referring to practice problem 3, part e, if you were to randomly select *one* respondent from this survey, what would be the probability the respondent would be male, given we already know the respondent voted favorable?

Solution The solution to practice problem 3, part e, can also be solved by a form of multiplication rule (to be discussed more fully later in the chapter), defined as

$$P(E_1, \text{ given } E_2) = \frac{P(E_1 \text{ and } E_2)}{P(E_2)}$$

$$P(\text{male, given favorable}) = \frac{P(\text{male and favorable})}{P(\text{favorable})}$$

Again, this is referred to as a conditional probability.

$$= \frac{360/1000}{760/1000} = \frac{360}{760}$$

3.3 More Complex Experiments: Tree Diagram

In the preceding examples, we selected *one* from a set of equally likely possibilities. For instance, we selected *one* person from a party of 20 people, and we selected *one* card from a deck of 52. However, in more complex problems, where we select *two* or more from a set of possibilities, it is often helpful to list all the equally likely outcomes of the experiment using a technique known as a **tree diagram.**

Example ———— Suppose we return to our party example. Only this time we attend a party of six people, of which two are famous TV celebrities. We shall use the numbers, 1, 2, 3, and 4 to identify the *non*-TV celebrities and the numbers 5 and 6 to identify the two famous TV celebrities.

   Now, what if a huge Green Giant were to walk up, lift the roof, and reach down into the party and randomly pluck up two people? What is the probability the two will both be famous TV celebrities?

Solution   If the Giant had selected one person from this party, this would have been a rather simple experiment. Selecting two is a little more complex as you will see.

To help us list all the equally likely ways we can select two people, we use a technique known as a *tree diagram,* as follows:

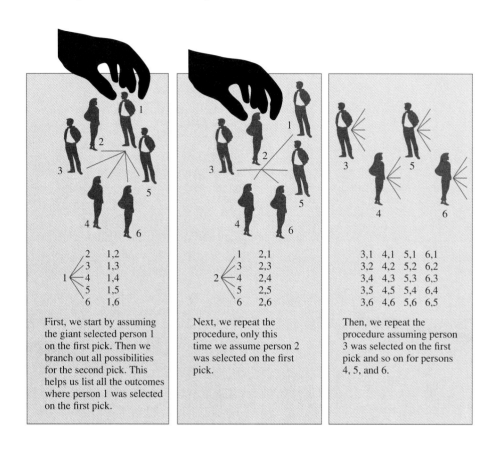

First, we start by assuming the giant selected person 1 on the first pick. Then we branch out all possibilities for the second pick. This helps us list all the outcomes where person 1 was selected on the first pick.

Next, we repeat the procedure, only this time we assume person 2 was selected on the first pick.

Then, we repeat the procedure assuming person 3 was selected on the first pick and so on for persons 4, 5, and 6.

Our complete listing would be as follows.

1,2	2,1	3,1	4,1	5,1	6,1	n = 30 equally likely outcomes
1,3	2,3	3,2	4,2	5,2	6,2	s = 2 chances for success (circled)
1,4	2,4	3,4	4,3	5,3	6,3	
1,5	2,5	3,5	4,5	5,4	6,4	
1,6	2,6	3,6	4,6	(5,6)	(6,5)	

This is called a **sample space** of equally likely outcomes. It represents all the ways the giant could have reached in and first selected one person, then reached in and selected a second person.

Note that outcome 5,6 is considered different from outcome 6,5. It is important to maintain the order of selection. This ensures that we have a complete listing of *all* equally likely outcomes.

Now we are ready to answer our question.

Since we have 2 chances for success (circled above) out of 30 equally likely outcomes,

$$P(\text{success}) = \frac{s}{n} \quad \begin{array}{l}\text{(number of chances for success)} \\ \text{(total equally likely outcomes*)}\end{array}$$

$$P(\substack{\text{celebrity}\\\text{1st pick}} \ and \ \substack{\text{celebrity}\\\text{2nd pick}}) = \frac{2}{30} \quad \text{(or 6.7\%)} \qquad \blacksquare$$

Now let's suppose we asked the same question but selected in a different manner.

Example ——————— Suppose at our party of six, of which two are famous TV celebrities, a Green Giant plucked up one person by the collar, *replaced* that one person into the party, then later returned and plucked up one person again.

What is the probability both picks would be famous TV celebrities?

Solution If we list our sample space of equally likely outcomes, we get

1,1	2,1	3,1	4,1	5,1	6,1	n = 36 equally likely outcomes
1,2	2,2	3,2	4,2	5,2	6,2	s = 4 chances for success
1,3	2,3	3,3	4,3	5,3	6,3	(circled)
1,4	2,4	3,4	4,4	5,4	6,4	
1,5	2,5	3,5	4,5	(5,5)	(6,5)	
1,6	2,6	3,6	4,6	(5,6)	(6,6)	

Note the addition of outcomes 1,1 and 2,2 and 3,3 and 4,4 and 5,5 and 6,6. Since we replaced the first person, we must include the possibility that the *same* person might be chosen twice.

Because we now have 4 chances for success (circled above) out of 36 equally likely outcomes,

$$P(\text{success}) = \frac{s}{n} \quad \begin{array}{l}\text{(number of chances for success)} \\ \text{(total equally likely outcomes)}\end{array}$$

$$P(\substack{\text{celebrity}\\\text{1st pick}} \ and \ \substack{\text{celebrity}\\\text{2nd pick}}) = \frac{4}{36} \quad \text{(or 11.1\%)} \qquad \blacksquare$$

We can also list sample spaces for three or more selections, as follows.

*Note: In more complex experiments, the word outcomes is used in place of possibilities.

Example ——————— In a woman's wardrobe of 3 blouses (white, beige, and tan), 1 scarf, and 2 skirts (navy blue and gray), a blouse and a skirt must be worn but a *scarf* is optional.

If we assume the woman *randomly* selected from each group, what is the probability the woman will be wearing a scarf *and* gray skirt?

Solution

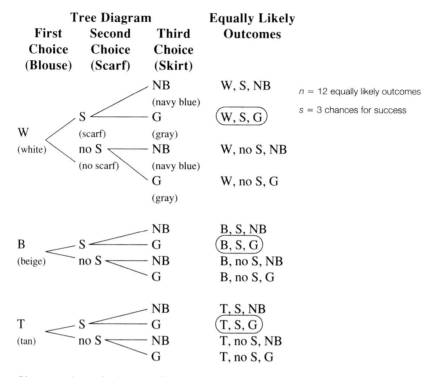

Tree Diagram

| **First Choice (Blouse)** | **Second Choice (Scarf)** | **Third Choice (Skirt)** | **Equally Likely Outcomes** |

W, S, NB
W, S, G
W, no S, NB
W, no S, G

B, S, NB
B, S, G
B, no S, NB
B, no S, G

T, S, NB
T, S, G
T, no S, NB
T, no S, G

n = 12 equally likely outcomes
s = 3 chances for success

Since we have 3 chances for success (circled) out of 12 equally likely ways an outfit can be put together,

$$P(\text{success}) = \frac{s}{n} \quad \begin{array}{l}\text{(number of chances for success)}\\ \text{(total equally likely outcomes)}\end{array}$$

$$P(\text{scarf and gray skirt}) = \frac{3}{12} \quad \text{(or 25\%)} \qquad \blacksquare$$

Note that we randomly sampled from each category. If color matching or personal taste were involved, the solution would be more complex.

Example ——————— Suppose we take a true–false quiz of four questions and we *randomly* guess on each answer, what is the probability of getting every question correct?

Solution

Let C = correct answer and X = incorrect answer, our sample space of equally likely outcomes would be as follows.

First Question	Second Question	Third Question	Fourth Question	Equally Likely Outcomes

n = 16 equally likely outcomes
s = 1 chance for success (circled)

Because we have 1 chance for success (circled) out of 16 equally likely outcomes,

$$P(\text{success}) = \frac{s}{n} \frac{\text{(number of chances for success)}}{\text{(total equally likely possibilities)}}$$

$$P(C \text{ and } C \text{ and } C \text{ and } C) = \frac{1}{16} \quad \text{(or approximately 6.3\%)}$$

3.4 More Complex Experiments: Multiplication Rules

Complex experiments (in which we select two or more from a set of possibilities) may also be solved by formula. However, we must be careful since each formula comes with restrictions that limits its use to a well-defined set of circumstances.

Dependent and Independent Events

In the case where two events are **dependent** (that is, the occurrence or nonoccurrence of one event affects the probability associated with the other event), we use the **general multiplication rule,** as follows.

For two dependent events

▼ *General Multiplication Rule*

If E_1 and E_2 are two defined events in a sample space, then

$$P(E_1 \text{ and } E_2) = P(E_1)P(E_2, \text{ given } E_1 \text{ has already occurred})$$

$P(E_2, \text{ given } E_1 \text{ has already occurred})$ is often symbolically expressed as $P(E_2 | E_1)$. $P(E_1 \text{ and } E_2)$ also equals $P(E_2)P(E_1, \text{ given } E_2 \text{ has already occurred})$.

In special cases, where the two events have no effect on each other's probability, we call the events **independent** (that is, the occurrence or nonoccurrence of one event has no effect on the probability of the other event). In which case, the multiplication rule is greatly simplified to:

▼ *Special Multiplication Rule*

For two independent events

If E_1 and E_2 are two independent events in a sample space, then,

$$P(E_1 \text{ and } E_2) = P(E_1)P(E_2)$$

Note that in $P(E_2)$ we eliminated the condition "given E_1 has already occurred."

For three or more independent events

If we can guarantee all events are independent, then we can expand our formula to three or more events as follows:

$$P(E_1 \text{ and } E_2 \text{ and } E_3 \ldots) = P(E_1)P(E_2)P(E_3) \ldots$$

Let's see how these rules apply to the examples we studied in the prior section.

Example ——————— We attend a party of 6 people, of which 2 are famous TV celebrities. If some Green Giant *randomly* plucked up 2 people by the collar, what is the probability both will be famous TV celebrities? (Note: this is the first example solved in section 3.3 using a tree diagram.)

Solution First we define "selecting a celebrity on the 1st pick" as *event one* (E_1) and "selecting a celebrity on the 2nd pick" as *event two* (E_2).

Since the two events are *dependent* (that is, whether or not we select a celebrity on the 1st pick affects the probability of selecting a celebrity on the 2nd pick), thus we use the general multiplication rule:

$$P(E_1 \text{ and } E_2) = P(E_1)P(E_2, \text{ given } E_1 \text{ has already occurred})$$

$$P\left(\substack{\text{celebrity} \\ \text{1st pick}} \text{ and } \substack{\text{celebrity} \\ \text{2nd pick}}\right) = P\left(\substack{\text{celebrity} \\ \text{1st pick}}\right) P\left(\substack{\text{celebrity} \\ \text{2nd pick}}, \text{ given we chose a celebrity on 1st pick}\right)$$

$$= \frac{2}{6} \cdot \frac{1}{5} = \frac{2}{30} \quad (6.7\%)$$

Notice that getting a celebrity on the first pick was $\frac{2}{6}$ as we would expect. However, now, we assume the first pick occurred, that is, we picked a celebrity out of the party. Thus, we have only 5 people left at the party with only 1 celebrity left, so the probability of selecting a celebrity on the second pick is 1 chance of success out of 5 possibilities or $\frac{1}{5}$. When we multiply $\frac{2}{6}$ and $\frac{1}{5}$, we get the same answer as we did in the prior section $\frac{2}{30}$ (or 6.7%). ■

One additional comment about *dependence*. Note in the above case, whether or not we selected a celebrity on the 1st pick affects the probability associated with selecting a celebrity on the 2nd pick. In other words, if a celebrity was chosen on the 1st pick, then P(celebrity on 2nd pick) equals $\frac{1}{5}$ (as stated above). However, if a celebrity was *not* chosen on the 1st pick, the P(celebrity on 2nd pick) equals $\frac{2}{5}$, which is quite different. This is what is meant when we say the occurrence or nonoccurrence of one event affects the probability associated with another event—that is, the two events are dependent.

The multiplication rule can also be expanded to include three or more *dependent* events, demonstrated in the following example.

Example ————————— Three cards are randomly selected from a 52-card deck. Calculate the probability all will be kings.

Solution We can expand the multiplication rule as follows: $P(E_1$ and E_2 and $E_3) = P(E_1)$ $P(E_2$, given E_1 occurred) $P(E_3$, given E_1 and E_2 occurred)

Let $\qquad\qquad K_1$ = selecting a king on 1st pick

$\qquad\qquad\qquad K_2$ = selecting a king on 2nd pick

$\qquad\qquad\qquad K_3$ = selecting a king on 3rd pick

$P(K_1$ and K_2 and $K_3) = P(K_1)P(K_2$, given K_1 occurred) $P(K_3$, given K_1 and K_2 occurred)

$$= \frac{4}{52} \cdot \frac{3}{51} \cdot \frac{2}{50} = \frac{24}{132,600} \quad \text{(near 0\%)}$$

Note that getting a king on the 1st pick was $\frac{4}{52}$, as we would expect. However, let's assume that this occurred—that is, we picked a king out of the deck. Thus, we have only 3 kings left out of 51 cards, so the probability of getting a king on the 2nd pick is $\frac{3}{51}$. Now, we assume the first two events occurred, that is, both kings were picked from the deck, so now we have 2 kings left out of 50 cards. Thus, the probability of getting a king on the 3rd pick is $\frac{2}{50}$. ■

As you can see, these formulas do save us time. In other words, we do not have to construct sample spaces of equally likely outcomes to calculate probabilities. For instance, in the example given above, we would have had to construct a sample space of 132,600 equally likely outcomes to solve this problem. A formidable task, indeed. But as easy as formulas are to use, they do come with restrictions that limit their use to a well-defined set of circumstances. And we must be careful to apply them exactly as presented.

Let's consider the second example from section 3.3 to demonstrate the conditions and restrictions for use of the *special* multiplication rule (for independent events).

Example ——————— Suppose at our party of 6, of which 2 are famous TV celebrities, a Green Giant plucked up one person by the collar, *replaced that one person* into the party, then later returned and plucked up one person again. (Note: This is the second example solved in section 3.3 using a tree diagram.)

What is the probability both picks would be famous TV celebrities?

Solution The two events are now *independent* since we replaced the first person. In other words, whether we select or do not select a celebrity on the 1st pick in no way influences the probability of selecting a celebrity on the 2nd pick.

$$P(E_1 \text{ and } E_2) = P(E_1)P(E_2)$$

$$P(\substack{\text{celebrity} \\ \text{1st pick}} \text{ and } \substack{\text{celebrity} \\ \text{2nd pick}}) = \frac{2}{6} \cdot \frac{2}{6} = \frac{4}{36} \quad (11.1\%)$$

Notice that the probability on the 1st pick was $\frac{2}{6}$. However, because we replaced the person back into the party, we still had six people at the party with two famous TV celebrities. So the probability of success on the 2nd pick was also $\frac{2}{6}$. When multiplied, the answer is $\frac{4}{36}$ (or 11.1%), which is the same answer we achieved when listing our sample space of equally likely outcomes in section 3.3. ■

One comment about *independence* in the preceding example: note the probability on the 2nd pick would have been $\frac{2}{6}$ no matter what the outcome of the 1st pick (whether we picked a celebrity or not on the 1st pick). This is what is meant by independence.

If we can guarantee all events are independent, that is, the outcome of one event in no way influences the probability of any other event, we can use the expanded form of the special multiplication rule, as follows:

$$P(E_1 \text{ and } E_2 \text{ and } E_3 \ldots) = P(E_1)P(E_2)P(E_3) \ldots$$

Example ——————— Suppose at our party of 6, of which 2 are famous TV celebrities, the Green Giant chooses 3 people, one at a time, *but replaces* each after the person was chosen. What is the probability all 3 will be famous TV celebrities?

Solution Because replacement in this case assures independence

$$P(\substack{\text{celebrity} \\ \text{1st pick}} \text{ and } \substack{\text{celebrity} \\ \text{2nd pick}} \text{ and } \substack{\text{celebrity} \\ \text{3rd pick}}) = \frac{2}{6} \cdot \frac{2}{6} \cdot \frac{2}{6} = \frac{8}{216} \quad (\text{or } 3.7\%) \quad ■$$

Example ———————— In a woman's wardrobe of 3 blouses (white, beige, and tan), 1 scarf, and 2 skirts (navy blue and gray), a blouse and a skirt must be worn but a *scarf* is optional. (Note: This is the third example solved in section 3.3 using a tree diagram.)

If we assume the woman randomly selected from each group, what is the probability the woman will be wearing a scarf *and* gray skirt?

Solution If we *randomly* select from each group, we essentially have three independent choices, since a choice from one group does not affect the probability of a choice from any other group. Thus

$$P(E_1 \text{ and } E_2 \text{ and } E_3) = P(E_1)P(E_2)P(E_3)$$

$$P(\text{any blouse and scarf and gray skirt}) = P(\text{any blouse}) \, P(\text{scarf}) \, P(\text{gray skirt})$$

$$= \frac{3}{3} \cdot \frac{1}{2} \cdot \frac{1}{2}$$

$$= \frac{3}{12} \quad (\text{or } 25\%)$$

Notice we must choose a blouse and all choices give us success. Because a scarf is optional, we have two choices (scarf or no scarf) of which one gives us success (scarf). We must choose a skirt and there is one chance for success (a gray skirt) out of two possibilities. ■

Note that we *randomly* selected from each category. If color matching or personal taste were involved, we would have lost our independence. In other words, say, if the woman selected a tan blouse, she may have personal preference for wearing the scarf and navy blue skirt, in which case the probabilities associated with the 2nd and 3rd choices would be greatly affected. These type of problems (where the choices are dependent) are much more difficult to solve, since they require knowledge of the extent one choice affects another.

Violation of independence is often the reason why statistical studies go awry since this requirement of independence is necessary for many of the formulas we use later in the text when we sample. If for some reason independence is violated, we may be obliged to use the general multiplication rule (which assesses the effect each event has on the probability of subsequent events). Unfortunately, in many experiments, especially those involving people such as in psychological or educational studies, these effects are quite difficult to assess and sometimes impossible. So it is important when we design a study to do our best to ensure independence when we can.

Let's examine this property of independence with one more example.

Example ———————— Suppose we take a true–false quiz of 4 questions and we *randomly* guess on each question, what is the probability of getting every question correct? (Note: This is the fourth example solved in section 3.3 using a tree diagram.)

Solution

If we guess randomly, this ensures *independence,* that is, whether we are correct or not correct on one guess in no way affects the probability of being correct on any other guess. Thus,

$$P(E_1 \text{ and } E_2 \text{ and } E_3 \text{ and } E_4) = P(E_1)P(E_2)P(E_3)P(E_4)$$

Let

$$C_1 = \text{correct on 1st question}$$
$$C_2 = \text{correct on 2nd question}$$
$$C_3 = \text{correct on 3rd question}$$
$$C_4 = \text{correct on 4th question}$$

$$P(C_1 \text{ and } C_2 \text{ and } C_3 \text{ and } C_4) = P(C_1)P(C_2)P(C_3)P(C_4)$$
$$= \frac{1}{2} \cdot \frac{1}{2} \cdot \frac{1}{2} \cdot \frac{1}{2} = \frac{1}{16} \quad (6.3\%)$$

Remember, for independence, the outcome on one event must in no way affect the probabilities associated with any other event.

Counting Principle

Although the counting principle can be used in many circumstances, one of its uses is to quickly count the total number of possible outcomes for an experiment.

▼ *Counting Principle*

> For a sequence of events, in which the first event can occur in *a* ways, the second in *b* ways, and the third in *c* ways, and so on, the total number of ways the events can occur together is
>
> $$a \cdot b \cdot c \ldots$$

Example

For our woman's wardrobe experiment of 3 blouses, scarf or no scarf, and 2 skirts, how many outcomes are possible?

Solution

Because the first event, picking a blouse, can occur in 3 ways; the second event, choosing a scarf, in 2 ways; and the third event, selecting a skirt, in 2 ways,

$$\text{Total number of possible outcomes} = 3 \cdot 2 \cdot 2$$
$$= 12$$ ■

Example

At our party of 6 people with 2 famous TV celebrities, how many ways can we select 2 people, given that

a. we replace the first person before we select the second?
b. we do not replace the first person?

Solution

a. Since the first event, selecting the first person, can occur in 6 ways, and the second event, selecting the second person, can occur in 6 ways, then

$$\text{Total number of possible outcomes} = 6 \cdot 6$$
$$= 36$$

b. Since the first event can occur in 6 ways and the second event can now only occur in 5 ways,

$$\text{Total number of possible outcomes} = 6 \cdot 5$$
$$= 30 \quad \blacksquare$$

Other circumstances in which we may use the counting principle are as follows.

Example ——————— In the 714 telephone area, how many different telephone numbers are possible?

Solution For a telephone number we have seven events:

____ ____ ____ - ____ ____ ____ ____

For the first event, selection of the first digit, we have 8 choices (2, 3, 4, 5, 6, 7, 8, and 9). Note that we cannot use the digit 0 or 1 as the first digit of a telephone number. For each other digit, we have 10 choices (0, 1, 2, 3, 4, 5, 6, 7, 8, and 9).

$$\text{Total number of ways these seven events can occur} = 8 \cdot 10 \cdot 10 \cdot 10 \cdot 10 \cdot 10 \cdot 10$$
$$= 8{,}000{,}000 \text{ telephone numbers} \quad \blacksquare$$

Example ——————— Suppose a particular state wishes to use 3 letters followed by 3 digits for an automobile license plate. How many different license plates are possible?

Solution For this license plate, we have six selections,

____ ____ ____ - ____ ____ ____

As long as there are no restrictions on which letters or digits can be used, we have 26 choices each for the first three selections and 10 choices each for the last three selections.

$$\text{Total number of ways these six events can occur} = 26 \cdot 26 \cdot 26 \cdot 10 \cdot 10 \cdot 10$$
$$= 17{,}576{,}000 \text{ license plates} \quad \blacksquare$$

In probability experiments, the counting formula is often used to count the total number of equally likely outcomes for an experiment, denoted by n in the probability formula,

$$P = \frac{s}{\boxed{n}} \quad \begin{array}{l} \text{(number of chances for success)} \\ \text{(total equally likely outcomes)} \end{array}$$

The counting formula can give you the total number of possible outcomes rather quickly, however the formula generally does not lend itself for use in calculating s, the number of chances for success.

3.5 Early Gambling Experiments Leading to Discovery of the Normal Curve

Early gambling experiments usually involving the tossing of coins and dice form the theoretical underpinning for many of our formulas and the statistical procedures we use today in statistics (such as, chi-square analysis and tests of proportions, which are discussed at length in chapters 10 and 11). However, these early experiments also paved the way for the original discovery of one of our most fundamental statistical tools, the normal curve, which is the subject of chapter 4. It is in these regards that we present the following.

Example

A fair coin is tossed 4 times. Bets are taken as to the number of heads that would turn up.

a. Use a tree diagram to list all equally likely outcomes.
b. Construct a histogram demonstrating how many times we would expect 0 heads, 1 head, 2 heads, 3 heads, and 4 heads to occur.
c. Find the probability of achieving 2 heads.
d. Would you bet even money on 2 heads? Explain your reason.

Rationale

Essentially we are sampling from a huge population of coin flips, where 50% possess the attribute of heads.

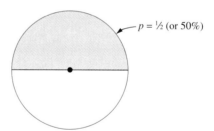

In other words, if we were to flip this coin millions and millions of times, 50% (or extremely close to 50%) would be heads.

Now, if we were to sample from this huge population of coin flips, in this case we are sampling 4 tosses (that is, we are selecting a sample size of $n = 4$), what may we expect to happen? We know from theory and a long history of experience, that if we were to randomly sample from such a population,

$$p_s \approx p$$

The sample proportion, p_s, will be approximately equal to the population proportion, p.

That is, since the population consists of 50% heads, the sample should consist of *approximately* 50% heads. In the case of $n = 4$ tosses, we should get *approximately* 2 heads. However, we can also get 1 head or 4 heads. How can we determine the percentage of times we can expect each of these outcomes to occur?

One way is to use a tree diagram to list all the equally likely outcomes that can occur when 4 coins are tossed and simply count the number of times we achieved zero heads, one head, two heads, three heads, and four heads, as follows:

Solution

a. We use a tree diagram to help us list the sample space of equally likely outcomes.

b. To construct the histogram,

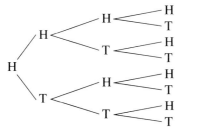

	First, we count how many
H, H, H, H	times we obtained 0 heads.
H, H, H, T	This occurred on one
H, H, T, H	occasion, circled at left,
H, H, T, T	so we indicate this with
———	one slash mark in our
H, T, H, H	tally.
H, T, H, T	
H, T, T, H	Tally
H, T, T, T	
———	0 Heads |
T, H, H, H	
T, H, H, T	1 Head
T, H, T, H	
T, H, T, T	2 Heads
———	
T, T, H, H	3 Heads
T, T, H, T	
T, T, T, H	4 Heads
(T, T, T, T)	

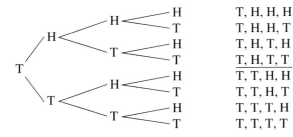

	Second, we count how many
H, H, H, H	times we obtained 1 head.
H, H, H, T	This occurred on four
H, H, T, H	occasions, circled at left,
H, H, T, T	so we indicate this with
———	four slash marks in our
H, T, H, H	tally.
H, T, H, T	
H, T, T, H	Tally
(H, T, T, T)	
———	0 Heads |
T, H, H, H	
T, H, H, T	1 Head ||||
T, H, T, H	
(T, H, T, T)	2 Heads
———	
(T, T, H, H)	3 Heads
(T, T, H, T)	
(T, T, T, H)	4 Heads
T, T, T, T	

Notice we have 16 equally likely outcomes.

Third, we continue for 2 heads, 3 heads, and 4 heads. The completed tally would appear as follows.

Tally

0 Heads |

1 Head ||||

2 Heads   卌 |

3 Heads ||||

4 Heads |

Notice, there are 16 slash marks, one for each different outcome.

Fourth, we encase the slashes in rectangular bars, and rotate the tally ¼ turn counterclockwise.

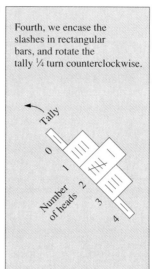

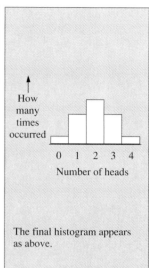

How many times occurred

0 1 2 3 4
Number of heads

The final histogram appears as above.

The histogram bars are usually labeled in one of three different ways in terms of

Frequency of occurrence

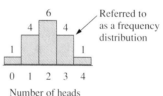

Referred to as a frequency distribution

A probability fraction

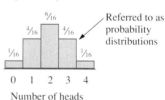

Referred to as probability distributions

A probability percentage

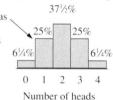

No matter which way the bars are labeled, however, they all give us the same information, as follows:

$$P(0 \text{ heads}) = 1 \text{ chance in } 16 = 1/16 = 6\frac{1}{4}\%$$
$$P(1 \text{ head}) = 4 \text{ chances in } 16 = 4/16 = 25\%$$
$$P(2 \text{ heads}) = 6 \text{ chances in } 16 = 6/16 = 37\frac{1}{2}\%$$
$$P(3 \text{ heads}) = 4 \text{ chances in } 16 = 4/16 = 25\%$$
$$P(4 \text{ heads}) = 1 \text{ chance in } 16 = 1/16 = 6\frac{1}{4}\%$$

c. So, to answer the question of the probability of 2 heads, we merely look at the results. You have *6 chances out of 16 = 37½%.*

d. If you did bet even money on 2 heads, unfortunately in the long run you would lose your money. Out of every 16 times you played the game, you can expect to win on 6 occasions and lose on 10. Sometimes this is expressed as odds, 6:10 (meaning, in the long run, you would average 6 wins and 10 losses out of each 16 plays).

Although the above probabilities were based on mathematical analysis (that is, by constructing sample spaces), experience has shown these to give reasonably accurate estimates of what we would expect to occur in the long run in actual practice. In other words, if we were to drop 4 coins on a table thousands and thousands of times and record the number of heads achieved on each drop, we would find we would get

0 heads approximately	$6\frac{1}{4}\%$ of the time
1 head approximately	25% of the time
2 heads approximately	$37\frac{1}{2}\%$ of the time
3 heads approximately	25% of the time
4 heads approximately	$6\frac{1}{4}\%$ of the time

Now, what if we decided to drop $n = 12$ coins or $n = 50$ coins on a table, what would we expect to get?

Well, we can determine this either mathematically, by constructing a sample space, or we can actually drop 12 coins (or 50 coins) thousands and thousands of times and tally the result.*

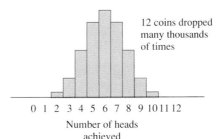

12 coins dropped many thousands of times

0 1 2 3 4 5 6 7 8 9 10 11 12

Number of heads achieved

Suppose we drop $n = 12$ coins on a table thousands and thousands of times and record the number of heads achieved on each drop, we would get a distribution something like this figure.

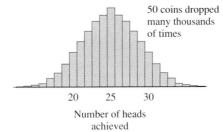

50 coins dropped many thousands of times

20   25   30

Number of heads achieved

Suppose we drop $n = 50$ coins on a table thousands and thousands of times and recorded the number of heads achieved on each drop, we would get a distribution something like this figure.

Look at the two histograms above. Both are symmetrical around the value we would most likely expect to occur. In the case of dropping $n = 12$ coins, we would most likely expect approximately 6 heads (50% of 12 = 6). And indeed we do most often get 6 heads. However on a great many occasions we get somewhat more than 6 heads, and on a great many occasions somewhat less, with the heights of the histogram bars seeming to fall off in the shape of a bell.

*Actually, simpler techniques and formulas are available to calculate these probabilities, which are discussed at the end of chapter 4 and in chapter 11.

Notice we have a similar situation in dropping $n = 50$ coins. The distribution is symmetrical around the value we would most likely expect, in this case 25 heads (50% of 50 = 25). However on a great many occasions we get somewhat more than 25 heads, and on a great many occasions somewhat less, with the heights of the histogram bars again falling off in the shape of a bell.

This bell-shaped pattern appears repeatedly with these coin experiments.

Now, you might ask, this may happen with coin tosses, where the probability of a head for a coin toss is $\frac{1}{2}$ (50%), but what if we sampled from a different population, say die tosses, where the probability of a particular face turning up is $\frac{1}{6}$ ($16\frac{2}{3}$%)? What happens then?

Okay, let's take 60 dice and paint one face on each blue (for identification purposes).

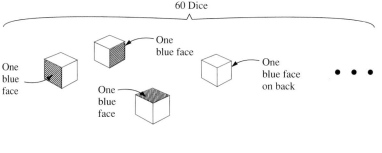

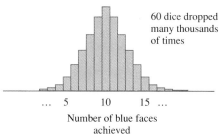

60 dice dropped many thousands of times

... 5   10   15 ...

Number of blue faces achieved

Say we drop $n = 60$ dice on a table thousands and thousands of times, and each time recorded the number of blue faces that turned up. If we tally the results into a histogram, we would get a distribution something like this figure.

Notice the shape of the distribution. It is symmetrical about the value we would most likely expect, in this case 10 blue faces. In other words, we would expect about 1 die in 6 to turn up blue, or 10 in 60. And indeed 10 blue faces would be our most frequently occurring value if we had constructed a sample space or in practice if we actually dropped 60 dice many thousands of times. However, on many occasions we get somewhat more than 10 blue faces and on many occasions somewhat less, again with the heights of the histogram bars falling off in the shape of a bell.

Bell-shaped distributions kept occurring with amazing regularity in these coin experiments and in the dice experiments when the number of dice dropped was large. This repetitive bell-shaped pattern in gambling experiments led De Moivre (in 1733) to the initial discovery of one of the most powerful predictive tools we have in all of statistics and the topic of our next chapter, the normal distribution.

3.6 Additional Probability Topics

Although the material presented thus far provides the main thrust of the techniques needed for future chapters, several additional topics are often encountered and are presented here. Specifically we introduce the mean and standard deviation of a discrete probability distribution, expected value, and permutations and combinations (combinations are further discussed in chapter 11, section 11.1).

Mean and Standard Deviation of a Discrete Probability Distribution

In section 3.5, we constructed a distribution that offered probabilities associated with achieving a given number of heads when a fair coin is tossed four times, as follows:

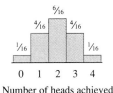

Number of heads achieved
in four tosses

This is called a discrete probability distribution. In such a distribution, each outcome for an experiment is associated with a specific probability of occurrence and the sum of all probabilities equals 1.00. (Note in the above case, $\frac{1}{16} + \frac{4}{16} + \frac{6}{16} + \frac{4}{16} + \frac{1}{16} = \frac{16}{16} = 1.00$.)

To calculate the mean and standard deviation of such a probability distribution, we use the following two formulas.

Mean

$$\mu = \Sigma \, xp \, (x) \qquad \text{x: one outcome}$$
$p(x)$: probability of achieving this outcome

For this example,
$\mu = \Sigma xp(x)$
$\mu = 0(1/16) + 1(4/16) + 2(6/16) + 3(4/16) + 4(1/16)$
$\mu = 2$

Standard Deviation

$$\sigma = \sqrt{\Sigma \, (x - \mu)^2 p \, (x)}$$

For this example,
$\sigma = \sqrt{\Sigma(x - \mu)^2 p \,(x)}$
$\sigma = \sqrt{(0 - 2)^2(1/16) + (1 - 2)^2(4/16) + (2 - 2)^2(6/16) + (3 - 2)^2(4/16) + (4 - 2)^2(1/16)}$
$\sigma = \sqrt{1} = 1$

> ### Probability Distribution
> A distribution that offers the probabilities associated with each possible outcome of an experiment, such that the sum of these probabilities always equals 1.00.

> ### Discrete Probability Distribution
> A probability distribution where each possible outcome of an experiment can only be one of a limited number of discrete values, that is, a value that when presented on a number line occupies only a distinct isolated point.

In other words, in the above experiment, we could only achieve 0, 1, 2, 3, or 4 heads as outcomes, a limited number of distinct isolated points on a number line. Note we could never achieve $1\frac{1}{4}$ heads or $3\frac{1}{2}$ heads or any values other than these limited isolated values of 0, 1, 2, 3, or 4.

The term **discrete** is discussed again in section 4.4 when we introduce one of the most frequently encountered discrete probability distributions, called the binomial sampling distribution.

Expected Value

In the example just presented, we calculated the mean and standard deviation (μ and σ) of a discrete probability distribution. This mean (μ) is often referred to as the **expected value.**

> ### Expected Value
> The long-range average of a repeated experiment, essentially the population mean, μ.
> $$\text{Expected value (EV)} = \mu = \Sigma\ xp(x)$$

In our experiment of tossing four coins, since $\mu = 2$ heads, this is the expected value. Note in this particular experiment, the expected value is also the most frequently occurring outcome (with probability $\frac{6}{16}$; refer to the histogram at the beginning of the section). However this is not always the case, as shown in the following discrete probability distributions:

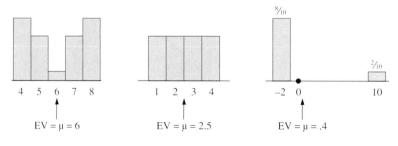

Note in the first example, the expected value, μ, has the *least* probability of occurring, and in the other two examples the expected value, μ, has *no* probability of occurring. Keep in mind, expected value is merely an average, the average value you would likely calculate if you repeated an experiment many many thousands of times, added up all the outcomes, and divided by *n*, the total number of values.

This concept of expected value (average value for a repeated experiment) is used extensively in business and scientific investigations but in honor of its origins in gambling, we introduce the following example:

Example ——————— Roulette is a game played by placing a bet that a ball will tumble into a particular pocket of a spinning wheel. There are 38 available pockets (numbered 1 to 36, 0, and 00), so your chance of winning is $\frac{1}{38}$, while your chance of losing is $\frac{37}{38}$. On a \$10 bet, say, a gambling house will return \$360 if you win, however the house will keep your \$10 if you lose. What is the EV, the expected value, for this experiment? In other words, if you played the game many thousands of times, added up all the winnings and losses, and divided by *n*, the number of times you played, what would be your expected average (the expected value)?

Solution This is solved as follows:

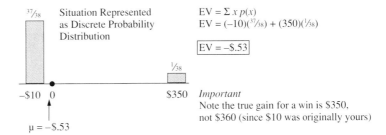

EV = Σ x p(x)
EV = (−10)($^{37}/_{38}$) + (350)($^1/_{38}$)

$$\boxed{\text{EV} = -\$.53}$$

Important
Note the true gain for a win is \$350,
not \$360 (since \$10 was originally yours)

Thus the expected value, μ, for this experiment is −\$.53, meaning if you repeated this experiment (betting \$10) many thousands of times, added up all your losses (−\$10's) and wins (+\$350), and divided by *n*, the number of times you placed a bet, the average would be about −\$.53. In other words, in the long run, over many thousands of bets, you will average losing 53¢ for each bet you place. Of course, from the point of view of the gambling house, they gain, *in the long run*, on average, 53¢ for each bet you place. ∎

Permutations and Combinations

In section 3.4, we introduced a useful method for quickly counting the total number of possible outcomes for an experiment, called the *counting principle*. In addition to the counting principle, other counting techniques are available, such as the **permutation** and **combination,** denoted $P_{n,s}$ and $C_{n,s}$ respectively.

A **permutation** counts the number of ways n *different* objects can be arranged, s at a time, where order of arrangement is important.

$$P_{n,s} = \frac{n!}{(n - s)!}$$

Factorial Symbol, !, is defined as

$$n! = n(n - 1)(n - 2) \ldots 1$$

For example,

$$5! = 5(4)(3)(2)(1) = 120$$
$$3! = 3(2)(1) = 6$$
$$1! = 1$$
$$0! = 1 \text{ by definition}$$

Let's look at permutations.

Example ——————— Suppose there are 5 books (A, B, C, D, and E) that are to be placed in 3 available positions on a shelf. How many different arrangements are possible?

Solution If order of arrangement is important, meaning book arrangement A, B, C is considered different from book arrangement A, C, B (even though the same books are used), we use the permutation formula, as follows:

$$P_{n,r} = \frac{n!}{(n - r)!}$$

$$P_{5,3} = \frac{5!}{(5 - 3)!} = \frac{5!}{2!}$$

$$P_{5,3} = \frac{5(4)(3)(2)(1)}{(2)(1)}$$

$$P_{5,3} = 60 \text{ ways}$$

60 Ways Listed

ABC	BAC	CAB	DAB	EAB
ABD	BAD	CAD	DAC	EAC
ABE	BAE	CAE	DAE	EAD
ACB	BCA	CBA	DBA	EBA
ACD	BCD	CBD	DBC	EBC
ACE	BCE	CBE	DBE	EBD
ADB	BDA	CDA	DCA	ECA
ADC	BDC	CDB	DCB	ECB
ADE	BDE	CDE	DCE	ECD
AEB	BEA	CEA	DEA	EDA
AEC	BEC	CEB	DEB	EDB
AED	BED	CED	DEC	EDC

In situations where order of arrangement is *not* important, say for instance, if from 5 books (A, B, C, D, and E) we select reading lists of 3 (demonstrated in the following example), we use one of the most popular counting devices in mathematics called the combination.

> A **combination** counts the number of ways n *different* objects can be formed into groups of size s (in these cases, order of arrangement or selection is *not* important).*
>
> $$C_{n,s} = \frac{n!}{(n-s)!\,s!}$$

Example ——————— From 5 books (A, B, C, D and E), how many reading lists of 3 different books can be made up?

Solution Well, if order of arrangement does *not* count, meaning reading list A, B, C is considered no different from reading list A, C, B (same books), we use the combination formula, as follows.

$$C_{n,s} = \frac{n!}{(n-s)!\,s!}$$

$$C_{5,3} = \frac{5!}{(5-3)!\,3!} = \frac{5!}{2!\,3!}$$

$$C_{5,3} = \frac{(5)(4)(3)(2)(1)}{(2)(1)\,(3)(2)(1)}$$

$$C_{5,3} = 10 \text{ ways}$$

10 Ways Listed

ABC	BCD	CDE
ABD	BCE	
ABE	BDE	
ACD		
ACE		
ADE		

In other words, only 10 reading lists are possible. Keep in mind, although 60 arrangements of three books are possible (if order counts), only 10 of these arrangements involve groups containing different books. ■

Summary

The study of probability grew out of the gambling dens of Europe over three centuries ago. Much of what we study today is still based on this early work.

In broad terms, we define probability as follows.

Probability: The proportion (or percentage) of times an event will occur in the long run, under similar circumstances.

Two methods we can use to obtain this probability percentage are the empirical approximation and classical approach.

Empirical approximation to probability: The fraction of times an event *has actually occurred* over a great many experiments conducted over a long period of time under similar circumstances, expressed as,

P(an event)

$$\approx \frac{\text{Number of Times Event Has Actually Occurred}}{\text{Total Number of Experiments (or Attempts)}}$$

As the number of experiments increases, this empirical fraction gets closer and closer to the true probability.

*Combinations may also be denoted as $_nC_s$, $C(n,s)$, C_s^n, and $\begin{bmatrix} n \\ s \end{bmatrix}$.

It is important to remember that use of the empirical fraction requires an experiment to be performed a great many times over a long period and under similar circumstances.

Subjective probability: Assigning a probability out of personal judgment. Such probabilities are not used in this text.

Probability defined classically: In an experiment of n different possibilities, *each having an equal chance of occurring*, the probability a particular event will occur is equal to the following fraction

$$\frac{\text{Number of Chances for Success}}{\text{Total Number of Equally Likely Possibilities}}$$

This can be expressed as

$$P(\text{an event}) = \frac{s}{n} \quad \begin{array}{l}\text{(number of chances for success)} \\ \text{(total equally likely possibilities)}\end{array}$$

Often the word, event, is replaced with the word, success, as follows:

$$P(\text{success}) = \frac{s}{n}$$

In dealing with probability fractions, there are two fundamental properties.

Property 1: The probability of an event occurring will always be a number between 0 and 1, inclusive.

$$0 \leq P(\text{an event}) \leq 1$$

No chance to occur (0%)   Certain to occur (100%)

Property 2: The probability of an event occurring *plus* the probability of an event *not* occurring $= 1$.

$$P(E) + P(\text{not } E) = 1$$

This property says there is a 100% probability that the event will either occur or not occur.

The words, *and* and *or* have very specific meanings in statistics.

AND: The word *and* between two or more conditions or events implies that *all* must be met for success.

OR: The word *or* between two or more conditions or events implies that *either* one or more may be met and that will give success.

Although many formulas are available to solve probability problems, each normally comes with restrictions and conditions that limit their use to a well-defined set of circumstances. It is best to first learn to solve simple experiments (where we select *one* from a set of possibilities) by the nonformula techniques described in the chapter.

However, in more complex problems, where we select *two* or more from a set of possibilities, shortcut techniques and formulas are often unavoidable. For such purposes we can use a tree diagram to help us list all equally likely outcomes in an experiment or we have a series of multiplication rules that are often necessary.

Tree diagram: A branching technique used to list equally likely outcomes in an experiment. This is used in conjunction with our classical probability formula,

$$P(\text{success}) = \frac{s}{n} \quad \begin{array}{l}\text{(number of chances for success)} \\ \text{(total equally likely outcomes)}\end{array}$$

Multiplication rules: Rules that allow us to calculate certain probabilities without having to list our sample space of equally likely outcomes.

General multiplication rule: If E_1 and E_2 are two events in a sample space, then

For two dependent events $P(E_1 \text{ and } E_2) = P(E_1)P(E_2,$ given E_1 has already occurred)

The rule can be expanded to three or more events.

For three or more dependent events $P(E_1 \text{ and } E_2 \text{ and } E_3 \ldots)$ $= P(E_1) \, P(E_2,$ given E_1 occurred) $P(E_3,$ given E_1 and E_2 occurred) $\ldots$

Special multiplication rule: If E_1 and E_2 are two independent events in a sample space, then

For two independent events $P(E_1 \text{ and } E_2) = P(E_1)P(E_2)$

The special multiplication rule can be expanded to three or more events

For three or more independent events $P(E_1 \text{ and } E_2 \text{ and } E_3 \ldots)$ $= P(E_1)P(E_2)P(E_3) \ldots$

Dependent events: The occurrence or nonoccurrence of one event affects the probability of one or more other events.

Independent events: The occurrence or nonoccurrence of one event in no way affects the probability of any other event.

Because many of the formulas we use later in the text when we sample require the condition of independence, it is important when we design a study to do our best to ensure independence of events when we can.

Counting principle: For a sequence of events, in which the first event can occur in a ways, the second event in b ways, the third event in c ways, and so on, the total number of ways the events can occur together is

$$a \cdot b \cdot c \ldots$$

Early gambling experiments leading to discovery of the normal curve: Early gambling experiments (usually involving the tossing of coins and dice) form the theoretical underpinning of many of the formulas and statistical procedures we use today in statistics (such as chi-square analysis and tests of proportions, which are discussed at length in chapters 10 and 11). However, these early experiments also paved the way for one of the most important discoveries in statistics and the subject of the next chapter, the normal curve.

Exercises

Note that full answers for exercises 1–5 and abbreviated answers for odd-numbered exercises thereafter are provided in the Answer Key.

3.1 Out of 200 tosses of a dart at a Friday night tournament, a player strikes the bull's-eye 50 times.

a. What is the true probability of a bull's-eye for this player?
b. Explain how one might empirically determine the true probability of a bull's-eye for this player.

3.2 Suppose you attend a party of 50 people (30 men and 20 women). If 10 of the men and 5 of the women have curly black hair and you were to *randomly* select one person from the party, what is the probability this one person would

a. have curly black hair?
b. be male?
c. be male and have curly black hair?
d. be male *or* have curly black hair?
e. be female and *not* have curly black hair?

Given we already know the person selected is a male, what is the probability the person selected

f. has curly black hair?
g. is a female?

3.3 Suppose we attend a party of 5 people, of which 2 are famous TV celebrities. What if we were to randomly select 2 people from this party.

a. Use a tree diagram to list all equally likely ways two people may be selected. Assume the first person is removed from the party prior to the second pick. Hint: use numbers 1, 2, and 3 to identify the *non*-TV celebrities and the numbers 4 and 5 to identify the TV celebrities. Then find the probability
  i. Both are TV celebrities.
  ii. Exactly one is a TV celebrity.
  iii. None are TV celebrities.
b. Assume the first person *is* replaced prior to the second pick. List the new sample space of equally likely outcomes, then find the probability
  i. Both are TV celebrities.
  ii. Exactly one is a TV celebrity.
  iii. None are TV celebrities.
c. Use the multiplication rule to calculate the probability *both* are TV celebrities if the first person is
  i. removed after the first pick.
  ii. replaced after the first pick.
d. Use the multiplication rule to calculate the probability *none* are TV celebrities if the first person is
  i. removed after the first pick.
  ii. replaced after the first pick.

Challenging Question:

e. Use the multiplication rule to calculate the probability *exactly one* is a TV celebrity if the first person is
  i. removed after the first pick.
  ii. replaced after the first pick.

3.4 Suppose a particular state wishes to use 2 letters followed by 4 digits for an automobile license plate, however the first digit cannot be 0. How many different license plates are possible?

3.5 If we toss a coin three times,

a. List the sample space of equally likely outcomes.
b. Tally the number of times 0, 1, 2, and 3 heads occurred, then construct a histogram to represent this.
c. What is the probability of 2 heads?
d. What is the probability of 0 heads?

3.6 In broad terms, probability is defined as the proportion (or fraction) of times an event will occur in the long run under similar conditions. More specifically, define probability

a. empirically.
b. classically.
c. subjectively.

Two fundamental properties of probability were presented. Property 1 states: the probability of an event occurring will range from 0 to 1, inclusive. Property 2 states: the probability of an event occurring plus the probability of an event not occurring is equal to 1.

d. Give an example of property 1 where P(an event) = 0 and P(an event) = 1.
e. Give an example of property 2.

Two events are said to be independent if the occurrence or nonoccurrence of one event . . .

f. Complete the definition.
g. Give an example of two independent events.
h. Give an example of two dependent events.

3.7 Over several months, the manager of a local pizza place noted that out of 8420 customers, 1455 spent over ten dollars. Use your empirical definition to estimate the probability a randomly selected person entering the pizza place will spend over ten dollars.

3.8 Out of 7325 fifth-graders tested over several years at a school in Louisville, Kentucky, 947 had some form of learning disorder. Estimate the probability a child selected from the fifth-grade in this school will have some form of learning disorder.

3.9 In a college study consisting of 12 psychology majors, 8 education majors, 11 business majors, and 5 pre-meds, if we select *one* person at random, what is the probability this person is

a. a psych major?
b. a business or education major?
c. not a business *and* not an education major?

3.10 The probabilities that a video store stocks a movie rated G, PG, or R are $P(G) = .26$, $P(PG) = .21$, and $P(R) = .38$, respectively. What is the probability a randomly selected movie will be rated

a. PG?
b. something other than G, PG, and R?
c. either G or PG?

3.11 A medication is known to cause the side effects of dryness of the mouth (D) or mild anxiety (A) with the following probabilities:

$$P(D) = .32$$
$$P(A) = .16$$

If the two side effects cannot occur together, find the probability that in any one person the medication will result in

a. dryness of the mouth *and* mild anxiety.
b. dryness of the mouth.
c. no dryness of the mouth.
d. no dryness of the mouth and no mild anxiety.

If the two side effects *can* occur together with probability, P(dryness and anxiety) = .10, find the probability that in any one person the medication will result in

e. dryness of the mouth only.
f. mild anxiety only.

3.12 A store rents out VCRs. Out of 7 available for rental, 3 have defects. If 2 VCRs are rented together, find the probability

a. both will be defective.
b. neither will be defective.

Challenging Question:

c. only one will be defective.

3.13 Out of three cards selected from a deck of 52, what is the probability all three will be aces? Solve without replacement, then solve with replacement.

3.14 Suppose we roll a pair of dice. List the sample space of equally likely outcomes, then find the probability of getting

a. snake eyes (two 1's).
b. a 5 on one die and a number other than 5 on the other die.
c. a sum of 7.

3.15 If we toss a coin 5 times, use the special multiplication rule to find the probability of getting 5 heads.

3.16 There are 3 roads from Allentown to Boynton. How many different round trips are possible if a person

a. *must* take a different road on the return trip?
b. can take either road either way?
c. *must* take the same road both ways?

3.17 A company employs 7 salespeople, 10 office workers, and 3 administrators. How many different committees can be formed consisting of a salesperson, an office worker, and an administrator?

3.18 Six coins are shaken in a tin can then dropped on a table and the number of heads counted. If we repeat the experiment a large number of times, how many times would we expect to get 0 heads, 1 head, 2 heads, 3 heads, 4 heads, 5 heads, and 6 heads? We can actually perform this experiment over and over again thousands of times to get the answer, or an easier way would be to use a tree diagram to help us list the sample space of equally likely outcomes.

Let's say the sample space was listed for us. This would result in 64 equally likely outcomes, which when tallied according to the number of heads would appear as follows:

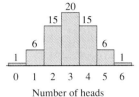

Expected Number of Heads Achieved if 6 Coins Are Tossed 64 Times

Based on the histogram, if you were to toss a coin six times, what is the probability of getting

a. 0 heads (in other words, all tails)?
b. 3 heads?
c. 4 or more heads?

3.19 It is known 25% of the women in a certain town in Minnesota possess the attribute of red hair and this might be expressed as follows:

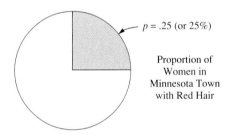

$p = .25$ (or 25%)

Proportion of Women in Minnesota Town with Red Hair

If we were to randomly sample 60 women from this town, how many would you expect to have red hair?

3.20 An experiment has outcomes 0, 1, 2, 3, and 4 with the probabilities as shown:

$$p(x): .3 \quad .3 \quad .2 \quad .1 \quad ?$$
$$x: \ 0 \quad 1 \quad 2 \quad 3 \quad 4$$

For this probability distribution,

a. Fill in the missing $p(x)$ value.
b. Construct the probability histogram.
c. Calculate μ and σ.

3.21   For an experiment with outcomes 3, 4, 5, and 6, the probabilities are as shown:

$$p(x): .1 \quad ? \quad .2 \quad .1$$
$$x: 3 \quad 4 \quad 5 \quad 6$$

For this probability distribution,

a. Fill in the missing $p(x)$ value.
b. Construct the probability histogram.
c. Calculate μ and σ.

3.22   Suppose a gambling bet of $5 will return $300 if you win (with probability $\frac{1}{80}$). Calculate the expected value for this experiment from the point of view of the person placing the bet. (Hint: a win actually returns $295.)

3.23   A precious stone is insured for $90,000 against theft by paying an annual premium of $$x$. If the insurance company wishes an expected value (gain) of $1200, and assumes the probability of theft is $\frac{1}{200}$ in a year period, at what rate, x, should they set their premium? (Hint: the loss to the company is $-$90,000 + $$x$ if the stone is stolen.)

3.24   Suppose fire insurance on a $120,000 house can be purchased for $800 per year. If the probability of total loss due to fire is $\frac{1}{400}$ and the probability of

50% loss due to fire is $\frac{1}{400}$, calculate the expected value for this experiment from the point of view of the insurance company. (Hint: total loss due to fire will cost the insurance company $119,200; they would give the home owner $120,000 but $800 was the home owner's original money; for 50% loss due to fire, the cost to the insurance company is $59,200.)

3.25   Suppose out of 6 staff members at a hospital (A, B, C, D, E, and F), a task force of 2 is to be chosen to evaluate procedures.

a. How many different task forces are possible?
b. Suppose first chosen is designated head of the task force, how many different task forces are possible now?

3.26   For financial purposes, a mining company divided its operations into 7 independent national regions. Suppose they wish to test a new financial procedure in 3 of these regions.

a. How many groupings of 3 regions are possible (if order is *not* important)?
b. How many groupings of 3 regions are possible (if order *is* important)?

CHAPTER 4

Normal Distribution

Carl Friedrich Gauss

4.0 Origins of the Concept

P erhaps the single most important distribution in all of statistics is the **normal distribution.** Its discovery dates back to the English mathematician, Abraham De Moivre[1] (1733), and his work on gambling experiments and is perhaps best illustrated with the following example.

Suppose a large number of well-balanced coins, say for instance 900 coins, are dropped on a table and the number of heads counted. How many heads would you expect? Many people would guess approximately 450 (half of 900) and, indeed, experience has shown that if this experiment were repeated thousands and thousands of times, most often you would get approximately 450 heads. However, on many occasions, you would get somewhat more than 450 heads and on many occasions somewhat less. If we were to actually record the results of these thousands and thousands of experiments into a histogram, it might appear as follows.

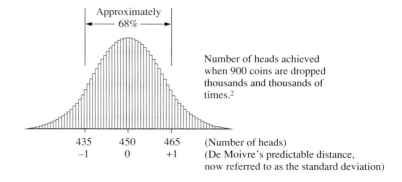

Number of heads achieved when 900 coins are dropped thousands and thousands of times.[2]

| | 435 | 450 | 465 | (Number of heads) |
| | −1 | 0 | +1 | (De Moivre's predictable distance, now referred to as the standard deviation) |

In studying these results, De Moivre noted that approximately 68% of the readings consistently fell within a predictable distance from the mean, denoted by the symbols −1 and +1. In other words, if you dropped 900 coins on a table and counted the number of heads, you would have a 68% probability there would be between 435 and 465 heads. De Moivre's predictable distance is now referred to as the **standard deviation.**[3]*

All information relevant to the understanding of the chapter is presented on each page as footnotes. However, certain information is presented at the end of the chapter as numbered *endnotes,* since they are mostly reference sources and historical fine points that tend to interfere with the flow of the material. It is not necessary to consult endnotes.

*De Moivre used the *inflection points* on the curve as his predictable distance. Inflection points are the points where the steep upward slope of the curve abruptly changes to a more gradual incline. This is useful when sketching the curve to properly estimate where the first standard deviation lines are located.

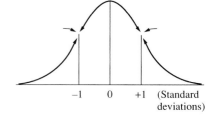

| | −1 | 0 | +1 | (Standard deviations) |

Furthermore, De Moivre noted that approximately 95% of the readings fell within −2 and +2 predictable distances (standard deviations) of the mean, as follows.

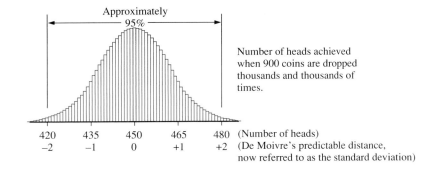

420	435	450	465	480	(Number of heads)
−2	−1	0	+1	+2	(De Moivre's predictable distance, now referred to as the standard deviation)

Number of heads achieved when 900 coins are dropped thousands and thousands of times.

De Moivre realized he had discovered something important (he spent over 12 years on it, ultimately deriving the equation for the normal distribution and calculating probabilities associated with its use that to this day would be considered quite accurate) but for reasons[4] was unable to interest others.

Half a century later, the famous French mathematician, Pierre Simon Laplace was working on a probability experiment similar to that of De Moivre, only Laplace's experiment (1781) concerned newborn infants.[5] Laplace was trying to prove that male babies were born with a higher frequency than female babies. Although Laplace's work is quite complex, let's consider the following simplified example.

Suppose we assume the probability of a male birth is $\frac{1}{2}$ (50%). If we equate the probability of having a male baby to the probability of achieving a head when a coin is tossed (which is also $\frac{1}{2}$),[6] then the resulting distribution of heads achieved when dropping 500,000 coins on a table, thousands and thousands of times, should be equivalent to the resulting distribution of male births achieved when 500,000 babies are born, in thousands and thousands of cities. In other words, we would expect approximately 250,000 heads (or approximately 250,000 male births) each time. However, on many occasions we would get somewhat more than 250,000 and on many occasions somewhat less. According to the laws of probability, the resulting distribution should look as follows.[7]

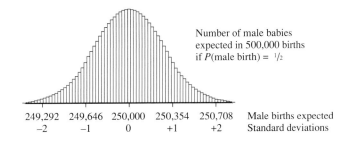

Number of male babies expected in 500,000 births if P(male birth) = $\frac{1}{2}$

249,292	249,646	250,000	250,354	250,708	Male births expected
−2	−1	0	+1	+2	Standard deviations

As it turned out, in the years 1745–1770 in Paris, there were equivalent to 254,856 male births out of 500,000.[8] According to the previous histogram, the likelihood of this occurring is almost 0. Just look at the histogram. The probability of getting over 251,000 male births is negligible. To get over 254,000 would be considered nearly impossible.

Laplace correctly reasoned that the probability of a male birth must then be greater than 50% and more likely closer to 51% (254,856/500,000 ≈ 51%). Laplace substantiated these findings using birth records from other European cities such as London and Naples, which also produced similar ratios with only minor variation that Laplace attributed to climate, food, or custom.[9] To this day, these probabilities hold true worldwide. The probability of a male birth is known to be nearly 51%, a female birth 49%.[10]

Although Laplace republished his findings in 1786, the work attracted only minor attention, probably due to the extreme complexity of his mathematical development.

The next stage in this unfolding discovery had to wait an additional 30 years for the work of the famous German mathematician, Carl Gauss. Interest in planetary motion dominated Europe in the late 1700s. Astronomers and mathematicians were encouraged with national grants and contests to correctly measure the position of certain stars and other celestial bodies, which were to be used to determine precise longitudinal measurements for sea navigation. However, imperfections in telescopic lenses (along with the imperfections and variations in the human eye) produced measurement errors that interfered with determining exact positions. These errors and how to deal with them perplexed astronomers for more than half a century until Carl Gauss in 1809 correctly reasoned that the errors of observation had to be distributed much like the heads in a coin experiment and thus created a minor revolution, at least in the field of astronomy.[11] Although the actual mathematics involve trigonometric equations, the underlying principle is quite simple. Suppose we use the following example.

Say the *true* position of the Moon's crater, Manilius,[12] at a certain place and time was known to be precisely 17°20′ from a known reference star. If we were to take thousands and thousands of measurements, the average of all these measurements might indeed be 17°20′, however many measurements would be greater than 17°20′ and many less. If we were to record all these thousands and thousands of measurements, the results might appear as follows.

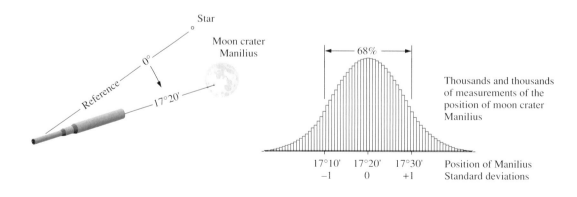

Notice how the distribution takes on that familiar bell-like shape, with the measurements symmetrical about the mean, and that approximately 68% of the errors fall within a certain predictable distance, which we now refer to as the standard deviation.[13]

The next two decades following Gauss's discovery were spent mostly gathering large bodies of data related to measurement error and working out mathematical theories associated with its use. However, outside the fields of astronomy and probability theory, the distribution we now call *normal* was relatively unknown.[14] Starting in the 1830s this started to change. In fact, much of the remainder of the 1800s was spent exploring applications to other fields.[15]

By the end of the century, the normal distribution had been successfully applied to such fields as experimental psychology (reaction times, stimuli-sensation measurement, memory), physics (molecular motion), biology (human height and chest measurements, size of fruit and other characteristics of plant and animal life), and education (talent and abilities as demonstrated by examination scores). By the 1930s, statistical techniques based on the normal distribution had become integral parts of the fields of biometric research, factory production, economics, and agriculture, and was on the verge of incorporation into numerous other fields.

> **Cautionary Note:** One of the myths of statistics is that most natural phenomenon, given enough observations, will take on a normal distribution. This is not so. However, still to this day, some of those trained in experimental research cling to this false belief. In fact, much natural phenomenon is skewed, bimodal, or exhibits a variety of distributive forms. Although *some* natural phenomenon can be closely estimated with the normal distribution, the normal distribution's importance derives more from its use in sampling theory, where this distribution reoccurs with uncanny repeatability, which is discussed in section 4.4 and in chapter 5.

Because a full understanding of the normal distribution is so vital for statistical inference (that is, use of samples to estimate population characteristics), we will spend the remainder of this chapter exploring its intricacies.

4.1 Idealized Normal Curve

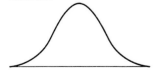

The concept of the idealized normal curve originally stems from the writings and mathematical methods of Laplace (1781, 1786)[16] and is based on the following underlying assumptions.

$n \to \infty$ **a.** The number of observations approaches infinity.   In other words, the number of observations (n) is enormously large (maybe millions or billions of measurements) on the same phenomenon.

$\Delta x \to 0$ **b.** The change in x approaches 0.

The change in x refers to the *width* of the histogram bars. To say the width "approaches zero" means the histogram bars are exceedingly narrow. In other words, if you were to measure adult height, the data must be grouped in exceedingly narrow categories. Say one histogram bar might represent all women $5'4\frac{1}{100}''$ while the next histogram bar represents all women $5'4\frac{2}{100}''$, the next $5'4\frac{3}{100}''$, which are exceedingly narrow groupings.

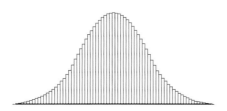

c. The resulting histogram contains thousands of histogram bars, which form into the shape of a bell.

If we connect the tops of the histogram bars, it would take on a smooth flowing appearance, which we refer to as the **normal curve.**

These conditions can be summarized pictorially as follows.

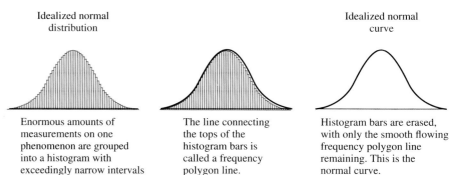

Idealized normal distribution

Idealized normal curve

Enormous amounts of measurements on one phenomenon are grouped into a histogram with exceedingly narrow intervals and form the shape of a bell.

The line connecting the tops of the histogram bars is called a frequency polygon line.

Histogram bars are erased, with only the smooth flowing frequency polygon line remaining. This is the normal curve.

Although the normal curve is somewhat of an idealized construction (it's rare that we can obtain millions or billions of measurements on one phenomenon), experience has shown it to be an indispensible tool in predicting probabilities associated with sampling. Note that the shape of the normal curve can vary slightly; however, certain characteristics are common to all normal curves, and these are presented next.

Characteristics of the Normal Curve

a. Bell-shaped, fading at tails. Theoretically, the distribution continues indefinitely in both directions, approaching but never touching the horizontal axis.

b. The total amount of data is 100%, symmetrical about the mean, μ, with 50% of the data above the value of μ, and 50% below.

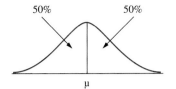

c. Approximately 68% of the data lies within -1 and $+1$ standard deviation of the mean, and approximately 95% of the data lies within -2 and $+2$ standard deviations of the mean.

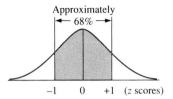

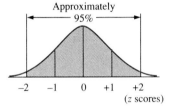

(Recall from chapter 2, section 2.6, a z score is defined as the number of standard deviations a value is away from the mean.)

d. The percentage of data between any two points is equal to the probability of randomly selecting a value between those two points.

   For example, if 28% of the data lies between points *a* and *b,* and if you randomly select one value from the entire population, the probability this one value will be between *a* and *b* is 28%.

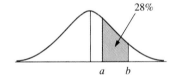

A brief word about terminology: we will often refer to the percentage of data in some part of the normal curve as an **area.** The terminology stems from calculus and pervades much of statistical writing, including this text. Just keep in mind, if we refer to the

Area in some shaded interval	this is equivalent to	% of data in that same interval

Now let's get down to specifics. How do we obtain precise percentages associated with the normal curve? We merely look them up in the normal curve table in the back of the text.

Use of the Normal Curve Table

Kramp was the first to tabulate the exact probabilities associated with the normal distribution, which appeared in 1799 in a book concerning the refraction of light. These tables were used for about 100 years and the tables in use today are only slight variations of the original.[17] Let's see how a contemporary table works.

Although contemporary tables vary slightly in structure, our table requires the understanding of three rules, two of which are presented below.

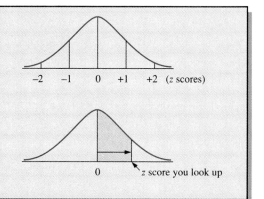

Rule 1
z scores (number of standard deviations from the mean) are used to locate position in the normal curve.

−2   −1   0   +1   +2 (*z* scores)

Rule 2
The table only gives the percentage of data *starting from the middle, z = 0*, out to the *z* score you look up.

0   *z* score you look up

At the upper right hand corner of the normal curve table (refer to "Statistical Tables" in back of book, or for quick reference, see inside cover) is a demonstration example. Let's use it as our first example.

Example   Find the percentage of data (or area) from $z = 0$ to $z = 1.28$.

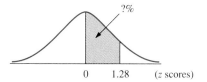

?%

0   1.28   (*z* scores)

Solution   Look under the *z* column to 1.2, then across to the .08 column (notice that $1.2 + .08 = 1.28$). Here we find the decimal .3997. To change .3997 to a percentage, we move the decimal two places to the right .39.97, to get 39.97%.

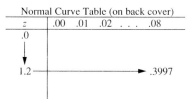

Normal Curve Table (on back cover)

z	.00	.01	.02	. . .	.08
.0					
1.2					.3997

Answer   39.97% of the data lies between $z = 0$ and $z = 1.28$.

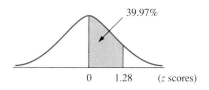

39.97%

0   1.28   (*z* scores)

See if you can solve the following two practice problems without looking at the answers.

Practice 1 ————— Find the % of data between $z = 0$ and $z = .93$.

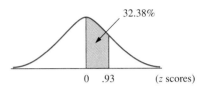

Answer   32.38%. Look under the z column for .9, then across to .03 (note: $.9 + .03 = .93$).

Practice 2 ————— Find the percentage of data between $z = 0$ and $z = 2.00$.

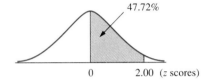

Answer   47.72%. Look under the z column for 2.0, then across to .00 (note: $2.0 + .00 = 2.00$).

Example ————— Find the percentage of data *above* $z = 1.28$.

Solution   Since the table reads from the middle ($z = 0$) out, when we look up $z = 1.28$, we get the percentage of data from $z = 0$ to $z = 1.28$, which is 39.97%. However, this is not the answer to our question. But, if we remember 50% of the data is in half the curve, then

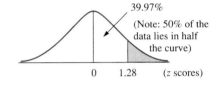

$$39.97\% + ? = 50.00\%$$

We solve this by subtracting 39.97% from both sides, to get

$$? = 50.00\% - 39.97\%$$
$$= 10.03\%$$

Answer   10.03% of the data lies *above* $z = 1.28$.

Practice 3 ————— Find the percentage of data *above* $z = .93$.

Answer   17.62% (50% minus 32.38%)

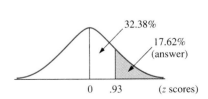

Practice 4 ——————— Find the percentage of data *above* $z = 2.00$.

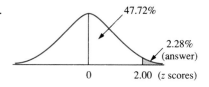

Answer 2.28% (50% minus 47.72%).
Remember: the two percentages (2.28% and 47.72%) must add up to 50%.

The table can also be used to get the percentage of data in any "slice" of the normal curve.

Example ——————— Find the percentage of data from $z = .93$ to $z = 1.28$.

Solution Find the percentage of data from $z = 0$ to 1.28. Then find the percentage of data from $z = 0$ to $z = .93$. Subtract the two percentages to get the answer. Perhaps a formula would be helpful.

$$A_{shaded} = A_2 \text{ minus } A_1$$
$$= 39.97\% - 32.38\%$$
$$= 7.59\%$$

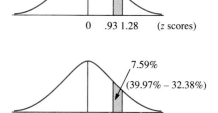

Answer 7.59% of the data lies between $z = .93$ and $z = 1.28$.

Practice 5 ——————— Find the percentage of data from $z = 1.28$ to $z = 2.00$.

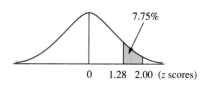

Answer 7.75% (47.72% minus 39.97%)

Practice 6 ——————— Find the percentage of data from $z = .10$ to $z = 1.28$.

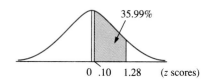

Answer 35.99% (39.97% minus 3.98%)

Note: To get the percentage of data for $z = .10$, we look down the z column to .1, then across to the .00 column (.1 + .00 = .10)

The third and last rule in using our normal curve table is as follows:

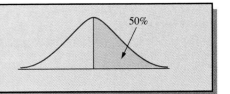

Rule 3
The table only gives percentages
for half the curve.

50%

However, since both halves are identical, the percentage of data from, say, $z = 0$ to $z = 1.28$ is identical to the percentage of data from $z = 0$ to $z = -1.28$. In other words, the percentage of data from $z = 0$ to $z = 1.28$ is 39.97% and the percentage of data from $z = 0$ to $z = -1.28$ is 39.97%. This is demonstrated in the following example.

Example

Find the percentage of data between $z = -1.28$ and $z = +1.28$.

Solution

First get the percentage of data from $z = 0$ to $z = +1.28$ by looking up $z = 1.28$. This gives us 39.97%. Next we get the percentage of data from $z = 0$ to $z = -1.28$ by looking up $z = 1.28$. This also gives us 39.97%. Notice in the diagram that we must *add* the two areas to get the total percentage of data from $z = -1.28$ to $z = +1.28$.

$$39.97\% + 39.97\% = 79.94\%$$

Answer

79.94% of the data lies between $z = -1.28$ and $z = +1.28$.

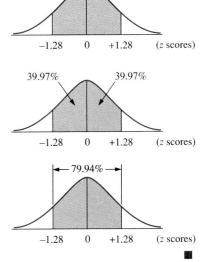

Practice 7

Find the percentage of data between $z = -1.28$ and $z = +.93$.

Answer

72.35% (39.97% + 32.38%)

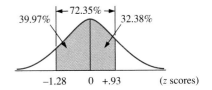

Practice 8

Find the percentage of data *below* $z = -1.85$.

Answer

3.22%

Note that the percentage of data from $z = 0$ to $z = -1.85$ is 46.78%. We then subtract this from 50% to get the answer: 50.00% minus 46.78% = 3.22%.

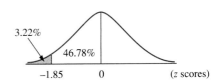

Notice that each case requires us to stop and think before using the normal curve table. For this reason, before we attempt to solve a problem, it is best to do the following.

> ### To Solve a Normal Curve Problem
> **1.** Draw a separate normal curve for each example.
> **2.** Shade the area asked for in the question.
> **3.** Determine the *z* score(s) at the cutoff(s).
> **4.** Stop and think a moment as to how the normal curve table can best be used to get the specific information we request.

4.2 Applications: Idealized Normal Curve

It was widely believed in the last century that once enough data is gathered almost all natural phenomenon will be shown to be normally distributed. Although today we know this not to be true, we do find much in nature and life that can be closely approximated with the idealized normal curve. Let's use the following example to demonstrate.

Suppose we measure the height of every male student enrolled at the Community College at Maxwell Airforce Base, Alabama,* and find the *average* height to be $\mu = 5'10''$. Now, what are the chances that every male student at this College will be $5'10''$? Of course, this is absurd. Although many will be in the vicinity of $5'10''$, the bulk of the students will probably be somewhat shorter than $5'10''$ or somewhat taller. Experience has shown that if we were to represent these height measurements in the form of a *histogram,* that chances are the histogram will build into the shape of a normal distribution and might look as follows.†

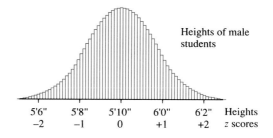

Heights of male students

| 5'6" | 5'8" | 5'10" | 6'0" | 6'2" | Heights |
| −2 | −1 | 0 | +1 | +2 | *z* scores |

*Although probably having the largest enrollment of any community college in the country (approximately 300,000 students), the College at Maxwell Airforce Base offers no on-campus courses. Instead, the college acts as a "clearing house" for incoming transfer credit from airforce personnel all over the world.

†Adolphe Quetelet (1846) was probably the first to demonstrate a population of male heights as closely fitting a normal distribution. For data, he used the heights of 100,000 French conscripts from the early 1800s. Can you guess the average height of a French soldier in those days? The answer is, $\mu = 5'0''$, for all males. Quetelet also showed chest measurements of nearly 6000 Scottish soldiers to be near normally distributed.

The spread of values would depend on a number of factors,* however let's say for this particular population, we calculated the standard deviation to be $\sigma = 2''$. If we fit an idealized normal curve over the data, the resulting representation would look as follows.

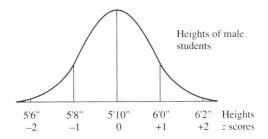

Just from knowing we have a normally distributed population with average height $\mu = 5'10''$ and standard deviation $\sigma = 2''$, we can immediately determine the following.

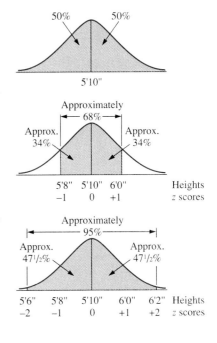

- 50% of the heights will be over $5'10''$, and 50% of the heights will be under $5'10''$.
- Approximately 68% of the heights will be within ±1 standard deviation of the mean, μ—that is, approximately 68% of the heights will be from $5'8''$ to $6'0''$.
  Since the curve is symmetrical, approximately 34% (half 68%) will be from $5'8''$ to $5'10''$ *and* approximately 34% will be from $5'10''$ to $6'0''$.
- Approximately 95% of the heights will be within ±2 standard deviation of the mean, μ—that is, approximately 95% of the heights will be from $5'6''$ to $6'2''$.
  Since the curve is symmetrical, approximately $47\frac{1}{2}\%$ (half of 95%) will be from $5'6''$ to $5'10''$ *and* approximately $47\frac{1}{2}\%$ will be from $5'10''$ to $6'2''$.

*Certain normal populations have been shown to be comprised of a number of smaller normal populations. In other words, several cultural groups in this case might mix (each with a different average height and normal distribution) into one larger composite normal distribution.

To determine more precise percentages, we must refer to the normal curve table, which requires z scores to be calculated to two decimal places. For this, we use the following formula.

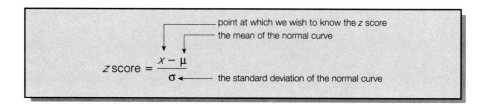

Recall, a z score is the number of standard deviations a value is away from the mean. Now let's look at an example.

Example

Suppose the heights of all male students at the Community College at Maxwell Airforce Base are known to be normally distributed with $\mu = 5'10''$ and $\sigma = 2''$, find the percentage of male students over $6'0''$.

Solution

We proceed in four steps.

a. Draw normal curve, listing real data and z scores for at least ±2 standard deviations.

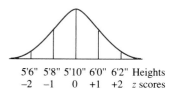

b. Shade the area in question, in this case, over $6'0''$.

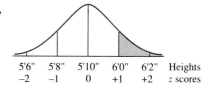

c. Calculate the z score at the cutoff ($6'0''$). This point is represented by the symbol, x, in the z formula.

$$Z = \frac{x - \mu}{\sigma} = \frac{6'0'' - 5'10''}{2''}$$

$$= \frac{2''}{2''} = +1.00$$

d. Now we must stop and think a moment as to how the normal curve table can be used to get the information we wish. If we look up $z = 1.00$, we get 34.13% (.3413), but this is *not* the answer.

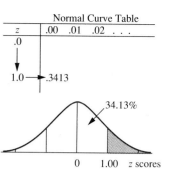

Normal Curve Table

z	.00	.01	.02	. . .
.0				
1.0	.3413			

However, if we subtract 34.13% from 50%, we get the percentage of data in the shaded region.

$$A_{shaded} = 50\% - 34.13\%$$
$$= 15.87\%$$

Note: 34.13% plus 15.87% equals 50% (half the curve).

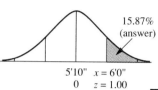

Answer

15.87% of the male students registered at the Community College at Maxwell Airforce Base are expected to be over 6′0″.

Example ————— Referring to the above problem: if we were to randomly select *one* male student from this Community College, what is the probability this one student would be over 6′0″?

Answer

Since 15.87% of the male students are over 6′0″ (according to the above problem), then the probability of randomly selecting a male student over 6′0″ is 15.87%.

Example ————— Again referring to the above problem: what percentage of the area under the curve is in the shaded region?

Answer

The words, area, percentage of data, and probability, have much the same meaning when discussing the normal curve. Thus, we can state,

$$\begin{array}{c}\text{Area of} \\ \text{shaded region}\end{array} = \begin{array}{c}\text{Percentage of data} \\ \text{in shaded region}\end{array} = \begin{array}{c}\text{Probability of} \\ \text{selecting a male} \\ \text{in shaded region}\end{array} = 15.87\%$$

Using the same population, now let's ask a different question.

Example ——————— Suppose the heights of male students at the Community College at Maxwell Airforce Base are known to be normally distributed with $\mu = 5'10''$ and $\sigma = 2''$. Find the percentage of male students who are over $5'7\frac{1}{2}''$.

Solution

Height

We proceed using the same four steps.

a. Draw the normal curve, listing real data and z scores for at least ± 2 standard deviations and
b. Shade the area in question.

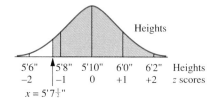

Heights

| 5'6" | 5'8" | 5'10" | 6'0" | 6'2" | Heights |
| -2 | -1 | 0 | +1 | +2 | z scores |

$x = 5'7\frac{1}{2}''$

c. Calculate the z score at the cutoff (in this case, at $5'7\frac{1}{2}''$).

$$z = \frac{x - \mu}{\sigma} = \frac{5'7\frac{1}{2} - 5'10}{2''} = \frac{-2\frac{1}{2}}{2}$$
$$= \frac{-2.5}{2} = -1.25$$

d. Look up $z = -1.25$, which gives us 39.44%. However this is not the complete answer. If we examine the diagram, we note that this is the percentage of males from $5'10''$ to $5'7\frac{1}{2}''$ (that is, from $z = 0$ to $z = -1.25$). To get the complete answer we must add 50%, the percentage of males over $5'10''$.

$$A_{\text{shaded}} = 39.44\% + 50\%$$
$$= 89.44\%$$

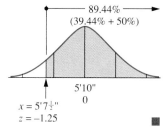

39.44%   50%

5'10"
0
$x = 5'7\frac{1}{2}''$
$z = -1.25$

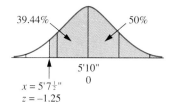

89.44%
(39.44% + 50%)

5'10"
0
$x = 5'7\frac{1}{2}''$
$z = -1.25$

Answer

89.44% of the male students registered at the Community College at Maxwell Airforce Base are expected to be over $5'7\frac{1}{2}''$.

Practice 1 ——————— For the problem above, find the percentage of male students who are $5'7\frac{1}{2}''$ to $6'0''$.

Answer

73.57% (39.44% + 34.13%)

Note: we must look up two %'s of data and then add the two together.

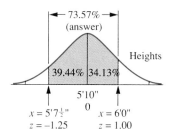

73.57%
(answer)

Heights

39.44% | 34.13%

5'10"
0
$x = 5'7\frac{1}{2}''$ $x = 6'0''$
$z = -1.25$ $z = 1.00$

Practice 2 ───────── For the problem above, find the percentage of male students who are $5'11\frac{1}{2}''$ to $6'1''$.

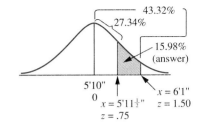

Answer 15.98% (43.32% − 27.34%)

Note: Again, we must look up two %'s of data, only this time we subtract.

4.3 Working Backward with the Normal Curve Table

The normal curve table can also be used in reverse. That is, if we already know the percentage of data in a certain region, we may be able to use the normal curve table to find the z score at the cutoff.

Example ───────── Suppose the percentage of data in the normal curve from $z = 0$ to $z = ?$ is known to be 30%. Find the missing z score.

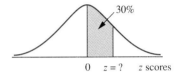

Solution Since the table reads from $z = 0$ out to $z = ?$ and we already know the percentage of data in this region is 30% (.3000 in table), we merely use the table in reverse. First, we find the percentage of data closest to .3000 (30%), which turns out to be .2995.

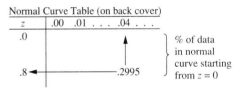

Next, we look across to the z score indicated, to get .8 and up to get .04. The z score is $z = .84$. (Note: if the percentage of data falls precisely midway between two values, we round to the higher z score.)

Answer The missing z score is $z = .84$ (in other words, approximately 30% of the data lies between $z = 0$ and $z = .84$).

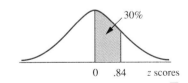

Practice 1 ───────── Find the z scores associated with the *middle 60%* of the data in the normal curve.

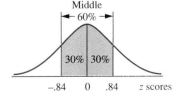

Answer $z = -.84$ to $z = +.84$

Note: We must split 60% into 30% plus 30%. When we look up 30% (closest value is .2995), we get $z = -.84$ and $z = +.84$.

Practice 2 ——————— Find the *z* score associated with the *upper 10%* of the data.

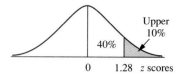

Answer ——————— *z* = +1.28

Note: We must look up 40% in the table, since the table starts reading from *z* = 0 outward. The closest value to 40% (.4000) is .3997, which gives us *z* = 1.28 at the cutoff.

Applications

It's long been known by experimental psychologists that different people react to the same stimuli in different times. For instance, an automobile driver responds to danger by jamming on the brakes but the precise time to react will vary from individual to individual. Similarly, the speed at which a student reacts to (or absorbs) facts in a classroom varies from individual to individual. Gerling (1838) was probably the first to demonstrate reaction times as normally distributed.[18]

Let's demonstrate with the following experiment.

Example ——————— Suppose a researcher concerned with measuring forms of intelligence in new-born infants sets up an experiment to electronically monitor the neurological reaction time when a tiny light is flashed into a baby's eye.

After testing thousands and thousands of newborn infants, it was found that the *average* reaction time was μ = 50 milliseconds (ms) with standard deviation σ = 10 ms. Assuming the reaction times are normally distributed, below what value would you expect to find the *fastest* 10% of the reaction times?

Solution ——————— This is a typical working-backward problem where the percentage of data is given (in this case, the fastest 10% of the values), and we use the normal curve table in reverse to determine the *z* score.

To start, we proceed in much the same way as solving any normal curve problem.

a. Draw the normal curve, listing real data and *z* scores for at least ±2 standard deviations.

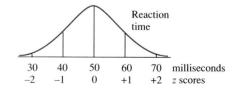

b. Shade the area in question. Since the fastest times would be *less than* 50 ms, we shade the extreme left of the normal curve, estimating 10%.

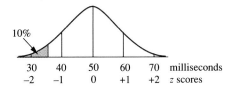

c. Next obtain the z score at the cutoff. To do this, we look up 40% (.4000) since the table reads data only from the *center* ($z = 0$) out.

Note in the table that the closest value to 40% (.4000) is .3997, which gives us a z score of $z = 1.28$. Since the z value is below μ, we must make the z value negative. Thus, $z = -1.28$.

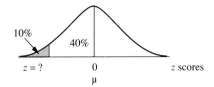

Normal Curve Table

z	.00	.01	. . .	.08
.0				
.				
.				
.				
1.2				.3997

d. Now use the z formula to solve for x, the real data value at the cutoff. Essentially we know z (-1.28), and we wish to solve for x in the formula:

$$z = \frac{x - \mu}{\sigma} \quad \begin{cases} \mu = 50 \\ \sigma = 10 \end{cases}$$

$$-1.28 = \frac{x - 50}{10}$$

$$(10)(-1.28) = x - 50$$

$$-12.8 = x - 50$$

$$37.2 = x$$

$$\text{or } x = 37.2 \text{ ms}$$

This calculation required some algebraic manipulation. First, we multiplied both sides of the equation by 10 to obtain $-12.8 = x - 50$. Second, we added $+50$ to both sides of the equation to get $37.2 = x$. In other words, at the cutoff, $x = 37.2$ ms.

Answer

Below 37.2 ms you would expect to find the fastest 10% of the reaction times.

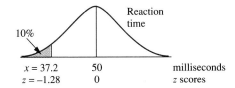

Practice 3 ———————— For the preceding problem, between what two values would you expect to find the *middle 95%* of the reaction times?

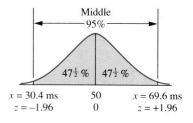

Answer Between 30.4 and 69.6 ms

Note: We must look up $47\frac{1}{2}$% (half of 95%), or in decimal form .4750, to obtain $z = 1.96$ on both sides. When we substitute $z = -1.96$ and $z = +1.96$ in our z formula, we obtain the following:

$$z = \frac{x - \mu}{\sigma} \qquad\qquad z = \frac{x - \mu}{\sigma}$$

$$-1.96 = \frac{x - 50}{10} \qquad +1.96 = \frac{x - 50}{10}$$

$$x = 30.4 \text{ ms} \qquad\qquad x = 69.6 \text{ ms}$$

(To solve, multiply both sides by 10, then add 50 to both sides.)

Practice 4 ———————— For the above problem, *above* what value would you expect to find the *slowest 70%* of the reaction times?

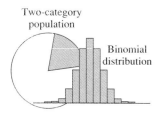

Answer Above 44.8 ms

Note: 50% of the data is above $\mu = 50$ ms so we must look up the remaining 20% (.2000). The closest value to .2000 is .1985, which is equivalent to $z = -.52$. Substituting $z = -.52$ in our z formula, we obtain the following:

$$z = \frac{x - \mu}{\sigma}$$

$$-.52 = \frac{x - 50}{10}$$

$$x = 44.8 \text{ ms}$$

4.4 Binomial Distribution: An Introduction to Sampling

Although some natural populations have distributions that can be approximated with the normal curve, the normal curve's importance is derived more from its consistent and uncanny ability to predict the outcomes when we *sample* from a population. Although different "types" of populations exist (from which we may sample), one of the most important in research is the two-category population.

> A **two-category population** is a population where every member is classified into exactly one of two categories.

Examples of two-category populations are as follows.

Medical Population
Many thousands of users of a new experimental drug designed to cure a specific form of bladder inflammation, classified into *users who were cured* and *users not cured*.

Educational Population
Hundreds of thousands of SAT verbal scores recorded over the past five years, classified into *scores 330 or less* and *scores above 330.*

Manufacturing Population
Millions of assembly-line batteries produced last month by a large manufacturer, classified into *batteries defective* and *batteries not defective.*

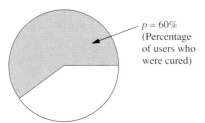
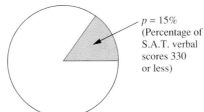
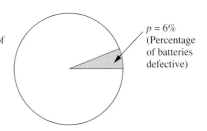

Psychological Population
Thousands of recipients of a new drug-free therapy, classified into *those who showed improvement* and *those with no improvement.*

Marketing Population
Hundreds of thousands of phone calls made to New Jersey residents last year by the Fullins Co. selling magazine subscriptions, classified into *calls resulting in a sale* and *calls resulting in no sale.*

Gambling Population
Billions of coin flips, classified into *those resulting in heads* and *those resulting in tails.*

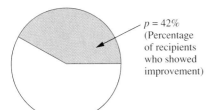
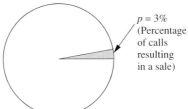
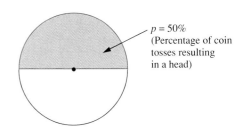

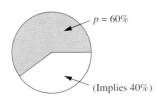

Notice that we may describe such two-category populations by the letter p, the proportion or percentage classified into *one* of the categories. Of course, once we know the percentage of the population in one category, we know the percentage in the other, since the sum of the two percentages must add to 100%. For instance, in the medical population, the first example above, if 60% of the users of this experimental drug were cured, indicated by the shaded region, this implies 40% were not cured. This 40% is represented by the *un*shaded region (note: 60% + 40% = 100%).

In case you were wondering, it doesn't matter which of the two categories in a two-category population we describe by p. For instance, in the manufacturing population, we described this population of assembly-line batteries as $p = 6\%$ defective, however a salesman for this company might describe this exact same population as $p = 94\%$ okay. Most often, we assign p to the particular category we are interested in.

One last important point: every member of a two-category population must fall into one or the other category. In other words, each battery in the manufacturing population must be classified as either defective or not defective. Each user of the experimental drug must be classified as either cured or not cured. Each telephone call in the marketing population must be classified as sale or no sale. There can be no borderline cases. Each member of the population uniquely fits into one or the other category.

Sampling from a Two-Category Population

Once we determine we have a two-category population and describe this population by p (the percentage of values in one category), then we may wish to know what we can expect when we sample from such a population.

Since the methodology for determining such sampling evolved from early gambling experiments, usually involving the tossing of coins or dice, we offer the following.*

Suppose we have the following two-category population:

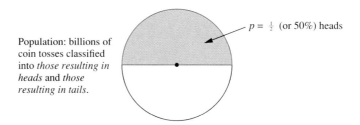

Population: billions of coin tosses classified into *those resulting in heads* and *those resulting in tails*.

$p = \frac{1}{2}$ (or 50%) heads

Now, if we were to randomly sample from this population, say for instance, we sample 12 coin flips, what may we expect to happen? We know from theory and a long history of experience, that if we were to *randomly* sample from any large two-category population,

$$p_s \approx p$$

The sample proportion, p_s, will be approximately equal to the population proportion, p.

*This topic was introduced at the end of chapter 3, section 3.5.

That is, since the population consists of 50% heads, then a random sample should consist of *approximately* 50% heads. In the case of $n = 12$ coin flips, we should get *approximately* 6 heads (50% of 12 = 6). However, we can also get 5 heads or perhaps even 9 heads. How can we determine the percentage of times we can expect each of these outcomes to occur?

One way is to actually perform this experiment a great many times, as follows.

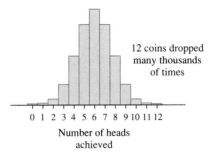

12 coins dropped many thousands of times

0 1 2 3 4 5 6 7 8 9 10 11 12

Number of heads achieved

Say we drop $n = 12$ coins on a table thousands and thousands of times and record the number of heads achieved on each drop, we would get something like the histogram shown here.

This is called a sampling distribution, defined as follows:

> A **sampling distribution** shows us what we can expect when we randomly select n values (a fixed number) repeatedly from a particular population.

In fact, the above sampling distribution shows us what we can expect when we randomly select $n = 12$ values repeatedly from a large two-category population described by $p = \frac{1}{2}$ (or 50%) heads.

These results can be summarized with the following diagrams:

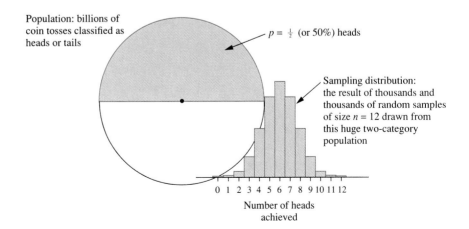

Population: billions of coin tosses classified as heads or tails

$p = \frac{1}{2}$ (or 50%) heads

Sampling distribution: the result of thousands and thousands of random samples of size $n = 12$ drawn from this huge two-category population

0 1 2 3 4 5 6 7 8 9 10 11 12

Number of heads achieved

Notice that this particular sampling distribution (the histogram) is symmetrical around the value we would most likely expect to occur. In the case of dropping $n = 12$ coins, we would most likely expect 50% heads or 6 heads. And indeed, in this instance, we do most often get 6 heads. This is called the **expected value** and can be calculated as follows.*

Expected value = np

Since our sample size is $n = 12$, and the population proportion is $p = \frac{1}{2}$ heads,

$$\begin{aligned} \text{Expected value} &= np \\ &= (12)(1/2 \text{ heads}) \\ &= 6 \text{ heads} \end{aligned}$$

Although we indeed most often get 6 heads, on a great many occasions we get somewhat more than 6 heads, and on a great many occasions somewhat less, with the heights of the histogram bars falling off in a shape strongly resembling that of a normal distribution.

Actually, this should not come as a surprise, since the initial discovery of the normal curve evolved from these same early coin experiments; recall De Moivre's and Laplace's work discussed at the beginning of this chapter.

In fact, these bell-shaped sampling distributions appear repeatedly in gambling experiments when n is large. For instance, the following:

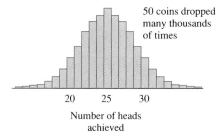

50 coins dropped many thousands of times

20   25   30

Number of heads achieved

Suppose we drop $n = 50$ coins on a table thousands and thousands of times and record the number of heads achieved on each drop, we would get a distribution something like this.

*Expected value was defined in section 3.6 using the general formula, expected value $= \Sigma xp(x)$. It can be shown for binomial experiments such as these, after algebraic manipulation, expected value $= np$.

Notice the sampling distribution is again symmetrical around the value we would most likely expect, which is 25 heads (since the population consists of $\frac{1}{2}$ heads, then any random sample should consist of *approximately* $\frac{1}{2}$ heads; $\frac{1}{2}$ of 50 = 25). Again, this may be calculated as follows:

$$\text{Expected value} = np$$
$$= (50)(1/2 \text{ heads})$$
$$= 25 \text{ heads}$$

And indeed we do most often get 25 heads (see histogram above); however, on a great many occasions we get somewhat more than 25 heads and on a great many occasions somewhat less, with the heights of the histogram bars again falling off in a shape resembling that of a normal distribution.

Okay, you might ask, this may happen with coin tosses, where the probability of a head for a coin toss is $\frac{1}{2}$ (50%), but what if we sampled a different population, say die tosses, where the probability of a particular face turning up is $\frac{1}{6}$ ($16\frac{2}{3}$%). What happens then?

Well, let's take 60 dice and paint one face on each blue (for identification purposes).

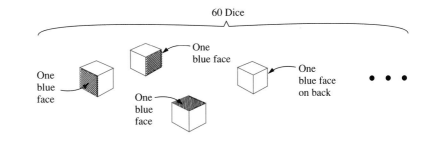

60 Dice

One blue face

One blue face

One blue face

One blue face on back

• • •

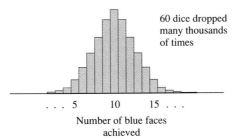

60 dice dropped many thousands of times

. . . 5   10   15 . . .

Number of blue faces achieved

Suppose we drop $n = 60$ dice on a table thousands and thousands of times, and each time record the number of blue faces that turn up. If we tally the results into a histogram, we would get a distribution something like this.

For a clearer picture of this, let's summarize the results with the following diagram.

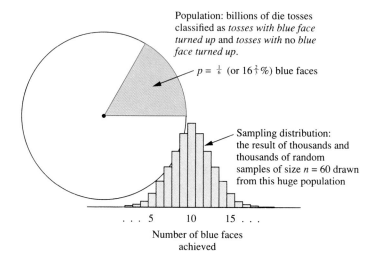

Population: billions of die tosses classified as *tosses with blue face turned up* and *tosses with* no *blue face turned up.*

$p = \frac{1}{6}$ (or $16\frac{2}{3}\%$) blue faces

Sampling distribution: the result of thousands and thousands of random samples of size $n = 60$ drawn from this huge population

. . . 5   10   15 . . .

Number of blue faces
achieved

Notice the shape of the sampling distribution (the histogram). It is symmetrical around the value we would most likely expect, in this case 10 blue faces. In other words, since the population consists of $\frac{1}{6}$ blue faces, any random sample should consist of *approximately* $\frac{1}{6}$ blue faces (note: $\frac{1}{6}$ of 60 = 10). Again, this expected value can be calculated as follows.

$$\begin{aligned} \text{Expected value} &= np \\ &= (60)(1/6 \text{ blue faces}) \\ &= 10 \text{ blue faces} \end{aligned}$$

And indeed 10 blue faces is our most frequently occurring value. However, on many occasions we get somewhat more than 10 blue faces and on many occasions somewhat less, again with the heights of the histogram bars falling off in a shape approximating that of a normal distribution.

Normal Curve Approximation to the Binomial Sampling Distribution

These bell-shaped sampling distributions kept occurring with amazing regularity in coin and dice experiments when n, the number of coins or dice dropped *was sufficiently large.* Of course, at this point you might ask, how large must n be to be considered ''sufficiently large''? Large enough, so when multiplied by p or $(1 - p)$, the result exceeds 5—which leads us to the following important rule.

In these types of sampling experiments, known as **binomial sampling experiments** (to be further defined in Section 4.5),

if expected value (np) > 5 *and* $n(1 - p)$ > 5

the sampling distribution (known as a **binomial sampling distribution**) will be approximately normally distributed with mean and standard deviation

$$\mu = \text{expected value } (np)$$
$$\sigma = \sqrt{np(1 - p)}$$

and a normal curve with these dimensions can be fitted over the distribution and used to estimate probabilities.

Does this imply that *if np* or $n(1 - p)$ is 5 or less, the normal curve cannot be used to estimate probabilities? Yes, for np or $n(1 - p)$ of 5 or less, the sampling distribution is often skewed or sloping and generally the normal curve cannot be depended on to give reliable estimates. For these special cases, other techniques are available, which are discussed in chapter 11.

For the remainder of this chapter, we will demonstrate only those situations where the sampling distribution can be approximated with the normal curve, namely when

$$\text{Expected value } (np) > 5 \quad \text{and} \quad n(1 - p) > 5$$

Let's see how this works in an example.

Example ———————— Out of 12 tosses of a coin, find the probability of achieving exactly 6 heads.

Solution Since the expected value $(np) = (12)(\frac{1}{2}) = 6$, which is greater than 5, and $n(1 - p) = (12)(1 - \frac{1}{2}) = 6$, which is greater than 5, we now know repeated samples of $n = 12$ will produce a sampling distribution approximately normally distributed such that a normal curve with mean and standard deviation as follows can be used to estimate probabilities.

$$\begin{aligned} \mu = \text{expected value} &= np \\ &= 12(1/2) \\ &= 6 \text{ heads} \end{aligned} \qquad \begin{aligned} \sigma &= \sqrt{np(1 - p)} \\ &= \sqrt{12(1/2)(1/2)} \\ &= 1.73 \end{aligned}$$

Now, to answer the question, what is the probability that out of 12 tosses we will achieve exactly 6 heads, we proceed as follows.

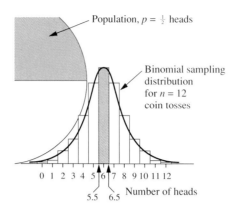

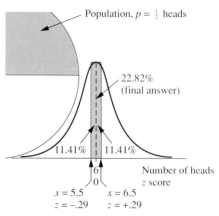

First, we shade the histogram bar representing exactly 6 heads. Note the shading must extend from 5.5 to 6.5 to include the entire width of the histogram bar representing 6 heads. This $\frac{1}{2}$-unit adjustment (referred to as a **continuity correction factor**) is necessary when the normal curve is used to estimate probabilities in the binomial sampling distribution. The term *continuity correction factor* is further defined at the end of the example. Now fit a normal curve over the histogram to estimate probabilities.

Resketch normal curve (for clarity) and shade area from 5.5 to 6.5. Using $\mu = 6$ and $\sigma = 1.73$, we solve as we would any normal curve problem by first calculating the z score at the cutoffs.

$$z = \frac{x - \mu}{\sigma} = \frac{5.5 - 6}{1.73} = \frac{-.5}{1.73} = -.29$$

The percentage of data from $z = 0$ to $z = -.29$ is 11.41%. Since there is an equal amount of data from $z = 0$ to $z = +.29$, we add 11.41% + 11.41% to get 22.82%.

Answer

Now we can say that the probability of achieving exactly 6 heads out of 12 tosses is *22.82%*. Visually, this can be represented as follows:

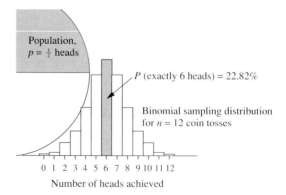

Terminology

> ### *Continuity Correction Factor*
> This refers to the $\frac{1}{2}$-unit shading adjustment(s) necessary to include the entire width of the histogram bar(s) in question.

This is necessary since it is the area occupied by the histogram bar that represents the probability that event will occur. So, remember, when using the normal curve to estimate binomial probabilities, we must shade the entire histogram bar(s) in question to get all the probability.

The *binomial sampling distribution* is sometimes referred to as a **discrete** data distribution, meaning the distribution contains only discrete values.

> ### *Discrete Values*
> Values that, when presented on a number line, occupy only distinct unconnected (or isolated) points.

Note in the histogram above, the data is classified only into values such as 0 heads, 1 head, 2 heads, 3 heads, etc. In other words, if 12 coins were dropped, you could never achieve $3\frac{1}{4}$ or $5\frac{1}{2}$ heads. When data can assume only isolated point values, such as in this case whole-number values, it is referred to as discrete.

4.5 Binomial Sampling Distribution: Applications

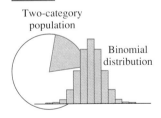

Two-category population

Binomial distribution

Of course, at this point, you may very well say, who cares about this; these are gambling experiments and I'm interested in business, psychology, medicine, education, or whatever.

Well, let's say, these binomial sampling distributions will form no matter what field of endeavor you apply them to, research in business, psychology, medicine, education, or whatever, provided you conform to the fundamental assumptions of binomial sampling, as follows.

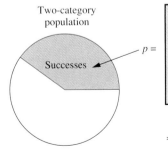

Two-category population

Successes

$p =$

> **Binomial sampling** assumes selection from a two-category population (with members in *one* category labeled "success"), such that
>
> **1.** there is a fixed number of selections, n (often referred to as "n trials"),
> **2.** with each selection (or trial) independent* and each having the same probability, p, a success will be chosen.

*__Independent__ means: whether or not we achieve a success on one selection in no way affects the probability of achieving a success on any other selection.

Binomial sampling can be used in a wide variety of contemporary applications, provided we conform to these fundamental conditions. These conditions are necessary to conform to the basic fundamentals that are innate to coin, dice, and other gambling experiments, on which the theory and mathematics is based.

Although we must evaluate every contemporary experiment on the above formal terms, in actual practice these conditions can often be satisfied by simply

---

Obtaining a Valid* Random Sample from a Large Two-Category Population

This will generally satisfy the above conditions for binomial sampling. A *large* population is defined as any population at least 20 times the size of the sample.

Let's see how all this applies to a contemporary experiment.

Example ——————— From many thousands of users of a new experimental drug designed to cure a specific form of bladder inflammation, it was found that 60% were cured.

Suppose we randomly select $n = 15$ individuals from this large two-category population, what percentage of the time (or with what probability) would we find 12 or more cured?

Solution Notice this is a binomial sampling experiment: there are 15 fixed selections from a two-category population, each independent and each having the same probability a cured individual (a success) will be chosen. *Random* selection from a *large* two-category population generally satisfies these conditions. Furthermore, since expected value $(np) = (15)(.60) = 9$ and $n(1 - p) = (15)(.40) = 6$, and both are greater than 5, the resulting binomial sampling distribution will be approximately normally distributed with mean and standard deviation calculated as follows:

Population: many thousands of users of a new experimental drug

$p = 60\%$ cured

Binomial sampling distribution: the result of many thousands of random samples of $n = 15$ drawn from this population

Number cured

$$\mu = np \qquad\qquad \sigma = \sqrt{np(1 - p)}$$
$$= 15(.60) \qquad\qquad = \sqrt{15(.60)(1 - .60)}$$
$$= 9 \qquad\qquad\qquad = \sqrt{15(.60)(.40)}$$
$$= 1.897$$
$$= 1.90$$

Now a normal curve with these dimensions ($\mu = 9$, $\sigma = 1.90$) can be fitted over the histogram to estimate probabilities in any portion.

---

*All sampling in this text assumes both internal and external validity, as discussed in section 1.1.

So, to answer our question, out of $n = 15$ randomly selected individuals what percentage of the time would we find 12 or more cured, we proceed as follows.

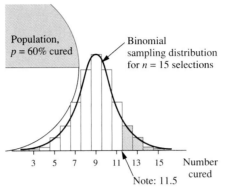

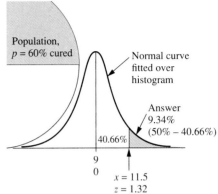

First, we shade the histogram bars representing 12 and above. Note the shading must extend to 11.5 to include the entire bar representing 12 cured. This half-unit adjustment is called your *continuity correction factor*. Now, we fit a normal curve over the histogram.

Second, we resketch the normal curve and shade the area 11.5 and above. Using $\mu = 9$ and $\sigma = 1.90$, we solve as we would solve any normal curve problem by first calculating the z score at the cutoff.

$$z = \frac{x - \mu}{\sigma} = \frac{11.5 - 9}{1.90} = \frac{2.5}{1.90} = 1.32$$

The percentage of data from $z = 0$ to $z = 1.32$ is 40.66%. Subtract this from 50% to get 9.34%.

Answer

Now we can say, 9.34% of the time we will achieve 12 or more cured when we randomly sample 15 from our population, and this can visually be represented as follows:

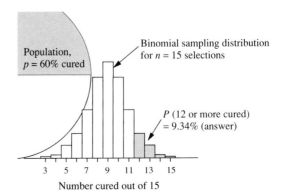

Let's summarize: since the population consists of 60% cured, any random sample will most likely consist of approximately 60% cured (60% of 15 = 9); thus, in this case, *approximately* 9 cured would be expected. Achieving 12 or more cured out of a sample of 15 is not very likely; in fact, this occurs only 9.34% of the time.   ■

It is important when we conduct a binomial sampling experiment that we maintain the conditions of *independence* and *a constant probability of success* from selection to selection. *Random* selection from a *large* population allows for this.

Importance of random selection

Note if selection in the above medical experiment were *not* random: let's say we used only members of the same family for our sample of 15. Family members may very well have similar genetic reactions to a drug. In this case, it would not be unlikely to get 100% of the sample, or even 0% of the sample cured. Generally, *non*random samples violate the prime conditions for binomial sampling, and will usually destroy our ability to predict probabilities. With *random* selection we can be assured of maintaining a constant probability of a success from selection to selection, and thus obtain a true representation of the population.

Importance of a large population

Second, if selection had been from a *small* population (under 20 times the size of your sample), this would violate our condition of independence. For instance, let's say our entire *population* in the medical experiment were not many thousands but instead merely 30 individuals, of which 18 were cured (60%). Now, if we were to randomly sample 15 from this very small population, how many cured individuals we selected, let's say, on the first few picks would greatly affect the probabilities associated with later picks.

Actually, sampling from small populations can be dealt with using other statistical tools, but not the binomial.

Remember, random selection from a large population allows us to maintain the conditions of the early coin and dice experiments, namely independence and a constant probability of success from selection to selection. Serious violation of these conditions can render your results valueless (and remember, for all sampling in this text, we assume internal and external validity has been assured, as discussed in chapter 1).

One more point before we continue. Keep in mind, the normal curve gives an approximation. The histogram bars are wide and the normal curve may fit well, but the fit is not perfect. For instance, the precise answer to the above problem is 9.05%. Our answer is 9.34%. Most would consider this quite close. Generally, for larger values of np and $n(1 - p)$ (for instance, when both np and $n(1 - p)$ exceed 14) the normal curve approximation for most purposes is almost exact. Of course, this leaves somewhat of a gap for np and $n(1 - p)$ between 5 and 14 in which we must exercise some professional judgment in evaluating probabilities. As a general rule, for np or $n(1 - p)$ between 5 and 14, the probabilities in the broad central region of the normal curve are considered reasonably accurate, while probabilities in the "very extreme" tails might best be verified with other methods. Other methods are available to get more precise answers, however these methods can be quite tedious to implement (again, more is discussed on these special cases in chapter 11). Now let's try another example.

This next example is presented not only for practice but to demonstrate that the approximating normal curve may peak at a value of μ that is not a whole number, even though the data in the binomial histogram is classified into discrete whole-number categories.

Example ———————— Out of millions of assembly-line batteries produced last month by a large manufacturer, 6% were known to be defective.

Out of 170 randomly selected batteries from this population, find the probability that 14 or less of these will be defective.

Solution This is binomial sampling since there are 170 fixed selections, each independent and each having the same 6% probability that a defective battery (a success) will be chosen. Random selection from a large two-category population generally satisfies the conditions for binomial sampling.

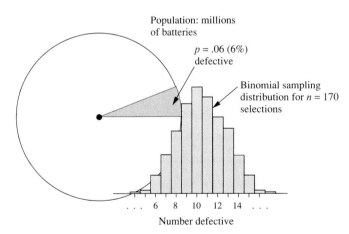

Population: millions of batteries

$p = .06$ (6%) defective

Binomial sampling distribution for $n = 170$ selections

Number defective

Since expected value $(np) = (170)(.06) = 10.2$ and $n(1 - p) = (170)(.94) = 159.8$, and both are greater than 5, the sampling distribution will be approximately normally distributed with mean and standard deviation calculated as follows:

$$\mu = np \qquad\qquad \sigma = \sqrt{np(1 - p)}$$
$$= 170(.06) \qquad\qquad = \sqrt{170(.06)(.94)}$$
$$= 10.2 \qquad\qquad = 3.096$$
$$\qquad\qquad\qquad\qquad = 3.1 \text{ (rounded)}$$

Note that the histogram shown is not quite symmetrical about any particular central value, thus the peak of the approximating normal curve will probably not be a whole number. In this case, the approximating normal curve peaks at $\mu = 10.2$, which is not a whole number. So, to answer the question, what is the probability that out of our sample of 170 we will find 14 or less defective, we proceed as follows.

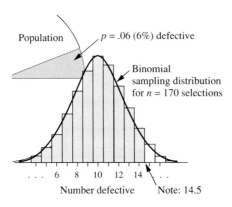

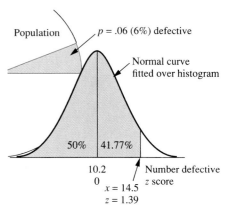

First, we shade the histogram bars representing 14 or less. Note the shading must extend to 14.5 to include the entire bar representing 14 defective. This half-unit adjustment is called your *continuity correction factor*. Now we fit a normal curve over the histogram.

Second, we resketch the normal curve and shade the area 14.5 and below. Using $\mu = 10.2$ and $\sigma = 3.1$, we solve as we would any normal curve problem by first calculating the z score at the cutoff.

$$z = \frac{x - \mu}{\sigma} = \frac{14.5 - 10.2}{3.1} = \frac{4.3}{3.1} = 1.39$$

The % of data from $z = 0$ to $z = 1.39$ is 41.77%. Add this to 50% to get 91.77% (answer).

Answer

Now we can say, 91.77% of the time (or with probability .9177) we will achieve 14 or less defective batteries when we randomly sample 170 from our population, and this can be visually represented as follows:

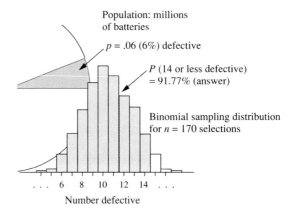

Note that the normal curve is merely a tool, a device we use to lay over the histogram to help us determine the percentage of data in some portion of the histogram. In these binomial sampling experiments, there can never be, in reality, 10.2 defective batteries or 14.5 defective batteries. You can get 10 or 11 or 13 or 15 or any whole number of defective batteries but never 10.2 or 14.5. These numbers are location points on our estimating device, the normal curve.

One more point concerning this problem: Say we were employed as a Quality Control manager on the assembly line that produces these batteries and from a month's production we randomly sampled $n = 170$ and found 22 defective batteries, what would you conclude? Look at the histogram.

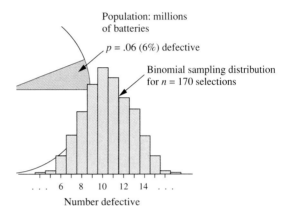

Population: millions of batteries

$p = .06$ (6%) defective

Binomial sampling distribution for $n = 170$ selections

Number defective

Certainly, if the population proportion were indeed $p = 6\%$ defective, then 22 defective batteries out of our sample of $n = 170$ would be an **extremely rare** event—in fact, nearly impossible. You can tell this just by looking at the histogram. The question is: did this *extremely rare* event occur or is the manufacturing process malfunctioning? In other words, is production out of control and no longer holding down the defective rate to $p = 6\%$? Certainly, a prudent quality control manager would investigate and would do so immediately before the process possibly degenerates further.   ■

One last point concerning $np > 5$ and $n(1 - p) > 5$ in a binomial sampling experiment:

np = Expected Number of Successes
$n(1 - p)$ = Expected Number of Failures

For example, in our battery experiment with $n = 170$ selections, we calculated $np = 10.2$ and $n(1 - p) = 159.8$. That is,

Expected Successes   +   Expected Failures   =   Total Selections
(10.2 defective batteries)   +   (159.8 okay batteries)   =   170 selections

So, instead of saying np and $n(1 - p)$ must each exceed 5 for the sampling distribution to be approximately normally distributed, we can say expected number of successes and expected number of failures must each exceed 5 for the sampling distribution to be approximately normally distributed, and the sum of these two numbers equals n, the total selections.

Summary

Perhaps the single most important distribution in all of statistics is the bell-shaped or normal distribution. The distribution was discovered seemingly under different circumstances by De Moivre (1733), Laplace (1781), and Gauss (1809) and encountered so frequently in experiments that sometime in the mid-to-late 1800s it adopted the name, *normal.*

Characteristics of the normal distribution:
Bell-shaped, fading at tails; symmetrical about μ, the mean, with 50% of the data in each half. Approximately 68% of the data lies within ±1

standard deviation of the mean, whereas approximately 95% of the data lies within ±2 standard deviations of the mean.

Normal curve table: This table offers the percentage of data in the normal curve between $z = 0$ (the position of μ) and any z score you look up. Recall, a z score is the number of standard deviations a value is away from the mean. To precisely calculate the z score of a value, x, we use the formula

$$z = \frac{x - \mu}{\sigma}$$

Normal Curve Table Usage

To Find Area, A

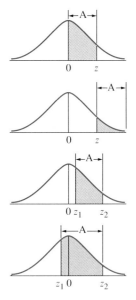

We Use the Following Procedure

A, the area between 0 and z can be found directly in the normal curve tables.

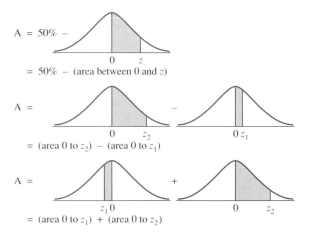

Working backward: The normal curve table can also be used in reverse. If we know the percentage of data between 0 and z, we look up this area (in decimal form) in the table and determine the closest z value. If the percentage of data falls precisely midway between two values, we round to the higher z score.

Sampling from a Two-Category Population

Two-category population: A two-category population is a population where every member is classified into exactly one of two categories.

Sampling distribution: A sampling distribution shows us what we can expect when we randomly select n values (a fixed number) repeatedly from a particular population.

Binomial sampling distribution: The resulting sampling distribution when we randomly select n values repeatedly from a large two-category population, where each selection is independent and each has the same probability, p, a success will be chosen.

Large-n binomial sampling: For np and $n(1 - p)$ greater than 5, the binomial sampling distribution can be approximated with a normal curve with the following dimensions:

$$\mu = np$$
$$\sigma = \sqrt{np(1 - p)}$$

Continuity correction: This refers to the $\frac{1}{2}$-unit shading adjustment(s) necessary to include the entire width of the histogram bar(s) in question.

Discrete values: The binomial sampling distribution is sometimes referred to as a discrete data distribution, discrete, meaning values that when presented on a number line occupy only distinct unconnected (or isolated) points. However to assess probabilities we represent these discrete values with histogram bars, where the area of a histogram bar at some value represents the probability of achieving that value.

Small-n binomial sampling: For np or $n(1 - p)$ of 5 or less, the binomial sampling distribution is often skewed or sloping and the normal curve cannot be depended on to give proper estimates. For these cases, other techniques can be employed that are discussed in chapter 11, section 11.1.

Exercises

Note that full answers for exercises 1–5 and abbreviated answers for odd-numbered exercises thereafter are provided in the Answer Key.

4.1

a. It was widely believed in the mid-1800s that given enough observations all natural phenomena, such as, heights, weights, reaction times, etc., from any common grouping will take on the shape of a normal distribution. Is this so? Explain.

b. In the construction of the idealized normal curve, three primary assumptions were presented. List each and explain.

c. The idealized normal curve has a number of characteristics. List four characteristics.

4.2 Use the normal curve table to determine the percentage of data in the normal curve

a. between $z = 0$ and $z = .82$.

b. above $z = 1.15$.

c. between $z = -1.09$ and $z = .47$.

d. between $z = 1.53$ and $z = 2.78$.

Work backward in the normal curve table to solve the following:

e. 32% of the data in the normal curve can be found between $z = 0$ and $z = ?$

f. Find the z score associated with the *lower* 5% of the data.

g. Find the z scores associated with the *middle* 98% of the data.

4.3 Suppose the heights of all female students at University of Maryland in College Park are known to be normally distributed with $\mu = 5'5''$ and $\sigma = 2''$, find the percentage of female students

a. under $5'2''$.
b. between $5'2\frac{1}{2}''$ and $5'8''$.
c. between $5'8\frac{1}{2}''$ and $5'9\frac{1}{2}''$.

4.4 Ebbinghaus in 1885, in a landmark experiment in Experimental Psychology, repeatedly measured the time necessary for an individual to memorize equal blocks of nonsense syllables (such as, zid, cuk, xot) and found the times to be normally distributed.

Using Ebbinghaus's data, suppose this individual takes an average of $\mu = 21.0$ minutes to complete the task of memorizing a block of nonsense syllables with standard deviation, $\sigma = 1.2$ minutes.

a. Below what value would you expect to find the fastest 10% of the times? (Note: the *fastest* times would be *less than* 21.0 minutes, thus we shade the extreme left of the normal curve, estimating 10%.)
b. Between what values would you expect to find the *middle* 50% of the times?

4.5 Selecting random samples of the same size *repeatedly* from a large two-category population creates a sampling distribution, known as a binomial sampling distribution, which is approximately normally distributed for expected value $(np) > 5$ and $n(1 - p) > 5$. Use this information to answer the following.

a. In a population of many thousands of users of a new experimental drug designed to cure a specific form of bladder inflammation, it was found that 60% were cured. Suppose we randomly select 15 users from this population, what is the probability we will find 7 or less cured?
b. Can we apply this population proportion ($p = 60\%$ cured) to future users, say in the case where the drug is to be distributed in another country? That is, can we expect about 60% cured? Discuss briefly.

4.6 Use the normal curve table to determine the percentage of data in the normal curve

a. between ± 1 standard deviation.
b. between ± 2 standard deviations.
c. between ± 3 standard deviations.

4.7 Use the normal curve table to determine the percentage of data in the normal curve

a. between $z = 0$ and $z = .38$.
b. above $z = -1.45$.
c. above $z = 1.45$.
d. between $z = .77$ and $z = 1.92$.
e. between $z = -.25$ and $z = 2.27$.
f. between $z = -1.63$ and $z = -2.89$.

Work backward in the normal curve table to solve the following.

g. 15% of the data in the normal curve can be found between $z = 0$ and $z = ?$
h. Find the z score associated with the *upper* 73.57% of the data.
i. Find the z scores associated with the *middle* 95% of the data.

4.8 Suppose standard IQ scores are known to be normally distributed with $\mu = 100$ and $\sigma = 15$. Find the percentage of individuals with IQ scores

a. above 125.
b. above 90.
c. between 62 and 72.
d. below 88.

Work backward in the normal curve table to solve the following.

e. Above what value would you expect to find the *upper* 30% of IQ scores?
f. Between what values would you expect to find the *middle* 75% of IQ scores?

4.9 Biological characteristics of a species are sometimes found to be near normally distributed. Suppose American anchovies, *Engraulis encrasicholus*, a species of herring commonly used on pizza, is known to have lengths that are normally distributed with μ = 10.2 centimeters (about 4′′) and σ = .68 centimeters (cm). Find the percentage of anchovies with lengths

a. below 9.0 cm.
b. below 10.7 cm.
c. between 9.5 cm and 10.8 cm.
d. between 11.0 cm and 11.4 cm.

Work backward in the normal curve table to solve the following.

e. Above what length would you expect to find the *longest* 15% of anchovies?
f. Between what lengths would you expect to find the *middle* 99% of anchovies?

4.10   Human characteristics are sometimes found to be near normally distributed. Quetelet in 1846 was probably the first to demonstrate this using the chest measurements of Scottish soldiers. He found the chest measurements to be normally distributed with μ = 39.5′′ and σ = 2.5′′. Find the probability of randomly selecting a measurement

a. between 36.5′′ and 38.5′′.
b. above 38.2′′.
c. between 39.2′′ and 40.6′′.
d. between 39.5′′ and 44.7′′.
e. Below what value would you expect to find the *smallest* 40% of the chest measurements?
f. Between what values would you expect to find the *middle* 96% of the chest measurements?

4.11   Galton demonstrated that large normal populations may, in fact, be comprised of several smaller normal populations. In 1875 he separated sweet pea seeds from the same parent by weight into several groups. Each group produced sweet peas with normally distributed weights but around different averages. When combined, these several smaller normal distributions formed into one large normal distribution centered around one common average.

Suppose the weights of a number of subspecies of Granny Smith apple combine to form one large normally distributed population of Granny Smith apple with μ = 6.9 ounces (oz) and σ = 1.1 oz.

a. What percentage of Granny Smith apples weigh more than 8.5 oz?
b. What percentage of apples weigh between 7.2 oz and 8.0 oz?
c. If you randomly select a Granny Smith apple, what is the probability the apple weighs less than 7.0 oz?
d. Above what weight would you find the *heaviest* 65% of the apples?
e. Between what weights would you find the *middle* 84% of the apples?

4.12   A binomial experiment is formally defined as a fixed number of trials (or selections), each independent and each having the same probability for success. Show how these conditions are met and solve the following.

a. Out of 20 tosses of a coin, what is the probability of getting 13 to 15 heads?
b. Out of n = 50 die tosses (one face of die is painted blue), what is the probability of turning up 10 or more blue faces?

4.13   The U.S. Military Academy at West Point is one of the nation's most selective colleges, accepting 11%* of applicants (according to the Insider's Guide to the Colleges). Out of n = 60 randomly selected applicants to the U.S. Military Academy,

a. how many would you *expect* to be accepted?
b. what is the probability 8 or less will be accepted?
c. what is the probability between 5 and 7 will be accepted?
d. what is the probability *exactly* 6 will be accepted?

4.14   67% of Americans feel secret files are being kept on them (based on data from *The Harper's Index*). Out of 25 randomly selected Americans, what is the probability 18 or more will feel secret files are being kept on them?

4.15   75% of those working in the visual, literary, or performing arts earn low wages from their art, under twelve thousand dollars per annum, based on data from Columbia's Research Center for Arts and

*Harvard accepts 15%.

Culture (*Columbia Magazine,* Summer 1990, p. 14). Out of 30 randomly selected artists,

a. how many would you *expect* to earn low wages?
b. what is the probability you will find at least 20 earning low wages?
c. what is the probability you will find 24 to 27 earning low wages?

4.16 In a marketing population of phone calls, 3% produced a sale. *If* this population proportion (p = 3%) can be applied to future phone calls, then out of 500 randomly monitored phone calls,

a. how many would you expect to produce a sale?
b. what is the probability of getting 11 to 14 sales?
c. what is the probability of getting 12 or less sales?

4.17 In a study on aggression, 23% of mice exposed to severe conditions of overcrowding resorted to bizarre social behavior, such as cannibalism. If this is representative of all mice, out of a randomly selected group of n = 100 mice exposed to these severe conditions,

a. find the probability you will get from 20 to 25 that exhibit bizarre social behavior.

b. find the probability you will get 18 or less that exhibit bizarre social behavior.
c. How valid is our assumption that p = 23% can be assigned to all mice? Discuss briefly.

4.18 88% of American high school students agree with their parents on the value of an education (according to studies from the University of Michigan, Institute for Social Research, ''Monitoring the Future'').* Out of n = 45 randomly selected American high school students,

a. find the probability that 35 or more will agree with their parents on the value of an education.
b. find the probability that 41 to 44 will agree with their parents on the value of an education.
c. If we were to randomly sample n = 45 American high school students ten years from now, can we expect about 88% of the sample to agree with their parents on the value of an education? Discuss briefly.

*The same studies also revealed only 47% agreed with their parents on what's permitted on a date.

Endnotes

1. De Moivre, although born in France, was obliged to move to England as a young man under the Edict of Nantes (which restricted religious and civil liberties to Huguenots), and in England De Moivre worked as a mathematics tutor and consultant for wealthy patrons.

2. Actually De Moivre simulated the number of heads expected when n coins are dropped by using the expansion of $(1 + 1)^n$. One can also use the coefficients of the expansion $(a + b)^n$.

3. De Moivre did not use the term, standard deviation. In fact, technically the concept of standard deviation was not to be fully recognized for at least another seven decades, until after Legendre's work on least squares (1805) in which he demonstrated $\Sigma (x - \mu)^2$ was minimum about the mean (refer to chapter 9, section 9.0, under ''Least-Squares Analysis'' for

further reading on this). De Moivre's predictable distance was calculated to be $\frac{1}{2}\sqrt{n}$, which we now know as the standard deviation in a binomial experiment when $p = \frac{1}{2}$, that is, $\sigma = \sqrt{np(1 - p)} = \sqrt{n\frac{1}{2}\frac{1}{2}} = \frac{1}{2}\sqrt{n}$. In this case, where n = 900, $\sigma = \frac{1}{2}\sqrt{900} = \frac{1}{2} \cdot 30 = 15$. This is further discussed in section 4.4. De Moivre arrived at $\frac{1}{2}\sqrt{n}$ (actually, $\frac{1}{2}\sqrt{n - 2}$, which is essentially equal to $\frac{1}{2}\sqrt{n}$ for large n) by determining the inflection points on the curve.

4. De Moivre's work at the time went relatively unnoticed and one can only speculate why. Perhaps the most probable reason is that mass statistical data was not as yet available, thus the practical application of De Moivre's discovery to social phenomenon could not be readily demonstrated—although De Moivre and a number of others felt it was only a matter of time until the laws of probability would be applied to a variety of social issues.

For an insightful discussion on this topic, refer to S. Stigler, *The History of Statistics* (Cambridge: Belknap Press, 1986), pp. 85–87.

5. It was unclear whether Laplace was familiar with De Moivre's work published 50 years earlier since he never mentioned De Moivre in his papers and his mathematical approach was quite different.

6. Laplace used the illustration of black and white tickets drawn from an urn.

7. At the time, Laplace (like De Moivre) was unaware of the concept of standard deviation. Laplace used a rather complex formulation to arrive at a suitable measure of spread. It was adequate for his purposes, but like much of Laplace's work exceedingly complex.

8. The actual figures were 251,527 males out of 493,472 births. All figures were scaled to 500,000 births for clarity.

9. The precise percentages were: Paris, 50.97% male births; London, 51.35%; Kingdom of Naples, 51.16% (Stigler, 1986).

10. According to *Newsweek* (April 16, 1990, p. 81), current averages worldwide are 50.6% male births, 49.4% female (102.5 males are born for every 100 females).

11. Gauss's reasoning essentially proceeded as follows: (1) it was generally acknowledged at the time that the arithmetic mean of several measurements was the best estimate of planetary position, (2) since the mean is most probable only if the errors are normally distributed, according to the method of least squares, then (3) errors must be normally distributed. Although one may find fault with Gauss's reasoning, the impact was monumental. Laplace seized on the argument giving it a solid base in logic based on his work with probability experiments.

Essentially, Laplace reasoned that a single observation must itself be an aggregate of more fundamental errors just like the outcome of 900 coins dropped on a table is the aggregate of many head-tail outcomes. It is surprising Laplace himself had not made the discovery, considering his intense involvement in both astronomy and probability theory.

12. Planets such as Jupiter and Saturn were not used at sea to measure longitude because of their relatively slow movement and other difficulties of measurement while at sea. Moon craters were highly visible and the Moon's motion relatively fast, offering more precise measurements.

13. Gauss and others in the 1800s used a variety of standard distances from the mean, however many were multiples of the standard deviation, such as .675 σ, which was referred to as the *probable error,* since 50% of the errors were expected to fall within ±.675 σ of μ. In 1893, Pearson coined the term *standard deviation* and advocated its universal use.

14. Use of the normal distribution was confined mostly to astronomy for several decades and, thus, throughout much of the 1800s was referred to as Gauss's law of error. Even to this day, the normal distribution is sometimes called the Gaussian distribution.

15. For further readings in this area, refer to H. Walker, *Studies in the History of Statistical Method* (Baltimore: Williams & Wilkins, 1929) and Stigler (1986).

16. In his newborn infant study (discussed in section 4.0), Laplace devised methods for calculating certain probabilities associated with the binomial distribution as $n \rightarrow \infty$, which Kramp in 1799 used to construct a full table of normal curve probabilities.

17. Kramp prepared the tables using $\sigma \sqrt{2}$ as the unit measure of dispersion, referred to as the *modulus.* Contemporary tables use σ, the standard deviation. Shepperd (1902) was the first to publish a table using σ, the standard deviation, as the unit measure.

18. E. W. Scripture, *The New Psychology* (1897), p. 443, as discussed and footnoted by Walker (1929), p. 24.

Central Limit Theorem

Not all populations are normally distributed and the reader should not assume them to be. However, for simplicity, the concepts in chapters 5 and 6 are presented using mostly normal or near normal populations.

*Non*normal populations are introduced in section 5.4, then demonstrated in chapters 7 and 8. ▼

5.1 Central Limit Theorem

One of the most remarkable theorems in statistics is called the **central limit theorem,** which is best explained through practical example.

Suppose a machine in a dress factory is set to cut pieces of silk material exactly to the length of 1000 mm. The pieces are then to be assembled into an outfit. Now, what are the chances the machine will cut every piece of silk material to precisely 1000 mm? Quite slim. Most of the cuts will be in the vicinity of 1000 mm, however many pieces will be shorter and many longer. Experience tells us that if a properly operating machine cuts millions and millions of pieces, the *histogram* representing the lengths of all these pieces of material may very well build into a shape closely resembling that of a normal distribution clustered around the average length of μ = 1000 mm, and might look as follows:*

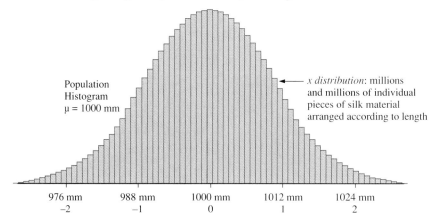

Population Histogram μ = 1000 mm

x distribution: millions and millions of individual pieces of silk material arranged according to length

| 976 mm | 988 mm | 1000 mm | 1012 mm | 1024 mm |
| −2 | −1 | 0 | 1 | 2 |

The standard deviation (σ) of a population such as this will vary from machine to machine. The speed of the cuts, even the length setting may affect the standard deviation, but let's say for our example we will "invent" a standard deviation of σ = 12 mm.

Now let us suppose, from these millions and millions of pieces we *randomly select* a sample of 36 pieces. Because our sample was *randomly* selected, we know from section 2.4 that:

$\bar{x} \approx \mu$ The sample average *is approximately equal to* the population average.

$s \approx \sigma$ The sample standard deviation *is approximately equal to* the population standard deviation.

*A single machine operating properly and uninterrupted will often produce goods whose measurement on a single characteristic, when recorded into a histogram, take on a shape strongly resembling that of a normal distribution.

So it should not be surprising that after we calculate $\bar{x}$, the average length of these 36 pieces in our sample, that this sample average ($\bar{x}$) might equal, say for instance, 998 mm, shown below.

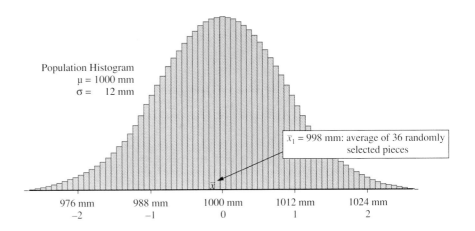

Since we know $\bar{x}$ should be *approximately equal to* μ, the question arises: would 998 mm be considered approximately equal to 1000 mm? In other words: how close must $\bar{x}$ be to be considered approximately equal to μ? For the answer to this question, we must take additional samples of 36 and actually calculate the values we get for $\bar{x}$. So, we randomly select a second sample of 36, then a third sample of 36, and even a fourth sample of 36. The new $\bar{x}$'s ($\bar{x}_2$, $\bar{x}_3$, and $\bar{x}_4$) are calculated and plotted along with $\bar{x}_1$ in the following diagram.

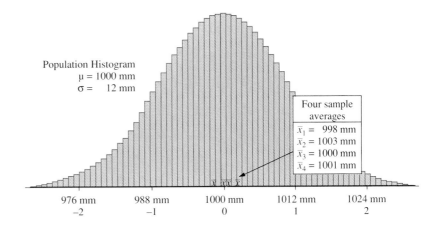

In the next sketch, we have added seven more sample averages ($\bar{x}$'s) to our plot. Note how the sample averages ($\bar{x}$'s) begin to "pile up" on the same readings, all in the vicinity of 1000 mm.

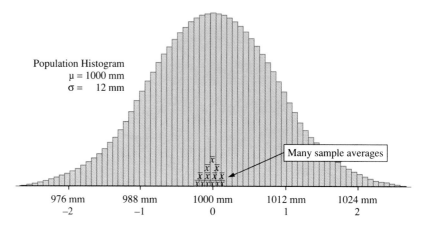

Population Histogram
μ = 1000 mm
σ = 12 mm

Many sample averages

| 976 mm | 988 mm | 1000 mm | 1012 mm | 1024 mm |
| -2 | -1 | 0 | 1 | 2 |

Now we run wild and randomly select thousands and thousands of samples, with each sample containing 36 pieces of material cut from the machine. For each sample of 36 pieces, we calculate the sample average such that, now, we have thousands and thousands of $\bar{x}$'s. Why on earth would anybody want to do this, you might ask? That's a difficult question to answer,* but somebody did and discovered something that, when put in combination with astute and sensible management, helped catapult numerous mid-sized businesses into gigantically successful worldwide empires. Two such empires are Proctor & Gamble and IBM. Management in these corporations use statistical techniques such as these in marketing research and quality control studies on a routine basis.

Okay, we now have thousands of $\bar{x}$'s. Now what? We group the results of all these thousands of *sample averages* and arrange them according to length into a **small histogram** (called an $\bar{x}$ distribution), which would look as follows:

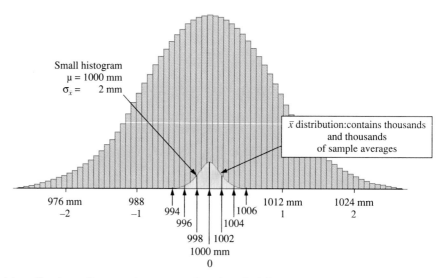

Small histogram
μ = 1000 mm
$\sigma_{\bar{x}} =$ 2 mm

$\bar{x}$ distribution:contains thousands and thousands of sample averages

976 mm	988	994		1006	1012 mm	1024 mm
-2	-1				1	2
		996		1004		
		998	1002			
		1000 mm				
		0				

*Actually, the earliest experiments were based on Carl Gauss's work on measurement error in the field of astronomy (1809, 1816, 1823; Encke, 1832, 1834). Laplace formulated the theoretical underpinnings (1810–1812) and, later, experiments in biological measurement, agriculture and machine output confirmed these early findings.

Perhaps now we can answer the question: how close does $\bar{x}$ have to be to be considered approximately equal to μ? We merely look at the results of the thousands and thousands of samples taken. Almost every sample average ($\bar{x}$) fell between 994 mm and 1006 mm. Are we saying it is impossible to get an $\bar{x}$ of, say, 1012 mm? No, not impossible, but highly improbable. In all the thousands of random samples actually taken, not one $\bar{x}$ even came near 1012 mm. In fact, most of the $\bar{x}$'s were clustered between 998 mm and 1002 mm with only a small few as far out as 994 mm or 1006 mm. Incredible!

And another thing. Did you notice the shape of the small histogram? Yes, a *normal distribution.* Which of course allows us to calculate areas that translate into probabilities. And this finally leads us to the central limit theorem.

▼ *Central Limit Theorem*

> If you continually select random samples of a fixed sample size greater than 30 from a normal population (or almost any population for that matter), the sample averages ($\bar{x}$'s) will gradually build into the shape of a normal distribution, called an $\bar{x}$ distribution, such that
>
> **1.** The mean of the $\bar{x}$ distribution is the same as μ, the mean of the population, and
> **2.** The standard deviation of the $\bar{x}$ distribution, called sigma x-bar* ($\sigma_{\bar{x}}$) is equal to the population standard deviation (σ) divided by the square root of the sample size (n). In other words,
>
> $$\sigma_{\bar{x}} = \sigma/\sqrt{n}.$$

In our "cutting machine" problem, since the population standard deviation (σ) is 12 mm and our sample size (n) is 36 pieces, $\sigma_{\bar{x}} = 12/\sqrt{36} = 2$ mm.

5.2 Applying the Central Limit Theorem

The central limit theorem is one of the most remarkable achievements in statistics. It brought statistics out of the Dark Ages. Until its discovery and widespread application (starting about 100 years ago) we had only been able to estimate population characteristics with very large bodies of data—data that often took months or years to gather and sort and was often outdated before it was analyzed. Now we have at our disposal a precise mathematical way of estimating what is happening now. And because of this we can make better decisions. Let's see how it works in the following three examples.

---

*$\sigma_{\bar{x}}$ is often called the **standard error of the mean;** however, we will simply refer to it as the standard deviation of the $\bar{x}$ distribution.

Example —————— A machine in a dress factory cuts pieces of silk material to **an average length of** $\mu = 1000$ mm with standard deviation $\sigma = 12$ mm. If we **continually take samples** of size $n = 36$, what percentage of the sample averages ($\bar{x}$'s) **would you expect** to fall between $\bar{x} = 995$ mm and $\bar{x} = 1005$ mm?

Solution   We do this problem as we would do any normal curve problem, **only now we are** dealing with a small normal curve fitted over the $\bar{x}$ histogram, **so we must first** calculate $\sigma_{\bar{x}}$, the standard deviation of the $\bar{x}$ histogram (or distribution), **and list** values for at least ±2 standard deviations in that distribution.

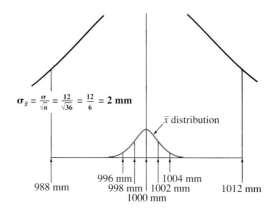

$$\sigma_{\bar{x}} = \frac{\sigma}{\sqrt{n}} = \frac{12}{\sqrt{36}} = \frac{12}{6} = 2 \text{ mm}$$

$\bar{x}$ distribution

996 mm | 1004 mm
988 mm | 998 mm | 1002 mm | 1012 mm
1000 mm

Next, we calculate the z scores at the cutoffs of $\bar{x} = 995$ mm and $\bar{x} = 1005$ mm. Notice the z formula is coded with the subscript, $\bar{x}$; this is to remind us we are working in the $\bar{x}$ distribution.

$$z = \frac{\bar{x} - \mu}{\sigma_{\bar{x}}}$$

$$z = \frac{995 - 1000}{2}$$

$$z = -2.50$$

$$z = \frac{\bar{x} - \mu}{\sigma_{\bar{x}}}$$

$$z = \frac{1005 - 1000}{2}$$

$$z = +2.50$$

Looking up the area for a z of $+2.50$, we get 49.38%, that is, 49.38% of all the $\bar{x}$'s fall between 1000 mm and 1005 mm (refer to the following diagram). Since the normal curve is symmetrical, a z of -2.50 will yield an additional 49.38%, that is, 49.38% of the $\bar{x}$'s fall between 1000 mm and 995 mm.

The completed solution would appear graphically as follows:

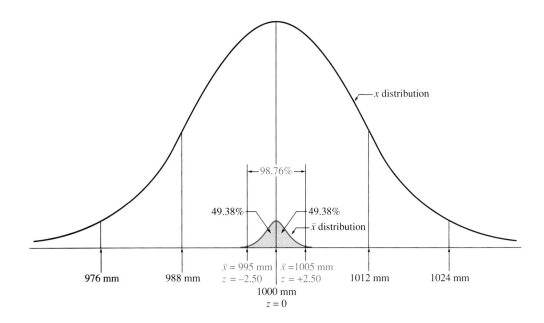

$\bar{x} = 995$ mm | $\bar{x} = 1005$ mm
$z = -2.50$ | $z = +2.50$
1000 mm
$z = 0$

Answer

If we continually take random samples of 36 pieces of material in each sample, and calculate the average length ($\bar{x}$) in each of these samples, then 98.76% of all the sample averages ($\bar{x}$'s) would be expected to fall between $\bar{x} = 995$ mm and $\bar{x} = 1005$ mm.

> ### Rounding Technique
> As a general rule, work in three decimal places throughout the entire problem. Only round final answers to two decimal places.
> This rounding technique is important to ensure that everyone arrives reasonably close to the same answer. Use of the technique will grow increasingly more critical as we proceed through the material.
> Exception: *z* scores will *always* be presented in two decimal places, even in calculations.

Note: Since 98.76% is such an overwhelmingly high percentage, if one day we turned on the machine, took a random sample of 36 pieces and calculated a sample average ($\bar{x}$) which fell *outside* this 995 mm to 1005 mm range, would you suspect a malfunction in the machine?

Certainly we would be suspicious, since any $\bar{x}$ outside this range would be considered quite an unlikely or rare occurrence and therefore we might very well ask: is the machine malfunctioning (cutting pieces to some other length?) or has that rare occurrence actually occurred? Generally, companies in this situation would explore the possibility of a malfunction.   ■

Example ————————— The National Institutes of Health agreed to supply active disease viruses, such as polio and AIDS, to research firms for the purpose of experimentation. A process is set up to automatically fill millions of small test tubes to an average of 9.00 milliliters of disease virus with standard deviation .35 milliliters. If we continually take random samples of 49 test tubes in each sample and calculate the average fill, $\bar{x}$, between what two values would you expect to find the middle 99% of all the sample averages ($\bar{x}$'s)?

Solution Since we are concerned with the "average" fill of 49 test tubes and not with the contents of one test tube, we use the $\bar{x}$ distribution.

This is a typical "working backward (given the area, find z)" problem for the normal curve, only now we are dealing with the $\bar{x}$ distribution so we must first calculate $\sigma_{\bar{x}}$, the standard deviation of the $\bar{x}$ distribution, and list values for at least ±2 standard deviations.

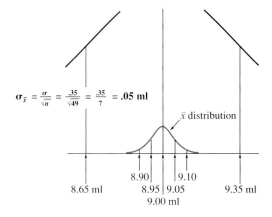

$$\sigma_{\bar{x}} = \frac{\sigma}{\sqrt{n}} = \frac{.35}{\sqrt{49}} = \frac{.35}{7} = .05 \text{ ml}$$

$\bar{x}$ distribution

8.90 9.10
8.65 ml   8.95 | 9.05   9.35 ml
9.00 ml

Since we know the area between the cutoffs is 99%, we merely look in the normal curve table for the corresponding z scores. *Remember:* the table reads "half" the normal curve, so we must look up an area of 49.5% ($\frac{1}{2}$ of 99%), which in decimal form is .4950.

According to the table, the corresponding z scores are -2.58 and $+2.58$ (note that .4950 fell precisely midway between two values in the table; in these cases, we round to the higher z score).

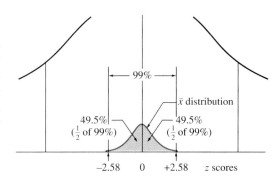

99%

$\bar{x}$ distribution

49.5%   49.5%
($\frac{1}{2}$ of 99%)   ($\frac{1}{2}$ of 99%)

-2.58 0 $+2.58$ z scores

Substituting -2.58 and $+2.58$ into our formula, we solve for the cutoffs:

$$z = \frac{\bar{x} - \mu}{\sigma_{\bar{x}}}$$

$$z = \frac{\bar{x} - \mu}{\sigma_{\bar{x}}}$$

$$-2.58 = \frac{\bar{x} - 9.00}{.05}$$

$$+2.58 = \frac{\bar{x} - 9.00}{.05}$$

Solving for $\bar{x}$:

Solving for $\bar{x}$:

$$\bar{x} = 8.87 \text{ ml}$$

$$\bar{x} = 9.13 \text{ ml}$$

Graphically, the solution would appear as follows:

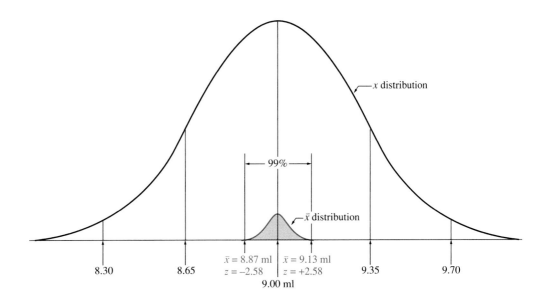

Answer

If we continually take random samples of 49 test tubes in each sample, and calculate the average fill ($\bar{x}$) in each of these samples, then 99% of all the sample averages ($\bar{x}$'s) would be expected to fall between $\bar{x} = 8.87$ ml and $\bar{x} = 9.13$ ml.

It is probably safe to say, if you randomly sample 49 test tubes from a properly functioning process, the average fill ($\bar{x}$) of this sample will fall between 8.87 ml and 9.13 ml. On any given day, if you obtain a sample average much *outside* this range, you might very well suspect the process is malfunctioning. Of course, we must keep in mind that 1% of the time (100% minus 99%), or approximately 1 out of every 100 times, a properly functioning process will produce a sample average ($\bar{x}$) outside this range, but since this is such a ''rare'' occurrence, it is probably wiser to check your filling operation for a malfunction.

Along with the value 95%, industry and research often use this value of 99% to establish a criteria for whether or not an operation may be malfunctioning. This is discussed at length in chapter 6.

Example ——————— Brell Shampoo, an "in-house" brand, is marketed along with various other shampoos through a large national chain of convenience stores. In these stores, Brell's market share has remained relatively constant at $\mu = 24.00$ (meaning: on average 24.00% of the shampoo sold in these stores is Brell) with standard deviation 3.20. If we continually take random samples, each consisting of 64 stores, and calculate the average market share ($\bar{x}$) for each sample, what percentage of the sample averages ($\bar{x}$'s) would you expect to have a market share of *less than* 23.80?

Solution Since we are concerned with the "average" market share in 64 stores and not the market share in an "individual" store, we use the $\bar{x}$ distribution. Remember: when using the $\bar{x}$ distribution we must first calculate $\sigma_{\bar{x}}$, the standard deviation of that distribution.

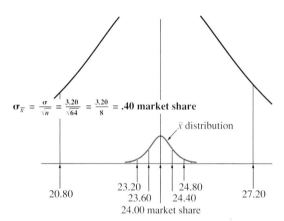

$$\sigma_{\bar{x}} = \frac{\sigma}{\sqrt{n}} = \frac{3.20}{\sqrt{64}} = \frac{3.20}{8} = .40 \text{ market share}$$

$\bar{x}$ distribution

20.80   23.20   24.80   27.20
         23.60   24.40
         24.00 market share

Next, we calculate the z score at the cutoff of $\bar{x} = 23.80$:

$$z = \frac{\bar{x} - \mu}{\sigma_{\bar{x}}} = \frac{23.80 - 24.00}{.40} = -.50$$

Since the area from $z = 0$ to $z = -.50$ is 19.15%, the area below $z = -.50$ must be 30.85% (50% minus 19.15%).

Graphically, the solution would appear as follows:

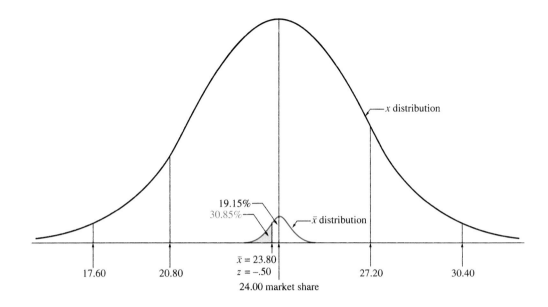

Answer

If we continually take random samples, each consisting of 64 stores, and calculate the average market share ($\bar{x}$) for each sample, then 30.85% of all the sample averages ($\bar{x}$'s) would be expected to have a market share of *less than* 23.80. ■

At this point a reminder may be needed: in the above problem concerning Brell Shampoo, we are not dealing with the market share in one outlet. We are dealing with the *average* market share in 64 randomly selected store outlets. The market share in one store can easily be, say, 19.00. However, the chances that the *average* market share in 64 randomly selected stores being 19.00 is nearly impossible.

Random Selection

Another reminder: the above mathematical procedure, or any mathematical procedure we discuss involving sampling, is based on the critically important process of random selection. Samples are chosen, in a way, similar to how lottery winners are selected. Each person (or store, in our case) has an equal chance of being selected on every pick. For instance, if you drive through Delaware and select 64 store outlets, this is *not* a random sample and these mathematical procedures cannot be used. For a random sample, you must have access to every store outlet in the country on each and every pick, say for instance, through a master list of all the stores. Then, each selection must be made as if "blindfolded," such that each store has an equal chance of being picked (see chapter 1 for further discussion).

5.3   How *n* and σ Affect $\sigma_{\bar{x}}$

The standard deviation of the $\bar{x}$ distribution, $\sigma_{\bar{x}}$, is influenced by two factors. The first is σ, the population standard deviation, and the second is *n,* the sample size, according to the formula $\sigma_{\bar{x}} = \sigma/\sqrt{n}$.

How σ Affects $\sigma_{\bar{x}}$

The relationship between σ and $\sigma_{\bar{x}}$ is direct. If σ increases, $\sigma_{\bar{x}}$ increases. If σ decreases, $\sigma_{\bar{x}}$ decreases. Furthermore, σ and $\sigma_{\bar{x}}$ increase or decrease in the same multiple. In other words, if σ doubles, $\sigma_{\bar{x}}$ doubles. If σ triples, $\sigma_{\bar{x}}$ triples. If σ and $\sigma_{\bar{x}}$ are respectively 12 mm and 2 mm, and σ quadruples to 48 mm, then $\sigma_{\bar{x}}$ quadruples to 8 mm.

In practical terms, however, σ is normally a fixed item. One normally cannot manipulate the population standard deviation, σ, to influence $\sigma_{\bar{x}}$. But its relationship to $\sigma_{\bar{x}}$ is still important for an understanding of advanced work.

How *n* Affects $\sigma_{\bar{x}}$

The relationship between your sample size, *n,* and $\sigma_{\bar{x}}$ is more complex and of more concern since we can often control *n.* Note that $\sigma_{\bar{x}}$ varies inversely as the square root of *n* according to the formula

$$\sigma_{\bar{x}} = \frac{\sigma}{\sqrt{n}}.$$

First, this means, as *n* increases, $\sigma_{\bar{x}}$ decreases.

Second, this means, as *n* increases to 4 times its original value, $\sigma_{\bar{x}}$ decreases to $\frac{1}{\sqrt{4}}$ or $\frac{1}{2}$ its original value.

If *n* increases to 9 times its original value, then $\sigma_{\bar{x}}$ decreases to $\frac{1}{\sqrt{9}}$ or $\frac{1}{3}$ its original value.

If *n* increases to 16 times its original value, then $\sigma_{\bar{x}}$ decreases to $\frac{1}{\sqrt{16}}$ or $\frac{1}{4}$ its original value.

And so on.

To understand the impact of increasing sample size, let us again use the cutting machine problem in the following example.

Example

Using the cutting machine problem with $\mu = 1000$ mm and $\sigma = 12$ mm, compare the $\bar{x}$ distribution when you change your sample size from 36 pieces in each sample to four times this value (144 pieces in each sample) and then again to sixteen times this value (576 pieces in each sample).

Solution

If we continually take random samples of size $n = 36$, then the resulting $\bar{x}$ distribution would have a standard deviation $\sigma_{\bar{x}} = \dfrac{12}{\sqrt{36}} = 2$ mm.

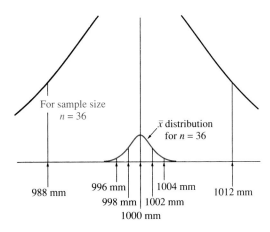

For sample size
$n = 36$

$\bar{x}$ distribution
for $n = 36$

988 mm   996 mm   1004 mm   1012 mm
998 mm | 1002 mm
1000 mm

If we continually take random samples of size $n = 144$, then the resulting $\bar{x}$ distribution would have a standard deviation $\sigma_{\bar{x}} = \dfrac{12}{\sqrt{144}} = 1$ mm.

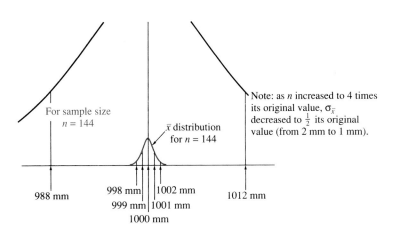

For sample size
$n = 144$

$\bar{x}$ distribution
for $n = 144$

Note: as *n* increased to 4 times its original value, $\sigma_{\bar{x}}$ decreased to $\frac{1}{2}$ its original value (from 2 mm to 1 mm).

988 mm   998 mm   1002 mm   1012 mm
999 mm | 1001 mm
1000 mm

If we continually take random samples of size $n = 576$, then the resulting $\bar{x}$ distribution would have a standard deviation $\sigma_{\bar{x}} = \dfrac{12}{\sqrt{576}} = \dfrac{1}{2}$ mm.

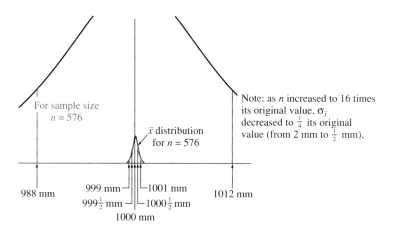

For sample size $n = 576$

$\bar{x}$ distribution for $n = 576$

Note: as n increased to 16 times its original value, $\sigma_{\bar{x}}$ decreased to $\frac{1}{4}$ its original value (from 2 mm to $\frac{1}{2}$ mm).

988 mm   999 mm — | — 1001 mm   1012 mm
$999\frac{1}{2}$ mm — | — $1000\frac{1}{2}$ mm
1000 mm

Note that a substantial increase in the sample size (n) is necessary to produce only a modest decrease in $\sigma_{\bar{x}}$. We had to increase our sample size to 16 times its original value (from 36 pieces to 576 pieces) to decrease $\sigma_{\bar{x}}$ to $\frac{1}{4}$ of its original value (from 2 mm to $\frac{1}{2}$ mm).

To fully understand the impact of sample size changes, let's see how it affects the location of the middle 95% of the $\bar{x}$'s in the next problem. ◼

Example ———————— Using the cutting machine problem with $\mu = 1000$ mm and $\sigma = 12$ mm, calculate where the middle 95% of the $\bar{x}$'s would be expected to fall for the three situations in the last example, that is, for $n = 36$, for $n = 144$, and for $n = 576$.

Solution The completed solutions are as follows:

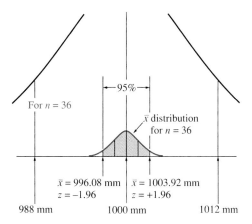

For *n* = 36

95%

$\bar{x}$ distribution
for *n* = 36

$\bar{x} = 996.08$ mm
$z = -1.96$

$\bar{x} = 1003.92$ mm
$z = +1.96$

988 mm   1000 mm   1012 mm

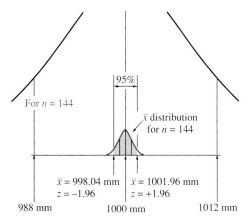

For *n* = 144

95%

$\bar{x}$ distribution
for *n* = 144

$\bar{x} = 998.04$ mm
$z = -1.96$

$\bar{x} = 1001.96$ mm
$z = +1.96$

988 mm   1000 mm   1012 mm

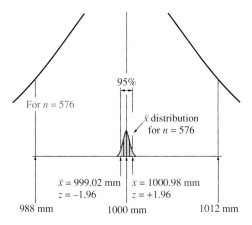

For *n* = 576

95%

$\bar{x}$ distribution
for *n* = 576

$\bar{x} = 999.02$ mm
$z = -1.96$

$\bar{x} = 1000.98$ mm
$z = +1.96$

988 mm   1000 mm   1012 mm

Notice the extreme compression of the $\bar{x}$'s when you increase your sample size to 16 times its original value. At $n = 36$, note that 95% of the $\bar{x}$'s clustered between 996.08 mm and 1003.92 mm. However, at $n = 576$, the sample averages ($\bar{x}$'s) drew in closer to 1000 mm, with 95% of the $\bar{x}$'s now clustered between 999.02 mm and 1000.98 mm.

The importance of this compression of $\bar{x}$'s when the sample size is increased will become apparent when we discuss controlling statistical errors in chapter 6.

5.4 Central Limit Theorem Applied to Nonnormal Populations

The amazing thing about the central limit theorem is that it applies to almost any shaped population (normal or nonnormal), provided your sample size is 30 or more ($n \geq 30$). That is,

> The $\bar{x}$'s will distribute normally around μ for almost any shaped population, provided
>
> $$n \geq 30$$

Let's look at some examples.

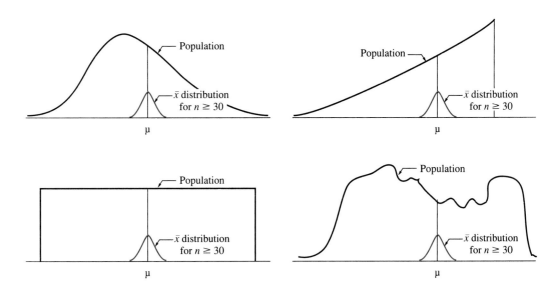

Note that in all cases, the sample size was 30 or more ($n \geq 30$) and the $\bar{x}$'s distributed normally around μ. And we can use the methods and techniques described in this chapter to predict where the $\bar{x}$'s will fall. Does this imply the $\bar{x}$'s will *not* distribute normally around μ if n is less than 30 ($n < 30$)? Yes and no. If your population is normal, the $\bar{x}$'s tend to distribute normally around μ no matter what the size of n (even for $n = 5$ or $n = 2$). However, for many *non*normal populations, the $\bar{x}$'s do *not* assume the normal shape unless the sample size is at least 20 or 25 or in some extreme cases much more. A good rule of thumb is, assume the $\bar{x}$'s normally distributed for sample sizes of $n \geq 30$, unless the population has a highly unusual shape (say for instance, an extraordinarily skewed distribution), in which case, n may have to be substantially more than 30 to ensure a normal $\bar{x}$ distribution. Further discussion on this is in sections 7.3 and 8.4.

Summary

Central limit theorem: For almost any shaped population, if you continually select random samples of a fixed sample size greater than 30, the sample averages ($\bar{x}$'s) will gradually build into the shape of a normal distribution, called an $\bar{x}$ distribution, such that:

1. The mean of the $\bar{x}$ distribution is μ, the mean of the population, and

2. The standard deviation of the $\bar{x}$ distribution, called sigma x-bar ($\sigma_{\bar{x}}$) is equal to the population standard deviation (σ) divided by the square root of your sample size (n). In other words,

$$\sigma_{\bar{x}} = \frac{\sigma}{\sqrt{n}}$$

Factors that affect $\sigma_{\bar{x}}$, the spread of the $\bar{x}$ distribution, are as follows.

σ affects the spread directly, that is, if σ *increases* to three times its value, $\sigma_{\bar{x}}$ will *increase* to three times its value.

n affects the spread inversely as its square root. If n *increases* to 16 times its value, then $\sigma_{\bar{x}}$ will *decrease* to $1/\sqrt{16}$ or 1/4 its value.

The amazing thing about the central limit theorem is that it applies to almost any shaped population (normal or nonnormal) provided the sample size is equal to or greater than 30. In other words, the $\bar{x}$'s will distribute normally around μ for almost any shaped population provided,

$$n \geq 30$$

Only in populations with quite unusual shapes (such as one with a highly extended skew) may we have to sample sometimes more than 30 to be assured of a normally distributed $\bar{x}$ distribution.

Exercises

Note that full answers for exercises 1–5 and abbreviated answers for odd-numbered exercises thereafter are provided in the Answer Key.

5.1 A machine cuts pieces of silk material to an average length of 1000 mm with standard deviation 12 mm. Between what two lengths would we expect to find the middle 95% of all the sample averages ($\bar{x}$'s)?

Assume sample size:

a. $n = 36$
b. $n = 144$
c. $n = 576$

5.2 A machine cuts pieces of silk material to an average length of 1000 mm with standard deviation 12 mm. Between what two lengths would we expect to find the middle 99% of all the sample averages ($\bar{x}$'s)?

Assume sample size:

a. $n = 36$
b. $n = 144$
c. $n = 576$

5.3 A national institute supplies active disease viruses for medical research. A process is set to fill small test tubes at an average of 9.00 ml with standard deviation of .35 ml. If we continually take random samples of 49 test tubes in each sample, what percentage of the $\bar{x}$'s would you expect to fall *below* 8.92 ml or *above* 9.08 ml?

5.4 Brell Shampoo, an "in-house" brand, is marketed through a large national chain of convenience stores. This chain also carries other national brands of shampoo. Brell's "in-house" market share is $\mu = 24.00$ (meaning: on average 24.00% of the shampoo sold in these stores is Brell) with standard deviation 3.20. If we continually take random samples, each sample consisting of 64 stores, and calculate the average market share ($\bar{x}$) in each sample, below what market share would we find the *lowest 1%* of all the sample averages ($\bar{x}$'s)?

5.5 The credit manager of a large sports shop made a statement at an important board meeting that the average age of their customers is 32 years old with standard deviation 4 years.

a. If the credit manager is correct and we were to continually take random samples of 100 customers each, what percentage of the $\bar{x}$'s would you expect to be between $\bar{x} = 31.5$ years old and $\bar{x} = 32.5$ years old?

b. If the credit manager is correct and we were to take *one* random sample of 100 customers, what is the probability the average of this one sample ($\bar{x}$) would be between 31.5 years old and 32.5 years old?

c. Between what two ages would you expect to find the middle 95% of all the sample averages ($\bar{x}$'s)?

5.6 Gaunt Health Farms, based on a survey of records of all visitors for the last five years, claim an average weight loss of 12.0 lb with standard deviation of 2.4 lb.

a. If you took a random sample of 36 from these records, what is the probability the sample average ($\bar{x}$) will be less than 11.0 lb?

b. If you took a random sample of 36 and your sample average, $\bar{x}$, was indeed less than 11.0 lb, would you be suspicious of their claim that the average visitor weight loss is 12.0 lb?

5.7 Bad-debt accounts are a serious source of profit drain for all businesses, but especially for the fashion industry, which deals with the risky whims of the public. Ralph Weetz Co., a distributor of women's blouses to small boutiques, was one such company. A computer tally of all bad-debt accounts of the past few decades reveals an accumulation of thousands of bad-debt accounts, with the average amount owed of $\mu = \$550.00$ and standard deviation $\sigma = \$75.90$.

a. Assuming a normal distribution, if we randomly selected one bad-debt account, what is the probability this one bad-debt account is between $538.00 and $562.00?

b. If we took a random sample of 40 bad-debt accounts and calculated the average amount owed ($\bar{x}$) in this sample, what is the probability this sample average ($\bar{x}$) will be between \$538.00 and \$562.00?

c. Assuming $n = 34$, with what probability can we assert a sample average ($\bar{x}$) will fall within \$20.00 of $\mu = \$550.00$?

5.8   A nationwide marketing study concluded the average age of horror film moviegoers is 17.4 years old with standard deviation 2.7 years.

a. Assuming a normal distribution, what percentage of horror film movie goers nationwide would you expect to be over 18.0 years old?

b. If we continually take random samples of 81 horror film movie goers nationwide and calculate the sample average ($\bar{x}$) for each sample, what percentage of the sample averages ($\bar{x}$'s) would you expect to be over 18.0 years?

c. If you took a random sample of 81 horror film movie goers, what is the probability the sample average ($\bar{x}$) would be over 18.0 years?

5.9   In a certain year, the nationwide SAT verbal score averaged $\mu = 430$ with standard deviation $\sigma = 96$. Answer the following assuming SAT scores are continuous over the scale 200 to 800.

a. Assuming a normal distribution, between what two values would you expect to find the middle 90% of SAT verbal scores?

b. If we continually take random samples of 144 students, and calculate the average SAT verbal score ($\bar{x}$) in each sample, between what two values would you expect to find the middle 90% of the sample averages ($\bar{x}$'s)?

c. If we continually take random samples of 42 students, and calculate the average SAT verbal score ($\bar{x}$) in each sample, between what two values would you expect to find the middle 90% of the sample averages ($\bar{x}$'s)?

5.10   In a certain year, the nationwide SAT mathematics score averaged $\mu = 470$ with standard deviation $\sigma = 96$. Answer the following assuming SAT scores are continuous over the scale 200 to 800.

a. Assuming a normal distribution, between what two values would you expect to find the middle 98% of SAT mathematics scores?

b. If we continually take random samples of 256 students, and calculate the average SAT mathematics score ($\bar{x}$) in each sample, between what two values would you expect to find the middle 98% of the sample averages ($\bar{x}$'s)?

c. If we continually take random samples of 30 students, and calculate the average SAT mathematics score ($\bar{x}$) in each sample, between what two values would you expect to find the middle 98% of the sample averages ($\bar{x}$'s)?

5.11   Medical doctors in Kansas City are known to work on average $\mu = 54.7$ hours per week with standard deviation $\sigma = 6.8$ hours.

a. Assuming a normal distribution, what is the probability a doctor will work less than 53.0 hr/wk?

b. What is the probability a random sample of 55 doctors will yield an average, $\bar{x}$, of less than 53.0 hr/wk?

c. With what probability can we assert a random sample average ($\bar{x}$) will be between 53.0 and 56.0 hr/wk, based on $n = 55$?

5.12   A survey indicated the average yearly salary of entry-level women managers in St. Paul to be $\mu = \$56,700$ with standard deviation $\sigma = \$7,200$.

a. Assuming a normal distribution, what is the probability a woman manager's entry-level salary will exceed \$58,000?

b. What is the probability a random sample of 50 women managers will yield an average entry-level salary ($\bar{x}$) exceeding \$58,000?

c. Assuming $n = 50$, with what probability can we assert a sample average ($\bar{x}$) will fall between \$55,000 and \$58,000?

d. Assuming $n = 42$, with what probability can we assert a sample average ($\bar{x}$) will fall within \$1500 of $\mu = \$56,700$?

Introduction to Hypothesis Testing

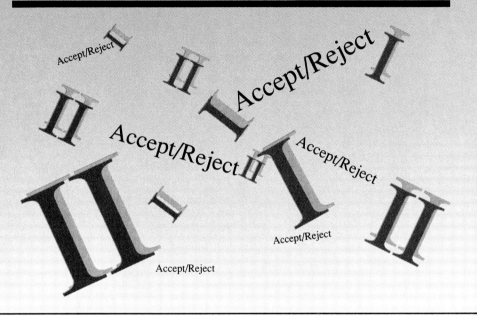

6.1 Basic Concepts of Hypothesis Testing

For a complete understanding of **hypothesis testing,** one must first understand three important concepts in statistics, namely, **accept/reject** decision making, the **Type I error,** and the **Type II error.** Let's see how these concepts interrelate using chapter 5's cutting machine problem. ▼

Suppose a machine in a dress factory cuts pieces of silk material to an average length of $\mu = 1000$ mm with standard deviation $\sigma = 12$ mm. If we were to continually take random samples, with 36 pieces of cut material in each sample, and calculate the average length ($\bar{x}$) of the 36 pieces in each of these samples, then experience tells us the resulting distribution of $\bar{x}$'s would take the form of the *small histogram* below.

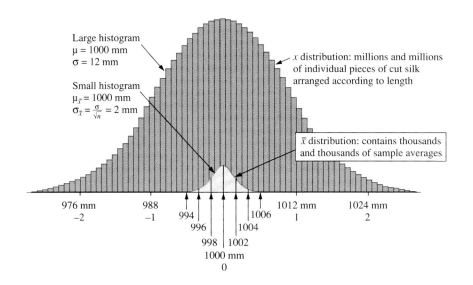

Large histogram
$\mu = 1000$ mm
$\sigma = 12$ mm

x distribution: millions and millions of individual pieces of cut silk arranged according to length

Small histogram
$\mu_{\bar{x}} = 1000$ mm
$\sigma_{\bar{x}} = \frac{\sigma}{\sqrt{n}} = 2$ mm

$\bar{x}$ distribution: contains thousands and thousands of sample averages

| 976 mm | 988 | 994 | 996 | 998 | 1000 mm | 1002 | 1004 | 1006 | 1012 mm | 1024 mm |
| -2 | -1 | | | | 0 | | | | 1 | 2 |

Now let us suppose the machine is turned off because it is late Friday afternoon and all the workers are leaving the dress factory for the weekend. The weekend passes and on Monday morning a sleepy-eyed operator starts up the cutting machine in preparation for the week's operation. The dial on the machine is set to cut at 1000 mm. We know from our experiment the prior week that *if* the machine is operating properly, the machine will be cutting pieces to an average length of 1000 mm, although some pieces will be shorter and some longer in accordance with the *large histogram* above.

But how can we be sure that in our absence someone has not tampered with the machine, that in closing down or starting up the machine critical parts have not vibrated loose, or that our sleepy-eyed operator has not accidentally moved the dial off its setting. The answer is: until we take some cuts, we usually have no way of knowing.

Certainly, the length of the first two or three pieces of material off the cutting machine will be measured. This acts as a quick and simple check for any gross malfunctioning. However, after you measure the first few pieces, let the machine run for a period to stabilize. Then take a true random sample, say for instance of 36 pieces. Measure the length of each piece in the sample and calculate $\bar{x}$, the average length.

With this completed, we refer to the small histogram reprinted below for your reference.

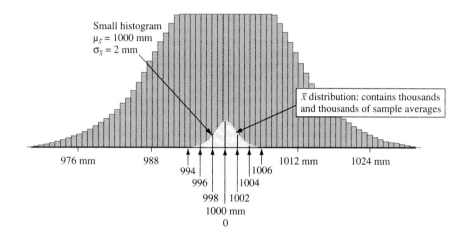

The small histogram shows us where sample averages ($\bar{x}$'s) should fall on a properly operating machine. Just about every sample average falls between 994 mm and 1006 mm. Certainly, on Monday morning, if you were to calculate an $\bar{x}$ *outside* this 994 mm to 1006 mm range, then most likely the machine is *not* operating properly, *not* cutting on average at 1000 mm. On the other hand, if you were to calculate a sample average between 998 mm and 1002 mm, you would feel the machine was probably cutting okay, since this is where you would expect to find most of the sample averages ($\bar{x}$'s) on a properly operating machine.

However, this leaves a gigantic borderline. What if you were to get a sample average of exactly 994 mm? Or 996 mm? Or 998 mm? At what point do we say the machine is cutting okay, or *not* cutting okay? Fortunately industry and research have been grappling with this question for decades and, from vast experience, have come up with some guidelines. One guideline establishes the middle 95% of the $\bar{x}$'s on a properly operating machine as a gauge in a cold-hearted *accept/reject* decision.

Accept/Reject Decision Making

Accept/reject decision making fundamentally proceeds in three steps, as follows:

1. The first step in an accept/reject decision is to set up some initial assumption. Although many initial assumptions are possible, in this case, the preferred initial assumption is $\mu = 1000$ mm since this is where we suspect (and hope) the machine is cutting on average. So, as first step, we state:

$$\text{Initial assumption: } \mu = 1000 \text{ mm}$$

2. The second step in an accept/reject decision is to establish some guideline for accepting or rejecting your initial assumption. In this case, we will choose the often used middle 95% of the $\bar{x}$'s guideline, as follows:

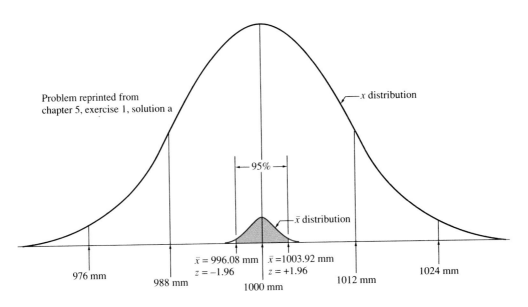

Problem reprinted from
chapter 5, exercise 1, solution a

As calculated in a prior example, the middle 95% of the $\bar{x}$'s on a properly operating machine would be expected to fall between 996.08 mm and 1003.92 mm, so your accept/reject decision would be as follows.

$$\text{Accept } \mu = 1000 \text{ mm}$$
if your sample $\bar{x}$ is between 996.08 mm and 1003.92 mm

$$\text{Reject } \mu = 1000 \text{ mm}$$
if your sample $\bar{x}$ is *outside* this range

3. The third step is a decision. If your sample average ($\bar{x}$) is between 996.08 mm and 1003.92 mm, you accept the machine as cutting properly at $\mu = 1000$ mm. If your sample $\bar{x}$ is outside this range, you assume the machine is cutting *improperly,* that is, *not* at $\mu = 1000$ mm. It's a simple *accept* $\mu = 1000$ mm or *reject* $\mu = 1000$ mm decision, no maybes, no in-betweens. You accept $\mu = 1000$ mm or reject $\mu = 1000$ mm. Period.

But aren't accept/reject decisions risky? Accept/reject decisions if properly thought out are one of the most powerful and efficient devices in statistical research, however, yes, they do come with risk. One of the risks is called the Type I error.

Type I Error

Let's continue with the cutting machine problem. If we adopt the middle 95% of the $\bar{x}$'s as our gauge for accepting or rejecting whether a machine is operating properly, then we must remember 5% of the $\bar{x}$'s will fall outside this 996.08 mm to 1003.92 mm range (100% minus 95%) when the machine is operating properly, as follows:

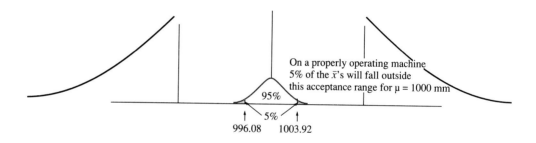

That means on a properly operating machine, 5% of the time you will reject the machine as operating properly. In other words, there is a 5% probability of rejecting a properly operating machine when, in fact, we should not reject it. Expressed in statistical terms,

> A Type I error is the probability of rejecting an initial assumption in error.

In this case, our initial assumption is that the machine is operating properly at $\mu = 1000$ mm. However, on a properly operating machine, 5% of the time you will reject this initial assumption in error. This is called a *Type I error.* This

Type I error is denoted by the symbol, α (alpha), traditionally labeled the **level of significance** and written in decimal form as follows:

Level of significance, $\alpha = .05$

Unfortunately we must live with some Type I error risk for the convenience and efficiency of an accept/reject decision.

Hold it a minute! What is he jabbering about? We don't have to live with this 5% risk, you might say. Why not establish a middle 99% of the $\bar{x}$'s for accepting the machine as operating properly. This lowers the Type I error risk to 1% (99% + 1% = 100%). In other words, only 1% of the $\bar{x}$'s on a properly operating machine will fall outside the 99% range. Written in statistical terms, you would say

Why not impose a level of significance, $\alpha = .01$?

The level of significance $\alpha = .01$ is also often used in industry and research. But let's see what happens when this is done.

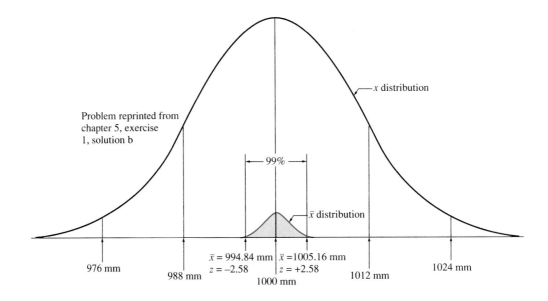

In this case, the middle 99% of the $\bar{x}$'s on a properly operating machine would be expected to fall between 994.84 mm and 1005.16 mm (as calculated in a prior exercise, noted above). Notice, if we lower the risk of a Type I error from 5% to 1%, we ''open up'' the range of sample averages where we would consider the machine operating properly, as follows:

$\alpha = .05$ Machine cutting okay if $\bar{x}$ is between 996.08 mm and 1003.92 mm
$\alpha = .01$ Machine cutting okay if $\bar{x}$ is between 994.84 mm and 1005.16 mm

By opening up the range for accepting the machine as operating properly, we reduce the risk of a Type I error from 5% to 1%. However, this leaves us more vulnerable to another form of risk, called the Type II error.

Type II Error

Okay, we have just decided to reduce our chances of a Type I error by opening up the range of sample averages where we would consider the machine as operating properly. We have adopted the following guideline.

$\alpha = .01$ Any sample average ($\bar{x}$) between 994.84 mm and 1005.16 mm is assumed to come from a properly operating machine, that is, a machine cutting okay at $\mu = 1000$ mm.

The situation now arises: what if we turn on the machine Monday morning, sample 36 pieces as usual, and obtain an $\bar{x}$ of say, 995.00 mm? You must conclude: machine okay, probably cutting at $\mu = 1000$ mm, since $\bar{x} = 995.00$ mm falls between 994.84 mm and 1005.16 mm, as follows:

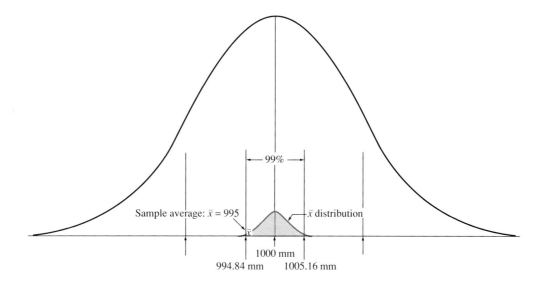

Now here's the problem. A sample average $\bar{x} = 995.00$ mm is also "typical" for a machine cutting at, say for instance, $\mu = 995$ mm. How do we know a pin is not stuck in the machine so that the machine is now, in reality, cutting on average to 995 mm? Or to 994 mm? Or to 996 mm? In other words, *Has there been a shift?* The answer to this question is: we don't know. Based on a sample average $\bar{x} = 995$ mm you accepted the fact of the machine cutting at $\mu = 1000$ mm. However if the machine in reality has actually shifted to $\mu = 995$ mm (or to any other value), we have just committed a Type II error. Written in statistical language,

A Type II error is the probability of accepting an initial assumption in error.

Again, our initial assumption is that the machine is operating properly at $\mu = 1000$ mm. If we accept $\mu = 1000$ mm, when in reality, $\mu \neq 1000$ mm, we have committed a Type II error. The probability of committing a Type II error is denoted by the symbol, β (beta).

Note, had we adopted the middle 95% of the $\bar{x}$'s for accepting a machine as operating properly ($\alpha = .05$) by accepting $\mu = 1000$ mm if $\bar{x}$ falls between 996.08 mm and 1003.92 mm, then a sample average ($\bar{x}$) of 995.00 mm would have alerted us to a possible malfunction since $\bar{x} = 995$ mm falls *outside* the 996.08 to 1003.92 range. In other words, had we accepted our original 5% Type I error risk, the Type II error above would not have occurred. Generally, decreasing the probability of a Type I error merely increases the probability of a Type II error.

To summarize

Type I error: Probability of rejecting an initial assumption in error.

Type II error: Probability of accepting an initial assumption in error.

Remember: decreasing the Type I error by imposing a lower α, say for instance going from .05 to .01, merely increases your Type II error risk.

Power

Statisticians will often evaluate a statistical test in terms of its **power.** Power simply means the probability of making the correct decision by avoiding a Type II error. If you calculate a Type II error risk of 10% the *power* of the test is 90%. If your Type II error risk is 30%, your power is 70%. Either you make the Type II error or you make the correct decision. The sum of the two must equal 100%, that is, β + Power = 100% *or* Power = 100% $-$ β. Written in decimal form, it is expressed as Power = $1 - \beta$. Remember: β is the probability of making a Type II error.

Precise calculations of the Type II error and power will be demonstrated in the following two examples. To summarize:

> *Power*
> Probability of making a correct decision by avoiding a Type II error
>
> Power = 100% $-$ β = $1 - \beta$

6.2 Applications

Accept/reject decision making is standard practice in statistical testing. However, accept/reject decisions do come with risk. Four examples that demonstrate this risk are presented.

Example ———————— In the cutting machine problem, suppose we establish the middle 92% of the $\bar{x}$'s as our *cutoffs* for accepting $\mu = 1000$ mm. Assuming $n = 36$ and $\sigma = 12$ mm,

    a. What is the probability of a Type I error?
    b. Between what $\bar{x}$ values would you accept the machine cutting at $\mu = 1000$ mm?
    c. Explain briefly how one might commit a Type I error.

Solution to (a) The probability of a Type I error is simply 8% ($100\% - 92\% = 8\%$). Written in statistical terms, you would state $\alpha = .08$. In other words, on a properly operating machine, 8% of the $\bar{x}$'s would fall *outside* your acceptance range for $\mu = 1000$ mm, as shown:

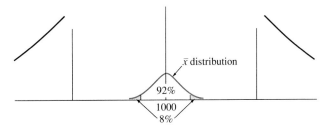

Solution to (b) This is a typical "working backward (given the area, find z)" problem for the normal curve, only now we are dealing with the normal curve of the $\bar{x}$ distribution. Since we know the area between the critical cutoffs is 92%, we merely look up the corresponding z scores. Doing this, we get $z = -1.75$ and $z = +1.75$. (Remember the table reads half the normal curve, so we must look up an area of 46% ($\frac{1}{2}$ of 92%) or in decimal form .4600.)

    Using $z = -1.75$ and $z = +1.75$, we calculate the values at the cutoffs as follows:

$$\sigma_{\bar{x}} = \frac{12}{\sqrt{36}} = 2 \qquad z = \frac{\bar{x} - \mu}{\sigma_{\bar{x}}} \qquad z = \frac{\bar{x} - \mu}{\sigma_{\bar{x}}}$$

$$-1.75 = \frac{\bar{x} - 1000}{2} \qquad +1.75 = \frac{\bar{x} - 1000}{2}$$

$$\bar{x} = 996.50 \text{ mm} \qquad \bar{x} = 1003.50 \text{ mm}$$

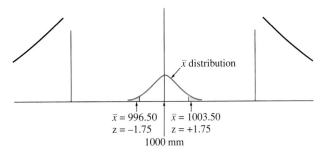

So, you would accept the machine as cutting properly, cutting at $\mu = 1000$ mm, if you obtained a sample average between $\bar{x} = 996.50$ mm and $\bar{x} = 1003.50$ mm.

Solution to (c)

The following might be a Type I error scenario: You random sample 36 pieces and your sample average falls *outside* this 996.50 mm to 1003.50 mm range, say for instance you obtain a sample average of $\bar{x} = 994$ mm. Based on this you reject the machine as operating properly and shut down production.

If indeed the machine is *not* operating properly, not cutting on average to $\mu = 1000$ mm, then you have made no error. Your decision was correct. However, if the machine is okay, cutting properly at $\mu = 1000$ mm, and you happened to have sampled one of those rare 8% occurrences, then you have made a Type I error. ■

Example ——————

In the cutting machine problem, suppose we arbitrarily establish $\bar{x} = 997$ mm to $\bar{x} = 1003$ mm as our cutoffs for accepting $\mu = 1000$ mm. Assuming $n = 36$ and $\sigma = 12$ mm,

  a. What is the probability of a Type I error?
  b. What is the probability of a Type II error if the machine shifts and is now cutting at $\mu = 995$ mm?
  c. What is the power of the test in part b?

Solution to (a)

The percentage of $\bar{x}$'s on a properly operating machine that fall *outside* the 997 mm to 1003 mm range is your Type I error risk. It is represented by the shaded regions in the diagram below. Calculating the percentage of data in the shaded regions we get

$$\sigma_{\bar{x}} = \frac{\sigma}{\sqrt{n}} = \frac{12}{\sqrt{36}} = 2 \text{ mm} \qquad z = \frac{\bar{x} - \mu}{\sigma_{\bar{x}}} \qquad z = \frac{\bar{x} - \mu}{\sigma_{\bar{x}}}$$

$$z = \frac{997 - 1000}{2} \qquad z = \frac{1003 - 1000}{2}$$

$$z = -1.50 \qquad z = +1.50$$

Looking up $z = 1.50$, we get 43.32%, with 6.68% in the tail.

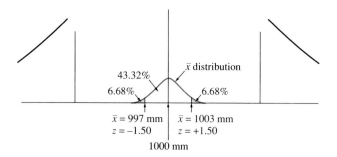

The probability of a Type I error is 13.36% (6.68% + 6.68%). In other words, on a properly operating machine the percentage of $\bar{x}$'s you would expect to fall *outside* the 997 mm to 1003 mm range is 13.36%. Written in decimal form (.1336) and rounded to two decimal places, you would say:

$$\alpha = .13$$

Solution to (b)

If a machine is indeed cutting at $\mu = 995$ mm, the sample averages ($\bar{x}$'s) now would cluster about $\mu = 995$ mm, as follows:

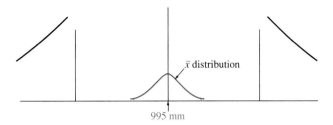

You would commit a Type II error if you accepted (in error) the machine cutting at $\mu = 1000$ mm. The only way you would accept a machine cutting at $\mu = 1000$ mm is if you took a random sample of 36 pieces and your sample average ($\bar{x}$) fell between 997 mm and 1003 mm. So, to calculate β, the probability of a Type II error, you must calculate the percentage of $\bar{x}$'s that would fall between 997 mm and 1003 mm from a machine cutting at $\mu = 995$ mm.

Remember: the machine is cutting at $\mu = 995$ mm but you are unaware of this. The only information available to you is your one sample average, $\bar{x}$.

To find the probability of a Type II error in this problem, we calculate the percentage of $\bar{x}$'s that we would expect to fall between 997 mm and 1003 mm, represented by the shaded region in the diagram below:

$$\sigma_{\bar{x}} = \frac{\sigma}{\sqrt{n}} = \frac{12}{\sqrt{36}} = 2 \text{ mm*} \qquad z = \frac{\bar{x} - \mu}{\sigma_{\bar{x}}} \qquad z = \frac{\bar{x} - \mu}{\sigma_{\bar{x}}}$$

$$z = \frac{997 - 995}{2} \qquad z = \frac{1003 - 995}{2}$$

$$z = +1.00 \qquad z = +4.00$$

(disregard since negligible data exists in this region)

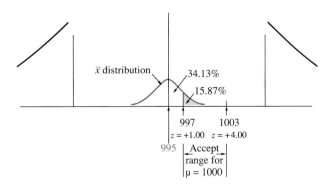

*Technical note: μ and σ are independent. A shift in μ will generally not affect σ. See endnote 10 in chapter 10 for further reading.

Looking up $z = 1.00$, we get 34.13%, thus our Type II error risk is 15.87% (50% minus 34.13%). Since 15.87% of the $\bar{x}$'s coming from a machine cutting at $\mu = 995$ mm would be expected to fall between 997 mm and 1003 mm, this 15.87% is the risk you will obtain one of these $\bar{x}$'s and thus conclude (erroneously) the machine was cutting at $\mu = 1000$ mm. So, your risk of a Type II error when μ shifts to 995 mm is 15.87%. Written in decimal form (.1587) and rounded to two decimal places, you would say:

$\beta = .16$ which is the probability you will accept your initial assumption, $\mu = 1000$ mm, in error. This is your Type II error risk for a shift to $\mu = 995$ mm.

Solution to (c)

Power is the probability of making a correct decision by avoiding a Type II error. For this test, Power = 84.13% (100% $-$ 15.87%). Written in decimal form,

$$\text{Power} = .84 \qquad \text{(Note: Power} = 1 - \beta$$
$$= 1.00 - .16$$
$$= .84)$$

Explained another way, since 84.13% of the $\bar{x}$'s (50% + 34.13%) will be lower than 997 mm, as shown on the previous diagram, you have an 84.13% chance your sample average will be less than 997 mm. In that case, you would reject $\mu = 1000$ mm, which would be the correct decision. In other words, we have an 84.13% chance of making the correct decision (rejecting $\mu = 1000$ mm) and a 15.87% chance of making the wrong decision (accepting $\mu = 1000$ mm). Power is the probability of making the correct decision in this situation and a Type II error is the probability of making the wrong decision.   ◼

Example ——————

The National Institutes of Health agreed to supply active disease viruses, such as polio and AIDS, to research firms for the purpose of experimentation. A process is set up to automatically fill millions of small test tubes to an average of 9.00 ml of disease virus with standard deviation .35 ml.

With sample sizes of $n = 49$ test tubes, it was calculated that 99% of the sample averages ($\bar{x}$'s) would fall between 8.87 ml and 9.13 ml (chapter 5, section 5.2, second example). If we use this 8.87 ml to 9.13 ml range of sample averages as our criterion to accept $\mu = 9.00$ ml,

a. What is the probability of a Type I error?
b. What is the probability of a Type II error if the process shifts to $\mu = 9.20$ ml?
c. What is the power of the test in part b?

Solution to (a)

Since 99% of the $\bar{x}$'s fall between 8.87 ml and 9.13 ml, 1% of the $\bar{x}$'s will fall outside this range. In other words, there is a 1% chance you will obtain an $\bar{x}$ *outside* this 8.87 ml to 9.13 ml range when the process is operating properly. So your probability of a Type I error is 1%. Written in decimal form you would say:

$\alpha = .01$ which is the probability you would reject your initial assumption ($\mu = 9.00$ ml) in error. This 1% is your Type I error risk.

Solution to (b)

The test tubes are now filling on average at μ = 9.20 ml, so your sample averages ($\bar{x}$'s) would now cluster about 9.20 ml, as follows:

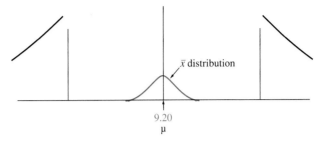

$\bar{x}$ distribution

9.20
μ

You would commit a Type II error if you accepted the process filling at μ = 9.00 ml. The only way you would accept the process filling at μ = 9.00 ml is if you took a random sample of 49 test tubes and your average ($\bar{x}$) fell between 8.87 ml and 9.13 ml. So, to calculate β, the probability of a Type II error, you must calculate the percentage of $\bar{x}$'s that would fall between 8.87 ml and 9.13 ml from a process filling at μ = 9.20 ml.

Remember: the process is filling at μ = 9.20 ml but you are unaware of this. The only information available to you is your one sample average, $\bar{x}$.

To find the probability of a Type II error in this problem, we calculate the percentage of $\bar{x}$'s that we would expect to fall between 8.87 ml and 9.13 ml, represented by the shaded region in the diagram that follows:

$$\sigma_{\bar{x}} = \frac{\sigma}{\sqrt{n}} = \frac{.35}{\sqrt{49}} = .05 \text{ ml} \qquad z = \frac{\bar{x} - \mu}{\sigma_{\bar{x}}} \qquad z = \frac{\bar{x} - \mu}{\sigma_{\bar{x}}}$$

$$z = \frac{8.87 - 9.20}{.05} \qquad z = \frac{9.13 - 9.20}{.05}$$

$$z = -6.60 \qquad z = -1.40$$

(disregard since negligible data exists in this region)

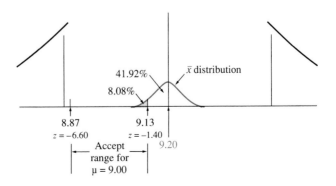

41.92%

8.08%

$\bar{x}$ distribution

8.87
$z = -6.60$

9.13
$z = -1.40$

9.20

Accept range for μ = 9.00

Looking up $z = -1.40$, we get 41.92%, thus our Type II error risk is 8.08%. Since 8.08% (50.00% minus 41.92%) of the $\bar{x}$'s coming from a process filling at $\mu = 9.20$ ml would be expected to fall between 8.87 ml and 9.13 ml, this 8.08% is the risk you will obtain one of these $\bar{x}$'s and thus conclude (erroneously) the process was filling at $\mu = 9.00$ ml. So, your risk of a Type II error when μ shifts to 9.20 ml is 8.08%. Written in decimal form (.0808) and rounded to two decimal places, you would say:

$\beta = .08$ which is the probability you will accept your initial assumption, $\mu = 9.00$ ml, in error; this is your Type II error risk for a shift to $\mu = 9.20$ ml.

Solution to (c)   Since power is the probability of making the correct decision by avoiding a Type II error, we get $100\% - 8.08\% = 91.92\%$

$$\text{Power} = 91.92\% \text{ or } .92$$

Put another way: since 91.92% (50% plus 41.92%) of the $\bar{x}$'s will be greater than 9.13 ml, you have a 91.92% chance your sample average will be greater than 9.13 ml. In that case, you would reject $\mu = 9.00$ ml, which would be the correct decision. In other words, we have a 91.92% chance of making the correct decision (rejecting $\mu = 9.00$ ml) and a 8.08% chance of making the wrong decision (accepting $\mu = 9.00$ ml).   ■

6.3 Controlling Error

Although a number of techniques can be used to decrease error risk, perhaps the most broadly preferred is increasing n, your sample size. This is best explained through practical example, as follows.

In our cutting machine problem (section 6.2, second example), we arbitrarily established $\bar{x} = 997$ mm to $\bar{x} = 1003$ mm as our cutoffs for accepting $\mu = 1000$ mm. Using $n = 36$ and $\sigma = 12$ mm, we calculated the probability of a Type I error to be 13.36% (6.68% + 6.68%), as follows:

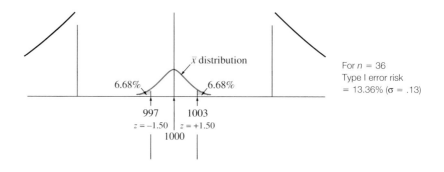

For the probability of a Type II error if μ shifts to 995 mm, we calculated 15.87%, as follows:

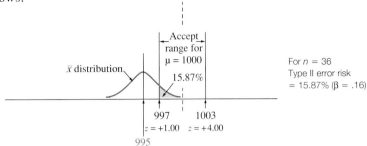

Now here's a problem: 13.36% is too high for a Type I error risk. It means, 13 times out of 100 you will reject a properly operating machine as malfunctioning. Checking out a properly operating machine for a malfunction that doesn't exist can be time consuming and expensive. Well, then, how do we lower this Type I error risk? There are a number of ways, but each come with drawbacks. We will discuss three.

In production or quality control situations, as in this example, most likely a **back up sample** would be taken, that is, a second random sample drawn from the machine's output. However, as mentioned above, this can be time consuming and expensive. Moreover, the use of back up samples in most industry and research situations is just not practical. Most statistical studies are exceedingly expensive (marketing studies often cost hundreds of thousands of dollars), exceedingly time consuming (scientific studies can easily range two to ten years), and exceedingly enervating.

A second approach is to arbitrarily set a lower Type I error risk, say from 13.36% to 1%, but as we discussed at length in prior sections, arbitrarily lowering your Type I error risk merely increases your Type II error risk (in fact, in this case lowering the Type I error risk from $\alpha = 13.36\%$ to $\alpha = 1\%$ would merely increase the β error, the Type II error risk, from 15.87% to over 53%) and a large Type II error risk means if the machine actually shifts, there is a high probability you won't be able to detect it.

A third approach, and perhaps the most preferred, is to increase your sample size. Let's see what happens when we increase our sample size to $n = 144$.

Example ——————— In our cutting machine problem, suppose we arbitrarily establish $\bar{x} = 997$ mm to $\bar{x} = 1003$ mm as our cutoffs for accepting $\mu = 1000$ mm, only this time we increase our sample size to $n = 144$. (Assume $\sigma = 12$ mm.)

a. Calculate the probability of making a Type I error.
b. Calculate the probability of making a Type II error if μ shifts to 995 mm.
c. Compare these results to the results when $n = 36$.

Solution to (a)

Since n, your sample size, has increased from 36 pieces of cut material to 144 pieces of cut material, the sample averages ($\bar{x}$'s) now cluster much closer to $\mu = 1000$ mm. In fact, calculating $\sigma_{\bar{x}}$ we get

$$\sigma_{\bar{x}} = \frac{12}{\sqrt{144}} = \frac{12}{12} = 1 \text{ mm}$$

The percentage of $\bar{x}$'s on a properly operating machine that fall outside the 997 mm to 1003 mm range is shown by the shaded regions in the following diagram. This is your Type I error risk. Notice how the shaded area can barely be seen. This is because, now, the $\bar{x}$'s have clustered much closer to $\mu = 1000$ mm. Calculating the percentage of data in the shaded region,

$$z = \frac{\bar{x} - \mu}{\sigma_{\bar{x}}} \qquad z = \frac{\bar{x} - \mu}{\sigma_{\bar{x}}}$$

$$z = \frac{997 - 1000}{1} \qquad z = \frac{1003 - 1000}{1}$$

$$z = -3.00 \qquad z = +3.00$$

Looking up $z = 3.00$, we get 49.87% with .13% in the tail.

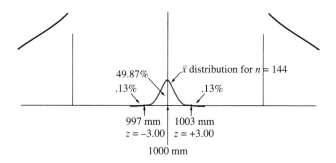

On a properly operating machine, the percentage of $\bar{x}$'s you would expect to fall outside the 997 mm to 1003 mm range is .26% (.13% + .13%), which is much less than 1%. This is your Type I error risk. Written in decimal form (.0026) and rounded to three decimal places, you would write $\alpha = .003$. In other words, there is less than 3 chances in 1000 of making a Type I error. Small indeed.

Solution to (b)

If the machine is indeed cutting at $\mu = 995$ mm, the sample averages ($\bar{x}$'s) would now cluster about $\mu = 995$ mm, as indicated in the following figure.

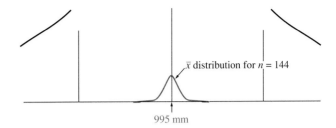

$\bar{x}$ distribution for $n = 144$

995 mm

You would commit a Type II error if you accepted (in error, of course) the machine cutting at $\mu = 1000$ mm. The only way you would accept the machine cutting at $\mu = 1000$ mm is if you took a random sample of 144 pieces and your sample average ($\bar{x}$) fell between 997 mm and 1003 mm. So, to calculate β, the probability of a Type II error, you must calculate the percentage of $\bar{x}$'s that would fall between 997 mm and 1003 mm from a machine operating at $\mu = 995$ mm.

Remember: the machine is cutting at $\mu = 995$ mm but you are unaware of this. The only information available to you is your one sample average, $\bar{x}$. To find the probability of a Type II error in this problem, we calculate the percentage of $\bar{x}$'s that we would expect to fall between 997 mm and 1003 mm, represented by the shaded region in the diagram below:

$$\sigma_{\bar{x}} = \frac{\sigma}{\sqrt{n}} = \frac{12}{\sqrt{144}} = \frac{12}{12} = 1 \text{ mm}$$

$$z = \frac{\bar{x} - \mu}{\sigma_{\bar{x}}} \qquad\qquad z = \frac{\bar{x} - \mu}{\sigma_{\bar{x}}}$$

$$z = \frac{997 - 995}{1} \qquad\qquad z = \frac{1003 - 995}{1}$$

$$z = +2.00 \qquad\qquad z = +8.00$$

(disregard since negligible data exists in this region)

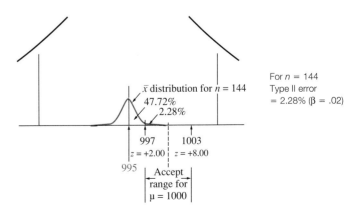

$\bar{x}$ distribution for $n = 144$
47.72%
2.28%

997   1003
$z = +2.00$ $z = +8.00$
995
Accept range for $\mu = 1000$

For $n = 144$
Type II error
= 2.28% ($\beta = .02$)

Looking up $z = +2.00$, we get 47.72%, thus our Type II error risk is 2.28%. Since 2.28% (50% − 47.72%) of the $\bar{x}$'s coming from a machine cutting at $\mu = 995$ mm would be expected to fall between $\bar{x} = 997$ mm and $\bar{x} = 1003$ mm, this is the risk you will obtain one of these $\bar{x}$'s and thus conclude (erroneously) the machine was cutting at $\mu = 1000$ mm. So, your risk of a Type II error when μ shifts to 995 mm is 2.28%. Written in decimal form (.0228) and rounded to two decimal places, you would say

$$\beta = .02$$

Solution to (c)   The following is a comparison of results:

For $n = 36$	For $n = 144$
$\alpha = 13.36\%$	$\alpha = $ less than 1%
$\beta = 15.87\%$	$\beta = 2.28\%$
(for μ shifting	(for μ shifting
to 995 mm)	to 995 mm)

When we increase our sample size from 36 pieces to 144 pieces, note the formidable drop in risk. The Type I error risk (α) drops from 13.36% to less than 1%. The Type II error risk (β) drops from 15.87% to 2.28%. Of course, in practical terms, increasing your sample size will add cost and inconvenience, but usually these are small prices to pay for the added protection. ■

In conclusion, controlling errors should be thought out at the initial stages of planning a statistical study. Although other methods are available, the preferred method for lowering Type I and Type II error risks is by increasing your sample size.

Summary

Three fundamental concepts were presented in this chapter as follows.

1. **Accept/reject decision making:** Accept/reject decision making proceeds as follows. The first step is to set up some initial assumption, say for instance that a population mean, μ, equals some specific value. Then we establish guidelines for accepting and rejecting this initial assumption. The final step is to accept or reject based on sample results.

Two popular guidelines for accept/reject decision making are as follows.

95% Guideline   We calculate where the middle 95% of the $\bar{x}$'s would be expected to fall if the initial assumption were true. If our sample $\bar{x}$ falls in this 95% range, we accept our initial assumption, otherwise reject (see sketch).

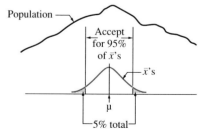

99% Guideline: We calculate where the middle 99% of the $\bar{x}$'s would be expected to fall if the initial assumption were true. If our sample $\bar{x}$ falls in this range, we accept our initial assumption, otherwise reject (see sketch).

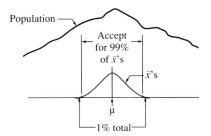

Accept/reject decision making is an efficient procedure well-suited for the cost-conscious needs of research and business, however it does come with risks, discussed as follows.

2. **Type I error (the α risk):** One risk is the probability of rejecting your initial assumption in error. For instance, say we establish the 95% guideline for accepting some initial assumption. Thus, if the initial assumption were true, 95% of the $\bar{x}$'s would fall in the accept zone and indeed we would make the correct decision (by accepting the initial assumption). However, this also implies that 5% of the $\bar{x}$'s will fall *outside* the accept zone even though the initial assumption is true, and we will reject the initial assumption in error.

This 5% is called the level of significance or Type I error risk for this experiment and is usually denoted as $\alpha = .05$.

3. **Type II error (the β risk):** Another risk is the probability of *accepting* the initial assumption in error. For instance, say we establish a 95% guideline for accepting some initial assumption, however this initial assumption is *not* true. Since we usually have no way of knowing the initial assumption is not true prior to sampling, we set up an accept zone and proceed with the experiment.

Now, by chance, we may actually get sample $\bar{x}$'s falling in this accept zone even though the initial assumption is incorrect.

The probability that your sample $\bar{x}$ will fall in this accept zone when the initial assumption is incorrect is called a Type II error risk and its probability will vary depending on several factors.

Power: Essentially power phrases the Type II error risk in positive terms. For instance, say under a particular set of conditions, we calculate the Type II error risk to be 12%, then the power of the experiment is 88% (100% minus 12%), meaning 88% of the time you will reject the initial assumption correctly and only 12% of the time will you commit the Type II error by accepting it erroneously (see sketch for detailed explanation of this example).

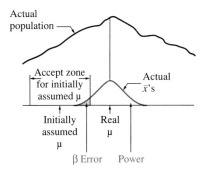

β Error: Say for instance, 12% of the actual $\bar{x}$'s fall in the accept zone for some initially assumed μ, thus 12% of the time you will accept the initially assumed μ in error (not realizing the real μ is in a different position).

Power: For this example the power is 88% (100% minus 12%), meaning 88% of the time the sample $\bar{x}$'s will fall outside the accept zone for the initially assumed μ and you will *reject* the initially assumed μ (which is the correct decision).

Controlling error in experiments: Generally in an experiment, we wish both risks (Type I and Type II) to be as low as possible. However, decreasing the Type I risk (say for instance, by lowering the α level of the experiment from .05 to .01) merely increases your Type II error risk.

Perhaps the best approach to reducing errors is to increase your sample size substantially, which lowers your Type II error risk, thus allowing for greater flexibility in setting a Type I error risk (α level) for your experiment. The negative side of increasing the sample size is that it may be costly, time-consuming, or in some cases not feasible.

Another approach is to conduct a second study, however this is often exceedingly costly, time consuming and may arouse questions as to why you didn't plan the initial study more carefully (say for instance, by using a larger sample size). And in many cases conducting a second study is simply not feasible.

Exercises

Note that full answers for exercises 1–5 and abbreviated answers for odd-numbered exercises thereafter are provided in the Answer Key.

6.1 In the cutting machine problem, suppose we establish the middle 98% of the $\bar{x}$'s as our guideline for accepting $\mu = 1000$ mm. Assuming $n = 36$ and $\sigma = 12$ mm,

a. What is the probability of a Type I error?
b. Between what $\bar{x}$ values would you accept the machine cutting at $\mu = 1000$ mm?
c. Explain briefly how one might commit a Type I error.

6.2 Referring to exercise 6.1,

a. What is the probability of a Type II error if the machine shifts to $\mu = 995$ mm?
b. Compare Type I and Type II error risks calculated in this question with those of the second example of section 6.2 where we calculated the Type I and II error risks to be 13.36% and 15.87%, respectively. What principle concerning Type I and Type II errors is demonstrated?

6.3 In the cutting machine problem, for $\mu = 1000$ mm, $\sigma = 12$ mm, and $n = 36$,

a. If you wish to establish a Type I error risk of 10%, find the $\bar{x}$ cutoffs for accepting $\mu = 1000$ mm.
b. Calculate the probability of a Type II error for a shift to $\mu = 1004$ mm.
c. What is the power of the test in part b?

6.4 In the National Institutes of Health problem, suppose we arbitrarily establish $\bar{x} = 8.90$ ml to $\bar{x} = 9.10$ ml as our cutoffs for accepting test tubes filling on average to $\mu = 9.00$ ml. Assuming $n = 49$ and $\sigma = .35$ ml,

a. What is the probability of making a Type I error?
b. Using this example, briefly define a Type I error and discuss the consequences of making a Type I error.
c. What is the probability of a Type II error for a shift to $\mu = 9.14$ ml?
d. Using this example, briefly define a Type II error and discuss the consequences of making a Type II error.

6.5 Referring to exercise 6.4,

a. What is the probability that when the process is "in control" (filling properly), you will believe the process malfunctioning?
b. What is the probability that when the process goes "out of control" (filling improperly), say filling at $\mu = 9.14$ ml, you will believe the process is filling correctly?

6.6 In the National Institutes of Health problem, for $\mu = 9.00$ ml, $\sigma = .35$ ml, and $n = 49$,

a. Establish a Type I error risk of 1% and find the $\bar{x}$ cutoffs for accepting $\mu = 9.00$ ml.
b. Calculate the probability of a Type II error for a shift to $\mu = 8.85$ ml.
c. What is the power of the test in part b?

6.7 In the cutting machine problem, for $\mu = 1000$ mm and $\sigma = 12$ mm, suppose we establish $\bar{x} = 997$ mm to $\bar{x} = 1003$ mm as our cutoffs for accepting $\mu = 1000$ mm, calculate your Type I error risk and your Type II error risk (for a shift to $\mu = 995$ mm) for,

a. $n = 30$.
b. $n = 100$.
c. Compare the results in parts a and b.

6.8 Brell shampoo, an "in-house" brand, is marketed through a large national chain of convenience stores. This chain also carries other national brands of shampoo. Brell's in-house market share is $\mu = 24.0$ (meaning: on average 24.0% of the shampoo sold in these stores is Brell) with standard deviation 3.2.

Suppose we arbitrarily establish $\bar{x} = 23.3$ to $\bar{x} = 24.7$ as our cutoffs for accepting $\mu = 24.0$; assuming sample size $n = 75$,

a. What is the probability of a Type I error?
b. What is the probability of a Type II error for a shift to $\mu = 23.0$?
c. What is the power of the test in part b?

6.9 Referring to exercise 6.8,

a. Suppose we establish $\bar{x} = 23.1$ to $\bar{x} = 24.9$ as our cutoffs for accepting $\mu = 24.0$, what effect would this have on our Type I and Type II error risks and on power?
b. Recommend a way to decrease both your Type I and Type II error risks.

6.10 In the horror film moviegoer problem, suppose we arbitrarily establish $\bar{x} = 17.0$ to $\bar{x} = 17.8$ years old as our cutoffs for accepting the average age of $\mu = 17.4$ years old. Assuming $\sigma = 2.7$ years, calculate the risks for a Type I error and for a Type II error (assuming a shift to $\mu = 16.8$ years old) for,

a. $n = 45$.
b. $n = 250$.
c. Compare the results in parts a and b.

Hypothesis Testing

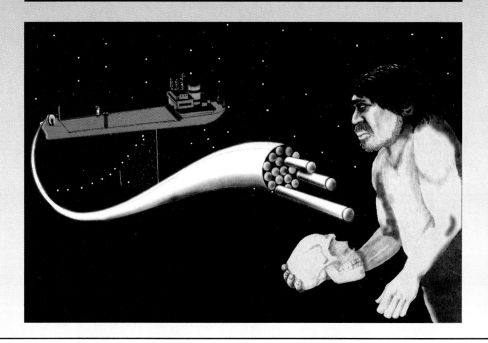

7.1 Two-Tailed Hypothesis Tests (Large Sample, $n \geq 30$)

Hypothesis testing is nothing more than a formalized approach to the central limit theorem incorporating the concepts of accept/reject decision making and Type I error. Let's see how it works in the following problem. ▼

Suppose the Fiche Company (a manufacturer of telephone cable) receives shipments of fiber optic thread, hair-thin strands of glass capable of transmitting hundreds of thousands of times more information than a copper wire. The Fiche Company will ultimately coat the fiber-optic threads with steel and plastic and bind several into cables to be laid on ocean floors for intercontinental communications. However, it is important for production purposes that the incoming shipments of hair-thin glass fiber thread maintain an average thickness of .560 mm. Of course the supplier of the thread claims this is so.

This is a typical situation in business. A supplier ships you goods and makes a claim with the expectation that you will believe that claim. In this case, the claim is: the average thickness of fiber optic thread in the shipment is .560 mm. In statistical terms, we call this a hypothesis.

> A **hypothesis,** then, is merely a claim put forth by someone. This hypothesis or claim is denoted by the symbol H, or H_0 (H-sub-zero) and referred to formally as the **null-hypothesis.**[*]

In this case, our claim or null hypothesis would be written

$$H_0: \mu = .560 \text{ mm}$$

This null hypothesis may or may not be true. The supplier may have documented evidence for making such a claim, or may simply be guessing. In fact, for all we know, the supplier may be lying outright, which of course obliges us as prudent individuals to test their claim. This test is referred to as a hypothesis test.

*Technical note: Actually the symbol H_0 originates from tests involving the comparison of two population means or proportions, however the symbol H_0 has now evolved to represent any hypothesis set up for the purposes of seeing if it can be rejected.

> ### Hypothesis Test
> A test designed to prove or disprove some initial claim, your null hypothesis, H_0.

When dealing with a hypothesis test, we always begin by assuming the claim or null hypothesis (H_0) is true, in this case that the supplier is correct, that indeed the average thickness is $\mu = .560$ mm for these shipments of fiber optic thread.

> We begin a hypothesis test by assuming H_0 is true.

Indeed, if we accept H_0: $\mu = .560$ mm as true (which we must to begin a hypothesis test), then we know from decades of experience a certain logic will necessarily follow, namely, if we were to measure the thickness of all the glass fiber in the shipment and arrange these measurements according to size into a histogram, these measurements would probably cluster about the average value of $\mu = .560$ mm, however many measurements would be less than .560 mm and many would be more, and the histogram *might* take on the following shape.

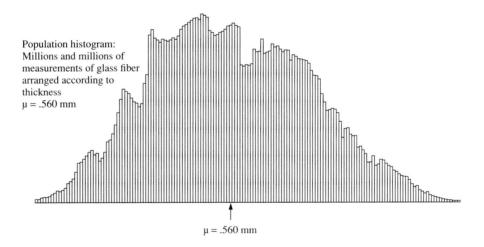

Population histogram:
Millions and millions of
measurements of glass fiber
arranged according to
thickness
$\mu = .560$ mm

$\mu = .560$ mm

Notice this population is somewhat ragged in shape with a slight skew. Although in real life we may not actually know the shape of the population prior to sampling, it would not be unusual for such a ragged skewed shape to appear. Although the output from one process or machine, properly operating and running uninterrupted, is often found to be normally or nearly normally distributed, an entire shipment may very well consist of output from several machines or processes over several periods of time and, thus, could vary considerably. When the output from various processes are mixed, a normal distribution may or may not form, depending on a number of factors. However, this should not make a difference in our analysis of μ, since whatever the shape of your population, as long as the sample size exceeds 30, the $\bar{x}$ distribution will be normally distributed, as follows:

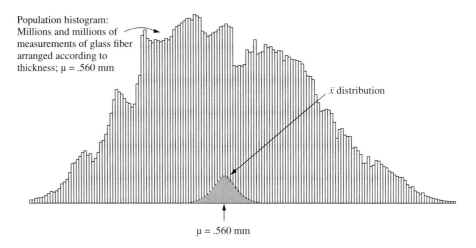

Population histogram:
Millions and millions of
measurements of glass fiber
arranged according to
thickness; μ = .560 mm

$\bar{x}$ distribution

μ = .560 mm

However, we do have another problem.

Noticeably absent from the above histogram is information concerning the standard deviation of this population, σ, which in real-life situations is often not supplied. In fact, more often than not, it is simply unknown. However, without σ we cannot calculate $\sigma_{\bar{x}}$.

$$\text{Remember: } \sigma_{\bar{x}} = \frac{\sigma}{\sqrt{n}}$$

And without $\sigma_{\bar{x}}$, we cannot estimate the spread of our $\bar{x}$ distribution, which tells us where we should expect sample averages ($\bar{x}$'s) to cluster—which of course forms the entire basis of our central limit theorem analysis. In other words, we are stuck!

But wait, the problem is not insurmountable. We have learned from prior exercises that when we randomly select 30 or more measurements from a population that

$\bar{x} \approx \mu$ the sample average, $\bar{x}$, is approximately equal to the population average, μ, and

$s \approx \sigma$ the sample standard deviation, s, is approximately equal to the population standard deviation.

If indeed $s \approx \sigma$, that is, the individual measurements in one sample are spread out in a manner similar to how the measurements in the entire population are spread out, we may be able to use the standard deviation of one sample, s, as an estimator of the standard deviation of the entire population, σ. Experience has confirmed that when your sample size is over 30, indeed the spread of measurements in one sample is a good estimator of the spread of measurements in the entire population—that is, s is a good estimator of σ, and this is precisely what is done in industry and research studies.

> s is used to estimate σ.

Since the standard deviation of one sample should give us what we want to know, namely, an approximation of σ, the standard deviation of the population, then the telephone cable manufacturer is obliged on receiving the shipment to take a *random sample*. Although many results are possible, let us say, for the purposes of this example that the manufacturer randomly samples 36 pieces of fiber-optic thread and calculates the following:

$$n = 36 \text{ measurements}$$
$$\bar{x} = .553 \text{ mm}$$
$$s = .030 \text{ mm}$$

If this is indeed a properly conducted random sample, the *spread* (standard deviation) of the 36 measurements should be similar to the spread (standard deviation) of the entire population. That is, if $s = .030$ mm (note sample results above) and if $s \approx \sigma$, then σ must be approximately equal to .030 mm. And we can use this estimate to calculate $\sigma_{\bar{x}}$, as follows:

$$\sigma_{\bar{x}} = \frac{\sigma}{\sqrt{n}} \approx \frac{s}{\sqrt{n}} \approx \frac{.030}{\sqrt{36}} \approx \frac{.030}{6}$$
$$\approx .005 \text{ mm}$$

Now that we know $\sigma_{\bar{x}}$ is approximately equal to .005 mm, we can now estimate the spread of the $\bar{x}$ distribution.

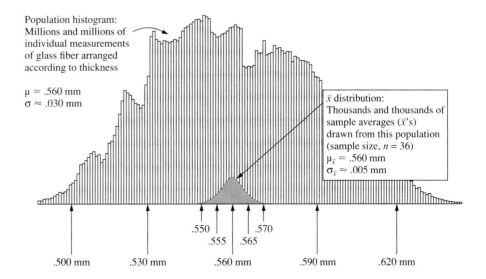

Population histogram: Millions and millions of individual measurements of glass fiber arranged according to thickness

$\mu = .560$ mm
$\sigma \approx .030$ mm

$\bar{x}$ distribution: Thousands and thousands of sample averages ($\bar{x}$'s) drawn from this population (sample size, $n = 36$)
$\mu_{\bar{x}} = .560$ mm
$\sigma_{\bar{x}} \approx .005$ mm

.550   .570
.555   .565

.500 mm   .530 mm   .560 mm   .590 mm   .620 mm

Keep in mind, what we have done so far is a make-believe construction based solely on the assumption that the supplier's claim $\mu = .560$ mm is true. We really do not know whether $\mu = .560$ mm is true or not. We are merely saying: ''if'' $\mu = .560$ mm is true, and ''if'' we were to measure every piece of fiber in the shipment, and ''if'' we continually took random samples of 36 measurements and calculated the sample average, $\bar{x}$, for each sample, then the central limit theorem tells us that the $\bar{x}$'s should form into a normally distributed $\bar{x}$ distribution, symmetrical about $\mu = .560$ mm and spread out as shown above.

Okay, now that we know what the $\bar{x}$ distribution should look like if the supplier's claim is true, how do we prove (or disprove) $\mu = .560$ mm? Simple. We take a random sample of 36 measurements from our shipment, calculate the sample average, $\bar{x}$, and observe if this $\bar{x}$ reasonably fits into the expected $\bar{x}$ distribution.

Wait a minute. We already took a random sample of 36 measurements. True. There's no point spending time and money on another sample. Let's use the $\bar{x}$ we observed from the earlier sample. If you recall, our sample results were as follows (reprinted here for convenience):

$$n = 36 \text{ measurements}$$
$$\bar{x} = .553 \text{ mm} \longleftarrow \text{(Now we are interested in this measurement)}$$
$$s = .030 \text{ mm}$$

Notice that, now, we are concerned with the $\bar{x}$ of the sample. In other words, does this $\bar{x}$ of .553 mm reasonably fit into our expected $\bar{x}$ distribution? And the answer is, yes. We can look at this sample average of .553 mm and look at the $\bar{x}$ distribution and see that this $\bar{x}$ of .553 mm is a reasonably likely occurrence. Observe:

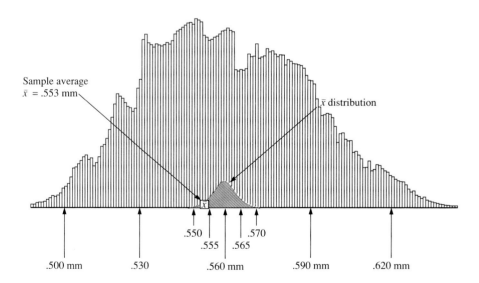

Since an $\bar{x}$ of .553 mm would be a reasonably likely occurrence, we conclude that the supplier's claim (μ = .560 mm) is quite possible. If we choose to make a firm *accept H_0* or *reject H_0* decision, then we

Accept H_0: μ = .560 mm

In reality, there is not enough evidence to prove μ is *precisely* .560 mm. The best we can show is that μ = .560 mm is reasonably possible given the evidence of this one sample. The concept of hypothesis testing is much like a jury trial: μ = .560 mm is innocent (accepted) unless *proven guilty*. Since a sample average of $\bar{x}$ = .553 mm does not prove the supplier's claim false, then we must assume the supplier's claim is true.

Professionally, this conclusion is written in a number of ways. Two of the most popular are:

The null hypothesis cannot be rejected

or

Results not significant*

Both statements say the same thing, that is, if we use the accept H_0–reject H_0 format, then we must accept the supplier's claim (μ = .560 mm), since we have no evidence to disprove the claim. My preference is to word the conclusion as follows:

Since the sample average of $\bar{x}$ = .553 mm reasonably fits into the expected $\bar{x}$ distribution for μ = .560 mm, we

Accept H_0: μ = .560 mm

Now you might feel a little uncomfortable accepting H_0 since your sample average (.553 mm) did not fall precisely on the claimed population value of .560 mm. And at this point you might say, why don't we continue sampling to be more positive of our decision? Unfortunately, in most areas of research, further sampling is not practical. It is usually expensive, time-consuming, and in some cases physically impossible (when test circumstances cannot be duplicated). Certainly in this production control experiment, another random sample can be taken with relative ease, however in most studies in marketing, medicine, sociology, economics, and other fields, we often must rely on the results of one and only

---

*The words *not significant* have a very special meaning in statistical testing. They mean the results may reasonably be attributed to "chance fluctuation." In other words, $\bar{x}$'s may very well vary, fluctuate by chance, between .550 mm and .570 mm when μ = .560 mm. Since we achieved an $\bar{x}$ (.553 mm) in this chance fluctuation range, we merely accept H_0. In broad terms, when sample results are,

Not significant: we accept H_0
Significant: we reject H_0

one sample. Even in this production control experiment, no one wants to absorb the added time and expense of further sampling unless absolutely necessary. In other words, in statistical studies,

we normally base our decision on one and only one sample.

And we will conform to this practice in this text. So, to sum up our experiment, if our one sample average, $\bar{x}$, is reasonably close to the claimed μ, we accept H_0 as true and therefore accept the shipment of fiber-optic thread as meeting our specification of $\mu = .560$ mm.

However, this may cause some questions, such as: at what point do we grow suspicious that our sample $\bar{x}$ is *not reasonably close* to μ? For instance, what if our sample average turned out to be .550 mm or .540 mm or .577? Clearly, these values are on the very fringe of the "expected" sample averages. Observe:

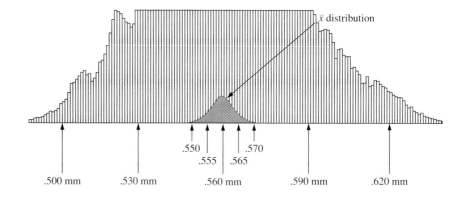

In other words, at what value of $\bar{x}$ do we begin to grow suspicious that maybe the supplier's claim is false? Fortunately, there are certain industry standards that have proven reliable over decades of use. Although a number of industry standards exist, one of the most popular is the

Level of significance, $\alpha = 5\%$ (.05)

Although discussed in the last chapter, a brief review here might be helpful. Essentially, a level of significance sets up the cutoffs, or boundaries for accepting or rejecting H_0. For instance,

For level of significance, $\alpha = 5\%$ (.05),* establish where the middle 95% of the $\bar{x}$'s are expected to fall if H_0 is true. Then, if the $\bar{x}$ you calculate

*Actually, many levels of significance are possible.

from your random sample falls inside (or exactly on the border) of this 95% range, accept H_0 as true. If the sample $\bar{x}$ falls outside, assume H_0 is false.

Visually we might present this $\alpha = .05$ hypothesis test as follows:

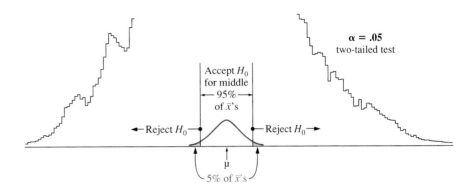

> ### Two-Tailed Test
> This is called a two-tailed hypothesis test since we have two tails of rejection (as shown shaded above). That is, we would reject the null hypothesis for any sample $\bar{x}$ falling in either of the *two* shaded tails.

To recap: if your sample $\bar{x}$ falls *inside* this 95% range (or on the border), accept H_0. If your sample $\bar{x}$ falls *outside* this range (that is, in the shaded tails), reject H_0. And this is precisely what is done in industry and research. Now let us repeat this problem as it would be worded and solved in practice.

Example ——————— A supplier claims the average thickness (diameter) of its fiber-optic thread is .560 mm. You receive a shipment and decide to test their claim at a .05 level of significance by taking a sample of 36 randomly selected measurements, with the following results:

$$n = 36 \text{ measurements}$$
$$\bar{x} = .553 \text{ mm}$$
$$s = .030 \text{ mm}$$

What can we conclude?

Solution

A hypothesis test consists of three fundamental sequences as follows.

Sequence   I. *Set up initial conditions: H_0, H_1, and level of significance*

In Our Example,
It Would Be

H_0:   State the null hypothesis, that is, H_0: $\mu = .560$ mm
the claim or assertion you wish to
test.

H_1:   State the alternative hypothesis. H_1: $\mu \neq .560$ mm
In other words, if H_0 proves false,
then what must we conclude?

α: State the level of significance, α, $\alpha = .05$ (5%)
that is, the risk of a Type I error
(the risk of rejecting H_0 in error).

Sequence   II. *Assume H_0 true, use α to establish cutoffs* as follows:

Calculate

$\sigma_{\bar{x}}$:   We must remember we are
dealing with $\bar{x}$'s and therefore
must first calculate $\sigma_{\bar{x}}$, the
standard deviation of the $\bar{x}$
distribution. Note in our formula
for $\sigma_{\bar{x}}$, we used s (the standard
deviation of the sample) as an
estimator of σ (the population
standard deviation).

$$\sigma_{\bar{x}} = \frac{\sigma}{\sqrt{n}} \approx \frac{s}{\sqrt{n}}$$
$$\approx \frac{.030}{\sqrt{36}} \approx \frac{.030}{6}$$
$$\approx .005 \text{ mm}$$

Draw Curves

Using our above calculation, $\sigma_{\bar{x}}$
$\approx .005$ mm, we estimate the
spread of the $\bar{x}$ distribution.

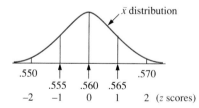

Establish Cutoffs (using α, the level of significance)
Our level of significance in this
case is $\alpha = .05$ (5%), which in
a two-tailed test implies we will
accept the middle 95% of the

$\bar{x}$'s as our boundary for accepting H_0 as true. We now look up the z scores corresponding to the middle 95% of the $\bar{x}$'s, which turn out to be $z = -1.96$ and $z = +1.96$.

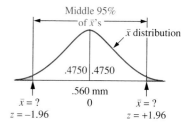

Remember: the normal curve table reads half the normal curve, starting from $z = 0$ out, so we look up $\frac{1}{2}$ of 95% or $47\frac{1}{2}\%$, which in decimal form is .4750 (as shown at right).

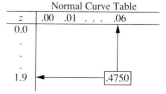

Substituting the z scores of -1.96 and $+1.96$ into our formula, we solve for the $\bar{x}$ at the cutoffs.

$$z = \frac{\bar{x} - \mu}{\sigma_{\bar{x}}}$$

$$-1.96 = \frac{\bar{x} - .560}{.005}$$

Solving for $\bar{x}$:
$$\bar{x} = .550 \text{ mm}$$

$$z = \frac{\bar{x} - \mu}{\sigma_{\bar{x}}}$$

$$+1.96 = \frac{\bar{x} - .560}{.005}$$

Solving for $\bar{x}$:
$$\bar{x} = .570 \text{ mm}$$

The completed solution would appear graphically as follows:

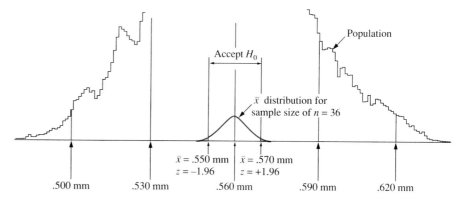

Note that the reject zones are shaded, that is, the zones where we would reject $\mu = .560$ mm as being true. This is your risk of a Type I error (5%).

Sequence III. *Accept or reject H_0 using your sample $\bar{x}$:* For this, two methods are available. Method One uses the actual value of the sample $\bar{x}$. Method Two uses the z score of the sample $\bar{x}$. Since each adds to understanding, we shall employ both.

METHOD ONE

This method uses the actual value of the sample $\bar{x}$ (.553) in the decision-making process.

Recall: Our sample results were as follows:

n = 36 measurements
$\bar{x}$ = .553 mm

Criteria: Accept H_0 (μ = .560 mm) if your sample $\bar{x}$ falls between the established $\bar{x}$ cutoffs of .550 mm and .570 mm, otherwise reject.

Decision: Since our sample $\bar{x}$ (.553 mm) fell in the acceptance zone for H_0, we accept H_0 (μ = .560 mm) as true.

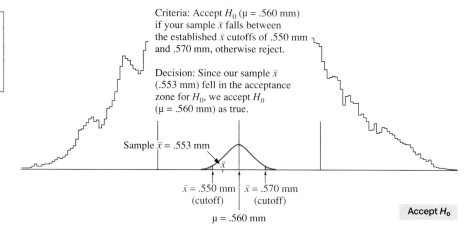

Sample $\bar{x}$ = .553 mm

$\bar{x}$ = .550 mm (cutoff) $\bar{x}$ = .570 mm (cutoff)

μ = .560 mm

Accept H_0

METHOD TWO

This method uses the z score of the sample $\bar{x}$ in the decision-making process. To use this method, however, we must first calculate the z score of our sample $\bar{x}$ (.553 mm), as follows.

$$z = \frac{\bar{x} - \mu}{\sigma_{\bar{x}}} = \frac{.553 - .560}{.005}$$

$$z = -1.40$$

Criteria: Accept H_0 (μ = .560 mm) if the z score of your sample $\bar{x}$ falls between the established z score cutoffs of -1.96 and $+1.96$, otherwise reject.

Decision: Since the z score of our sample $\bar{x}$ (-1.40) fell in the acceptance zone for H_0, we accept H_0 (μ = .560 mm) as true.

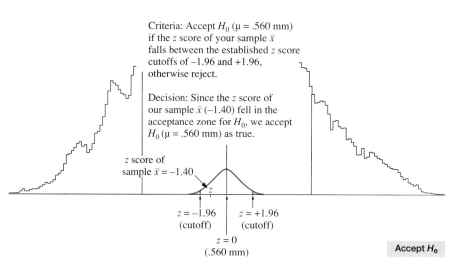

z score of sample $\bar{x}$ = -1.40

$z = -1.96$ (cutoff) $z = +1.96$ (cutoff)

$z = 0$ (.560 mm)

Accept H_0

Whether we use the actual value of the sample $\bar{x}$ or the z score of the sample $\bar{x}$, we will always make the same decision. In this case, we accept H_0. Generally, the z score is preferred by those most familiar with statistical technique since the z score is a more informative measure. Note that we better understand the position of the sample $\bar{x}$ if we say it is -1.40 standard deviations from the claimed μ than if we merely presented its actual value of .553 mm.*

Answer

The final answer may be presented in a number of ways, depending on the technical expertise of those reading the report.

a. If the report is to be presented to individuals unfamiliar with statistical technique, perhaps the following offers a clear approach:

Since the sample average we obtained from the shipment (.553 mm) falls inside the range (.550 to .570) where we would most likely expect sample averages to fall if H_0 were true, we accept H_0: $\mu = .560$ mm, and therefore accept the shipment.

> **Accept H_0**

b. However, this same answer may very well appear in a technical report worded in terms of z scores, as follows:

Since our sample z of -1.40 is not less than -1.96, the null hypothesis cannot be rejected. The difference between .553 mm and .560 mm is not large enough to provide evidence at the .05 level of significance that the shipment does not meet supplier's specification.

> **Null hypothesis cannot be rejected**

---

*P-value approach: Actually a third method is also used. This method calculates the probability of achieving a result *at least* as many standard deviations from the expected value as your sample result.

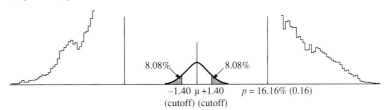

Let's reconsider the above example. Since we achieved a sample result 1.40 standard deviations from the expected value, μ (calculated in Method Two), we shade all the area that is *at least* 1.40 standard deviations from μ. Note in a two-tailed test, we shade *both* tails. Next we look up the probability of achieving a sample result in this shaded area, which is 16.16% (8.08% in each tail). This is our *p*-value. This is usually expressed in technical reports and computer software printouts as either $p = .16$ or $p > .05$ (meaning the probability of achieving this sample $\bar{x}$ is greater than the α level of the test).

> **For $p \geq \alpha$, Accept H_0, otherwise reject**

Since in our case, $.16 \geq .05$, we Accept H_0.

c. Then again, many reports simply present the results as

$z = -1.40$ (not significant).

Results not significant*

Believe it or not, all three answers say the same thing. Try to understand the technical explanations using z scores, since this is typical of how research reports are presented. ■

Control Charts

In production studies and occasionally in marketing, medical, and other studies, the same hypothesis test may be repeated a number of times. For instance, what if this telephone cable manufacturer in the prior problem were to accept this shipment of fiber-optic thread and then ordered additional fiber-optic thread under the same specifications, to be delivered once a month for several months? Each monthly shipment may very well be tested in an identical manner. When essentially the same test must be repeated on a periodic basis, a **control chart** can be set up as follows:

Construction of Control Chart

1. On a graph, establish cutoffs for a given hypothesis test. In industrial production, cutoffs are usually referred to as *control limits*.

2. Rotate graph $\frac{1}{4}$ turn counterclockwise, extending the cutoff lines to the right. Shade rejection zone.

3. Plot each sample $\bar{x}$ sequentially to the right. Connect each $\bar{x}$ to prior result with a line segment.

In a control chart, you may choose to use either actual values or z scores to represent the readings. For instance, say we use actual values, we would proceed (using our fiber-optic thread example) as follows.

---

*Again, the words *not significant* have a very special meaning in statistical testing. Essentially, *not significant* means: the sample results (in this case, $\bar{x} = .553$ or $z = -1.40$) are considered ''chance fluctuation.'' In other words, we would expect to find $\bar{x}$'s between ± 1.96 standard deviations of the mean if H_0 were true. Since the z score (-1.40) of our sample $\bar{x}$ was in this chance fluctuation range between ± 1.96 standard deviations, it is deemed not significant and we accept H_0.

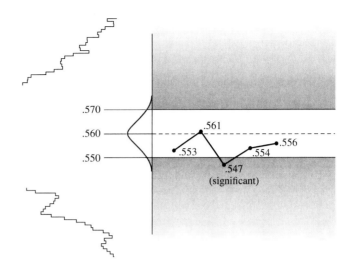

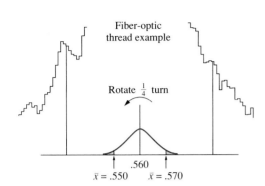

Cutoffs established, taken from prior example.

Rotate $\frac{1}{4}$ turn counterclockwise, extending cutoff lines to the right and shading rejection zone (as shown in next diagram).

Now let's say we receive 5 shipments over several months and calculate the sample $\bar{x}$ for each as follows.

$\bar{x} = .553$ mm $\bar{x} = .554$ mm

$\bar{x} = .561$ mm $\bar{x} = .556$ mm

$\bar{x} = .547$ mm (significant)

Each sample $\bar{x}$ is plotted sequentially as the shipment comes in and connected with a line segment to prior result (as shown above).

Note that one sample $\bar{x}$ (.547 mm) was marked "significant." This means, based on this one sample average, we would reject this particular shipment as not meeting specifications. At this point, the production supervisor would likely be called in. After verifying results, the supervisor may very well call the manufacturer of the fiber-optic thread to inform them that their process was not meeting specification, and most likely "out of control." A process is deemed out of control when sample $\bar{x}$'s fall outside the control limits for acceptance of H_0 and we suspect a possible deterioration of the process.

Note that a control chart provides a clear visual history of this hypothesis test. Often we learn more about a process by keeping this kind of record. Sometimes we can spot a trend, a process going out of control *before* a significant sample $\bar{x}$ is achieved. Or we may be able to pick up slight shifts in the value of μ, even though sample $\bar{x}$'s are in control. For a process in control, the sample $\bar{x}$'s should fluctuate (usually in a ragged pattern) around the value of μ. Notice that the $\bar{x}$'s we calculated, .553, .561, .547, .554, and .556, seem to fluctuate more around the value of .555 (than the value .560). If this trend continues for future shipments, we may very well suspect the thickness of the fiber-optic thread shipped may be on average, $\mu = .555$ mm. Of course, whether or not this slight shift makes a difference in our production would have to be assessed.

> A *control chart** provides a clear visual history of a repetitive test.

7.2 One-Tailed Hypothesis Tests (Large Sample, $n \geq 30$)

A one-tailed hypothesis test is quite similar in method to a two-tailed hypothesis test, except in a one-tailed test, the Type I error risk (α) is assigned to only *one* tail of the $\bar{x}$ distribution.

> **One-Tailed Hypothesis Test**[†]
> All the Type I error risk, α, is assigned to *one* tail of the $\bar{x}$ distribution, and we reject H_0 for any sample $\bar{x}$ falling in this *one* tail only.

The α risk may be assigned to either the right or left tail, depending on the hypothesis you wish to test. The following two examples demonstrate this.

---

*Historical note: Walter Shewhart first developed control charts in 1924, which were tested and developed within the Bell Telephone System, 1926–1931. For further historical reading on this topic, refer to, W. Peters, *Counting for Something* (New York: Springer-Verlag, 1987), Chapter 16, ''Quality Control,'' pp. 151–162.

[†]Actually, some controversy surrounds the use of one-tailed hypothesis testing. Refer to D. Howell, *Statistical Methods for Psychology* (Boston: PWS Publishers, 1982, pp. 64–66) for a discussion of one- and two-tailed tests. Essentially, Howell argues that an investigator may start with a one-tailed test, yet reject in two tails, thus inadvertently increasing the α level of the experiment. Howell also states, ''A number of empirical studies have shown that the common statistical tests . . . are remarkably robust when they are run as two-tailed tests, but are not always so robust when run as one-tailed tests.'' **Robustness** is the degree to which you can violate the assumptions of a test and yet leave the validity more or less unaffected.

To Test the Hypothesis

H_0: μ = 150 *or more,* we would assign the total α risk to the left tail in the $\bar{x}$ distribution and reject H_0 for any sample $\bar{x}$ in this left tail, as shown shaded below.

To Test the Hypothesis

H_0: μ = 150 *or less,* we would assign the total α risk to the right tail in the $\bar{x}$ distribution and reject H_0 for any sample $\bar{x}$ in this right tail, as shown shaded below.

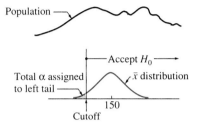

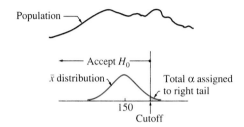

Notice in each example above, the total Type I error risk, α, was assigned to only *one* tail in the $\bar{x}$ distribution. This will affect the determination of the z score at the cutoff. Other than this, a one-tailed test is conducted in almost an identical manner as a two-tailed test. For instance, suppose we wish to test the following null hypothesis:

$$H_0\text{: } \mu = 150 \text{ or more}$$

at an α = .05 level of significance; we would proceed as follows:

First, assign the total α risk (.05 or 5%) to the left tail of the $\bar{x}$ distribution. Why the left tail? Well, because we reject H_0 only if our sample average falls *significantly below* 150. (Note: you would not reject the hypothesis, μ = 150 or more, for a sample $\bar{x}$ greater than 150.)

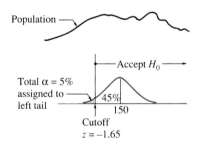

Second, to find the z score at the cutoff, we look up in the normal curve table, 45% (50% − 5% = 45%). Remember, the table reads from the center of the normal curve out. (Note: .4500 falls midway between .4495 and .4505 in the table, thus, round to the higher number, .4505, which is z = 1.65.) Since the cutoff is *below* μ, we apply a negative sign to the z score; thus z_{cutoff} = −1.65.

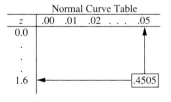

The decision-making process, in this case, is: Accept H_0 for any sample $\bar{x}$ to the right of the cutoff, otherwise reject. That is, we reject H_0 for any sample $\bar{x}$ in the shaded tail.

At this point you might ask, why don't we make sure μ is 150 or more by shading all of the values below 150 as follows?

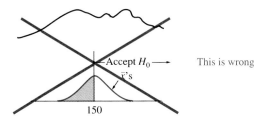

Remember, we are talking about sample averages, $\bar{x}$'s, and $\bar{x}$'s tend to fluctuate around the population average, μ. That is, μ may very well be exactly 150 and still you could get sample $\bar{x}$'s *below* this value. So, we must leave some margin below μ for the $\bar{x}$'s to fluctuate, as follows:

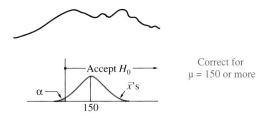

To recap: in a one-tailed hypothesis test, you assign the total α risk to *one* tail of the $\bar{x}$ distribution (which we shade), and you reject H_0 for any sample $\bar{x}$ in this shaded tail. Other than this, a one-tailed hypothesis test is conducted in almost an identical manner as a two-tailed hypothesis test.

At this point, I have found the following two reminders helpful.

Keep in mind for all hypothesis tests

1. Use α, the level of significance, to establish the cutoff(s), and
2. Shade where you would reject H_0.

Applications

Now let's see how this works in a study in psychiatric medicine.

Example

Elavil,* a powerful sedating drug prescribed by psychiatrists, has been proven effective over decades of use in the treatment of depression, however Elavil can have side effects (causing high blood pressure, dry mouth and impotence problems, blurred vision, etc.), so daily dosages must be minimized. However, one patient may need 75 mg per day to relieve depression while another patient may need 250 mg per day to have the same effect, depending on the individual.

Suppose a leading trade journal makes the claim that the average effective dosage nationwide for patients is *at least* 150 mg per day. Concerned such an article may influence psychiatrists to unnecessarily increase dosages, suppose the National Institute of Mental Health in Bethesda, Md., conducts a test by randomly sampling 400 patients nationwide, with the following result:

$$n = 400 \text{ patients}$$
$$\bar{x} = 141.6 \text{ mg/day minimum effective dosage}$$
$$s = 48.2 \text{ mg/day}$$

Use this sample result to test at a level of significance of $\alpha = .03$, the trade journal's claim

$$\mu \text{ is } \textit{at least} \text{ 150 mg/day.}$$

Solution

This is a one-tailed hypothesis test since in effect the trade journal's claim is $\mu = 150$ mg/day *or more*. In other words, you would reject the claim only if your sample $\bar{x}$ was unreasonably *below* 150 mg/day. That is, we reject only in one direction.

A hypothesis test consists of three fundamental sequences as follows:

Sequence   I. Set up initial conditions

State null hypothesis, the initial claim you wish to test:	H_0: $\mu = 150$ mg/day or more ($\mu \geq 150$)
State alternative hypothesis. If H_0 proves false, what must we conclude?	H_1: μ is less than 150 mg/day ($\mu < 150$)
State the risk of rejecting H_0 in error, the level of significance, α.	$\alpha = .03$ (3%)

*Elavil is part of the tricyclic family of antidepressant drugs, along with Sinequan, Tofranil, and Norpramin. Each relieves depression with varying degrees of sedating effect. Elavil is one of the more powerful sedating drugs, often used when agitation or sleeplessness accompany depression.

Sequence   II. Assume H_0 true, use α to establish cutoff(s)

Calculate

$\sigma_{\bar{x}}$:

$$\sigma_{\bar{x}} = \frac{\sigma}{\sqrt{n}} \approx \frac{s}{\sqrt{n}} \approx \frac{48.2}{\sqrt{400}} \approx \frac{48.2}{20} \approx 2.41$$

(Note s was used to estimate σ.)

Draw curves
Using our above calculation, $\sigma_{\bar{x}}$ = 2.4, we estimate the spread of the $\bar{x}$ distribution.

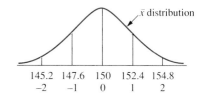

Establish cutoffs using α
Assign the total α risk (.03 or 3%) to the *left* tail of the $\bar{x}$ distribution because we would reject H_0 only if our sample $\bar{x}$ falls significantly below 150 mg/day.

Next, to find the z score at the cutoff, we look up 47% in the normal curve table (50% − 3% = 47%). Remember, the table reads from the center of the normal curve out. (47% in decimal form is .4700; the closest value is .4699.)

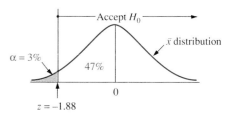

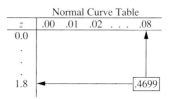

Since the cutoff is *below* μ = 150, the z score will be negative. Thus, $z = -1.88$. Substituting the z score of -1.88 into our formula, we solve for the $\bar{x}$ value at the cutoff.

$$z = \frac{\bar{x} - \mu}{\sigma_{\bar{x}}}$$

$$-1.88 = \frac{\bar{x} - 150}{2.4}$$

Solving for $\bar{x}$:

$$\bar{x} = 145.5 \text{ mg/day at the cutoff}$$

Thus, the values at the cutoff might be represented as follows:

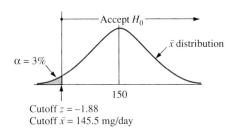

Cutoff $z = -1.88$
Cutoff $\bar{x} = 145.5$ mg/day

Note that in the diagram above there is no indication of the population, such as population standard deviation lines. This is because the sample size is so large ($n = 400$), causing the $\bar{x}$'s to cluster so tightly around μ that the population standard deviation lines are far out of view.

Sequence

III. Accept or reject H_0 using your sample $\bar{x}$

METHOD ONE
This method uses the actual value of the sample $\bar{x}$ (141.6 mg/day) in the decision-making process.

RECALL: our sample results were as follows:
$n = 400$ patients
$\bar{x} = 141.6$ mg/day.

Criteria: Accept H_0 ($\mu = 150$ mg/day or more) if the sample $\bar{x}$ falls *above* (or on border of) the $\bar{x}$ cutoff of 145.5 mg/day, otherwise reject.

Decision: Since our sample $\bar{x}$ (141.6 mg/day) fell in the rejection zone, we reject H_0 and accept H_1, the **alternative hypothesis** (μ is *less than* 150 mg/day).

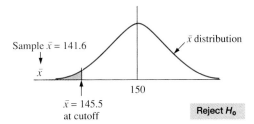

METHOD TWO

This method uses the z score of the sample $\bar{x}$ in the decision-making process. To use this method, however, we must first calculate the z score of our sample $\bar{x}$ (141.6 mg/day), as follows:

$$z = \frac{\bar{x} - \mu}{\sigma_{\bar{x}}} = \frac{141.6 - 150}{2.4}$$

$$z = -3.50$$

Criteria: Accept H_0 (μ = 150 mg/day or more) if the z score of the sample $\bar{x}$ falls *above* (or on border of) the z score cutoff of -1.88, otherwise reject.

Decision: Since the z score of our sample $\bar{x}$ (-3.50) fell in the rejection zone, we reject H_0 and accept H_1, the alternative hypothesis (μ is *less than* 150 mg/day).

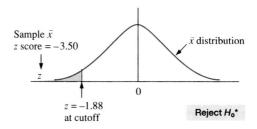

Whether we use the actual value or z score of the sample $\bar{x}$, we will always make the same decision. In this case, we reject H_0. This implies we accept H_1 (μ is less than 150 mg/day).

Answer

The final answer may be presented using actual values or z scores. We will use both.

Actual values   Since the sample average (141.6 mg/day) was below the cutoff of 145.5 mg/day, we reject H_0. Therefore we

Accept H_1: μ is *less than* 150 mg/day

z scores   Since the sample average z score (-3.50) was below the cutoff of -1.88, we reject H_0. Therefore we

Accept H_1: μ is *less than* 150 mg/day

*P-value approach: Actually, as mentioned in the prior section, a third method is also used. This method calculates the probability of achieving a result *at least* as many standard deviations from the expected value as your sample result. Let's consider this using the above example. Since we achieved a sample result of -3.50 standard deviations from the expected value, μ, we shade all the area that is *at least* -3.50 standard deviations from μ. Note in a one-tailed test, we shade in *one* tail only. Next we look up the probability of achieving a sample result in this shaded area, which is .02% (.0002), a negligible amount. This is our *p*-value. This is usually expressed in research reports and computer software printouts as either p = .0002 or $p < .03$ (meaning the probability of achieving this sample $\bar{x}$ is less than the α level of the test).

.02% (.0002)

−3.50   μ

For $p \geq \alpha$, Accept H_0, otherwise reject

Since in our case, .0002 < .03, we reject H_0.

Other ways the answer may be expressed:

The null hypothesis is rejected,

or

$$z = -3.50 \text{ (significant)}$$ ∎

Control Charts

Besides this sample result, $\bar{x} = 141.6$ mg, as presented in the previous example, suppose two identical studies (same hypothesis, same sample size and level of significance) yielded $\bar{x} = 145.8$ mg and $\bar{x} = 142.7$ mg. Plot these three results into a *control chart* and indicate significant findings.

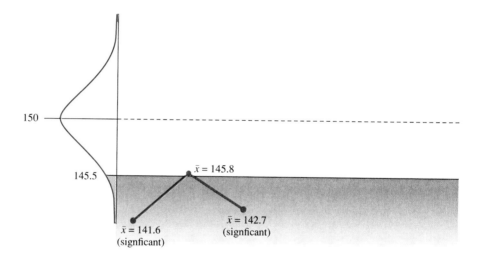

7.3 Small-Sample Hypothesis Tests ($n < 30$)

In statistical testing, small-sample sizes ($n < 30$) may also be used effectively, but only when the following conditions are satisfied.

When using small samples ($n < 30$)

1. Your population should be normally distributed or at least somewhat mound shaped, *and*

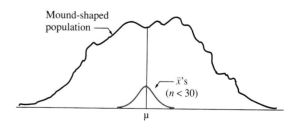

2. If we use s to estimate σ in the calculation $\sigma_{\bar{x}} = \sigma/\sqrt{n}$, we must use a t score, not a z score, to define the number of standard deviations the $\bar{x}$'s would be expected to fall from μ.

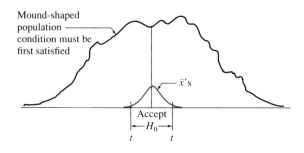

Essentially, the first condition (population normal or at least mound shaped) must be satisfied to ensure that the $\bar{x}$'s cluster close enough to μ in a reasonably normal shape such that accurate predictions can be made. When your population is *not* mound shaped, the $\bar{x}$'s spread out in a variety of patterns, often quite far from μ, as illustrated below:

Three Population Types

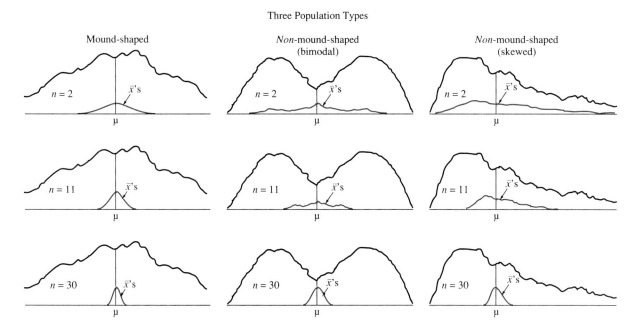

Notice it is only in the mound-shaped population that the $\bar{x}$'s cluster close to μ in a normal distribution for *all* sample sizes. For this reason, small-sample sizes can be used when your population is mound shaped. Notice in *non*-mound-shaped populations, a small-sample size ($n < 30$) will often produce $\bar{x}$'s that are quite far from μ, forming a variety of *non*normal patterns, making predictions about where the $\bar{x}$'s will fall quite perilous. Thus, when sampling from *non*-mound-shaped populations, small-sample sizes should be avoided. Of course, when your sample size grows sufficiently large (usually $n \geq 30$ is considered sufficiently large), the $\bar{x}$'s draw in quite close to μ for almost any shaped population, even for *non*-mound-shaped populations. Thus, for $n \geq 30$, the population shape has little effect on the $\bar{x}$ distribution shape; the $\bar{x}$'s will distribute normally about μ for nearly any shaped population, thus assuring reliable predictions.*

Assuming we have a mound-shaped population, a second condition must also be satisfied. If we chose to use s to estimate σ (which we most often do) in the formula, $\sigma_{\bar{x}} = \sigma/\sqrt{n}$, this necessitates a correction factor in our calculations—which we shall call the t score adjustment. Although this is more fully discussed in chapter 8, section 8.4, an overview here might be helpful.

Whereas, a large sample s is a reasonably good estimator of σ, meaning that s-values cluster quite close to σ. More specifically, if we were to take thousands of samples of size, say, $n = 35$ and calculated the standard deviation, s, for each sample and plotted these thousands of s's, the resulting distribution would be clustered relatively close around the true value of σ, as follows:

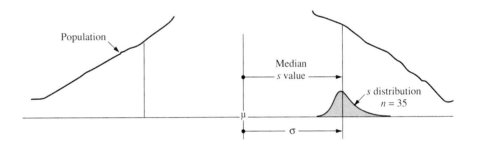

*Technical note: For *highly* unusual population shapes (such as, a population with an extraordinary skew), n may have to be larger than 30 to be assured of a normally distributed $\bar{x}$ distribution. An example of this might be the distribution of annual salaries of workers in lower Manhattan, which includes the highly skewed million-dollar-plus salaries of many Wall Street executives.

This is not the case for small-sample s's. Small-sample s's tend to *under*estimate σ; in fact, small-sample s's tend to underestimate σ more and more as n (the sample size) decreases,* illustrated as follows:

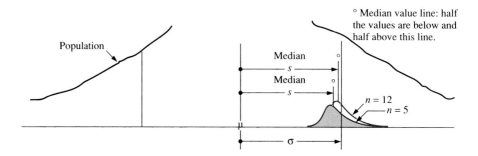

Notice that as your sample size decreases ($n = 35$, $n = 12$, $n = 5$), the s's tend to shift to lesser values, such that the median s value falls farther and farther below the true value of σ. This means, more often than not, when you calculate the s of your sample, it will be *less* in value than σ.

Now, since we use s in place of σ in the formula,

$$\sigma_{\bar{x}} = \frac{\sigma}{\sqrt{n}} \approx \frac{s}{\sqrt{n}}$$

If s underestimates σ, then $\sigma_{\bar{x}}$ will also be *under*estimated. In other words, more often than not, we will calculate a $\sigma_{\bar{x}}$ that is less in value than it actually is—because of the *under*estimated s. Visually, this might be represented as follows:

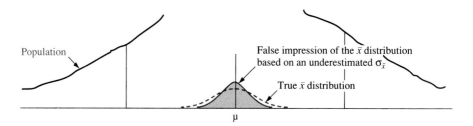

*Technical note: s^2 distributes around σ^2 in a chi-square-shaped distribution and on average $s^2 = \sigma^2$ for all sample sizes, however the median s^2 value drops below σ^2 as n decreases. The distribution of s is similar and can be found by compressing the base line suitably, to paraphrase W. Gossett (the statistician who originally published these findings in *Biometrika,* VI. pp. 1–25, 1908, under the pen name, Student), the distribution of s^2 has a direct linear relationship to the distribution of χ^2, chi-square, specifically, $s^2 = (\frac{\sigma^2}{n-1})\chi^2$.

Fortunately, we can compensate for this. Say, for instance, we conduct a two-tailed hypothesis test at $\alpha = .05$, using a sample size of $n = 5$. We know, for a large sample, $\alpha = .05$ implies an interval bounded by ±1.96 standard deviations. In other words, 95% of the $\bar{x}$'s are expected to fall within ±1.96 standard deviations of μ. However, in the case of $n = 5$, we must open up the interval to ±2.78 standard deviations, use the letter t and say, 95% of the $\bar{x}$'s will fall in the interval between ±2.78 standard deviations, illustrated as follows:

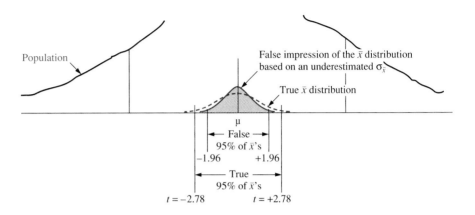

In other words, to ensure we have enclosed the *true* 95% of the $\bar{x}$'s when using small sample size $n = 5$, we must go out farther, to ±2.78 standard deviations (*not* ±1.96). And instead of using the letter z to represent the number of standard deviations, we use the letter,* t.

Of course, at this point, you might ask, where do we locate this t score adjustment of ±2.78? The answer is simple; we look it up in the t tables in the

---

*Technical note: Actually, a number of liberties were taken with this explanation. In reality, t-values were derived empirically by W. Gossett. He started with a known mound-shaped population of values, then literally took thousands of random samples of size $n = 2$. For each, he calculated the number of standard deviations (z score) the sample $\bar{x}$ *appeared* to fall from μ (using the s of the sample for each calculation). These z scores (many of which were distorted because of the *under*estimations of $\sigma_{\bar{x}}$) were plotted into a histogram that he called the t distribution for $n = 2$. This distribution resembles a normal curve, but is more flat on top and spread out in the tails. He repeated this process for $n = 3$, $n = 4$, etc. In the case above, for instance, where $n = 5$, he noted 95% of the $\bar{x}$'s appear to fall within ±2.78 standard deviations of μ, based on a number of distorted estimates of $\sigma_{\bar{x}}$. He had also derived equations for the distributions from fundamental theory and used these empirical results to validate these equations. (For further details, refer to endnote 1.)

  Further technical note: W. Gossett in his original 1908 *Biometrika* article claimed the population shape can deviate quite far from normal before this would influence the predictive value of these t distributions.

back of the text. Since we are conducting a two-tailed hypothesis test at $\alpha = .05$, we look under *that* particular column, then down to degrees of freedom (df) of 4 (our sample size minus one, $5 - 1 = 4$), as follows:

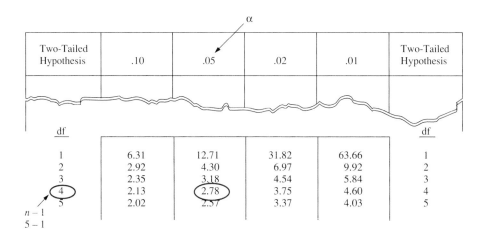

Degrees of freedom (or df): Although difficult to define completely,* let us say for our purposes, df defines a value that allows us to identify the correct sampling distribution and thus the proper cutoff value(s) for a given experiment. Essentially each df value represents a different sampling distribution.

Applications

Now let's see how small-sample testing works in two experiments. The first is a classroom experiment concerning social intelligence in children.

*Technical note: df is more formally defined as: "For any problem the number of degrees of freedom is the number of variables reduced by the number of independent restrictions on those variables," H. Walker and J. Lev, *Elementary Statistical Methods.* (New York: Holt, Rinehart, and Winston, 1969), p. 276. The concept was known to Carl Gauss (1826) but it wasn't until Sir Ronald Fisher's 1915 paper, "Frequency distribution of the values of the correlation coefficient in samples from an indefinitely large population," *Biometrika,* Vol. x, pp. 507–521, where Fisher applied n-dimensional geometry to its application that it gained widespread attention. The concept is quite advanced, but for those who wish a deeper understanding, refer to H. Walker, "Degrees of Freedom," *Journal of Educational Psychology* 31 (1940): pp. 253–269.

Example

Social IQ in young children, that is, a child's ability to properly assess nonverbal messages of teachers and peers (such as, tone of voice, body gesture, and facial expression) and properly assess social boundaries (such as, sensing how close to stand, responding at proper intervals, and smoothness in entering groups and in communicating emotions like anger and happiness) may be a more accurate measure than mental IQ in predicting later academic achievement and overall success throughout life, according to researchers at Harvard, University of Illinois, University of North Carolina, and other institutions.

Suppose Ms. Peach has her Lake County, Illinois, elementary school class of 9 slow students (underachiever class) tested on a social IQ scale, with the following results.

$$n = 9 \text{ underachiever students}$$
$$\bar{x} = 85.3 \text{ social IQ}$$
$$s = 11.7 \qquad \text{(assume a normal population)}$$

a. Test the hypothesis at $\alpha = .02$ that these students came from a population with social IQ, $\mu = 100$ (which is average for children of this age). Are the results significant?

b. If the data constitutes a valid random sample of underachiever students, what conclusions can be drawn? Briefly discuss validity.

Solution

Since a small-sample size, $n = 9$, was used, we must be careful that two conditions are first satisfied: (1) the population is at least somewhat mound shaped (it was stated, assume a normal population, so this condition is satisfied) and (2) if s is used to estimate σ, t scores must be used, not z scores (notice: we were *not* given σ, so s must be used to estimate σ, thus we must use t scores).

A hypothesis test consists of three fundamental sequences as follows.

Sequence   I. Set up initial conditions

State null hypothesis, the initial claim you H_0: $\mu = 100$ social IQ
wish to test.

State the alternative hypothesis. If H_0 proves H_1: $\mu \neq 100$ social IQ
false, what must we conclude?

State the risk of rejecting H_0 in error, the $\alpha = .02$ (2%)
level of significance, σ.

Sequence   II. Assume H_0 true, use α to establish cutoff(s)

Calculate

$\sigma_{\bar{x}}$: $\sigma_{\bar{x}} = \dfrac{\sigma}{\sqrt{n}} \approx \dfrac{s}{\sqrt{n}} \approx \dfrac{11.7}{\sqrt{9}} \approx \dfrac{11.7}{3} \approx 3.9$

(Note s was used to estimate σ above; a small-sample s will tend to underestimate σ, causing $\sigma_{\bar{x}}$ to be underestimated, too.)

Draw Curves

Using our above calculation, $\sigma_{\bar{x}} \approx$ 3.9, we estimate the spread of the $\bar{x}$ distribution. (Keep in mind, $\sigma_{\bar{x}}$ is probably *under*estimated, however we will compensate for this by using t scores.)

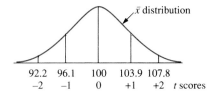

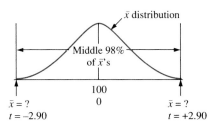

Establish Cutoffs using α

Our level of significance in this case is $\alpha = .02$ (2%) which in a two-tailed test implies we will accept the middle 98% of the $\bar{x}$'s as our boundary for accepting H_0 as true. In our t tables, we look up: two-tailed hypothesis, $\alpha = .02$, to obtain $t = \pm 2.90$ standard deviations. Remember: Look down the df (degrees of freedom) column to 8 (df $= n - 1 = 9 - 1 = 8$).

Note: Had this been a *large* sample, the boundaries would have been ±2.33 standard deviations instead of ±2.90.

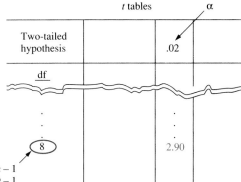

We solve using the z formula, only now we use the letter t in place of z. We can view the t score* as an adjustment to the z score (to compensate for the underestimations of $\sigma_{\bar{x}}$).

*Technical note: Actually, we are sampling from the t distribution (for $n = 9$), which is, essentially, a distribution of distorted z values. More specifically, it is a distribution of thousands and thousands of z scores describing how far in standard deviations the $\bar{x}$'s *appear* to fall from μ, based on distortions created by using small-sample s's to estimate $\sigma_{\bar{x}}$. In this example, the middle 98% of the $\bar{x}$'s appear to fall in the interval between ±2.90 standard deviations of μ. In reality, the $\bar{x}$'s are not falling ±2.90 standard deviations from μ, but the underestimations of $\sigma_{\bar{x}}$ make them appear so. These distorted z scores are called t scores. Explanation for math whizzes: Rearranging the z formula, we have $z\sigma_{\bar{x}} = \bar{x} - \mu$. If $\sigma_{\bar{x}}$ is underestimated, z must increase to keep $\bar{x} - \mu$ constant. This might be written: $z\uparrow$ $\sigma_{\bar{x}}\downarrow = \bar{x} - \mu$ (a constant value). Thus, $z\uparrow = t$. Put thousands of these distorted z's into a distribution = t distribution. Each sample size has a separate t distribution.

$$t_{\bar{x}} = \frac{\bar{x} - \mu}{\sigma_{\bar{x}}} \qquad\qquad t_{\bar{x}} = \frac{\bar{x} - \mu}{\sigma_{\bar{x}}}$$

$$-2.90 = \frac{\bar{x} - 100}{3.9} \qquad\qquad +2.90 = \frac{\bar{x} - 100}{3.9}$$

Solving for $\bar{x}$: $\bar{x} = 88.69$ $\qquad\qquad\qquad \bar{x} = 111.31$

The complete solution showing the cutoffs might appear as follows:

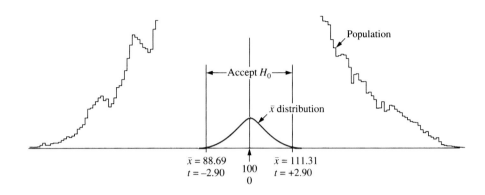

III. Accept or reject H_0 using your sample $\bar{x}$

Sequence

METHOD ONE
This method uses the actual value of the sample $\bar{x}$ (85.3 social IQ) in the decision-making process.

Recall: our sample results were as follows:
$n = 9$ students
$\bar{x} = 85.3$ social IQ

Criteria: Accept H_0 ($\mu = 100$ social I.Q.) if your sample $\bar{x}$ falls between the established $\bar{x}$ cutoffs of 88.69 and 111.31, otherwise reject.

Decision: Since our sample $\bar{x}$ (85.3) fell in the rejection zone, we reject H_0 and accept H_1, the alternative hypothesis ($\mu \ne 100$ social I.Q.).

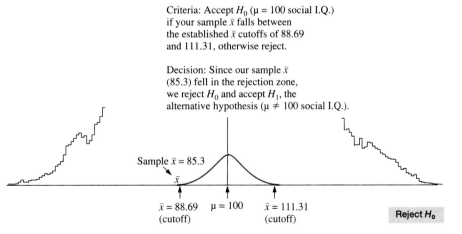

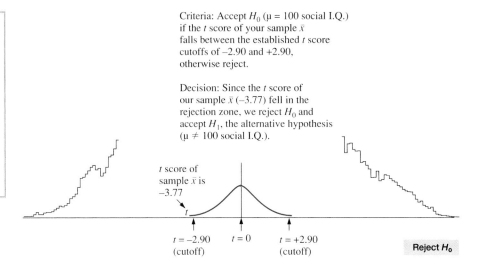

METHOD TWO

This method uses the t score of the sample $\bar{x}$ in the decision-making process. To use this method, however, we must first calculate the t score of our sample $\bar{x}$ (85.3 social IQ), as follows:

$$t = \frac{\bar{x} - \mu}{\sigma_{\bar{x}}} = \frac{85.3 - 100}{3.9}$$

$$t = -3.77$$

Criteria: Accept H_0 ($\mu = 100$ social I.Q.) if the t score of your sample $\bar{x}$ falls between the established t score cutoffs of -2.90 and $+2.90$, otherwise reject.

Decision: Since the t score of our sample $\bar{x}$ (-3.77) fell in the rejection zone, we reject H_0 and accept H_1, the alternative hypothesis ($\mu \neq 100$ social I.Q.).

Whether we use the actual value of the sample $\bar{x}$ or the equivalent t score, we will always make the same decision. In this case, we reject H_0. This implies we accept H_1: ($\mu \neq 100$).

Answer

a. The final answer may be presented using actual values or t scores.

Actual values   Since the sample average (sample $\bar{x} = 85.3$ social IQ) was outside the range (88.31 to 113.31) where we would most likely expect sample averages to fall if H_0 were true, we reject H_0. Therefore, we must

Accept H_1: $\mu \neq 100$ social IQ

t scores   Since the t score of the sample average (-3.77) was outside the range (-2.90 to $+2.90$) where we would most likely expect t scores of sample averages to fall if H_0 were true, we reject H_0. Therefore, we must

Accept H_1: $\mu \neq 100$ social IQ

The answer may also be expressed simply as:

The null hypothesis is rejected,

or

$t = -3.77$ (significant)

So, to answer part a, yes, the results are significant.

b. If the data constitutes a valid random sample of underachievers, the results provide evidence that on average underachievers score significantly below other children in their age group in the social skills observed in the experiment, which we labeled social IQ.

However, like many experiments in the social sciences and education, this experiment is replete with potential violations to validity, as follows.

Random selection: First, the study violates our most basic tenet of sampling: random selection. Intact groups (such as, classes, clubs, and residents of a building) often have common interests or ties and cannot be assumed to represent a cross section of the target population. For instance, this particular underachiever class may be located in a high-family-income district and all nine students may be from privileged families. In other words, this one underachiever class may very well *not* represent the general population of underachievers.*

Unfortunately, a good many statistical studies in education, marketing, psychology, medicine, and other fields still employ this method of using intact groups, which is often why experiments in these fields that measure the same phenomenon vary so widely in results. Random selection must be assured if we are to use a sample as representative of a population. Intact groups do not constitute random selection (refer to chapter 1 for a more detailed discussion on random selection).

Other aspects of validity: Of course, even if random selection from the underachiever population can be assured, the potential for violations to other aspects of validity in such experiments is enormous. Essentially, we must guarantee internal and external validity, defined in regards to this experiment as follows:

Internal validity: the certainty that our observations were *accurate* and *reliable* measures of the social characteristics we set out to measure.

External validity: the certainty that our *methods and presence* in no way affected the true social behavior of the children.

Assuring internal and external validity in experiments of this nature is no easy task. Actually, this experiment and its inherent questions of validity open up the whole topic of problems besetting scientific investigation in the social sciences. Although much too broad a topic to address here, I will discuss it briefly, then recommend two books for further reference.

To begin with, let us say, "labels" are dangerous in any scientific study. In this experiment, we called a certain set of social characteristics, social IQ, when, in fact, we really do not know what we measured, except for a collection of social characteristics at a certain point in a child's life. In reality, what we *may* be measuring is merely adaptability to white middle-class behavior in America rather than a universal set of social intelligence that crosses all cultures and classes. And this social intelligence may be "learned" behavior rather than "inborn," thus the phrase IQ, which implies a natural capacity, may be misleading. But, be that as it may, whatever we label these characteristics, they first must be precisely defined and set forth at the beginning of the experiment so future researchers (or any reader for that matter) can properly assess and criticize your work.

---

*Intact groups may more effectively be used as *one* element in a random sample, say for instance, if we randomly selected 36 classes nationwide, in which case, one class offers one result in a random sample of $n = 36$ classes.

Now, once the characteristics we wish to measure are set forth, we must determine the best way to get an *accurate and reliable* measure of these characteristics. When personal judgment of an observer is involved, this is a difficult task. For instance, let's use how close a child should stand while talking. What objective scale can be used? Is the observer in any way biased? Would a child be rated the same from observer to observer? Does a rating of 120 on this characteristic mean twice the social savvy as a rating of 60? And what does "twice" the social savvy mean? Obstacles abound in this type of experiment.

Even if accurate and reliable measures can be attained (thus, ensuring internal validity), did the *methods* of the experiment or *presence* of the observer in any way alter the true social behavior of the child? Were the children observed in secret in their natural environment? Or were artificial situations enacted? How were the times chosen to observe the child—convenience to the observer, when the child is at play, when the child is upset? These are all variables that may substantially influence results.

For further reading on proper experimental technique for testing in psychology, education, and other fields where similar difficulties exist in obtaining random selection and controlling risks to validity, refer to D. Campbell and J. Stanley, *Experimental and Quasi-Experimental Designs for Research* (Boston: Houghton Mifflin Co., 1963) and L. Tyler and W. B. Walsh, *Tests and Measurements* (Englewood Cliffs, N.J.: Prentice-Hall, 1979). ■

Starting from Raw Data

Our second example concerns the cranial capacity (brain size) of Neanderthal man. The example is presented not only to demonstrate small-sample testing, but to demonstrate how a hypothesis test is performed starting from raw data.

Example ——————— Much of the 1800s was spent by certain medical doctors and scientists trying to prove the cranial capacity of modern Caucasian man was larger (and, by implication, of greater intellect) than earlier Caucasian groups or other cultural groups, such as, Mongolian, Semitic, Malay, American, African, etc. Many educated people in the 1800s accepted this without evidence, but nothing was more illuminating than the fossil skulls first dug up in 1856 of Neanderthal man (a cousin to our Cro-Magnum species but much older and more primitive, who roamed Europe and the Middle East from 200,000 B.C. to 30,000 B.C.). Suppose 11 such Neanderthal skulls yielded the following:

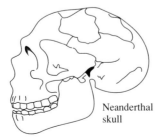
Neanderthal skull

**Neanderthal Skull
Cranial Capacity
(in cubic inches)**

85	91
89	94
93	90
90	88
91	86
	93

(assume a normal population)

a. Calculate $\bar{x}$ and s for this sample.
b. At a .05 level of significance, test the claim the sample came from a population with mean of 87.0 cubic inches *or less* (87.0 cubic inches is the average cranial capacity of modern Caucasian man). Are the results significant?
c. If the data constitutes a valid random sample from the Neanderthal population, what conclusions can be drawn? Briefly discuss validity.

Solution   a. We calculate $\bar{x}$ and s for this data, as follows:

Neanderthal Skull
Cranial Capacity

(in cubic inches)	$\bar{x}$	$x - \bar{x}$	$(x - \bar{x})^2$
85	90	-5	25
89	90	-1	1
93	90	3	9
90	90	0	0
91	90	1	1
91	90	1	1
94	90	4	16
90	90	0	0
88	90	-2	4
86	90	-4	16
93	90	3	9
$\Sigma x = 990$		$\Sigma(x - \bar{x})^2 = 82$	

$$\bar{x} = \frac{\Sigma x}{n} = \frac{990}{11}$$
$$\bar{x} = 90.0 \text{ cubic inches}$$

$$s = \sqrt{\frac{\Sigma(x - \bar{x})^2}{n - 1}}$$
$$s = \sqrt{\frac{82}{11 - 1}} = \sqrt{8.2}$$
$$s = 2.864 \text{ cubic inches}$$

We summarize the results for part a as follows:

$$n = 11 \text{ Neanderthal skulls}$$
$$\bar{x} = 90.0 \text{ cubic inches (cranial capacity)}$$
$$s = 2.86 \text{ cubic inches}$$

b. Since a small-sample size was used ($n = 11$), we must be careful that two conditions are first satisfied: (1) the population is at least somewhat mound shaped (it was stated, assume a normal population, so this condition is satisfied) and (2) since σ is not given and we must use s to estimate σ, we use t scores, not z scores.

A hypothesis test consists of three fundamental sequences as follows.

Sequence   I. Set up initial conditions

State null hypothesis, the initial claim you wish to test:	H_0: $\mu = 87.0$ *or less* ($\mu \leq 87.0$)
State alternative hypothesis. If H_0 proves false, what must we conclude?	H_1: μ is more than 87.0 ($\mu > 87.0$)
State the risk of rejecting H_0 in error, the level of significance, α.	$\alpha = .05$ (5%)

Sequence   II. Assume H_0 true, use α to establish cutoff(s)

Calculate

$\sigma_{\bar{x}}$: $\sigma_{\bar{x}} = \dfrac{\sigma}{\sqrt{n}} \approx \dfrac{s}{\sqrt{n}} \approx \dfrac{2.864}{\sqrt{11}} \approx \dfrac{2.864}{3.317} \approx .863$

Draw curves

Using our above calculation, $\sigma_{\bar{x}} \approx$ .86, we estimate the spread of the $\bar{x}$ distribution.

(Keep in mind, $\sigma_{\bar{x}}$ is probably *under*estimated, however we will compensate for this by using *t* scores.)

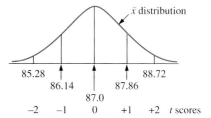

Since we are dealing with a one-tailed test (Note: $\mu = 87.0$ *or less*), we must assign the total α risk to one tail, in this case, to the right tail. Why the right tail? Because we reject H_0 only if our sample $\bar{x}$ falls significantly above 87.0 cubic inches.

Establish cutoffs using α

Our level of significance in this case is $\alpha = .05$ (5%), establishing the lower 95% of the $\bar{x}$'s as our region for accepting H_0 as true. In our t table, we look up a one-tailed hypothesis, $\alpha = .05$, to obtain $t = +1.81$. Remember: look down the df (degrees of freedom) column to 10 (df $= n - 1 = 11 - 1 = 10$).

Note: Had this been a large sample, the boundary would have been 1.65 standard deviations instead of 1.81.

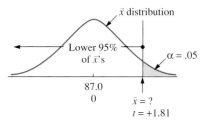

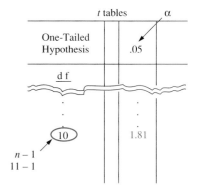

We solve using the z formula, only now we use the letter t in place of z. We can view the t score as simply an adjustment to the z score (to compensate for the uncertainty created by using small-sample s's in the calculation of $\sigma_{\bar{x}}$).*

$$t = \frac{\bar{x} - \mu}{\sigma_{\bar{x}}}$$

$$+1.81 = \frac{\bar{x} - 87.0}{.863}$$

Note: $(.863)(1.81) = 1.56$ (rounded); adding this to 87.0 gives *88.56* cubic inches

Solving for $\bar{x}$: $\bar{x} = 88.56$

The completed solution might appear graphically as follows:

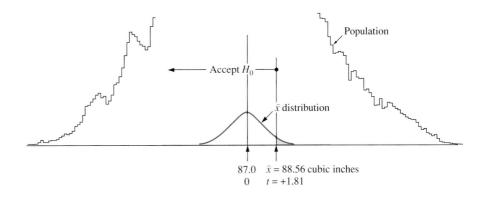

Sequence

III. Accept or reject H_0 using your sample $\bar{x}$

METHOD ONE
This method uses the actual value of the sample $\bar{x}$ (90.0) in the decision-making process.

RECALL: our sample results were as follows:
$n = 11$ skulls
$\bar{x} = 90.0$ cubic inches

Criteria: Accept H_0 ($\mu = 87.0$ or less) if your sample $\bar{x}$ falls below the established cutoff of 88.56 cubic inches, otherwise reject.

Decision: Since our sample $\bar{x}$ (90.0) fell in the rejection zone, we reject H_0 and accept H_1, the alternative hypothesis (μ is more than 87.0 cubic inches).

sample $\bar{x} = 90.0$

$\mu = 87.0$ $\bar{x} = 88.56$
(cutoff)

Reject H_0

*In statistics, we view ourselves as sampling values from a precisely defined distribution. In this case, we are sampling from the t distribution (for $n = 11$), which is, in essence, a histogram of thousands and thousands of z scores describing how far in standard deviations the $\bar{x}$'s *appear* to fall from μ, based on distortions created by using small-sample s's to estimate $\sigma_{\bar{x}}$. In this example, the lower 95% of the $\bar{x}$'s appear to fall below $+1.81$ standard deviations from μ.

METHOD TWO

This method uses the t score of the sample $\bar{x}$ in the decision-making process. To use this method, however, we must first calculate the t score of our sample $\bar{x}$ (90.0 cubic inches), as follows:

$$t = \frac{\bar{x} - \mu}{\sigma_{\bar{x}}} = \frac{90 - 87}{.863}$$

$$t = +3.48$$

Criteria: Accept H_0 ($\mu = 87.0$ or less) if the t score of your sample $\bar{x}$ falls below the established t score cutoff of $+1.81$, otherwise reject.

Decision: Since the t score of our sample $\bar{x}$ ($+3.48$) fell in the rejection zone, we reject H_0 and accept H_1, the alternative hypothesis (μ is more than 87.0 cubic inches).

t score of sample $\bar{x}$ is $+3.48$

$t = 0$ $t = +1.81$
(cutoff)

Reject H_0

Whether we use the actual value of the sample $\bar{x}$ or its equivalent t score, we will always make the same decision, in this case, we reject H_0. This implies we accept H_1 (μ is more than 87.0 cubic inches).

Answer

a. $\bar{x} = 90.0$; $s = 2.864$

b. The final answer may be presented using actual values or t scores.

Actual values   Since the sample average (90.0 cubic inches) was *above* the range where we would expect sample averages to fall if H_0 were true (up to 88.56 cubic inches), we reject H_0. Therefore, we must

> Accept H_1: μ is more than an 87.0 cubic inch cranial capacity

t scores   Since the t score of the sample $\bar{x}$ ($+3.48$) was *above* the range where we would expect t scores to fall if H_0 were true (up to $+1.81$), we reject H_0. Therefore, we must

> Accept H_1: μ is more than an 87.0 cubic inch cranial capacity

The answer may also be expressed simply as:

> *The null hypothesis is rejected,*

or

> $t = +3.48$ (significant)

So, to answer part b, yes, the results are significant.

c. If the data constitutes a valid random sample of Neanderthal skulls, the results indicate Neanderthal man had a larger brain size than modern Caucasian man. This, by the way, is true according to numerous skull findings over the past century and a half. Cro-Magnum man (our direct ancestors), according to accumulated evidence, also had larger cranial capacity.

Validity: The primary issue here seems to be, again, random selection. If all eleven skulls came from the same fossil site, this does *not* constitute random selection. Coming from one fossil site, all the skulls may have been from members of one family or clan and it is not unusual for members of the same family or clan to have similar biological traits (e.g., large heads). In other words, validity would be more assured if the skulls were randomly selected from, let's say, a great many Neanderthal skulls discovered at several widely scattered sites.

Other risks to validity seem minimal. *Accurate and reliable* measures of cranial capacity can be achieved with properly calibrated scientific instruments, of course barring researcher mistakes, shoddy technique, or questions of honesty.

Historical discussion: This and other evidence led scientists in the late 1800s to conclude that brain size does not determine intelligence. Some other evidence was women on average have 5 to 10 cubic inches less cranial capacity than men and larger people tend to have larger brains. Perhaps one final deciding factor was the death of Carl Gauss in 1855 (universally acclaimed mathematician, scientist, and genius who, by the way, is responsible for the discovery and validation of much of the work you've studied in the last few chapters, specifically, the normal curve, standard deviation, and the central limit theorem). Upon autopsy, it was discovered Gauss's brain was near average in size. The theory that a large brain produces great intellect soon thereafter began to crumble. So, even in his death, the great Carl Gauss contributed to the advancement of knowledge. Although of near average brain size, he was a true giant among men.

One last word in Gauss's defense: Upon autopsy of his brain, it was noted, however, the surface of Gauss's brain was more richly textured than the average man's brain, with many more folds and crevices, as illustrated below.

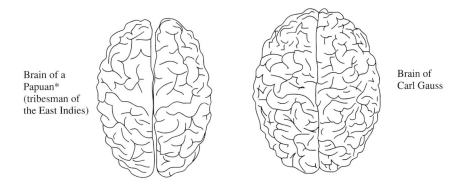

Brain of a Papuan* (tribesman of the East Indies)

Brain of Carl Gauss

*From E. A. Spitzka, *Transactions of the American Philosophical Society* 21 (1907): 175–308, as presented in S. J. Gould, *The Mismeasurement of Man* (New York: W. W. Norton Co., 1981).

Perhaps a brain is like a radiator. A radiator with more folds and crevices (thus, more surface area) radiates more heat. Perhaps a brain with more folds and crevices radiates more intellect. For further reading on these topics of intelligence and cranial measurement, refer to S. J. Gould, *The Mismeasurement of Man.*

Summary

Hypothesis Test

A test designed to prove or disprove some initial claim. This initial claim is referred to as the null hypothesis and denoted as H_0.

   We begin any hypothesis test by assuming the initial claim, the null hypothesis (H_0) is true. The second step is to establish a range of values where we would expect sample results (in this case, $\bar{x}$'s) to fall if H_0 were true. If the sample $\bar{x}$ of your experiment falls in this range, we merely accept H_0, otherwise we reject.

Physical Layout for Hypothesis Test (Large Sample, n ≥ 30)

Both the population and $\bar{x}$ distributions have the same mean, μ, established by the null hypothesis, H_0.

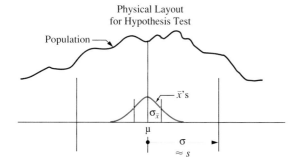

Physical Layout
for Hypothesis Test

   The spread of the population, σ, can be estimated by s, the spread of the sample data. And the spread of the $\bar{x}$ distribution, $\sigma_{\bar{x}}$, is calculated using the central limit theorem formula,

$$\sigma_{\bar{x}} = \frac{\sigma}{\sqrt{n}} \approx \frac{s}{\sqrt{n}}$$

where s, the spread of the sample data, can be used to estimate σ, the spread of the data in the population.

Establishing Cutoffs for Accepting H_0

The cutoffs are established using the level of significance (α risk) you are willing to accept in the experiment. For instance, if you establish $\alpha = .05$, you are willing to accept a 5% risk that when H_0 is true, you will reject it in error (refer to chapter 6 for a full discussion of errors).

   For a two-tailed test, $\alpha = .05$ implies $2\frac{1}{2}\%$ of the risk is placed in each of the two extreme tails, establishing regions where you would reject H_0.

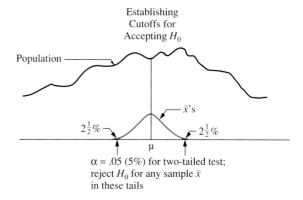

Establishing
Cutoffs for
Accepting H_0

$\alpha = .05$ (5%) for two-tailed test; reject H_0 for any sample $\bar{x}$ in these tails

For a one-tailed test, the entire α risk is placed in one tail and we reject H_0 if our sample $\bar{x}$ falls in that one tail.

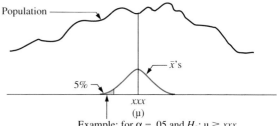

Example: for $\alpha = .05$ and H_0: $\mu \geq xxx$, we would reject H_0 for any $\bar{x}$ in the extreme left (shaded) tail

Determining which tail in which to place the α risk is dependent on how H_0 is stated. For example, if H_0 is stated, $\mu = xxx$ *or more,* this requires the entire α risk be placed in the extreme left tail (where you would reject H_0, shown in the previous sketch).

Control Charts

A control chart provides a clear visual history of a repetitive test. Essentially, cutoffs are established in a hypothesis test (using actual values or z scores) and the graph rotated $\frac{1}{4}$-turn counterclockwise, extending the cutoff lines to the right. This provides a clear on-going space to plot several sample $\bar{x}$'s. These $\bar{x}$'s, represented as dots in the sketch below, are often connected to each other by line segments as shown.

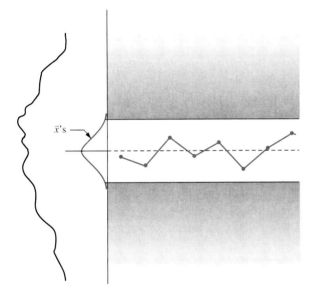

Small-Sample Hypothesis Testing (n < 30)

Small samples ($n < 30$) can be effectively used in hypothesis testing provided two conditions are satisfied, namely

1. The population from which you sample is normally distributed or at least somewhat mound shaped. Generally, the smaller your sample size, the more critical this restriction, and

2. If we use s to estimate σ (which we almost always do) in the calculation of $\sigma_{\bar{x}}$, we must use a t score and not a z score to define the number of standard deviations the $\bar{x}$'s would be expected to fall from μ.

Discussion of condition 1: Although mild to moderate violation of this normal-population condition can often be tolerated with little effect on the validity of the test, severe departure (such as when a population is extremely skewed) can seriously compromise the test's validity.

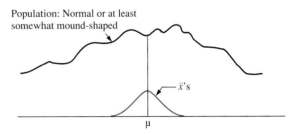

Discussion of condition 2: Because of the tendency of small-sample s's to underestimate σ, we use a t score at the cutoffs (and not a z score), which essentially compensates for this tendency and allows us to maintain the validity level (α risk) of the test.

The t score values can be obtained from the t tables in back of the text (df $= n - 1$). Once these two conditions are satisfied, the small-sample hypothesis test is conducted in a manner much like any hypothesis test. For a two-tailed hypothesis test, the layout would be as follows.

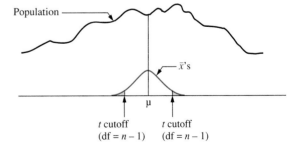

Exercises

Note that full answers for exercises 1–5 and abbreviated answers for odd-numbered exercises thereafter are provided in the Answer Key.

7.1   A supplier claims the average thickness (diameter) of its fiber-optic thread is .560 mm (no more, no less), per your specifications. You receive a shipment but before you accept it, you take a random sample, with the following results:

$$n = 225 \text{ measurements}$$
$$\bar{x} = .553 \text{ mm}$$
$$s = .030 \text{ mm}$$

a. Test the supplier's claim at a .01 level of significance.
b. Are the results "significant"? Would you accept or reject the shipment?

7.2   Elavil is a powerful sedating drug prescribed by psychiatrists for the treatment of depression; however, dosages must be minimized to reduce side effects. A leading health organization claims the minimum effective dosage nationwide is on average 140.0 mg/day *or less*. A manufacturer of the drug decides to test the claim with the following sample results:

$$n = 900 \text{ patients}$$
$$\bar{x} = 141.6 \text{ mg/day minimum effective dosage}$$
$$s = 48.2 \text{ mg/day}$$

a. Test the health organization's claim ($\mu \leq 140.0$ mg/day) at $\alpha = .04$.
b. Are the results "significant"? Do we have evidence to refute the health organization's claim?

7.3
a. For exercise 7.1, set up a control chart.
b. On this control chart, plot the shipment result from exercise 7.1, $\bar{x} = .553$ mm, along with additional shipment results: $\bar{x} = .558$ mm, $\bar{x} = .551$ mm, and $\bar{x} = .559$ mm. Indicate the results that are significant.
c. For exercise 7.2, set up a control chart.

d. On this control chart, plot the sample results from exercise 7.2, $\bar{x} = 141.6$ mg/day, along with the results of subsequently conducted studies: $\bar{x} = 138.7$ mg/day, $\bar{x} = 137.1$ mg/day, and $\bar{x} = 140.9$ mg/day. Indicate the results that are significant.

7.4   Social IQ in children (the ability to read subtle, nonverbal clues and accurately assess social boundaries) may be a better predictor of success in life than mental IQ according to recent studies.

Suppose Mrs. Berry has her Raleigh-Durham elementary school class of 12 gifted (high mental IQ) students tested on a social IQ scale, with the following results:

$$n = 12 \text{ students}$$
$$\bar{x} = 110.8 \text{ social IQ}$$
$$s = 16.3 \qquad \text{(assume a normal population)}$$

a. Test the hypothesis at $\alpha = .02$ that these students came from a population with social IQ, $\mu = 100$ (which is average for children of this age). Are the results significant?
b. If the data constitutes a valid random sample of gifted students, what conclusions can be drawn? Briefly discuss validity.

7.5   Anthropologists have long claimed it is not necesarily the size of the brain that determines intelligence. To prove their point, measurements are taken of the cranial capacity of $n = 10$ Neanderthal skulls (a species of primitive man living during the period from 200,000 B.C. to 30,000 B.C.) with the following results:

(In cubic inches)

87	98
88	87
95	84
91	95
89	96 (assume a normal population)

a. Calculate $\bar{x}$ and s for this sample.
b. At a .05 level of significance, test the claim that the sample group came from a population with a mean

of 87.0 cubic inches *or less* (87.0 cubic inches is the approximate average cranial capacity of modern man). Are the results significant?

c. If the data constitutes a valid random sample of Neanderthal skulls, what conclusions can be drawn? Briefly discuss validity.

7.6 Bloomindorfs Department Store in Manhattan hired a new specialist to redo their autumn fashion windows. The specialist is known for her exquisite coordinations of clothing, antique furniture, and accessories (all of which are sold at Bloomindorfs).

After setting up the new displays, Bloomindorfs wished to test whether any change had occurred in the number of customers entering the store per hour. Suppose it is known through electronic counters that the average number of customers for this store is 212 per hour and that a random sample now produces the following data.

$$n = 30 \text{ one-hour intervals}$$
$$\bar{x} = 231 \text{ customers per hour}$$
$$s = 82 \text{ customers per hour}$$

At a .05 level of significance, does the above data imply the new window displays have affected the average number of customers per hour entering the store? In other words, test the hypothesis, $\mu = 212$ per hour.

7.7 Excessive weight gain is a fear of those who wish to quit smoking, according to an article in the *New England Journal of Medicine*. Data gathered from the 1970s and 1980s indicate that after two years men gain on average 6.0 lb and women on average 8.0 lb more than those who continue to smoke.

Suppose a group of epidemiologists at the University of Rochester conducted the following research to see if these results still hold true in the 1990s.

Men: $n = 110$
        $\bar{x} = 6.8 \text{ lb increase}$
        $s = 2.7 \text{ lb}$

Women: $n = 90$
        $\bar{x} = 7.4 \text{ lb increase}$
        $s = 3.5 \text{ lb}$

a. At a .02 level of significance, test the claim that men still gain on average 6.0 lb.
b. At a .04 level of significance, test the claim that women still gain on average 8.0 lb.
c. Refer to part a. Suppose three such studies were conducted in the 1990s, yielding $\bar{x} = 6.8$, $\bar{x} = 6.23$, and $\bar{x} = 6.47$. Set up a control chart demonstrating this.
d. Briefly discuss validity.

7.8 Pig farmers in Canada are trying to sell pork as the "other white meat besides chicken" (according to an article in *Western Producer*), claiming to have evolved a leaner, more efficient species of pig. One of the claimed advantages of this new species is that it takes substantially less time for a pig to reach final market weight (thus saving a farmer money in feeding and tending).

Suppose researchers at Texas Agriculture decided to test the claim, with the following results:

$$n = 80 \text{ randomly selected new-species pigs}$$
$$\bar{x} = 157 \text{ days (birth to final market weight)}$$
$$s = 11.5 \text{ days}$$

The old species took on average 170 days to reach final market weight. At a .03 level of significance, test the hypothesis that the average number of days to reach final weight for this new species of pig is the same as the old species, $\mu = 170$ days. (Notice we test the sample $\bar{x}$ against the established norm, in this case, that pigs currently take on average, $\mu = 170$ days to reach final market weight; it is common to use the established norm to set the hypothesis.)

7.9 Growth hormone administered to short children is thought to increase height beyond expected levels, according to published data (*Lancet* 336:1331–1334). Suppose research yielded the following results.

$$n = 38 \text{ children}$$
$$\bar{x} = 1.2 \text{ inch growth (beyond expected)}$$
$$s = .84 \text{ inches}$$

Generally in such experiments we assume a null hypothesis of "no change," in this case, 0'' growth (beyond expected). In other words, it is customary in scientific investigations to start with the assumption that the treatment is ineffective. In this case, that a child taking the hormone will grow, on average, no more or no less than expected for that time period. We must *prove* otherwise.

At $\alpha = .01$, test the claim that this sample was taken from a population of $\mu = 0''$ growth (beyond expected).

7.10 A toiletry manufacturer claims their bottles contain *at least* 9.00 oz. of bath lotion (as stamped on the label). Suppose the Federal Trade Commission (FTC) investigated by randomly sampling 49 bottles, with the following results:

$$n = 49 \text{ bottles}$$
$$\bar{x} = 8.94 \text{ oz.}$$
$$s = .12 \text{ oz.}$$

At a .02 level of significance, does the FTC have legal grounds to proceed against the company on the unfair practice of short selling? In other words, test the claim, μ is at least 9.00 oz. ($\mu \geq 9.00$).

7.11 Certain nutritionists are outraged over the apparent lack of concern among physicians about so-called moderate cholesterol levels, claiming the average heart attack victim's cholesterol level is 230 mg/dl *or less*.

Suppose this prompted two studies, yielding the following results:

First Study: $n = 80$ heart attack victims
$\bar{x} = 226.0$ mg/dl cholesterol level prior to heart attack
$s = 37.3$ mg/dl

Second Study: $n = 67$ heart attack victims
$\bar{x} = 241.1$ mg/dl cholesterol level prior to heart attack
$s = 32.5$ mg/dl

a. Use the data in the first study to test the claim, $\mu \leq 230$ mg/dl ($\alpha = .01$).
b. Use the data in the second study to test the claim, $\mu \leq 230$ mg/dl ($\alpha = .05$).
c. What might account for the marked difference in results from the two studies?

7.12 Bypass surgery for obesity is increasingly becoming an option for those who have failed more moderate weight controlling strategies. In the procedure, vast tracts of the stomach are stapled off leaving a small pouch and narrow pathway leading to the intestines. Claims of weight losses have averaged over 120 lb (monitored two years after surgery).

Suppose researchers at the University of California tested this claim with the following results:

$n = 73$ patients monitored two years after surgery
$\bar{x} = 108$ lb weight loss
$s = 23$ lb

At $\alpha = .01$, test the claim, $\mu \geq 120.0$ lb weight loss.

7.13 To qualify for poverty funds, legislators in a particular district in Philadelphia had to show average household income for a family of four was $11,809 *or below*. A study was commissioned yielding the following:

$n = 53$ households in district
$\bar{x} = \$12,053$ annual household income
$s = \$4,320$

a. At $\alpha = .03$, does this particular legislative district qualify for poverty funds?
b. To reduce the possibility of falsely disqualifying a district, should a .03 or .01 level of significance be used? Briefly explain.

7.14 After the theft of several masterpieces, a New England Art Museum was quite concerned that the night security guard was diligent in performing his rounds. One way to monitor this was to ensure it took on average 21.0 minutes (no more, no less) to complete a known series of checks. The security guard was randomly clocked, yielding the following:

$n = 12$ observations
$\bar{x} = 18.6$ minutes
$s = 4.3$ minutes   (assume a normal population)

a. At $\alpha = .01$, test the claim that the average time it takes for the security guard to make the rounds is 21.0 minutes ($\mu = 21.0$).

b. At α = .10, test the same claim, μ = 21.0.

c. To reduce the possibility of falsely accusing the guard of not adequately performing rounds, should a .01 or .10 α level be set? Briefly explain.

7.15 Jessica, a new recruit at a local army base in Georgia, claims her average time to clean her M16 rifle to pass inspection is "a cool 11.7 min." Her buddies, suspecting an unsubstantiated brag, decide to test her claim and randomly clocked her (in secret) on 6 attempts, yielding:

n = 6 rifle cleanings
$\bar{x}$ = 13.57 minutes
s = 3.2 minutes   (assume a normal population)

a. At α = .05, test the claim that the average time it takes Jessica to clean her M16 rifle to pass inspection is at most 11.7 minutes ($\mu \leq 11.7$).

b. At α = .025, test the same claim, $\mu \leq 11.7$.

7.16 A rare Assyrian coin minted well over 2000 years ago was claimed to have contained *at least* 3.2 grams of gold, on average. A museum curator managed to locate 8 such coins and assessed their gold content at:

2.9, 3.5, 3.0, 3.2, 3.3, 3.0, 2.7, 3.2   (assume a normal population)

a. At α = .01 level of significance, test the claim $\mu \geq 3.2$.

b. If the data constitutes a valid random sample of these Assyrian coins, what conclusions can be drawn? Briefly discuss validity.

7.17 Children who are abused or severely neglected have lower intelligence and an increased risk of depression, drug abuse, and suicide, according to researchers at State University of New York/ Albany and at University of Minnesota (*New York Times*, February 18, 1991, p. A11).

Suppose the American Association for the Advancement of Science, in an effort to substantiate these claims, sponsored a study in the Albany area, which monitored several randomly selected children. Of these, seven turned out to be abused or severely neglected, yielding the following:

Change in IQ: -6, -17, -12, -15, -9, -7, and -11 points
(assume a normal population)

a. In such experiments, we assume μ = 0 change in IQ. In other words, it is customary in scientific investigations to start with the assumption the factors being studied (abuse or neglect) do not affect our measured variable (IQ). We must *prove* otherwise.

At α = .02, test the hypothesis of μ = 0 change in IQ.

b. If the data constitutes a valid random sample, what conclusions can be drawn? Briefly discuss validity.

7.18 Bone loss from prolonged space travel can be a serious problem, according to researchers at Pennsylvania State University, especially in the leg and spine regions. Suppose some estimate, on average, a 3% *or more* bone loss from extended space travel and the National Aeronautic and Space Administration (NASA) decided to conduct research on nine astronauts, yielding the following:

Bone Loss in Leg and Spinal Regions From Extended Space Travel (In Percentage Loss)

1.2	2.0	5.0
2.9	2.4	1.7
3.2	0.0	4.1

(assume a normal population)

a. At α = .05, test the claim of μ = 3.0% *or more* bone loss.

b. If the data constitutes a valid random sample, what conclusions can be drawn? Briefly discuss validity.

Endnotes

1. Historical endnote on W. Gossett: Gossett was part of a group of young university scientists in 1899 appointed to the brewing staff at the Guinness Breweries in Dublin. He tried to apply statistical techniques to experiments at the brewery and soon grew concerned using small-sample s's to estimate σ in experiments involving the quality of raw material (barley, hops, etc.), and in production tests and in finished-product tests.

This and other concerns led to a meeting with Karl Pearson in 1905. Gossett spent part of the academic year (1906–1907) at Pearson's Biometric Laboratory in London. From this stay, came a number of original papers, including the famous "Probable Error of a Mean" paper in 1908, in which he presented the t distributions for small-sample testing, which Fisher later incorporated into his ANOVA tables. We owe much to the original work of W. Gossett who published under the pen name, Student (assumedly referring to being a student of Karl Pearson).

Some early history: The t score distributions were derived empirically by W. Gossett and first published in an article, "The Probable Error of a Mean," which appeared in *Biometrika*, VI, pp. 1–25, in 1908, under the pen name, Student. Gossett was a British statistician and advisor to the Guinness Breweries in Dublin, Ireland. The Guinness Company forbade employees from publishing the results of research, however, the firm relaxed this ruling in Gossett's case to allow him to publish under a pen name.

Gossett derived the t distributions empirically using published data of body measurements (height and left middle finger) of 3000 criminals, from which he repeatedly selected small random samples. He successfully applied the results to other published data taken from (1) the *Journal of Physiology* (1904), which showed the effects of optical isomers of hyoscyamine hydrobromide in producing sleep, and data taken from (2) the *Journal of the Agricultural Society* concerning the causes which lead to the production of hard (glutinous) wheat and soft (starchy) wheat.

Gossett's article was a ground-breaking achievement. In many experiments, only small samples can be used, which Gossett cited as, "some chemical, many biological and most agricultural and large scale" experiments, and these had been outside the range of statistical enquiry, that is, up until his research.

See endnotes 1 and 10 in chapter 10 for more on W. Gossett.

Confidence Intervals

T he remarkable thing about the central limit theorem is that it works even for *non*normal populations. Say for instance your population is skewed, bimodal, or any shape that's not too uncommon, the $\bar{x}$'s will distribute normally around μ—*as long as your sample size is 30 or more ($n \geq 30$).*

Visually, it might appear as follows.

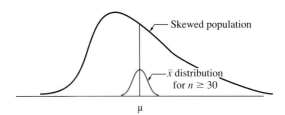

8.1 Confidence Interval Estimate of μ

When little or nothing is known about μ, the population average, a random sample may be drawn from the population and the sample average, $\bar{x}$, used as a central estimator in establishing a range of values believed to contain μ. This range of values is called the confidence interval estimate of μ and is accompanied by a probability that μ will indeed be within this range. Let's explore the rationale.

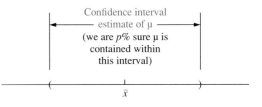

Suppose it is known through comprehensive studies that American women, ages 18–34, heights 5'3''–5'7'', weigh on average $\mu = 130.0$ lb with standard deviation $\sigma = 12.0$ lb and the shape of the population is unknown.

According to the central limit theorem, then, if we were to continually take random samples, with 144 women in each sample, and calculate the average weight ($\bar{x}$) for each of these samples, the sample averages ($\bar{x}$'s) would gradually build into the shape of a normal or near normal distribution around μ (no matter what the shape of your population), with standard deviation, $\sigma_{\bar{x}} = \sigma/\sqrt{n} = 12.0/\sqrt{144} = 1.00$ lb, as follows.

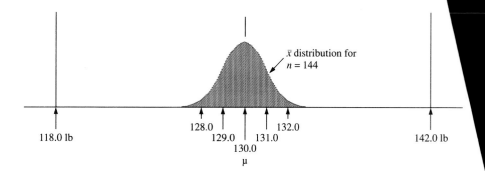

Now, to fully understand confidence intervals, let's play an imaginary game together.* For a moment, pretend to be a sample average, $\bar{x}$. I know this sounds silly. But just for a moment, pretend you are the average weight of 144 randomly selected women and your calculated value is, say for example, 129.0 lb. Refer to the small histogram above. From *your viewpoint,* as $\bar{x} = 129.0$ lb, μ would appear on which side of you, how many units away? Don't read on. Try it.

The answer is: μ would appear on your right-hand side, exactly 1.0 lb away, and might be depicted as follows:

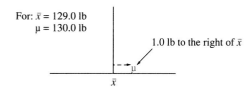

Now let's pretend again. This time, you are a second $\bar{x}$, a different sample average of 144 women, only this time you are calculated to be $\bar{x} = 131.5$ lb. Again, we wish to locate μ *from your viewpoint,* so now μ would appear on which side of you and how many units away? Refer to the small histogram at the beginning of this section.

*Although formally defined in a somewhat different manner by J. Neyman (1937), we will choose to model confidence intervals on a more intuitive footing. Neyman's argument is presented later in this section.

The answer is: μ would appear on your left-hand side, 1.5 lb away, as depicted below.

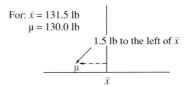

Wait a minute. How could μ be 1.0 lb to the right in the first example and now 1.5 lb to the left? μ doesn't move around. That's true, μ is fixed. It's the $\bar{x}$'s that are moving around. But remember, we are playing an imaginary game; so just imagine yourself, one $\bar{x}$, *at the center of the page,* recording the position of μ from your viewpoint.

Okay, now we repeat the experiment a third time. This time you are $\bar{x} = 129.6$ lb, again *at the center of the page.* Guess where μ will appear now? Refer to the small histogram presented at the beginning of this section. Don't read on.

The answer is: μ would be 0.4 lb away, on your right-hand side, as depicted below.

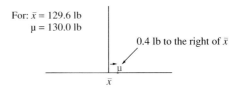

Important: note that this distance, the +0.4 just derived, can be easily calculated using the following formula:

$$\boxed{\text{Distance} = \mu - \bar{x}}$$

$$
\begin{aligned}
\text{Distance} &= \mu - \bar{x} \\
&= 130.0 - 129.6 \\
&= +0.4
\end{aligned}
$$

Now where is this all leading? Well, suppose we put the results of all three experiments above on *one* graph, as follows:

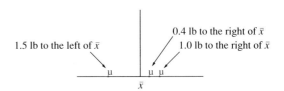

Next we run the experiment several more times, each time recording the location of μ (from $\bar{x}$'s viewpoint).

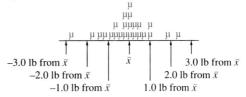

Notice how these locations of μ begin to "pile up" on the same values, mostly within 1.0 lb of $\bar{x}$, the sample average. Why is that? Well, since most $\bar{x}$'s are within 1.0 lb of μ (according to the small histogram at the beginning of this section), most likely you would select one of these $\bar{x}$'s. In which case, μ would most likely be within 1 lb of you.

Yes, some $\bar{x}$'s are 2.0 lb from μ. Not many, however. And almost none are 3.0 lb from μ. So, μ is most likely to be within 1.0 lb, far less likely to be 2.0 lb away and almost never 3.0 lb away. Which leads us to the following.

If we run wild and calculate thousands and thousands of $\bar{x}$'s, with each $\bar{x}$ you look around and record the location of μ, the resultant distribution would build into the following:

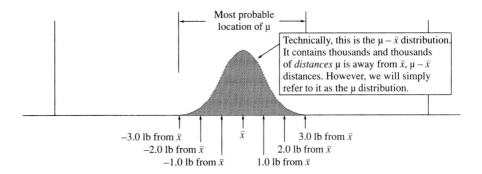

Are you saying, we can predict that μ will be between -3.0 lb and $+3.0$ lb of whatever sample average ($\bar{x}$) we obtain? Yes, and we can predict it with 99.7% certainty.* Do you mean it's impossible for μ to be 5.0 lb away? Not impossible, but highly improbable. In all the thousands of random samples actually taken, not one $\mu - \bar{x}$ distance came even near 5.0 lb away. In fact, most $\mu - \bar{x}$ distances were within 1.0 lb of the sample average.

Examine for a moment this histogram, which we will label for convenience the μ distribution. Did you notice its shape? Yes, it is a *normal distribution*. Only this time with $\bar{x}$ at the center.

---

*± 1 lb represents ± 1 standard deviation for this problem, as we will see, thus ± 3 lb represents ± 3 standard deviations, which includes 99.7% of the values.

Did you also notice the standard deviation. Yes, 1.0 lb, which is precisely the standard deviation of the original $\bar{x}$ distribution (shown at the beginning of this section). In fact,

> The $\mu - \bar{x}$ (or μ) distribution is identical in shape and size to the $\bar{x}$ distribution.

So it should not be a surprise when I say, to calculate the standard deviation of the μ distribution, σ_μ, we use the same formula we used to calculate the standard deviation of the $\bar{x}$ distribution, as follows:

$$\sigma_\mu = \sigma_{\bar{x}} = \frac{\sigma}{\sqrt{n}} = \frac{12.0}{\sqrt{144}} = \frac{12.0}{12} = 1.0 \text{ lb}$$

> To calculate the standard deviation of the μ distribution, we use
> $$\sigma_\mu = \sigma_{\bar{x}} = \frac{\sigma}{\sqrt{n}}$$
> Keep in mind, it's the *distances* from $\bar{x}$ to μ that are recorded into the μ distribution (not the value of μ itself, which is a fixed quantity).

To recap: confidence interval estimates are used when little or nothing is known about μ, the population average. Now we have a way of establishing a range of values believed to contain μ by using $\bar{x}$, the average of one sample drawn from that population, as a central estimate, and $\sigma_{\bar{x}}$ as a measure of spread.*

Let's see how it works.

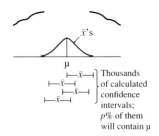

Thousands of calculated confidence intervals; $p\%$ of them will contain μ

*Jerzy Neyman (1937) was probably the first to present a complete argument for confidence intervals, although, clearly, the concepts of point and interval estimation were in common practice at least as far back as Laplace and Poisson, prior to 1840 (S. Stigler, *History of Statistics*, Cambridge: Belknap Press of Harvard University Press, 1986, p. 187). In fact, several statisticians prior to 1937 (Sir Ronald Fisher included) seem to assume the existence of confidence intervals intuitively, seemingly in a manner much like that demonstrated above. Neyman's formal argument was presented in "Outline of a Theory of Statistical Estimation Based on the Classical Theory of Probability," *Phil. Trans. Royal Society (London)*, Series A 236, pp. 336–380, 1937. Essentially he demonstrated that repeated samples of size n from a population could be used to produce intervals with different lower and upper limits, such that $p\%$ of the intervals would contain μ (actually Neyman's argument was quite general. He referred to estimating parameters which are unknown and constant; in our case the parameter is μ). This is demonstrated in the diagram. Essentially, if thousands and thousands of samples are drawn from a population and, say for instance, a 95% confidence interval calculated for each, then even though we get different upper and lower confidence limits, 95% of the intervals would contain the value of μ. However, the remaining 5% of the intervals would not contain μ.

8.2 Applications

Although broadly applied in research, the confidence interval estimate of μ is also one of the most frequently used statistical procedures in everyday life. Let's see how it works in the following two examples.

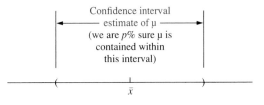

Confidence interval
estimate of μ
(we are p% sure μ is
contained within
this interval)

$\bar{x}$

Example

Suppose *Sturm*, a European magazine, is running a series of articles comparing European women to American women. It was already known through comprehensive studies that the average weight of American women (ages 18–34, 5′3″–5′7″) was μ = 130.0 lb, so no further testing was needed for American women.

However, the editors were dissatisfied with existing data for European women and decided to take a random sample of 144 European women (ages 18–34, 5′3″–5′7″) to *estimate* their average weight.

Sample results yielded: $\bar{x}$ = 128.6 lb

Find the 95% confidence interval estimate of μ. That is, between what two values can we be 95% sure of locating μ, the true average weight of these European women? (Assume σ = 12.0 lb.)

Solution

Notice that μ, the average weight of European women, is unknown. This is a classic case in which confidence interval estimates are used. Your population μ is unknown and we will use our sample average, $\bar{x}$ = 128.6 lb, to construct an interval in which μ might be found. Calculating a confidence interval proceeds in three relatively easy steps, as follows:

First, use your sample average, $\bar{x}$ = 128.6 lb as the center of your μ − $\bar{x}$ (or μ) distribution. We know that in *any* normal distribution, 95% of the data is between $z = -1.96$ and $z = +1.96$, so we shade this interval as follows:

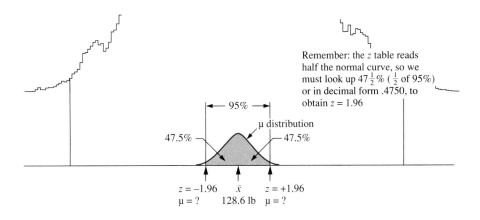

Remember: the *z* table reads half the normal curve, so we must look up $47\frac{1}{2}\%$ ($\frac{1}{2}$ of 95%) or in decimal form .4750, to obtain $z = 1.96$

95%

μ distribution

47.5%   47.5%

$z = -1.96$ $\bar{x}$ $z = +1.96$
μ = ?   128.6 lb   μ = ?

There is a 95% probability μ will be contained in the shaded interval. Notice that the precise location of μ is unknown. However, we do know an *interval* where μ is most likely to be found.

Second, calculate the standard deviation of the μ distribution using the formula,

$$\sigma_\mu = \sigma_{\bar{x}} = \frac{\sigma}{\sqrt{n}} = \frac{12.0}{\sqrt{144}} = \frac{12.0}{12} = 1.0 \text{ lb}$$

Third, we solve for the lower and upper limits of μ using the z formula. However, since $\bar{x}$ has replaced μ as the center of the distribution, to be technically correct, we must reverse $\bar{x}$ and μ in the z formula. In other words, $\bar{x} - \mu$ becomes $\mu - \bar{x}$. This gives us the correct minus and plus signs for our calculations, as follows:

$$z = \frac{\mu - \bar{x}}{\sigma_{\bar{x}}} \qquad\qquad z = \frac{\mu - \bar{x}}{\sigma_{\bar{x}}}$$

$$-1.96 = \frac{\mu - 128.6}{1.0} \qquad 1.96 = \frac{\mu - 128.6}{1.0}$$

Solving for μ (and rounding to one decimal place), we get*

$$\mu = 126.6 \text{ lb} \qquad\qquad \mu = 130.6 \text{ lb}$$

Graphically, the solution would appear as follows:

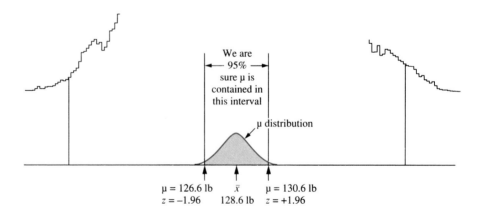

We are
← 95% →
sure μ is
contained in
this interval

μ distribution

$\mu = 126.6$ lb $\bar{x}$ $\mu = 130.6$ lb
$z = -1.96$ 128.6 lb $z = +1.96$

---

*Neyman (1937) referred to these as the lower and upper confidence limits. In our case, then,

$$\mu_{\text{lower}} = 126.6 \text{ lb} \qquad \mu_{\text{upper}} = 130.6 \text{ lb}$$

and we are $p\%$ sure (in our case, 95% sure) the true population mean, μ, is contained within these limits. We will forego use of the subscripts, lower and upper, since it is apparent from the values obtained. The lower and upper limits can also be calculated using the formulas:

$$\mu_{\text{lower}} = \bar{x} - z\,\sigma_{\bar{x}} \qquad \mu_{\text{upper}} = \bar{x} + z\,\sigma_{\bar{x}}$$

Answer

Based on the sample taken, $\bar{x} = 128.6$ lb, we are 95% sure the average weight of European women (ages 18–34, heights 5′3″–5′7″) is contained in the interval from 126.6 lb to 130.6 lb.   ▨

Notice that μ, the true average weight of these European women, is really unknown. All we can state is that μ is probably somewhere between 126.6 lb and 130.6 lb. But even of this, we are not 100% sure. However, we are 95% sure. Could μ = 132 lb? Yes, although it is not likely. Note that sample results allow us to construct an interval where μ is most likely to be found, but the interval does not pinpoint μ's precise location. But this is still a rather valuable piece of information, since prior to the study we may have had little or no idea of the true average weight of European women. At least now we can confidently state, we are 95% sure the *average* weight of European women is somewhere between 126.6 lb and 130.6 lb.

> ### Confidence Interval Estimate of μ
> A range of values believed to contain μ, and stated with a specific degree (or probability) of certainty.

Let me ask another question. Since it is known that American women (ages 18–34, heights 5′3″–5′7″) weigh on average μ = 130.0 lb, can the editors of *Sturm* state European women weigh less on average than equivalent American women? Based on the above study, the answer is no. According to the above, μ is probably somewhere between 126.6 lb and 130.6 lb, so European women (for these ages and heights) could very well weigh on average μ = 130.0 lb, the same as American women. But, then again, these European women could weigh on average μ = 129.0 lb or μ = 128.0 lb or μ = 127.0 lb. In other words, we are unsure of the precise location of μ. All we know is the average weight of European women (for these ages and heights) is most likely in the interval between 126.6 lb and 130.6 lb, and we are 95% sure.

But how can we avoid these "unsure" situations? We can't, unless of course we weigh every individual woman in Europe and calculate the real μ. However, we can make the estimation far more precise. For instance, had the editors of *Sturm* magazine used a much larger sample size, say for instance, $n = 2304$ women, then the $\bar{x}$'s would have clustered much closer to μ. In this situation, $\sigma_{\bar{x}}$ is reduced to $\frac{1}{4}$ lb (the original $\sigma_{\bar{x}}$ was 1 lb) and now the 95% confidence interval estimate of μ would calculate out to be 128.1 lb to 129.1 lb. In other words, for $n = 2304$ women and $\bar{x} = 128.6$ lb, we would be 95% sure the *true* average weight is located in the interval between 128.1 lb and 129.1 lb.

So, now with $n = 2304$ women and $\bar{x} = 128.6$ lb, the probability of μ being 130 lb is nearly impossible and *Sturm* could state with strong statistical evidence that European women (for these ages and heights) do indeed weigh less on average than American women. But unfortunately *Sturm*'s original study contained too few women ($n = 144$) to make this determination.

> Lesson to be learned: larger sample sizes more precisely pinpoint the
> location of μ. It is best to keep your sample size as large as possible.

Example ———————— A sociological society wished to *estimate* the average number of hours Japanese teenagers put into class and home study per week. A randomly selected sample of Japanese teenagers yielded the following results:

$$n = 400 \text{ Japanese teenagers}$$
$$\bar{x} = 57.7 \text{ hours of class and home study per week}$$
$$s = 6.0 \text{ hours per week}$$

Construct a 98% confidence interval for the *true* average number of hours Japanese teenagers put into class and home study per week (round final answer to one decimal place).

Solution Again we proceed in three steps.

First, use your sample average $\bar{x} = 57.7$ hours as the center of the $\mu - \bar{x}$ (or μ) distribution. We know that in *any* normal distribution 98% of the data is between $z = -2.33$ and $z = +2.33$, so we shade this interval as follows:

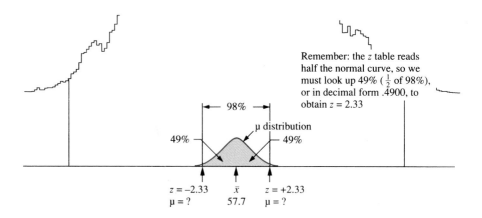

Remember: the z table reads half the normal curve, so we must look up 49% ($\frac{1}{2}$ of 98%), or in decimal form .4900, to obtain $z = 2.33$

There is a 98% probability μ will be contained in the shaded interval. Notice that the precise location of μ is unknown. However, we do know the interval where μ is most likely to be found.

Second, calculate the standard deviation of the μ distribution using the formula,

$$\sigma_{\bar{x}} = \frac{\sigma}{\sqrt{n}}$$

Often in real-life situations, σ, the standard deviation of the population is either not supplied or unknown, as in this example. In this case, we use s, the standard deviation of the sample, in place of σ, since

$$s \approx \sigma$$

In other words, the individual measurements in one sample are spread out in a manner similar to how the measurements in the entire population are spread out. If this is true, and a century of experience has confirmed this to be true, then we can use s, the standard deviation of this sample, as a reasonably good estimator of σ, the standard deviation of the population. So now we calculate,

$$\sigma_{\bar{x}} = \frac{\sigma}{\sqrt{n}} \approx \frac{s}{\sqrt{n}} \approx \frac{6.0}{\sqrt{400}} \approx \frac{6.0}{20} \approx .3 \text{ hours}$$

Third, we solve for μ using the z formula. Remember, since $\bar{x}$ has replaced μ as the center of the distribution, to get the correct plus and minus signs for our calculations, we must replace $\bar{x} - \mu$ with $\mu - \bar{x}$, as follows:

$$z = \frac{\mu - \bar{x}}{\sigma_{\bar{x}}} \qquad\qquad z = \frac{\mu - \bar{x}}{\sigma_{\bar{x}}}$$

$$-2.33 = \frac{\mu - 57.7}{.3} \qquad\qquad +2.33 = \frac{\mu - 57.7}{.3}$$

Solving for μ (and rounding to one decimal place), we get,

$$\mu = 57.0 \text{ hours} \qquad\qquad \mu = 58.4 \text{ hours}$$

These are formally referred to as lower and upper confidence limits.

Graphically, the solution would appear as follows:

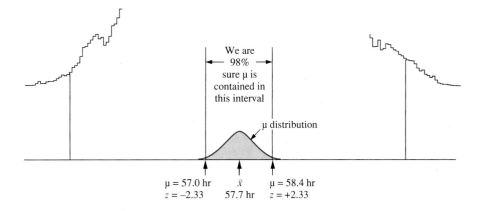

Answer

Based on the sample taken ($\bar{x}$ = 57.7 hours per week and s = 6.0 hours), we are 98% sure the true average number of hours Japanese teenagers put into class and home study per week will be contained in the interval between 57.0 hr and 58.4 hr. ∎

At this point a reminder might be helpful: keep in mind that the μ distribution is comprised of *distances* μ might be from $\bar{x}$ but not the actual value of μ itself, which is a fixed (and unknown) quantity. μ does not vary. Its distance to $\bar{x}$ varies (because $\bar{x}$ varies), and the μ distribution is a recording of all these possible distances.

8.3 Special Needs

Sometimes when working with confidence intervals, we cater to special needs. In the first example, we will calculate the probability that μ is within a specific given distance of $\bar{x}$. In the second example, the sample size, *n,* is calculated to meet certain criteria.

Example

A leading medical journal wished to determine the average starting salary for recently graduated MDs (after internship and residency). A nationwide random sample yielded:

$$n = 62 \text{ recently graduated MDs}$$
$$\bar{x} = \$91,300 \text{ first year salary}$$
$$s = \$6,000$$

The journal wanted to guarantee the readership *with a certain probability* that μ, the *real* average annual salary earned by recently graduated MDs, is within $1,000 of $\bar{x}$, the sample average. With what probability can they make this assertion?

Solution

Essentially, the question asks: with what assurance do we know μ, the *true* average salary earned by recently graduated MDs is between *$91,300 minus $1,000* and *$91,300 plus $1,000*. In other words, what is the probability that μ is between $90,300 and $92,300?

This is a "typical" normal curve problem. We merely calculate the percentage of data between $90,300 and $92,300. Let's proceed using the same three steps we used in the prior examples.

First, use your sample average $\bar{x}$ = $91,300 as the center of the $\mu - \bar{x}$ (or μ) distribution. Now, we shade the interval between $90,300 and $92,300, which we approximate, as follows.

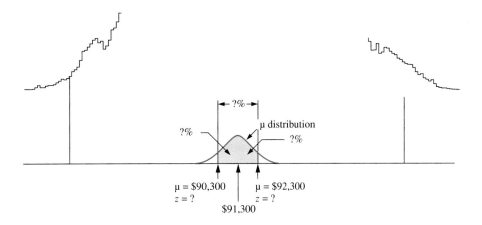

Second, calculate the standard deviation of the μ distribution using the formula,

$$\sigma_{\bar{x}} = \frac{\sigma}{\sqrt{n}}$$

Since σ is unknown, we use s, the standard deviation of the sample, as an estimator, as follows:

$$\sigma_{\bar{x}} = \frac{\sigma}{\sqrt{n}} \approx \frac{s}{\sqrt{n}} \approx \frac{\$6,000}{\sqrt{62}} \approx \frac{\$6,000}{7.874} \approx \$762$$

Third, we solve for the two z scores at the cutoffs. Remember to replace $\bar{x} - \mu$ with $\mu - \bar{x}$.

$$z = \frac{\mu - \bar{x}}{\sigma_{\bar{x}}} \qquad\qquad z = \frac{\mu - \bar{x}}{\sigma_{\bar{x}}}$$

$$z = \frac{\$90,300 - \$91,300}{\$762} \qquad z = \frac{\$92,300 - \$91,300}{\$762}$$

$$z = -1.31 \qquad\qquad z = +1.31$$

Looking up the area for $z = -1.31$, we get 40.49%. Since the normal curve is symmetrical, a z of +1.31 will yield an additional 40.49%. So, there is an 80.98% probability (40.49% + 40.49%), that μ will be contained in the interval from $90,300 to $92,300.

The completed solution would appear graphically as shown on the following page.

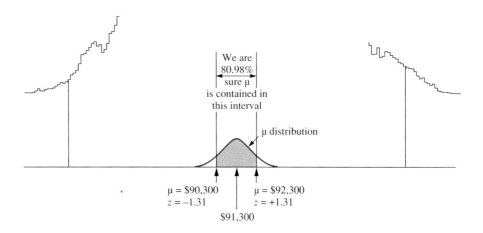

We are 80.98% sure μ is contained in this interval

μ distribution

μ = $90,300   μ = $92,300
z = −1.31   z = +1.31

$91,300

Answer

We can guarantee our readership with 80.98% confidence (or probability) that μ, the *true* average starting salary of recently graduated MDs will be within ±$1,000 of our sample average, $\bar{x}$ = $91,300, that is, between $90,300 and $92,300. ■

Example ——————

To properly assess costs before bidding on a contract, a Toy Design Outfit needed to *estimate* the average time necessary to assemble a new Walkie Doll. For a precise measurement, the financial department had to be 90% sure the *true* average time needed to assemble a new Walkie Doll, μ, lies within 2.0 minutes of the sample estimate, $\bar{x}$.

Calculate n, the sample size needed to be 90% sure μ lies within 2.0 minutes of $\bar{x}$ (assume σ = 18.0 minutes).

Solution

To calculate n, the sample size necessary to meet these conditions, we proceed somewhat differently, as follows.

The first step is basically the same. Use $\bar{x}$ as the center of the μ − $\bar{x}$ (or μ) distribution. Since we know that in *any* normal distribution, 90% of the data falls between $z = -1.65$ and $z = +1.65$, we shade this interval as follows:

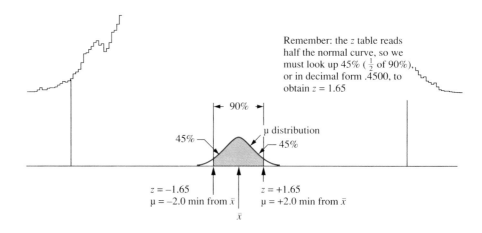

Remember: the z table reads half the normal curve, so we must look up 45% ($\frac{1}{2}$ of 90%), or in decimal form .4500, to obtain z = 1.65

90%

45%   μ distribution
                   45%

z = −1.65   z = +1.65
μ = −2.0 min from $\bar{x}$ μ = +2.0 min from $\bar{x}$

$\bar{x}$

Notice that at the cutoff μ must be exactly 2.0 minutes from $\bar{x}$. This was stated in the problem: we wish to be ''90% sure μ lies within 2.0 minutes of $\bar{x}$.'' Now, even though μ and $\bar{x}$ are unknown in this problem, we *do* know the distance from μ to $\bar{x}$ is 2.0 minutes. Another way of saying this, is

$$\mu - \bar{x} = 2.0 \text{ minutes*}$$

In other words, we did not need to know the value of μ or the value of $\bar{x}$ to know $\mu - \bar{x}$.

The second step involves algebraically manipulating the two formulas $z = (\mu - \bar{x})/\sigma_{\bar{x}}$ and $\sigma_{\bar{x}} = \sigma/\sqrt{n}$ into one combined formula for n (the derivation is presented after completion of this example), resulting in the following:

$$n = \left(\frac{z\,\sigma}{\mu - \bar{x}}\right)^2$$

Substituting in $z = 1.65$, $\sigma = 18.0$, and $\mu - \bar{x} = 2.0$:

$$n = \left(\frac{z\,\sigma}{\mu - \bar{x}}\right)^2 = \left[\frac{(1.65)(18.0)}{2.0}\right]^2 = (14.85)^2$$
$$n = 220.52, \quad \text{or}$$
$$n = 221 \text{ (rounded to a whole number)}$$

In the above formula, the quantity $\mu - \bar{x}$ is traditionally referred to as the **maximum error** in estimating μ.† In our case, the financial department wanted the maximum error in estimating μ limited to 2.0 minutes, probably for an accurate cost analysis to bid properly on the job.

Notice we had to know the value of σ, the standard deviation of the population, for this calculation. In real-life situations, σ may be quite difficult to determine, since we have not as yet taken a sample. Without a sample we have no s to estimate σ. However, we might guess ''past experience'' in assembling similar items may have allowed the toy designers to estimate σ at 18.0 minutes. In practice (when feasible) it is wise to actually take a prior sample ($n \geq 30$) and use the obtained s to estimate σ.

Answer

Thus, using a random selection of workers and conditions, we must assemble the new Walkie Doll 221 times to be 90% sure the *true* average time it takes to assemble the doll will be within 2.0 minutes of $\bar{x}$.   ■

---

*Actually, $\mu - \bar{x}$ could also equal -2.0 minutes, however this will not affect subsequent calculations since the value will ultimately be squared.

†Actually it's the quantity $|\mu - \bar{x}|$ that is referred to as the maximum error in estimating μ. This quantity is also referred to as the error of estimate, the maximum error of estimate, the maximum allowable error, and other terms.

Too often statistical studies are performed yielding inconclusive results because too small a sample size was used. It's best to first determine your needs, then estimate a sample size necessary to meet those needs prior to conducting a study.

▼ Derivation of Formula for n

Start with: $z = \dfrac{\mu - \bar{x}}{\sigma_{\bar{x}}}$

Replace $\sigma_{\bar{x}}$ with $\sigma/\sqrt{n}$; invert $\sigma/\sqrt{n}$ and multiply as follows:

A. $z = \dfrac{(\mu - \bar{x})}{\sigma_{\bar{x}}} = \dfrac{(\mu - \bar{x})}{\sigma/\sqrt{n}} = \dfrac{\sqrt{n}(\mu - \bar{x})}{\sigma} = z \longleftarrow \left[\begin{array}{l}\text{put } z \text{ on right} \\ \text{side of equation}\end{array}\right]$

resulting in: $\dfrac{\sqrt{n}(\mu - \bar{x})}{\sigma} = z$

B. Multiply both sides by $[\sigma/(\mu - \bar{x})]$ and cancel

$$\left(\frac{\sigma}{(\mu - \bar{x})}\right)\frac{\sqrt{n}(\mu - \bar{x})}{\sigma} = z\left(\frac{\sigma}{(\mu - \bar{x})}\right)$$

resulting in: $\sqrt{n} = \dfrac{z\sigma}{(\mu - \bar{x})}$

C. Square both sides to get:

$$n = \left[\frac{z\sigma}{(\mu - \bar{x})}\right]^2$$

8.4 Confidence Intervals Using Small Samples ($n < 30$)

Smaller samples are undesirable for a number of reasons. Many of the reasons (increased Type II error risk in hypothesis testing and added uncertainty in locating the true value of μ in confidence intervals) have already been discussed in prior sections.

However, if your sample size actually drops below 30 ($n < 30$), two additional problems are posed, as follows: (1) the need for assurance that your population is normally distributed or, at least, somewhat mound shaped and (2) the t score adjustment. Although discussed in chapter 7, section 7.3, we will address these two issues more fully here.

The first and perhaps most formidable of the two additional problems posed when using a small sample ($n < 30$) is the necessity that *your population be normally distributed or at least somewhat mound shaped.*

Assurance of Normal or Near Normal Population*

When your sample size is small, *non*normal populations affect the $\bar{x}$ distribution in a somewhat erratic and unpredictable manner. Results ($\bar{x}$ and s) from such studies can often be misleading and should be used only with caution. And, in general, the smaller the sample size, the more untrustworthy you should view your results. Let's see how it works.

Example —————— Suppose we return to the editors of *Sturm* magazine and their quest to *estimate* the average weight of European women (ages 18–34, heights 5′3″–5′7″). Only now, what if, instead of using a large sample, 144, *Sturm* chose to use a small sample ($n < 30$) and obtained the following results,

$$n = 7 \text{ European women}$$
$$\bar{x} = 154.3 \text{ lb}$$
$$s = 69.4 \text{ lb}$$

Calculate the 95% confidence interval estimate of μ, that is, the interval where we can expect to be 95% sure of finding the *true* average weight of these European women.

Solution No solution. Data unreliable.

Answer Since the sample size was small ($n < 30$) and *Sturm* gave us no assurance the population was normally distributed, we must consider the data unreliable and not worthy of further analysis—this is especially true in view of the fact the sample size was exceedingly small, only $n = 7$. Had the sample size been $n = 25$ or thereabouts (a number close to 30), maybe we might "risk it." ∎

This cold-hearted decision to invalidate the data had to be made. Small samples can produce a highly unpredictable array of $\bar{x}$'s if drawn from *non*normal populations, and this is especially true the smaller your sample size.

But why is this so? Let's look at an example. What if the distribution of weight of European women in the population was actually skewed (which, in

*Technical note: Wm. Gossett in his original 1908 *Biometrika* article claimed the population shape can deviate quite far from normal before this would influence the predictive value of the t distributions.

reality, may well be the case for such weight distributions) say for instance with $\mu = 130$ lb and looked as follows:

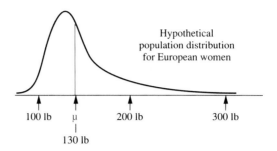

If we choose to use a small sample to estimate μ, the following scenario might occur.

Say we use sample size $n = 7$, that is, we randomly select seven European women and use their average weight, $\bar{x}$, to estimate μ, the average weight of all European women. Now, what if we select seven women, six weighing approximately 130 lb and one, by sheer chance, weighing 300 lb. Look at the skewed distribution above. This is not at all unlikely since approximately 5% to 8% of the women, according to the diagram, seem to weigh over 200 lb. In which case, the one woman who weighed 300 lb would dramatically affect our sample average, $\bar{x}$—which is precisely how the data in the preceding example was calculated. Six women weighing 130 lb each were averaged with one weighing 300 lb, with the following result:

$$\text{Sample average: } \bar{x} = 154.3 \text{ lb} \quad n = 7$$

Look for a moment at this bizarre result. Surely, $\bar{x}$ cannot be trusted to properly estimate μ—because of the inclusion of one skewed weight of 300 lb. Remember: $\mu = 130$ lb.

Had we used a large sample, say $n = 144$, the inclusion of one skewed weight of 300 lb (with the remaining 143 women weighing approximately 130 lb each) would have produced the following result:

$$\text{Sample average: } \bar{x} = 131.2 \text{ lb} \quad n = 144$$

Even if we include four women in our sample of $n = 144$, each weighing 300 lb (which, by the way, would be highly unlikely if we were to *randomly* select), even then, the results would be

$$\text{Sample average: } \bar{x} = 134.7 \text{ lb} \quad n = 144$$

In both cases using $n = 144$, $\bar{x}$ is a far better estimator of the population μ. Notice the impact of one extreme reading on the results when $n = 7$.

To sum up: when using a small sample ($n < 30$), we must be assured that our population is normally distributed (or at least somewhat mound shaped) and generally, the smaller the sample size, the more important this becomes.

However for large samples ($n \geq 30$), experience has shown that $\bar{x}$ can be used as a reliable estimator of μ even though your population is *non*normal and, thus, for large samples we generally need *no* assurances about the shape of our population.* (In chapter 10, we will use the chi-square goodness-of-fit test to assess normality in a population; specifically this is demonstrated in homework example 10.10.)

The second of the two problems posed when using a small-sample size ($n < 30$) is that when s is used to estimate σ, a t score adjustment is necessary.

t Score Adjustment When s Is Used to Estimate σ

Even if we could overcome the first problem, that is, assurance that our population is normal or at least somewhat mound shaped, a second obstacle arises. If we choose to use s to estimate σ (which we most always do) in the formula $\sigma_{\bar{x}} = \sigma/\sqrt{n}$, this necessitates a correction factor in our calculations—which we will call the t score adjustment.

Actually, t scores are not adjustments but rather a series of distributions derived empirically.† However, it is easier to view them as adjustments, because this is precisely their function—to adjust for the uncertainty created when s is used to estimate σ in the formula

$$\sigma_{\bar{x}} = \frac{\sigma}{\sqrt{n}}$$

Since $s \approx \sigma$, we say,

$$\sigma_{\bar{x}} \approx \frac{s}{\sqrt{n}}$$

Now, in large samples ($n \geq 30$), s is a reasonably good estimator of σ, and therefore the t adjustment is minimal and for this reason we have, thus far, ignored it for large-sample testing ($n \geq 30$). However, we cannot ignore it for small

---

*Actually for highly unusual population shapes (say for instance, one with an extraordinarily long skew), n may have to be considerably more than 30 to ensure that $\bar{x}$ is a reliable estimator of μ.

†More will be discussed on this shortly.

samples. As your sample size drops farther and farther below 30 ($n < 30$), s becomes noticeably a poorer and poorer estimator of σ and we must deal with this error. Let's see how this works.

Say we use a "large" sample size, for instance, $n = 50$, and take thousands and thousands of samples and calculated s for each sample. The thousands of s values would for the most part cluster quite close to the true value of σ and if we were to tally these thousands of s's into a small histogram, which we will call an s distribution, it might look as follows:

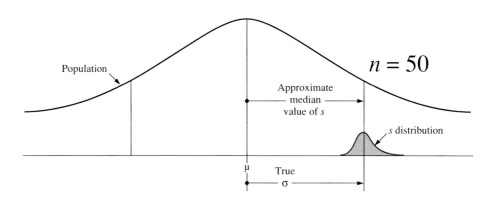

Notice how the s's cluster around the value of σ in a slightly skewed manner.* However, for large-sample sizes ($n \geq 30$), the skew is minor and the larger the sample size, the more normal in shape it becomes. So, basically we are saying, for large samples ($n \geq 30$), s distributes near normally around σ in a reasonably tight cluster, such that, approximately half the values of s will be above σ, and half below, and we ignore the t score adjustment.

Now let's see what happens with small samples ($n < 30$).

Two cases are presented. First, if we were to use a small-sample size, say $n = 15$, and took thousands and thousands of samples, calculating s for each sample, *more than* half the s's would fall below the value of σ. This is shown in the diagram below. A **median** value line (the vertical line) is drawn for reference: half the s's are above this line, and half are below. Note that the bulk of the s's in the distribution fall short of the true value of σ.

*For $n \geq 30$, the mean and median s value will be near equal. This is not true for $n < 30$.

Second, if we were to use a smaller sample, say $n = 4$, and took thousands of samples, calculating s for each sample, the value of most of the s's would now cluster *far below* σ. Visually, the two cases, $n = 15$ and $n = 4$, might look as follows:

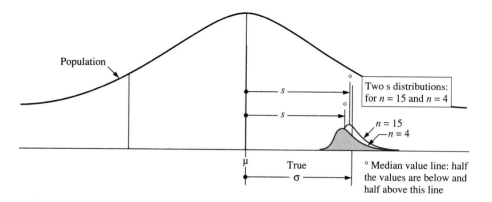

Notice that as your sample size decreases, a higher percentage of s's fall below the true value of σ, in an increasingly skewed distribution.

The reason s falls short of σ centers around the fact that when a small sample ($n < 30$) is drawn from a *normal* population, we are less likely to include ''extreme'' values. This is because most values in a *normal* population cluster fairly close to μ, with only a small percentage of extreme values (values far away from μ). Because extreme values are less likely to be included in your calculation of s when using a small sample, the value of s tends to be less than the value of σ. And, in general, the smaller the sample size, the less likely we are to include extreme values and therefore the more s will tend to underestimate σ. And, as we can see from the diagram,

s tends to underestimate σ more and more as your sample size decreases.*

Let's see what effect all this has on our calculations for $\sigma_{\bar{x}}$.

Since s replaces σ in the formula $\sigma_{\bar{x}} = \sigma/\sqrt{n} \approx s/\sqrt{n}$, and since s tends to be smaller in value than σ, we tend to calculate a $\sigma_{\bar{x}}$ that is smaller in value than it actually is. In other words, we will tend to get a false impression as to how

---

*Technical note: Actually, s^2 distributes around σ^2 in a chi-square-shaped distribution that is skewed (although near normal for large-sample size, $n \geq 30$). On average, $s^2 = \sigma^2$ for all sample sizes, however, when the sample size is small, the skew on the chi-square distribution becomes especially pronounced and a greater number of s^2 values will fall below σ^2 than above it, even though still, on average, $s^2 = \sigma^2$. This is typical of skewed distributions. The median and mean values tend to grow farther apart as the skew becomes more extreme. In other words, even though s^2 is still an unbiased estimator of σ^2 (meaning: on average equal to σ^2), *more than* half the s^2 readings will fall below σ^2.

close the $\bar{x}$'s are clustered about μ. In most cases, we will falsely believe the $\bar{x}$'s are clustered closer to μ than they actually are. Visually, we might present this as follows:

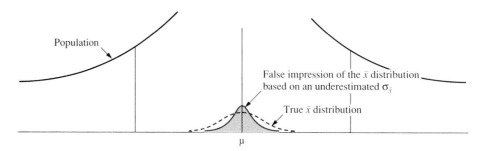

Of course at this point you might ask, if s doesn't always give us an accurate estimate of $\sigma_{\bar{x}}$, why use it? For two reasons: (1) in research studies, s is often the only estimator of σ we have, and (2) the error can be compensated for by using a t score—which finally brings us to a demonstration problem.

Say we wish to calculate a 95% confidence interval estimate of μ, using sample size, $n = 4$. Since a small sample is used, we must keep in mind the $\bar{x}$ distribution (and also the μ distribution, since it is identical) is probably underestimated. To compensate for this uncertainty, we must ''open up'' the interval to ensure we have a true 95% confidence interval estimate of μ.

In other words, if we were to use a large-sample size ($n \geq 30$), a 95% confidence interval would imply an interval bounded by ± 1.96 standard deviations (as shown in the shaded interval below, marked ''false 95%''). However, in the case of $n = 4$, we open up the interval to ± 3.18 standard deviations, and now use the letter t as follows:

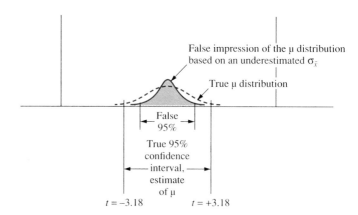

Thus, to ensure we have a true 95% confidence interval estimate of μ when using small-sample size $n = 4$, we must go out to ±3.18 standard deviations (*not* ±1.96). And instead of using the letter z to represent the number of standard deviations, we use the letter t.

Now, how did we obtain the t score adjustment of ±3.18? That's the easy part. We merely looked it up in our t tables in the back of the text. Simply look under the column 95% for 95% confidence interval, then down to degrees of freedom, df, of 3 (our sample size minus one: $4 - 1 = 3$), to obtain a t score adjustment of ±3.18, as follows:

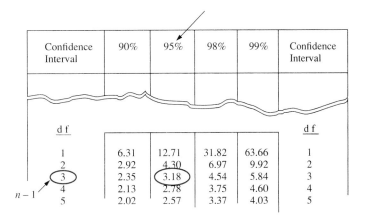

Confidence Interval	90%	95%	98%	99%	Confidence Interval
d f					d f
1	6.31	12.71	31.82	63.66	1
2	2.92	4.30	6.97	9.92	2
3	2.35	3.18	4.54	5.84	3
4	2.13	2.78	3.75	4.60	4
5	2.02	2.57	3.37	4.03	5

Remember, for this test,

$$df = \text{sample size} - 1$$
$$df = n - 1$$

To recap: although t scores are a series of distributions in their own right derived empirically,* for simplicity we may consider them adjustments since t scores, in effect, adjust for the uncertainty created when we use s to estimate σ.

But enough of this. Let's see how the t score adjustment works in the following two examples.

*Refer to endnote 1 in chapter 7.

Example ——————— A business publication wished to determine the average starting salary for recently graduated MBAs (with nontechnical undergraduate degrees and minimal,
or no, work experience). A nationwide random sample yielded:

$$n = 16 \text{ recently graduated MBAs}$$
$$\bar{x} = \$39,600$$
$$s = \$5600$$

Assuming a normally distributed population, calculate the 95% confidence interval estimate of μ. In other words, calculate an interval where we would expect
to be 95% sure of finding the *true* average salary of recently graduated MBAs.

Solution Notice the question states, "assuming a normally distributed population." If this
is a frivolous assumption (not based on fact) the sample data should be viewed
with suspicion. However, we will assume the business publication has good
reason to believe the population is normally distributed. In which case, we proceed with the same three steps we would use to calculate any confidence interval.

First, use your sample average, $\bar{x} = \$39,600$, as the center of the μ distri-
bution. We know when using a large sample, 95% of the data would be expected
to fall between $z = \pm1.96$. However, now, since we are using a *small sample,* we
must adjust for the fact s most likely is causing $\sigma_{\bar{x}}$ to be underestimated in the
formula $\sigma_{\bar{x}} = \sigma/\sqrt{n} \approx s/\sqrt{n}$. To compensate, we look up the t score adjustment,
to obtain $t = \pm2.13$, as follows:

Confidence Interval	90%	95%	98%	99%	Confidence Interval
	1.77	2.16	2.65	3.01	13
14	1.76	2.14	2.62	2.98	14
15	1.75	2.13	2.60	2.95	15
16	1.75	2.12	2.58	2.92	16
17	1.74	2.11	2.57	3.00	17

$n - 1$

Remember: look down the 95% column to a df of 15 (your sample
size minus one) $16 - 1 = 15$.

We now know that 95% of the time μ would be expected to be contained within ± 2.13 standard deviations of $\bar{x}$, that is, between $t = -2.13$ and $t = +2.13$, so we shade as follows:

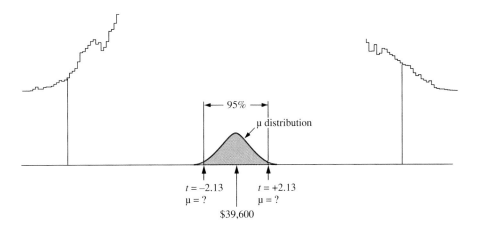

Second, we calculate $\sigma_{\bar{x}}$, as follows:

$$\sigma_{\bar{x}} = \frac{\sigma}{\sqrt{n}} \approx \frac{s}{\sqrt{n}} \approx \frac{\$5600}{\sqrt{16}} \approx \frac{\$5600}{4} \approx \$1400$$

Third, we solve using the z formula, only now we use the letter t in place of z, noting that

a. s was used to estimate σ, and
b. a small sample ($n < 30$) was probably used

Technically a t score is needed for all sample sizes when s is used to estimate σ. However, for large samples, the adjustment is considered negligible and for this reason we will ignore it for sample sizes $n \geq 30$.

$$\cancel{z} = \frac{\mu - \bar{x}}{\sigma_{\bar{x}}} \qquad\qquad \cancel{z} = \frac{\mu - \bar{x}}{\sigma_{\bar{x}}}$$

$$t = \frac{\mu - \bar{x}}{\sigma_{\bar{x}}} \qquad\qquad t = \frac{\mu - \bar{x}}{\sigma_{\bar{x}}}$$

$$-2.13 = \frac{\mu - \$39,600}{\$1400} \qquad\qquad +2.13 = \frac{\mu - \$39,600}{\$1400}$$

Solving for μ,

$$\mu = \$36,618 \qquad\qquad\qquad \mu = \$42,582$$

Graphically, the solution would appear as follows:

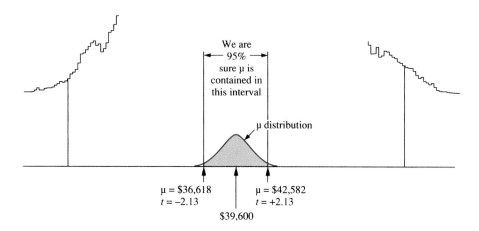

We are
← 95% →
sure μ is
contained in
this interval

μ distribution

$\mu = \$36,618$ $\mu = \$42,582$
$t = -2.13$ $t = +2.13$

$\$39,600$

Answer

Based on the sample taken ($\bar{x} = \$39,600$), we are 95% confident the true *average* annual salary for recently graduated MBAs is contained in the interval between $\$36,618$ and $\$42,582$. ■

Starting from Raw Data

Example

Suppose a medical research group in Darien, Connecticut, used a random sample of hospital patients to test a new drug reputed to lower cholesterol levels in the blood. After six months of use, the drug produced the following decreases in cholesterol in seven patients (in mg/dl blood): 24.6, 45.7, 32.0, 42.0, 23.8, 20.3, and 21.6. Assuming a normal population, construct a 90% confidence interval estimate of μ.

Solution

To get started, we must calculate $\bar{x}$ and s for this data, as follows:

Decrease in Cholesterol (mg/dl blood)	$\bar{x}$	$x - \bar{x}$	$(x - \bar{x})^2$
24.6	30.0	−5.4	29.16
45.7	30.0	15.7	246.49
32.0	30.0	2.0	4.00
42.0	30.0	12.0	144.00
23.8	30.0	−6.2	38.44
20.3	30.0	−9.7	94.09
21.6	30.0	−8.4	70.56
$\Sigma x = 210.0$		$\Sigma (x - \bar{x})^2 = 626.74$	

$$\bar{x} = \frac{\Sigma x}{n} = \frac{210.0}{7}$$
$$= 30.0 \text{ mg/dl}$$

$$s = \sqrt{\frac{\Sigma(x - \bar{x})^2}{n - 1}}$$
$$= \sqrt{\frac{626.74}{7 - 1}}$$
$$= \sqrt{104.457}$$
$$= 10.220 \text{ mg/dl}$$

Knowing $\bar{x} = 30.0$ and $s = 10.220$, we can now proceed using our three-step procedure for confidence intervals.

First, use your sample average, $\bar{x} = 30.0$, as the center of your μ distribution. We know using a large sample, 90% of the data would be expected to fall between $z = \pm 1.65$. However, now, since we are using a *small sample*, we must adjust for the fact s most likely is causing $\sigma_{\bar{x}}$ to be underestimated in the formula, $\sigma_{\bar{x}} = \sigma/\sqrt{n} \approx s/\sqrt{n}$. To compensate, we look up the t score and obtain $t = \pm 1.94$, as follows:

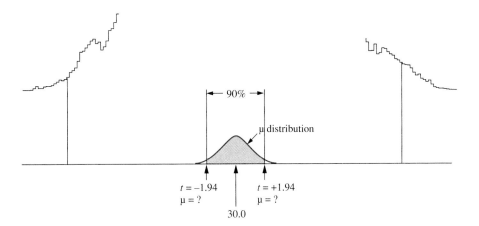

Confidence Interval	90%	95%	98%	99%	Confidence Interval
	2.13	2.78	3.75	4.60	4
5	2.02	2.57	3.37	4.03	5
6	1.94	2.45	3.14	3.71	6
7	1.89	2.36	3.00	3.50	7
8	1.86	2.31	2.90	3.36	8

$n - 1$

Remember: look down the 90% column to df of 6 (your sample size minus one).

We now know, 90% of the time μ would be expected to be contained within ± 1.94 standard deviations of $\bar{x}$, that is, between $t = -1.94$ and $t = +1.94$, as follows:

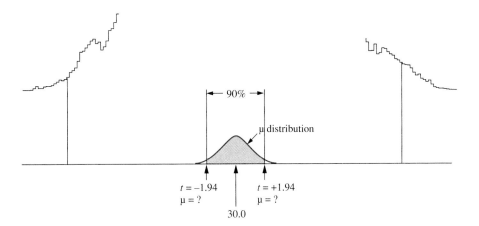

Second, we calculate $\sigma_{\bar{x}}$, as follows:

$$\sigma_{\bar{x}} = \frac{\sigma}{\sqrt{n}} \approx \frac{s}{\sqrt{n}} \approx \frac{10.220}{\sqrt{7}} \approx \frac{10.220}{2.646} \approx 3.862 \text{ mg/dl}$$

Third, we solve using the z formula, only now we use the letter t in place of z, noting that

a. s was used to estimate σ, and
b. a small sample ($n < 30$) was probably used.

Technically, a t score is needed for all sample sizes when s is used to estimate σ. However, for large samples, the adjustment is considered negligible and for this reason we will ignore it for sample sizes $n \geq 30$.

$$\cancel{z} = \frac{\mu - \bar{x}}{\sigma_{\bar{x}}} \qquad\qquad \cancel{z} = \frac{\mu - \bar{x}}{\sigma_{\bar{x}}}$$

$$t = \frac{\mu - \bar{x}}{\sigma_{\bar{x}}} \qquad\qquad t = \frac{\mu - \bar{x}}{\sigma_{\bar{x}}}$$

$$-1.94 = \frac{\mu - 30.0}{3.862} \qquad\qquad +1.94 = \frac{\mu - 30.0}{3.862}$$

Solving for μ,

$$\mu = 22.51 \qquad\qquad \mu = 37.49$$

Graphically, the solution would appear as follows:

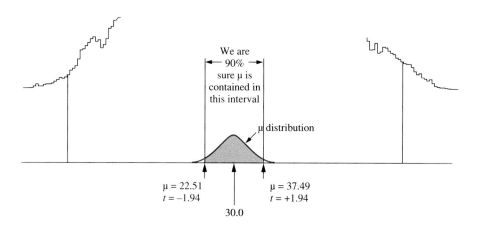

Answer   Based on the sample taken ($\bar{x} = 30.0$ mg/dl decrease in cholesterol), we are 90% confident the true *average* decrease in cholesterol is contained in the interval between 22.51 mg/dl and 37.49 mg/dl.

Summary

Confidence Interval Estimate of μ (n ≥ 30)

A confidence interval is defined as a range of values believed to contain μ and stated with a specific degree (or probability) of certainty, as follows:

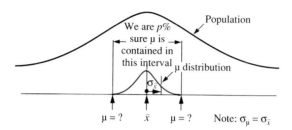

$\mu = ?$ $\bar{x}$ $\mu = ?$ Note: $\sigma_\mu = \sigma_{\bar{x}}$

To construct a confidence interval, the $\bar{x}$ is used as the center of the interval with $\sigma_{\bar{x}}$ as a measure of spread.

    For instance, if we wish to be 95% sure μ is contained in the interval, the interval would span ±1.96 standard deviations.

    The interval can be calculated using the following formula.

$$z = \frac{\mu - \bar{x}}{\sigma_{\bar{x}}}$$

Generally, a confidence interval is used when little or nothing is known about μ and we wish an estimation of its location.*

Special Needs

Two special cases were introduced.

1. Finding the percentage confidence that μ is within a specific distance of $\bar{x}$: In a way this can be thought of as a working backward problem. In the previous case, we were given the percentage

confidence and asked to calculate the interval. In this case, we are given the interval and asked to calculate the percentage confidence.

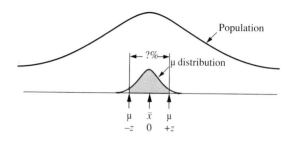

μ $\bar{x}$ μ
$-z$ 0 $+z$

In both instances we use the same formula:

$$z = \frac{\mu - \bar{x}}{\sigma_{\bar{x}}}$$

Only now we solve for the z score (instead of μ) and determine the percentage of data between ±z. The percentage data is essentially the percentage confidence of μ being within this interval.

2. Finding n, the sample size needed to meet certain criteria: To determine the sample size needed for specific requirements, we use the following formula:

$$n = \left(\frac{z\,\sigma}{\mu - \bar{x}}\right)^2$$

For instance, say we wish to be 95% sure μ is within three units of $\bar{x}$, we substitute into the formula the following values,

$$n = \left(\frac{1.96(\sigma)}{3}\right)^2$$

σ is generally estimated using data from prior studies. Large-sample s's from these studies will tend to give reasonably good estimates of σ.

*An alternative formula used to calculate a confidence interval is as follows:
$\bar{x} \pm z\,\sigma_{\bar{x}}$

terval Estimate

< 30) can be effectively used to establish confidence intervals provided two conditions are satisfied, namely

1. The population from which you sample is normally distributed or at least somewhat mound shaped. Generally, the smaller your sample size, the more critical this restriction.

2. If we use s to estimate σ (which we almost always do) in the calculation of $\sigma_{\bar{x}}$, we must use a t score and not a z score to define the number of standard deviations the $\bar{x}$'s would be expected to fall from μ.

For further summary discussion, see chapter 7 Summary (under small-sample hypothesis testing).

Exercises

Note that full answers for exercises 1–5 and abbreviated answers for odd-numbered exercises thereafter are provided in the Answer Key.

8.1 *Sturm,* a European magazine, is running a series of articles comparing European women to American women. The editors were dissatisfied with existing data and decided to take a random sample of 900 European women (ages 18–34, heights 5'3''–5'7'') to *estimate* their average weight.

Sample results yielded: $\bar{x}$ = 128.6 lb
(assume σ = 12.0 lb)

a. Find the 95% confidence interval estimate of μ (round final answers to one decimal place).
b. Find the 99% confidence interval estimate of μ (round final answers to one decimal place).
c. Based on the above confidence intervals, can we state European women, on average, weigh less than equivalent American women? (Assume equivalent American women weigh μ = 130.0 lb.)
d. What *dis*advantage can you see in using a 99% confidence interval rather than a 95% confidence interval? Explain.

8.2 A sociological society wished to *estimate* the average number of hours Japanese teenagers put into class and home study per week. A randomly selected sample of Japanese teenagers yielded the following results (round final answers to one decimal place):

n = 400 Japanese teenagers
$\bar{x}$ = 57.7 hours per week
s = 6.0 hours

a. Construct a 90% confidence interval for μ, the *true* average number of hours Japanese teenagers put into class and home study per week.
b. Explain the meaning of your answer in part a.
c. Construct a 95% confidence interval for μ, the *true* average number of hours Japanese teenagers put into class and home study per week.
d. Explain the meaning of your answer in part c.
e. With what probability can we assert that μ, the *true* average number of hours Japanese teenagers put into class and home study, is within .2 hours of $\bar{x}$, our sample average?

8.3 To properly assess costs before bidding on a contract, a Toy Design Outfit needed to *estimate* the average time necessary to assemble a new Walkie Doll. For a precise assessment, the financial department had to be 90% sure the sample estimate, $\bar{x}$, would be within 2.0 minutes of μ, the *true* average time needed to assemble a new Walkie Doll.

Calculate n, the sample size needed to be 90% sure μ is within 2.0 minutes of $\bar{x}$ (assume σ = 12.0 minutes).

8.4 A leading medical journal wished to determine the average starting salary for recently graduated MDs (after internship and residency). A randomly selected nationwide sample yielded,

$$n = 6 \text{ recently graduated MDs}$$
$$\bar{x} = \$100,000 \text{ yearly salary}$$
$$s = \$30,000$$

Calculate the 95% confidence interval estimate of μ.

8.5 Two independent nationwide medical studies using a random selection of patients tested a new drug to lower cholesterol levels in the blood, with the following results:

a. Study 1: $n = 9$ patients
$\bar{x} = 38.3$ mg/dl decrease in cholesterol
$s = 7.83$ mg/dl
Assuming a normally distributed population, construct a 95% confidence interval estimate of μ.
b. Study 2: A random selection of $n = 5$ patients produced the following decreases in cholesterol: 33.0, 38.2, 28.0, 30.6, and 30.2. Assuming a normally distributed population, construct a 95% confidence interval estimate of μ.
c. Explain your findings.
d. Since both experiments were performed using the same drug, what might we conclude about μ, the *true* average decrease in cholesterol produced by the drug? What might we conclude about σ, the population standard deviation?
e. From the wording of the problem above, describe the nature of who might be our target population.

8.6 Prompted by complaints, a federal agency decided to investigate the Twiggy Diet Farm located in central Pennsylvania. Twiggy brochures had testimonials from past customers claiming outrageous losses of weight. The federal agency conducted a random sampling of 49 past customers with the following results: $\bar{x} = 21.0$ lb weight loss for a three-week stay with $s = 8.4$ lb.

a. Construct a 90% confidence interval estimate of μ.
b. Construct a 95% confidence interval estimate of μ.
c. Construct a 99% confidence interval estimate of μ.

d. With what probability can we assert μ, the *true* average weight loss, will be within 1.0 lb of $\bar{x}$, our sample average?
e. What sample size (n) is needed to be 95% confident μ, the *true* average weight loss, is within 1.0 lb of $\bar{x}$, our sample average? (Assume $\sigma = 8.4$ lb.)

8.7 A Michigan State campus newsletter wished to compare the average height of male students on campus (ages 18–24) to the average height of equivalent males nationwide. A random sample on campus yielded the following results,

$$n = 64 \text{ male students}$$
$$\bar{x} = 71.0 \text{ inches } (5'11'')$$
$$s = 2.24 \text{ inches}$$

a. Construct an 80% confidence interval estimate of μ.
b. Construct a 95% confidence interval estimate of μ.
c. Construct a 99% confidence interval estimate of μ.
d. With what probability can we assert μ, the *true* average height of Michigan State male students, will be within .25 inch of $\bar{x}$, our sample average.
e. What sample size (n) is needed to be 90% confident μ, the *true* average height of Michigan State male students, is within .25 inch of $\bar{x}$, our sample average (assume $\sigma = 2.24$ inches).
f. If the average height of equivalent males nationwide is $\mu = 70.0$ inches (5'10''), can the campus newsletter, in good conscience, claim male students at Michigan State on average are taller than equivalent male students nationwide?

8.8 Midwest Gas and Electric wanted to estimate the average amount owed on delinquent accounts. A random sample from their vast files yielded the following results,

$$n = 45 \text{ delinquent accounts}$$
$$\bar{x} = \$32.40 \text{ owed}$$
$$s = \$8.62$$

a. Construct an 85% confidence interval estimate of μ.
b. Construct a 95% confidence interval estimate of μ.

c. With what probability can we assert μ, the *true* average amount owed by delinquent accounts is within $1.00 of $\bar{x}$, our sample average.

d. What sample size (n) is needed to be 99% confident μ, the *true* average amount owed by delinquent accounts is within $1.00 of $\bar{x}$, our sample average (assume σ = $8.62).

8.9 A marketing consulting firm wanted to estimate the average amount spent by shoppers entering a particular supermarket. A randomly selected sample yielded the following results:

$$n = 130 \text{ shoppers}$$
$$\bar{x} = \$24.50$$
$$s = \$4.80$$

a. Construct a 92% confidence interval estimate of μ.

b. Construct a 98% confidence interval estimate of μ.

c. With what probability can we assert μ, the *true* average amount spent by shoppers is within $0.50 of $\bar{x}$, our sample average?

d. What sample size (n) is needed to be 95% sure μ, the *true* average amount spent by shoppers is within $0.50 of $\bar{x}$, our sample average? (Assume σ = $4.80.)

8.10 *Hollywood Journal* magazine ran three nationwide studies to compare IQ levels of actors, artists, and TV/film directors, and obtained these results (note: average IQ = 100, genius levels, above 150) (assume a normal population for each group):

Study 1: Actors

$$n = 14 \text{ actors}$$
$$\bar{x} = 108.0$$
$$s = 10.5$$

Study 2: Artists

$$n = 12 \text{ artists}$$
$$\bar{x} = 122.3$$
$$s = 9.2$$

Study 3: TV/Film Directors

$$n = 9 \text{ TV/film directors}$$
$$\bar{x} = 118.2$$
$$s = 8.4$$

a. Construct a 95% confidence interval for actors.

b. Construct a 95% confidence interval for artists.

c. Construct a 95% confidence interval for TV/film directors.

d. Based on the above results, can we state artists have a higher IQ than TV/film directors?

e. What else might be concluded based on the above results?

8.11 ''Scalper'' tickets for a concert of The Vengeful Red, a popular recording group, were purchased in huge blocks and sold at prices far in excess of their original cost. Suppose two popular magazines, *Gathering Moss* and *Fever*, ran independent studies to estimate the *average* price paid for a scalper ticket as follows:

Gathering Moss magazine: a random sample yielded (assume a normal population for both studies)

$$n = 10 \text{ tickets}$$
$$\bar{x} = \$40.50 \text{ cost per scalper ticket}$$
$$s = \$6.20$$

Fever magazine: a random sample yielded

$$n = 8 \text{ tickets}$$
$$\bar{x} = \$34.75 \text{ cost per scalper ticket}$$
$$s = \$5.60$$

a. Construct a 90% confidence interval estimate of μ, the average amount paid for a scalper ticket, based on the *Gathering Moss* study.

b. Construct a 90% confidence interval estimate of μ, the average amount paid for a scalper ticket, based on the *Fever* study.

c. What do you think would account for the substantial difference in the sample averages ($40.50 versus $34.75)?

d. What might we conclude about μ, the true average price paid for a scalper ticket, based on the above results?

Regression-Correlation

9.0 Origins of the Concept

What Gauss's discovery of the normal curve was to the early 1800s, Galton's discovery of **regression-correlation** was to the late 1800s; both caused major revolutions precipitating a flurry of subsequent exploration, excitement, and useful application. ▼

Although by 1870 the Gaussian curve of errors (a common name for the normal curve in that period) had proven useful in a number of applications, it grew increasingly apparent that application to experiments in the social sciences was inadequate. As powerful as Gauss's curve and the central limit theorem techniques proved to be in astronomy, experimental psychology, and, later, in industrial quality control and other areas, these *one*-dimensional methods fell short of the *multi*dimensional needs of biologists, economists, sociologists, and other social scientists. Let's explore why this is so.

Until now, each example in this text has concerned itself with the measurement of *one* factor, say for instance, *height*.

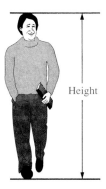

Height

For a population, then, we might plot many such measurements into a histogram or perhaps randomly sample a few such measurements to estimate population characteristics. However interesting as this is, it is far too simplistic for the needs of most social scientists who, by necessity, are most often concerned with the extent one factor influences another factor in a world of competing influences.* For instance,

To what extent does parent height influence the full-grown height of their offspring

Note: other competing influences affecting offspring height might be: nutrition, grandparent height, effects of drug or trauma in early childhood, etc.

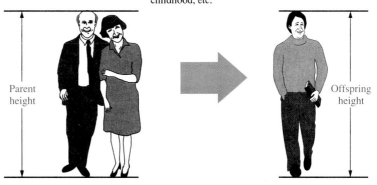

Parent height

Offspring height

Experiments of such a nature are concerned with the relationship between *two* measurements, in this case, parent height *and* offspring height. This type of data is referred to as **bivariate** (meaning having two variables). Now each *pair*

All endnotes (referenced by numbers) are presented at the end of the chapter. Note that all information relevant to the understanding of the chapter is presented on each page as sidenotes or footnotes. However, certain information is presented at the end of the chapter as *endnotes* since they are mostly reference sources and historical fine points, which tend to interfere with the flow of the material. It is *not* necessary to consult endnotes.

*The inability of social scientists to control competing factors in an experiment is probably the key factor that led to the development of regression-correlation, and later to the evolution of chi-square and ANOVA (topics to be presented in chapter 10). Whereas researchers in the hard sciences (physics, chemistry, etc.) can very often control competing influences using a laboratory, researchers in the social sciences usually do not have this option and must rely on real-world (nonlaboratory) data, which often has numerous factors simultaneously influencing results. The question is how to isolate these influences so we can assess the effect one single factor has on that which we are measuring.

of measurements is thought of as a single entity, often represented with the notation (x,y), where x = average parent height and y = offspring height. A population of many such entities might look as follows:

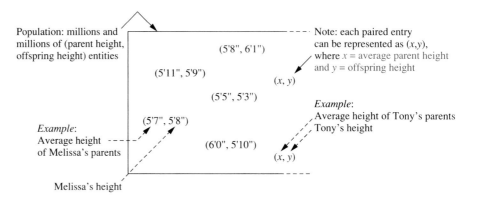

As you can see, *two*-dimensional data is far more difficult to represent than just taking millions of one-dimensional height measurements and constructing a histogram. How can we construct a histogram for millions of *pairs* of data? And even if we could, how do we statistically analyze the two separate normal curves that would probably result (one for average parent height, one for offspring height), presented perhaps, for lack of a better visual, as follows:

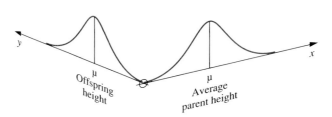

Note: all offspring heights if grouped together would most likely be normally distributed and all average parent heights if grouped together would most likely be normally distributed. The question is: how can we treat them as connected values to reveal some relationship?

This presented a whole new dimension of complexity that took numerous decades to breach. The first successful advance came in 1885 from an unexpected quarter—Sir Francis Galton and his studies in heredity.

Let's see what he did.

Galton's Parent-Offspring Height Experiment

As an experimentor, Galton (who, by the way, was first cousin to Charles Darwin)[1] was interested in showing how talent runs in families. This led him to conduct a number of experiments, most notably the very one described above concerning the relationship between parent height and offspring height. Galton elicited a sample of 928 (parent height, offspring height) measurements from the general British population of the day and presented the results in the following table. Let's take a moment to consider the table.[2]

Note: all female heights are multiplied by 1.08 to equate them to male heights.

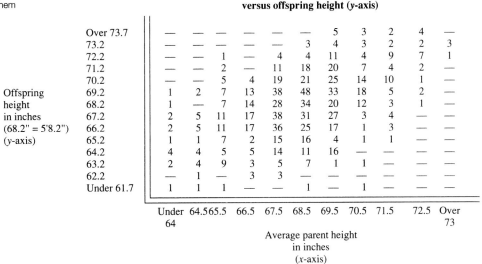

Table of parent height (averaged) (*x*-axis) versus offspring height (*y*-axis)

Offspring height in inches (68.2″ = 5'8.2″) (*y*-axis)

Offspring height	Under 64	64.5	65.5	66.5	67.5	68.5	69.5	70.5	71.5	72.5	Over 73
Over 73.7	—	—	—	—	—	—	5	3	2	4	—
73.2	—	—	—	—	—	3	4	3	2	2	3
72.2	—	—	1	—	4	4	11	4	9	7	1
71.2	—	—	2	—	11	18	20	7	4	2	—
70.2	—	—	5	4	19	21	25	14	10	1	—
69.2	1	2	7	13	38	48	33	18	5	2	—
68.2	1	—	7	14	28	34	20	12	3	1	—
67.2	2	5	11	17	38	31	27	3	4	—	—
66.2	2	5	11	17	36	25	17	1	3	—	—
65.2	1	1	7	2	15	16	4	1	1	—	—
64.2	4	4	5	5	14	11	16	—	—	—	—
63.2	2	4	9	3	5	7	1	1	—	—	—
62.2	—	1	—	3	3	—	—	—	—	—	—
Under 61.7	1	1	1	—	—	1	—	1	—	—	—

Average parent height in inches (*x*-axis)

First, Galton multiplied *all* female heights by 1.08 to bring them up to equivalent male heights. For instance, a 64.5″ female would be recorded in the above table as 69.5″ (64.5 × 1.08 ≈ 69.5). In other words, a $5'4\frac{1}{2}$ ″ female is recorded as $5'9\frac{1}{2}$ ″, her equivalent male height.

Second, look at parent-height, 71.5″. This means that the father and mother's *average* height is 71.5″ ($5'11\frac{1}{2}$ ″). Keep in mind, the mother's height had been multiplied by 1.08 before averaging. Now, *look* up the 71.5 column (that is, parent height of $5'11\frac{1}{2}$ ″ column). Up this column, Galton recorded 43 offspring (1 + 3 + 4 + 3 + 5 + 10 + 4 + 9 + 2 + 2 = 43). Of these 43 offspring, note that 10 are 70.2″ tall (obtained from the extreme left margin). This means there were 10 offspring of height 70.2″ (5'10.2″) for these parents. Also note that 9 offspring were 72.2″.

Now Galton was ready to address the question: To what extent does parent height influence offspring height? In an attempt to answer this, Galton rearranged the data in a number of ways, performing various mathematical analyses until he stumbled on one arrangement that seemed to give him the beginnings of a solution. We will not go through these detailed rearrangements,[3] however the essense of his analysis was as follows.

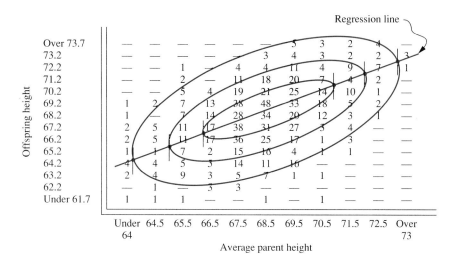

First, after some regrouping of the data, Galton constructed an array of concentric ellipses tracing paths of similar frequency. He then drew vertical lines tipping each ellipse. When he connected the tips, he found a perfectly *straight* upward-tilting line would form, which Galton felt gave the best estimate of offspring height—for any particular parent height. He called this the **regression line.**

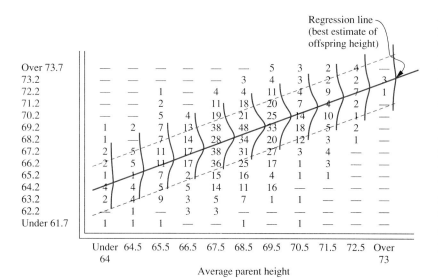

Now, whereas the regression line provided the best estimate of offspring height for each parent height, Galton noted small normal distributions of offspring heights would form around each point on the line, and that each of these small normal curves were spread out to a similar extent. In other words, each of these little normal curves seemed to have the same standard deviation (represented by the dotted line).[4]

In 1885, Galton first delivered the results of his analysis to the British Association for the Advancement of Science and, in one stroke, advanced knowledge on two fronts: one in the area of statistics, the other in the study of heredity. Let's take heredity first. In his 1886 article,[5] Galton summarized his findings by using a parent-height line slicing through the regression line, drawn something like the following:

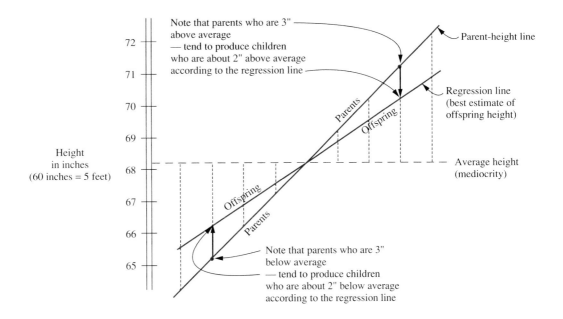

Effectively, the above sketch shows that *tall* parents tend to produce tall children, although children shorter than they are. And *short* parents tend to produce short children, although children taller than they are. For instance (see examples given in the above sketch): parents about 3'' taller than average tend to produce children who are only about 2'' taller than average,[6] and parents 3'' shorter than average tend to produce children only about 2'' shorter than average. Note that in each case, offspring height was *closer* to the population average than that of their parents.

Galton called the phenomenon **regression.** That is, offspring tend to "regress" toward mediocrity (average). This concept of regression is not only found in heredity studies, but has been confirmed in numerous unexpected areas of inquiry.* To this day, the concept of **regression toward the mean** is a serious

---

*Regression "is the tendency of the ideal mean filial type to depart from the parental type, reverting to what may be roughly and perhaps fairly described as the average ancestral type. If family variability had been the only process . . . the dispersion of the race from its mean ideal type would indefinitely increase with the number of generations, but (regression) checks this increase, and brings it to a standstill" (Walker, 1929, quoting Galton, p. 104).

The phenomenon of regression toward the mean can also be found in many studies unrelated to heredity. For instance, in education, high IQ children "regress" toward mediocrity in reading ability. In other words, higher IQ children do not read at the same proportionally high level as their IQ would indicate. This is an important concept that many teachers are unaware of. And failing to recognize this, they may set unrealistic goals for their pupils. A clearer understanding of the concept of regression may very well improve the effectiveness of teachers (Walker, 1969).[7]

concern to every researcher performing statistical analyses, and especially to researchers in the biological and social sciences.

Application to Statistics

As important as Galton's work was to the field of heredity, it was probably more important to the field of statistics.[8]

First, Galton made it clear that a large normal distribution of values (say for instance, offspring height) may indeed be broken out into a number of smaller normal distributions, and each of these smaller normal distributions can be separately assessed across a second variable (in this case, parent-height).

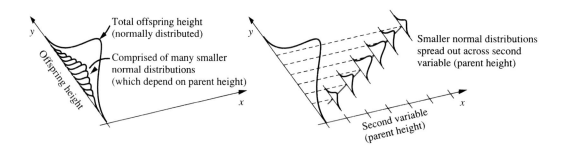

Second, he showed that a rather complex multidimensional distribution of two variables (in this case, the bivariate normal distribution)* may be represented by a simple *straight* line, and this straight line used to *predict* the value of one variable, given the other.

*__Bivariate normal distribution:__ if x and y are normally distributed, when plotted as paired measures on a three-dimensional graph, the surface created may or may not form into the bivariate normal distribution (which is presented in the diagram below). In the above case, it does. With a bivariate normal distribution, linear regression necessarily follows. Yule showed that the linear model can also be used even when the distribution is not bivariate normal but the regression is linear. Essentially, this means: if a straight line best describes the data, the shape of the multidimensional surface is not material, and the linear regression model will work. He also showed there are situations where the regression could differ from linear, but the linear model can still be used (because of its simplicity) without great loss to accuracy.[9]

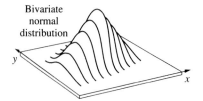

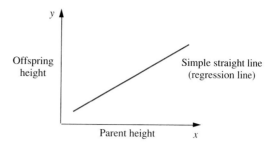

Third, Galton developed a measure to assess the *strength of the relationship* between the two variables, which in one paper he described as an index of co-relation (now called, **correlation**). The existence of co-relation implies that an increase in one variable is accompanied by a simultaneous increase (or decrease) in the other variable. Visually, this might appear as follows:

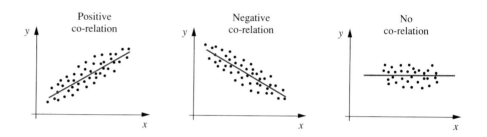

In the first two diagrams notice how closely the data points follow the sloping trend of the regression line. As the *x*-values increase, the *y*-values also tend to increase (called *positive correlation*) or decrease (called *negative correlation*). The last diagram presents an example of *no* correlation, that is, as *x* increases, the *y*-values show no upward or downward tendency. Galton assigned a value to his index of co-relation by measuring the slope of the regression line in standard units (that is, when the *x,y*-pairs are plotted in *z* scores from their respective $\bar{x}$ and $\bar{y}$).*

---

*Actually, the measure of correlation is defined in several ways.[10] The definition closest to Galton's original formulation of co-relation (which is still in use today) is defined as the slope of the regression line when each *x,y*-value is calculated and plotted in standard units (that is, if each data point is plotted in standard deviations [*z* scores] from its respective $\bar{x}$ and $\bar{y}$ value). For instance, if $r = .7$ (.7/1), this means that a .7 standard deviation increase in *y* is associated with a 1 standard deviation increase in *x*.

  $r = 1$ (1/1) is the strongest positive co-relation possible. And $r = -1$ (−1/1) is the strongest negative co-relation possible. This means that a 1 standard deviation change in *y* is associated with a 1 standard deviation change in *x*.

But with all this, probably the most important contribution Galton made was that he inspired others to continue his work and, thus, set the stage for the development of a good many of the advanced techniques we use today (multiple regression, chi-square, and ANOVA), all of which were initially developed on the heels of his groundbreaking work (mostly in the period from 1893 to 1918).

Although in 1885 the importance of Galton's work was not immediately recognized, this soon began to change.[11] In 1892, W. F. R. Welton attempted to apply the analysis to data collected on marine life, demonstrating a co-relation among body organs in shrimp (and, later, in shore crabs). This and other factors seem to have ignited the enthusiasm of Professor Karl Pearson, a mathematician and scholar with extraordinary drive and curiosity, who was to propel Galton's work to undreamed of heights. In fact, it was probably Karl Pearson's ability to enflame followers with his own zeal that set the stage for elevating Galton's analysis from a simple tool in heredity to an entire field of academic inquiry (see endnote 11 for further discussion).

Although it was W. F. R. Welton and Karl Pearson's initial force that vaulted the analysis into national prominence (circa 1892–1914 and later), the emphasis of the analysis was still clearly on biology (shrimps, shore crabs, theories of evolution, and human organs). Finally, a tool was available to link biological traits by mathematical formulation and the name of this new exclusive tool for biology was to be called *regression-correlation analysis.*

What was needed at this point was someone to break the analysis out of this limited biology-oriented mold and elevate it to all manner of social and scientific experimentation. However, for 200 years researchers and mathematicians have been trying to do just this, to apply statistical and probability techniques to other fields—with limited success. And even with Galton's formulation, it was a thorny problem. But Professor Pearson's 26-year-old research assistant, G. Udny Yule came up with the missing ingredient. Ironically, the ingredient had been around for 92 years.

Least-Squares Analysis

G. Udny Yule in 1897 supplied the key missing ingredient that was to indelibly link Galton's work to established statistical theory and open the way for application to all manner of social experiment, in economics, education, sociology, medical research, marketing research, etc. It was to be called **least-squares analysis.** Let's see how it works.

Like Galton, Yule sought the line that best represented the central flow of the data. Whereas Galton would provide a visual solution to determine this central-flow line (recall: Galton used vertical lines tipping ellipses to establish his regression line), Yule wanted to work out a more precise mathematical formulation tied in with existing statistical theory.[12] For this, he borrowed a nearly century old technique from Legendre. Before we can discuss Yule's groundbreaking achievement, however, we must first give a brief summary of what Legendre did 92 years before (in fact, it was Legendre's work that was essential to Carl Gauss and his formulation of the standard deviation, the normal distribution, and the central limit theorem, see chapters 2, 4, and 5). Now this same technique was to be the guiding inspiration for Yule. Let's see how it works.

Legendre, 1805 ("Sur la méthode des moindres quarrés") *["On the method of least squares"]*

In 1805, in a paper concerning the orbit of comets,[13] Legendre had described an ingenious method for determining the true position of a celestial body when only a few faulty observations were available, described as follows:

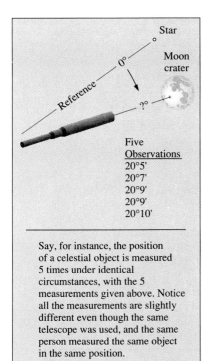

Five
Observations
20°5'
20°7'
20°9'
20°9'
20°10'

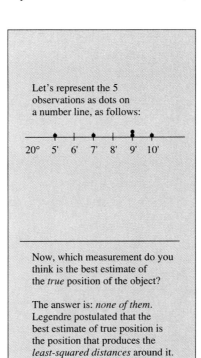

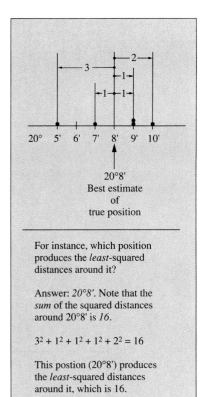

20°8'
Best estimate
of
true position

Say, for instance, the position of a celestial object is measured 5 times under identical circumstances, with the 5 measurements given above. Notice all the measurements are slightly different even though the same telescope was used, and the same person measured the same object in the same position.

Now, which measurement do you think is the best estimate of the *true* position of the object?

The answer is: *none of them.* Legendre postulated that the best estimate of true position is the position that produces the *least-squared distances* around it.

For instance, which position produces the *least*-squared distances around it?

Answer: *20°8'*. Note that the *sum* of the squared distances around 20°8' is *16*.

$$3^2 + 1^2 + 1^2 + 1^2 + 2^2 = 16$$

This postion (20°8') produces the *least*-squared distances around it, which is 16.

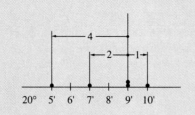

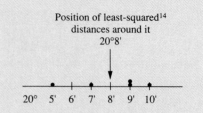

Position of least-squared[14]
distances around it
20°8'

To further understand this, suppose we take another value, say 20°9', and calculate the squared distances around this value. We get

$$4^2 + 2^2 + 0^2 + 0^2 + 1^2 = 21$$

Note the above sum of squared distances, 21, is *greater than* the sum of squared distances we calculated around 20°8', which was 16. In fact, choose any other position above and try it. Any position other than 20°8' will give *more* sum of squared distances around it.

So, 20°8' gives us the *least*-squared distances around it,* and is thus the best estimate of the Moon crater's true position, so postulated Legendre. And this has been confirmed by decades of use. It was an important finding then in astronomy, it was important to Gauss's subsequent formulation of the standard deviation,† normal distribution and central limit theorem, and now over ninety years later, it will prove important to G. Undy Yule. In fact, we will be using this method throughout much of the remainder of the text. The method is called:

Least-squares analysis

*The actual calculation of this position of least-squared distances can be found by algebraic manipulation or by calculus,[15] however it can also be found by simply adding the observations and dividing by n, which is precisely $\bar{x}$, the sample average. Thus, the position of least-squared distances is $\bar{x}$, the arithmetic mean of the observations. Note: $\bar{x} = (5 + 7 + 9 + 9 + 10)/5 = 8'$ (the position of least-squared distances).

†Since the position of least-squared distances (or $\bar{x}$) is the best estimate of central position, Gauss felt a natural link existed between $\bar{x}$ and the *sum* of the squared distances around it, that the sum of squared distances (which is a unique value, minimum around $\bar{x}$) could be used to formulate a measure of spread for the observations, thus resulting in his formulation of the standard deviation.[16] Recall, for a sample,

$$s = \sqrt{\frac{\Sigma(x - \bar{x})^2}{n - 1}}$$

Note: Sum of squared distances around $\bar{x}$ (a unique minimum value) is used in the standard deviation formulation.

It was Legendre's work that lead Gauss to reason (actually, speculate) that observational errors were, indeed, normally distributed around μ, the true position of the celestial object (see chapter 4, especially endnote 11 in that chapter for further detail).

In the previous example, the calculation would then be:

$$s = \sqrt{\frac{16}{5 - 1}} = \sqrt{4} = 2$$

Thus

$$\begin{cases} \bar{x} = 20°\ 8' \\ s = 2' \end{cases}$$

Now, how did Yule apply this analysis to (x,y) pairs? Easy. Yule simply said: sample several (x,y)-observations from a population, plot them on a graph, and it is the *line* that produces the least-squared vertical distances around it that best estimates the true central position of the population data, as follows:[17]

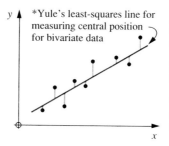

*Yule's least-squares line for measuring central position for bivariate data

Rationale
dotted path = vertical distance from point to line

Square each vertical distance, then add the squared distances.

The line that produces the *least* sum of squared vertical distances best represents the population of x, y observations.

Of course, at this point, you might ask: how do we actually calculate this line? That's easy. Yule worked out simple equations to calculate the least-squares regression line, but we will save these equations for section 9.2.

At this point, it is more important for us to understand the far-reaching impact of Yule's work. Let's consider it.

In inferential statistics, we are primarily concerned that samples give consistently close approximations to important population characteristics. And the *sample* least-squares regression line does precisely this. It gives consistently close approximations to the *population* least-squares regression line. Of course, we might ask, is the least-squares regression line itself an "important" population characteristic? And the answer is, yes. Essentially this regression line measures the central balance between two variables and does so using existing theory, the same theory in fact used by Carl Gauss to justify the adoption of the average and standard deviation for single-variate data. Essentially the regression line tracks a *changing central value of y*, continually increasing (or decreasing†) for each new value of x. Just as a simple *average* measures central position for single-variate data, a *regression line* measures changing central position for bivariate data.

*Technical note: More precisely, the above regression line best represents the expected y-value for any given x (denoted: y on x). A second regression line can be drawn through the data that represents the best expected x-value for any given y (denoted: x on y).

†In the case of no correlation, the regression line is horizontal (on average, the y-values remain the same).

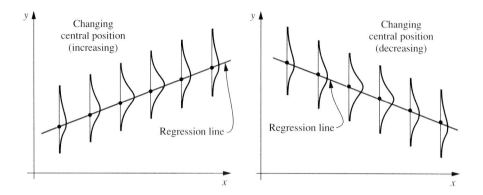

So, to recap: the least-squares regression line itself is an excellent measure of central position or central flow for bivariate data, plus the *sample* least-squares regression line generally gives close approximations to the true population least-squares regression line.*

Now, after nearly a century of continual use, challenges, and further development, Yule's least-squares regression line for defining central position for bivariate data is still with us, firmly rooted in advanced statistical techniques, such as, multiple regression, ANOVA, and numerous other advanced tools. Least-squares analysis, although not completely understood nor accepted in the 1800s (except by astronomers and a few others), now finds its permanent place in the history of scientific investigation. Least-squares analysis is certainly discussed further, because much of advanced statistics is dependent on its use and understanding.

Yule's Application to Social Issues

Actually, Yule's intent in his original 1897 analysis was simply to break Galton's work out of the limited arena of biological studies (to which Galton, Pearson, and Welton had assigned it) and apply the analysis to social concerns, marriage, poverty, divorce, crime, and so on—subjects in which Yule seemed to have a strong personal interest.

Yule's first application in 1897 concerned welfare. Although the actual study is a bit complex (Yule had data for hundreds of welfare districts and the British system of the day was somewhat different from the contemporary American system), a modified situation might proceed as follows:

*The larger the sample size generally the closer the approximation (discussed later in the chapter).

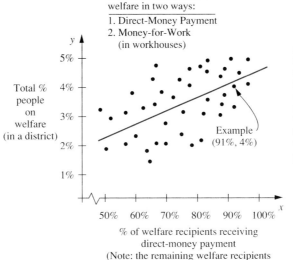

The British system gave
welfare in two ways:
1. Direct-Money Payment
2. Money-for-Work
(in workhouses)

Explanation of Graph [18]
Each (x,y) dot represents
one district. For example,
the arrow is pointing to
one welfare district with an
(x,y) value of (91%,4%).
This means 4% of the district
is on welfare and, of those on
welfare, 91% receive direct-money
payment.

Total % people on welfare (in a district)

Example
(91%, 4%)

% of welfare recipients receiving
direct-money payment
(Note: the remaining welfare recipients
receive money for their labor in workhouses)

Note the y-axis represents *total poverty* in a district (the percentage of people on some form of welfare) and the x-axis represents the percentage of recipients receiving direct-money welfare (no work).

Yule was trying to show that the more direct-money a district gives (versus money-for-work), the higher the percentage of welfare in that district. In other words, by implication, the more generous a district is in giving direct-money payments (versus giving jobs-in-workhouses for money), the more people you may expect on the welfare roles.[19] Yule effectively demonstrated a positive correlation between[20] x and y. However, the question arises, does his analysis prove cause and effect? In other words, is a district's generosity in giving direct-money payments *causing* the increase in number of people on the welfare roles?

Correlation Does Not Prove Cause and Effect

This example demonstrates the great dilemma of regression-correlation analysis which, to this day, exists. If an obvious correlation is demonstrated by plotting (x,y)-observations,* does this mean an increase in the x-variable actually *causes* an increase (or decrease) in the y-variable? Or are other factors, perhaps one or more unknown influences, causing both x and y to *appear* to vary simultaneously.

*Actually, we will learn precise mathematical methods for determining whether an actual correlation exists in a population, based on sample results.

It was pointed out to Yule (see endnote 19) that an unmeasured, lurking variable may indeed be causing x and y to appear to vary together. In this case, it was cited that, quality of administration in districts was causing what appeared to be a correlation, when indeed no true correlation existed. The argument being: (1) poor administrators tend to be in poor districts; (2) poor administrators give more direct-money payments, i.e., they are lax; and (3) it is the low quality of administrators that correlates to high direct-money payment, not the level of poverty in a district.[21]

Perhaps a more contemporary example would better demonstrate this concept of a lurking, unmeasured, variable influencing results. Suppose a random sample of several young schoolchildren were measured as to shoe size and reading ability, with the following result:

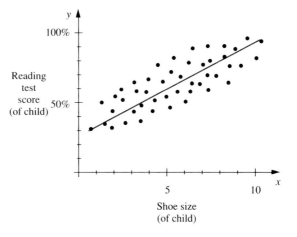

Clearly, big-footed children read better. Quite a strong correlation is apparent from the data above. Just look at it. In general, the larger the shoe size, the more advanced the reading ability of the child. Could this be? Yes. Try it yourself. Randomly select several schoolchildren from your local elementary school, measure their shoe size, and test their reading ability. You will get a strong positive correlation. As shoe size increases, reading ability tends to increase markedly.

Can you identify the unmeasured or lurking variable causing x and y to appear to vary together? Hint: What would cause both shoe size *and* reading ability to improve? Answer: *age*. The higher dots on the graph represent *older* children who have larger feet and read better than younger children, represented by the dots in the lower portion of the graph, who have smaller feet and, as yet, have poorer reading skills. So, it is not the shoe size causing changes in reading ability, it is the unmeasured variable *age* causing both *to appear* to vary together. In conclusion, let me repeat, even a strong correlation does not prove cause and effect.

9.1 Graphing: A Brief Review

This section offers a brief review of the graphing and analyzing techniques needed for this chapter. Much of it should look familiar. If you find the review too elementary, simply proceed to section 9.2.

Essentially, there are four techniques that should be mastered before proceeding in this chapter. They are as follows:

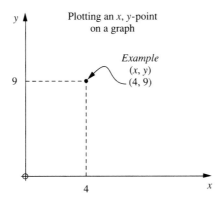

An x,y-point is located at the intersection of its two values. For example, let's take the point (4,9). Its location is at the intersection of $x = 4$, $y = 9$, as shown in this illustration. If you forget which comes first, the x or the y, remember, it's alphabetical: $x–y–z$.

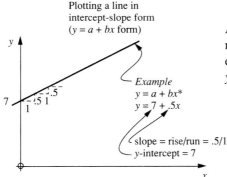

A continuous stream of points all moving in one direction is a line. The equation of a line may be represented as $y = a + bx$, such that,

a = the y-value at the point where the line crosses the y-axis.

b = a measure of steepness, called the **slope.***

*Don't get confused. In high school algebra, the letter b may have been used to represent the y-intercept. Here, the letter b represents the slope. This is done for reasons related to advanced work. Just remember, for our purposes,

$$\boxed{b = \text{slope}}$$

We can think of the slope as the dimensions of a step, as illustrated.

$$\text{slope} = \frac{\text{rise of step}}{\text{run of step}}$$

In our example above, where the slope is .5,

$$\text{slope} = \frac{\text{rise of step}}{\text{run of step}} = \frac{.5}{1} = .5$$

Essentially, the slope tells us how many units the line rises (or descends) for 1 unit increase in x.

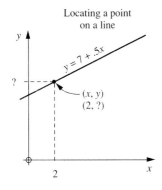

Locating a point on a line

To locate an x,y-point on line $y = 7 + .5x$, we must know either its x-value or its y-value. For example, say we know its x-value is 2. To get its y-value, we substitute 2 into the line equation for x, as follows:

$$
\begin{aligned}
y &= a + bx \\
&= 7 + .5x \\
&= 7 + .5(2) \\
&= 7 + 1 \\
&= 8
\end{aligned}
$$

This yields 8. Thus, when $x = 2$, $y = 8$. This point on the line has the value

$$(2,8)$$

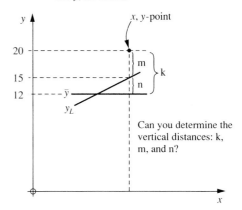

Calculating vertical distances (y-distances) between points and lines
Note: in the following example, one x, y-point and two lines, y_L and $\bar{y}$, are drawn.

Can you determine the vertical distances: k, m, and n?

Throughout the chapter, we will be measuring vertical distances (y-distances) between points and lines. For the vertical or y-distances referenced above,

$$
\begin{array}{lll}
k = y - \bar{y} & m = y - y_L & n = y_L - \bar{y} \\
\ \ = 20 - 12 & \ \ = 20 - 15 & \ \ = 15 - 12 \\
\ \ = 8 & \ \ = 5 & \ \ = 3
\end{array}
$$

The following practice problems are presented with the answers at the right.

Practice

a. Plot point (1,7).
b. Plot line: $y = 5 - x$
c. On line $y = 5 - x$, locate the point whose x-value is 4. What is the point's y-value?
d. Find m, the vertical distance from the x,y-point to line, y_L.

Answers

a.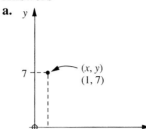

b.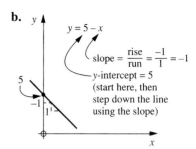

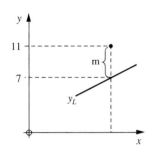

c.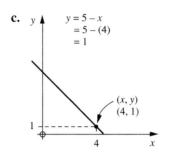

d.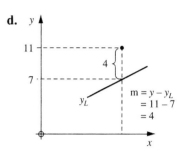

9.2 Organizing and Analyzing Bivariate Data

Overview

What chapter 2 does for single-variable data, sections 9.2–9.4 do for bivariate (two-variable) data. The purpose is to offer techniques for organizing and analyzing data for descriptive purposes. Specifically, we will study the most commonly used:

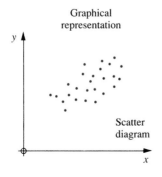

Graphical representation

Scatter diagram

x,y-observations (bivariate data) can be clearly and concisely represented by a *scatter diagram*, shown above. The scatter diagram consists of an x- and y-axis with the x,y-observations plotted according to their values. In the sketch, 26 x,y-observations were plotted.

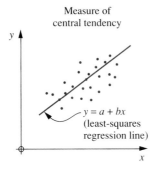

Measure of
central tendency

$y = a + bx$
(least-squares
regression line)

No measure of central tendency (or central flow) has gained such far-reaching acceptance and full integration into advanced technique as Yule's *least-squares regression line*. Specifically, we will learn how to calculate a and b (the y-intercept and slope) of the least-squares regression line for specific data.

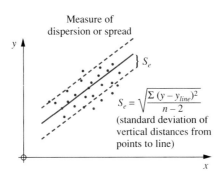

Measure of
dispersion or spread

$$S_e = \sqrt{\frac{\Sigma (y - y_{line})^2}{n - 2}}$$

(standard deviation of
vertical distances from
points to line)

Although a number of measures of spread can be derived, one of the most useful for inferential purposes (using samples to estimate population characteristics) is S_e, the standard deviation of the vertical distances from the points to the line.* Unfortunately, as useful as S_e is in helping us estimate population characteristics, it does not provide us with an adequate measure of the ability of one variable to predict the value of another, referred to as **correlation** and defined as follows.

▼ Correlation

The ability of one variable to *predict* the value of another variable.

To see why S_e does not provide us with an adequate measure of predictability or correlation, let's look at the following three cases:

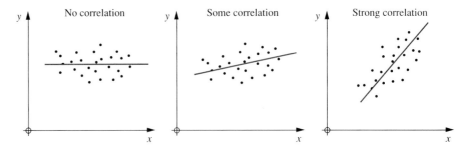

*S_e is commonly referred to as the **standard error of estimate.** Like any standard deviation, S_e is a sort of average distance a point is away from some central value, in this case, a sort of average *vertical* distance a point is away from the regression line.

Notice in each case, the data points are clustered rather similarly about the regression line, with approximately the same S_e, standard deviation of vertical distances about the line. However, in the first case, no predictive ability exists; that is, as x increases, y on average remains the same. Thus, knowledge of x gives us no knowledge of the value of y. In the second case, some predictive ability exists (although slight). But in the last case, strong predictive ability exists (knowledge of x offers us strong indications as to the value of y; note that larger x's are generally associated with larger y's). In other words, the value of S_e was approximately the same in all three cases, yet the three cases exhibited decidedly different degrees of correlation (predictive ability).

To measure this correlation (the ability of one variable to predict the value of another), we need a measure that takes into account not only the closeness of points about the regression line, as does S_e, but also the relative steepness or tilt of the slope. Specifically, we will study,

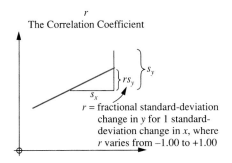

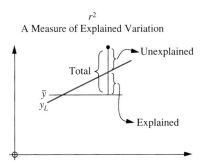

The correlation coefficient, r, is probably the most popular gauge to assess predictability or correlation (sometimes referred to as the strength of relationship between x and y). Actually, the concept of r can be explained in a number of ways. We will use two: first, as simply a measure of predictability and, second, as the slope of the regression line expressed in standard deviational changes, shown above.

Explained variation, r^2, attempts to quantify how much better the regression line, y_L, is in predicting y-values, rather than simply using $\bar{y}$, the average y value. Specifically, explained variation measures the extent sum of squared distances (from points to line) are reduced by using y_L rather than $\bar{y}$.

Once we master the descriptive techniques presented above we will present, in section 9.4, methods and conditions by which sample results may be used to estimate population characteristics.* To help us better understand the interrelationship of the concepts, we use one primary demonstration example throughout the chapter. Because the example involves standardized testing and is used so extensively, let's offer a brief introduction, as follows.

*One exception is the significance test for r, which is presented earlier because of its importance.

Regression-correlation, which developed initially in the 1890s, rapidly found its way into numerous disciplines by 1925,[22] but none perhaps so fully as in the area of standardized testing. In the 1920s and 1930s, explosive growth was witnessed in this area, to some extent, because of need,[23] but perhaps to a greater extent because now objective statistical tools were available that, for the first time, offered the promise of bringing scientific objectivity to the assessment of a test.

One of the prime considerations in developing a standardized test is whether the test indeed measures what it is supposed to measure (such as, does the test truly measure some personality trait or artistic talent?) or predicts what it is supposed to predict (such as, does the test actually predict future military success or future job or college success?). The extent to which a test measures what it purports to measure is referred to as the **test's validity.** Some tests have high validity (meaning they very accurately measure or predict something), others have medium or low validity.

Often a test is used in conjunction with other information to increase its validity, that is, to increase its power to identify traits, talent, or to predict. When used for the purposes of prediction, these tests (combined with other information) are often referred to as **predictor models.**

For instance, the SAT predictor model combines a student's high school SAT scores with other high school records to arrive at a numerical value for that student. These numerical values are then matched against a student's performance in college to see if there is a trend; for instance, does a greater high school numerical value predict, on average, a greater college freshmen GPA? If such a trend exists (in other words, if the predictor model has some *predictive validity*), then a college may want to use the predictor model to evaluate potential incoming freshmen.*

Linear Regression

Let's consider these ideas using the following example.

Example ——————— Suppose Middletown College in Salina, Kansas (a fictitious school) wished to test whether a particular predictor model (SAT scores combined with high school GPA and achievement test scores) could be used to predict success at their school and records of $n = 17$ current students were randomly selected, yielding the following.†

---

*Actual models are often more complex, weighing predictive factors with advanced statistical techniques. The example presented has been simplified to demonstrate the techniques presented in this chapter.[24]

†The first predictive validity studies for the SAT were conducted in the late 1920s and colleges and universities across the nation have conducted thousands of such studies since. In the 1964–1984 period alone, the College Board says they have helped 725 colleges prepare more than 2000 validity studies.[25] Note: Many colleges use the American College Test (ACT) for similar predictive purposes.

x (HS predictor score: 20–80 scale)	y (College freshmen GPA: 0–4 scale)
42	3.2
55	2.9
61	3.3
32	2.5
67	2.9
47	2.4
52	3.5
71	3.7
35	2.1
42	2.0
38	2.6
43	2.5
46	2.8
50	2.4
52	2.6
56	2.7
63	2.5

a. Construct a scatter diagram.

b. Calculate a and b, and use these to write the regression line equation in $y = a + bx$ form.* With this equation, predict freshmen GPA for a high school (HS) predictor score of: (i) 37 and (ii) 62.

c. Use the two (x,y) pairs, calculated in the prior question, to plot the least-squares regression line.

In this section, we will use this example to demonstrate linear regression, and again use the example throughout the chapter to demonstrate the remaining concepts. A summary is presented at the end of section 9.3.

Solution

a. To construct a scatter diagram, we do the following.

Note the first student above had high school predictor score 42, with college GPA 3.2. This (42, 3.2)-student we plot as an x,y-dot on a graph, as follows:

If we plot all 17 students, we get the scatter diagram below:

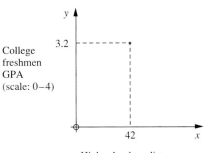

High school predictor score (scale: 20–80)

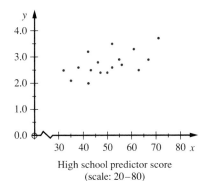

High school predictor score (scale: 20–80)

*Technical note: Actually, two regression lines are possible: one where $\Sigma (x - x_L)^2$ is minimized and one where $\Sigma (y - y_L)^2$ is minimized. Because our goal is to use x to predict y, we will minimize $\Sigma (y - y_L)^2$, which is the traditional regression model. Both, however, result in the same correlation coefficient, r (presented in section 9.3).

b. To calculate the regression-line equation in $y = a + bx$ form, we must consider Yule's criterion: Yule sought to determine the line that created the least-squared vertical distances from points to line (for a full discussion, see section 9.0 under heading, Least-Squares Analysis; also endnotes 12 and 17 describe the mathematical reasoning). To calculate a and b (the y-intercept and slope) for the line satisfying these conditions, the following formulas are used:

$$a = \frac{\Sigma y \, \Sigma x^2 - \Sigma x \, \Sigma xy}{n\Sigma x^2 - (\Sigma x)^2}$$

The formulas require the following column calculations.

$$a = \frac{(46.6)(44{,}684) - (852)(2383.5)}{17(44{,}684) - (852)^2}$$

$$= \frac{2{,}082{,}274.4 - 2{,}030{,}742}{33{,}724}$$

$$= \frac{51{,}532.4}{33{,}724} = 1.5280631$$

Note: The equations for a and b both have the same denominator, thus it is not necessary to recalculate 33,724.

$$\boxed{a = 1.528}$$

$$b = \frac{n \, \Sigma xy - \Sigma x \Sigma y}{n\Sigma x^2 - (\Sigma x)^2}$$

$$b = \frac{17(2383.5) - (852)(46.6)}{33{,}724}$$

$$= \frac{40{,}519.5 - 39{,}703.2}{33{,}724}$$

$$= \frac{816.3}{33{,}724} = .0242053$$

$$\boxed{b = .024}$$

HS score	College GPA			
x	y	x^2	xy	y^2
42	3.2	1764	134.4	10.24
55	2.9	3025	159.5	8.41
61	3.3	3721	201.3	10.89
32	2.5	1024	80.0	6.25
67	2.9	4489	194.3	8.41
47	2.4	2209	112.8	5.76
52	3.5	2704	182.0	12.25
71	3.7	5041	262.7	13.69
35	2.1	1225	73.5	4.41
42	2.0	1764	84.0	4.00
38	2.6	1444	98.8	6.76
43	2.5	1849	107.5	6.25
46	2.8	2116	128.8	7.84
50	2.4	2500	120.0	5.76
52	2.6	2704	135.2	6.76
56	2.7	3136	151.2	7.29
63	2.5	3969	157.5	6.25
$\Sigma x =$ 852	$\Sigma y =$ 46.6	$\Sigma x^2 =$ 44,684	$\Sigma xy =$ 2383.5	$\Sigma y^2 =$ 131.22

Note that the Σy^2 value will not be used until section 9.3.

Substituting $a = 1.528$ and $b = .024$ into the regression-line equation

Regression line equation

$$\boxed{y_L = a + bx}$$

we get

$$y_L = 1.528 + .024x \qquad \text{(GPA = 1.528 + .024 [HS score])}$$

i. To predict freshmen GPA for an x-value of 37 (HS score), we substitute 37 into the equation for x, as follows:

$$y_L = 1.528 + .024x$$
$$= 1.528 + .024(37)$$
$$= 2.4$$

ii. To predict freshmen GPA for an x-value of 62 (HS score), we substitute 62 into the equation for x, as follows:

$$y_L = 1.528 + .024(62)$$
$$= 3.0$$

Summary
Based on the regression-line equation, we would predict a

i. Freshmen GPA of 2.4 for a HS score of 37, representing point (37,2.4) on the regression line.
ii. Freshmen GPA of 3.0 for a HS score of 62, representing point (62,3.0) on the regression line.

c. Plot the least-squares regression-line. In the prior answer, we used the regression-line equation to determine two x,y-points as follows:

(37,2.4)   (62,3.0)

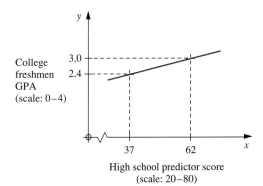

Because these points are on the regression line, we can use these to locate the line's exact position, demonstrated above. The line drawn through the original data would appear as follows:

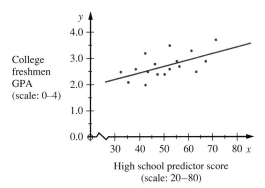

To recap, the regression line provides us with the best-fit line for the data using Yule's least-squares criterion. In other words, if we were to actually measure the vertical distances from each point to the regression line, square this quantity, and add them together, this sum (called the *sum of squared distances about the regression line*) is the *least* sum that can be achieved. Any other line gives us a higher sum of squared distances.

But this still doesn't address the question of how good the line is in predicting. Certainly, a least-squares line can be drawn through any collection of points, such as the two examples shown here.

However, we would feel a lot more comfortable using the regression line in the right diagram to predict y-values (given some x-value) rather than using the regression line in the left diagram. In other words, we need some measure of how well the line predicts before we can logically use it for such purposes. To evaluate how well the line predicts, two popular methods, or measures, have emerged: r, the **correlation coefficient,** and r^2, the **coefficient of determination.** Although both are discussed in the next section, keep in mind that r^2 is the preferred measure. ∎

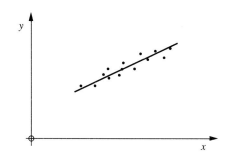

9.3 Organizing and Analyzing Bivariate Data (Correlation and Related Topics)

In this section, we introduce the linear correlation coefficient, r, defined in two ways, a test of significance to determine if linear correlation exists in the population, and the coefficient of determination, r^2 (a measure of explained variation).

r, the Linear Correlation Coefficient

Defined as a Measure of Predictability

> If we define correlation as the ability of one variable to predict the value of another, then *r* is defined as *a measure* of this ability. This measure, *r*, will vary between −1.00 and +1.00, and the closer *r* draws to one of these values, generally the more reliable the prediction.*

*Technical note: We normally view x as a fixed independent variable (in regression) predicting a random variable, y. However, regression can be carried out even if x is random. This flexibility is not true for correlation. For correlation, both x and y must be random. Because we wish to demonstrate regression and correlation together in a single example, we will use situations where both x and y are random. However, we will still approach the example in the classical regression sense as some independent x predicting a suspected dependent y (even though, technically, for correlation either x or y can be used as the independent variable).

To calculate r, we use the following formula.*

Formula to calculate correlation
coefficient r

$$r = \frac{n \Sigma\, xy - \Sigma\, x \Sigma\, y}{\sqrt{n \Sigma\, x^2 - (\Sigma\, x)^2} \; \sqrt{n \Sigma\, y^2 - (\Sigma\, y)^2}}$$

Many of the computations necessary for this formula have already been calculated, as indicated below.

Same as the numerator in the b calculation
(section 9.2 demonstration example, solution)

$$r = \frac{816.3}{\sqrt{33{,}724} \; \sqrt{17(131.22) - (46.6)^2}}$$

Same as the denominator in the b calculation
(section 9.2 demonstration example, solution)

$$r = \frac{816.3}{183.641 \; \sqrt{2{,}230.74 - 2171.56}} = \frac{816.3}{(183.641)(7.693)}$$

$$= \frac{816.3}{1412.750} = .5778$$

Thus,
$r = .5778$
$r = .58$ (rounded)

*Technical note: The formula we will use stems from Karl Pearson's original work (circa 1892–1896), defining r as a product-moment coefficient, as follows.

$$r = \frac{[\Sigma\, (x - \bar{x})(y - \bar{y})]/(n - 1)}{s_x s_y}$$

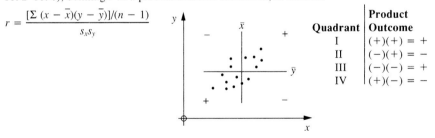

Quadrant	Product Outcome
I	$(+)(+) = +$
II	$(-)(+) = -$
III	$(-)(-) = +$
IV	$(+)(-) = -$

Explanation: From each point, we measure the distance to the $\bar{x}$ line and then the distance to the $\bar{y}$ line. These two distances are multiplied together. This is repeated for all points. Then, these products are added. Notice, on the products, the resulting signs. For points mostly in Quadrants I and III, the products will be mostly positive, and the closer the points are to a 45° line, the greater the positive value (points in the II and IV Quadrants will be negative, indicating negative correlation). Finally, the product-sum over $(n - 1)$ is divided by $s_x s_y$, the standard deviation of the x-values times the standard deviation of the y-values, to reduce the units to a dimensionless scale from -1 to $+1$. The formula we use in this text is just an algebraic rearrangement of Pearson's product-moment coefficient above.

We defined r as a measure of the ability of one variable to predict another. In this case, we are interested in the ability of high school scores to predict college GPAs. But what exactly does this value $r = .58$ calculated above mean? Well, we know that as r draws closer to -1.00 or $+1.00$, the data points generally cluster tighter around a distinctly sloping regression line, and we have greater certainty one variable (for instance, high school scores) will give reliable predictions of the other variable (in this case, college GPAs), however beyond this we know little more.*

r Defined as the Standard Slope

Another way of defining r, which offers a more visual interpretation and is actually the way originally envisioned by Sir Francis Galton (who initially developed the concept), defines r as the standard slope.†

Essentially, Galton plotted the x,y-observations in standard units (as standard deviations from their respective $\bar{x}$-and $\bar{y}$-values) and defined the slope of the resulting regression line as r. Essentially then, r, which we now refer to as the correlation coefficient, defines the expected fractional standard deviation increase in y associated with one standard deviation increase in x, as follows:

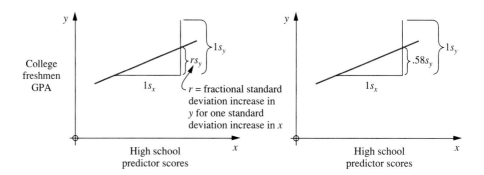

In other words, if you calculate $\bar{x}$ and s_x (the mean and standard deviation) for all the high school predictor scores, and calculate $\bar{y}$ and s_y (the mean and standard deviation) for all the college freshmen GPA scores, then one standard deviation increase in x is associated with r standard deviation increases in y. In our case,

---

*Actually, the value r^2 provides us with additional information (discussed later in this section).

†Technical note: Plotted in standard units, both regression lines (y on x and x on y) have the same standard slope, r, which Galton originally identified as a regression coefficient. However, in his writings he also used terminology such as closeness of co-relation and index of co-relation.

since $r = .58$, one standard deviation increase in x (HS scores) is associated with a .58 standard deviation increase in y (College GPA).* This relationship is expressed as standard slope, derived as follows:

$$\text{Slope} = \frac{\text{change in } y}{\text{change in } x} = \frac{.58 s_y}{1 s_x}$$

If we set $s_y = 1$ and $s_x = 1$ (as we do in z score notation), we get the **standard slope,** r, as follows:

$$\text{standard slope, } r = \frac{.58(1)}{1(1)} = \frac{.58}{1} \text{ or } .58$$

Thus, r is the slope of the regression line expressed in z score or standard deviational changes.

By expressing slope in z score changes, we lend some degree of objectivity to the results, that is, the results will not be influenced by the units or scale a researcher might choose. For instance, if the HS predictor scale 20–80 was changed to 200–800, the standard slope, r, would still be the same.†

To recap: although r (the correlation coefficient) can be interpreted in various ways, essentially r is a measure of the ability of one variable to predict the value of another and is commonly referred to as a measure of the strength of the relationship between x and y. In general, the closer the data points cluster to a sharply sloping regression line, the closer the r-value will draw to $+1.00$ or -1.00, and the more confident we are that for any given x the regression line will offer reliable predictions of y. Different r-values are demonstrated as follows:

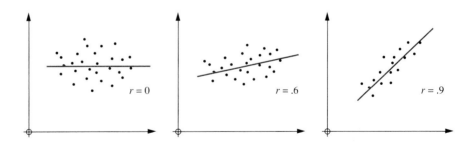

*Note that $+1.00$ is the maximum possible standard deviation change in y (or -1.00 the minimum possible standard deviation change in y) for one standard deviation change in x. This has been proven mathematically.

†Provided, of course, the interval proportionality is maintained.

However, as important as r is in measuring correlation in sample data, it does not guarantee we have correlation in the population from which the sample was drawn. In other words, we could indeed have *no* correlation in the population and yet sample 17 x,y-observations and "by chance" obtain a sizable correlation in the sample. This problem is addressed next.

Test of Significance for r

Before we begin, keep in mind, sample results only estimate population characteristics, so in this case,

$r \approx \rho$ the sample linear correlation coefficient, r, is approximately equal to ρ (rho), the population linear correlation coefficient.*

The problem arises in that ρ (rho) could very well equal zero (*no* correlation in the population), and we may very well get a sizable correlation in the sample. This is especially true when we use small-sample sizes.

So, to ensure that ρ does not equal zero, we set up the following test.

$H_0: \rho = 0$ Our null hypothesis (initial assumption) is: ρ equals zero (*no* correlation exists in the population).

$H_1: \rho \neq 0$ The alternative hypothesis, $\rho \neq 0$, indicates that if H_0 proves false, then some correlation must exist in the population.

$\alpha = .05$ Tests in this chapter will be limited to alpha values .05 and .01. In this case, we will use .05 (meaning: a 5% chance exists that when H_0 is true, you will reject it in error).

---

*Note: $r \approx \rho$ only provided certain population assumptions are satisfied. When samples are used to estimate population characteristics, population assumptions are usually necessary, and violation of these assumptions may result in faulty predictions. Generally, to use r as a (relatively) unbiased estimator of ρ, we assume random selection from a *bivariate normal distribution*, discussed in section 9.0 footnote. Such a distribution has normality of arrays and homogeneity of variance across both variables (further discussion in section 9.4) and the separate distributions of x and y are normal. For $\rho = 0$, r will distribute normally about 0. For ρ other than 0, the distribution of r is skewed, however r may be transformed to Fisher's Z [$Z = \frac{1}{2} \ln(1 + r)/(1 - r)$], which nearly normally distributes, and Fisher's Z can be used for reasonable estimates. (Actually, for $n \geq 30$, r distributes nearly normally about 0 when $\rho = 0$. For $n < 30$, the t distributions are used to test significance.)

Any hypothesis test begins by assuming H_0 true. Thus, if H_0 is true and indeed the population correlation coefficient equals zero ($\rho = 0$), and we were to take thousands of samples of size $n = 17$ from this population, the following would occur:

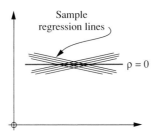

If indeed the population correlation coefficient equals zero ($\rho = 0$), then the population regression line would lie flat (horizontal) and the thousands of sample regression lines would cluster about this $\rho = 0$ line, as shown.*

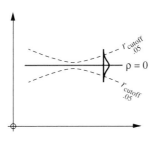

If we were to draw boundaries containing the middle 95% of the sample regression lines, denoted by $r_{cutoff,.05}$, it might appear as shown here. In fact, the standard slopes (correlations) are normally distributed around 0. In the above case, for $n = 17$ observations, the boundaries would be $\pm.48$, denoted as r_{cutoff}. This boundary or cutoff value is obtained in the r table in back of text (df $= n - 2$). This will be demonstrated shortly.

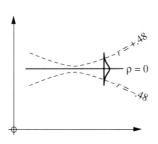

In other words, if 17 bivariate observations are randomly selected from a bivariate normal population with *no* correlation ($\rho = 0$), 95% of the sample regression lines would have standard slopes (correlations) anywhere from $r = -.48$ to $r = +.48$.

So, if we obtain from our data an r-value between $-.48$ and $+.48$, ρ could indeed be 0; in these cases, we accept $\rho = 0$. In other words, for

$$-.48 \leq r \leq .48$$
$$\text{Accept: } H_0 \colon \rho = 0$$

*Actually, the sample regression lines seem to give the appearance of pivoting about some central value, which indeed they do, the x and y averages of the population, which we will denote as (μ_x, μ_y). This is because all the sample regression lines go through their own central value, their own $(\bar{x}, \bar{y})$ value, and since sample averages tend to approximate population averages [recall: $\bar{x} \approx \mu_x$, in our case, $(\bar{x}, \bar{y}) \approx (\mu_x, \mu_y)$], the centers of the sample regression lines tend to cluster close to the population center, (μ_x, μ_y).

Because our sample data gave us an r of .5778 (.58 rounded), we reject H_0 and accept the alternative hypothesis, $H_1: \rho \neq 0$, which means we are reasonably sure some correlation exists in the population. The above procedure can be summarized by using the following illustration:

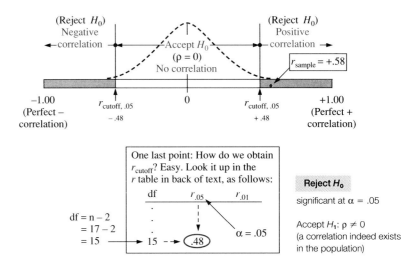

But as useful as r is in assessing the strength of correlation (strength of prediction), it does not truly reflect proportionality. In other words, we cannot say, a correlation at $r = .8$ is twice as good as one at $r = .4$. Another measure, measuring the extent of explained variation and called r^2, allows us a more proportional comparison and thus is the preferred measure. This concept of explained variation and use of r^2 is the topic of our next discussion.

r^2, a Measure of Explained Variation*

At this point, we may have forgotten our original example. Let's consider it again.

Example ——————— Suppose Middletown College in Salina, Kansas (a fictitious school), wished to test whether a particular predictor model (SAT scores combined with high school GPA and achievement test scores) could be used to predict success at their school and records of $n = 17$ current students were randomly selected, yielding the following:

*r^2 is commonly referred to as the **coefficient of determination**.

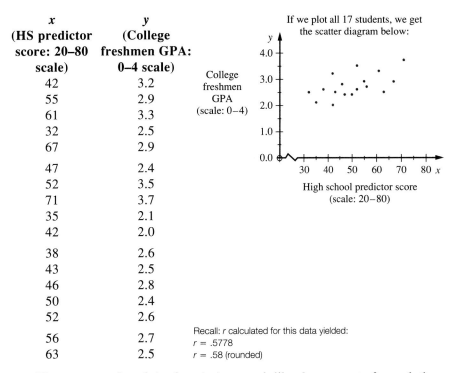

x (HS predictor score: 20–80 scale)	y (College freshmen GPA: 0–4 scale)
42	3.2
55	2.9
61	3.3
32	2.5
67	2.9
47	2.4
52	3.5
71	3.7
35	2.1
42	2.0
38	2.6
43	2.5
46	2.8
50	2.4
52	2.6
56	2.7
63	2.5

Recall: r calculated for this data yielded:
$r = .5778$
$r = .58$ (rounded)

The concept of *explained variation,* much like the concept of correlation, attempts to measure the extent one variable can be used to predict another, however explained variation does it in a way that opens the door to more advanced analysis and integration into other techniques.

The actual calculation of explained variation is quite easy. It is simply r^2. And if we wish to convert this to a percentage, we multiply by 100, thus,

% Explained variation = 100 r^2.

For our example,

$$\text{Percentage explained variation} = 100\ r^2$$
$$= 100(.5778)^2$$
$$= 100(.3339)$$
$$= 33.39\%$$

Note: this is the r we calculated from our sample data: $r = .5778$ (which we had also rounded to .58).

However, the above formula offers little by way of understanding. The actual concept is explained as follows.

The relationship between x and y can be modeled a number of ways. One model (important for advanced work) uses the sum of squared distances as a gauge in measuring the *improvement* the regression line, y_L, offers in predicting y-values over merely using $\bar{y}$, the average y-value. This is explained in the following:

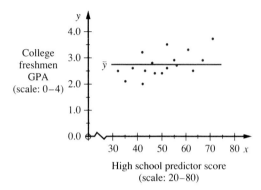

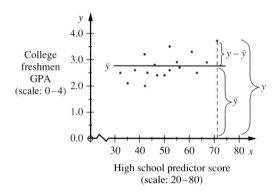

Of the 17 students originally chosen, say we had no knowledge of a student's high school predictor score or any other information. Then the best predictor of freshmen GPA for any randomly chosen student in that group would be $\bar{y}$, the average GPA, calculated as

$$\bar{y} = \frac{\Sigma y}{n}$$
$$= \frac{46.6}{17}$$
$$= 2.74$$

The sum Σy is obtained from the calculations in section 9.2.

In other words, if we had to predict a freshmen GPA (given no further information about the student), our best guess would be 2.74, the average freshmen GPA of the sample, and indeed most of the data points do cluster rather close to this average of 2.74.

In fact, to measure exactly "how close" the data points cluster, we calculate the sum of squared distances the points are away from the $\bar{y}$-line. For each point, we subtract $y - \bar{y}$ (the y-value of the point minus the y-value of the line), demonstrated for one point above. Each of these $(y - \bar{y})$ distances is then squared, and these squares are added together, shown as follows.

y	$\bar{y}$	$y - \bar{y}$	$(y - \bar{y})^2$
3.2	2.74	.46	.2116
2.9	2.74	.16	.0256
3.3	2.74	.56	.3136
2.5	2.74	−.24	.0576
.	.	.	.
.	.	.	.
.	.	.	.
2.7	2.74	−.04	.0016
2.5	2.74	−.24	.0576

$$\Sigma (y - \bar{y})^2 = 3.4812$$

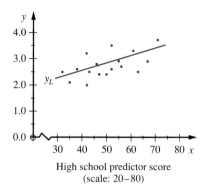

High school predictor score
(scale: 20–80)

To recap:

$y - \bar{y}$ gives the vertical distance from one point to the $\bar{y}$-line. If we square all the vertical distances and add these up, we get 3.4812 (sum of squared distances around $\bar{y}$), which is referred to as total variation.

> Total variation $= \Sigma (y - \bar{y})^2$
> $= 3.4812$

Recall: this would be the first step in calculating the standard deviation of the y-values. If we divide 3.4812 by $n - 1$ and take the square root, we have the standard deviation of values about $\bar{y}$. However, it is preferable to use the sum of squared distances in assessing spread rather than the standard deviation.*

Now what if we say we have devised a better predictor of the y-values, say, using the regression line, y_L. Certainly, then, if the regression line were a better predictor, the points in some sense should lie closer to the regression line than to the $\bar{y}$-line. And this should be reflected in a *lessening* of the sum of squared distances around y_L.

So, now, if the sum of squared distances from the points-to-regression line are calculated, we get what is referred to as unexplained variation.†

> Unexplained variation $= \Sigma (y - y_L)^2$
> $= 2.3189$
> (unexplained variation is also referred to as **residual variation**)

Essentially this is the remaining sum of squared distances after the regression line is used to describe the points.

---

*The reason the sum of squared distances is used rather than the standard deviation is that standard deviations, s's, are not additive (nor are s^2's without the same df). In advanced work, we often have to add or combine the spread measures for a number of samples. It is far easier to have a measure of spread we can readily add, especially when many of these computations are performed. Also, these calculations are highly sensitive to rounding errors. The less averaging and square rooting we do, the better our estimates. In other words, the sums of squared distances are much easier to work with than standard deviations, however they should be viewed in the same way, as simply measures of spread.

†Actually, a number of simplifying formulas are available for these calculations, such as,

Total variation $= \Sigma y^2 - \dfrac{(\Sigma y)^2}{n} = 131.22 - \dfrac{(46.6)^2}{17} = 3.4812$ (same as above)

Unexplained variation $= \Sigma y^2 - a \Sigma y - b \Sigma xy$
$\qquad = 131.22 - (1.5280631)(46.6) - (.0242053)(2383.5)$
$\qquad = 2.3189$

Note, these calculations are very sensitive to rounding and several decimal places are necessary for accuracy. The full points-to-line calculation for unexplained variation is demonstrated in section 9.4.

So, as you can see, the regression line reduces the *total sum of squared distances* from 3.4812 to 2.3189. This is a *reduction* of 1.1623 squared units (3.4812 − 2.3189 = 1.1623) and referred to as *explained variation*. Essentially, total variation = explained variation plus unexplained variation; thus, rearranging, we state

> Explained variation = Total variation − Unexplained variation

In our case, explained variation = 3.4812 − 2.3189 =

> 1.1623

To express this as a percentage of the total, we do the following:

> $$\text{Percentage explained variation} = \frac{\text{Explained variation}}{\text{Total variation}} \times 100$$

$$= \frac{1.1623}{3.4812} \times 100$$
$$= 33.39\%$$

Note: using 100 r^2 earlier, we had arrived at the same answer:

> Percentage explained variation = 100 r^2
> = 100(.5778)^2
> = 33.39%

In other words, if the squared distances are reduced 33.39%, as in the above case, we say the regression line ''explains'' 33.39% of the variation. And it is common for researchers to phrase this as, 33.39% of the variation in y is explained by its association with x (meaning HS predictor scores explain 33.39% of the variation in college freshmen GPA). Of course, this implies that 66.61% (100% − 33.39%) is *un*explained or remaining.

A common model used to represent this relationship is the following:

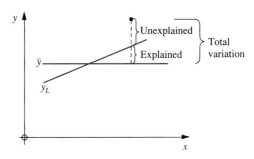

	Total variation		Explained variation		Unexplained variation
	$(y - \bar{y})$	$=$	$(y_L - \bar{y})$	$+$	$(y - y_L)$

From this, it can be proven for all points:

$$\Sigma(y - \bar{y})^2 = \Sigma(y_L - \bar{y})^2 + \Sigma(y - y_L)^2$$

3.4812	$=$	1.1623	$+$	2.3189	
100%	$=$	33.39%	$+$	66.61%	
Total variation	$=$	Explained variation	$+$	Unexplained variation	

To recap: Without any prior knowledge of the student, $\bar{y}$ gives us the best predictor of GPA. However, by using the regression line, we may get closer predictions to what students actually achieve.

How much closer? Well, one way to measure this is to say the regression line reduces the sum of squared distances by 33.39%. In other words, the line better fits the points by reducing the sum of squared distances. Of course, the 33.39% reduction implies 66.61% of sum of the squared distances still remain. It is quite understandable that *un*explained factors would enter into college freshmen GPA achievement, such as, hard work, motivation, extracurricular influences, course load, difficulty of courses, and so on.

Summary and Discussion

A summary of the results for the calculations presented in sections 9.2 and 9.3 is as follows.

The sample of $n = 17$ x,y-observations yielded the following regression line:

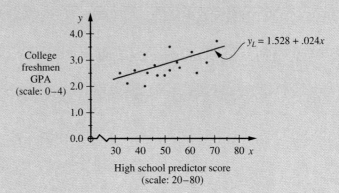

The ability of one variable to predict the value of another is measured by r, the correlation coefficient, and for this example yields $+.58$. Generally the closer r approaches -1.00 or $+1.00$, the more reliable the prediction. A second way of viewing r is as the slope of the regression line expressed in standard units (z scores). That is, one standard deviation change in x is associated with an r standard deviation change in y. In other words, one standard deviation change on the regression line in high school predictor score is associated with a .58 standard deviation increase in college GPA.

    Since our r value (.5778, rounded to .58) exceeded the cutoff value of .48, we conclude at least some positive correlation exists in the population (that is, $\rho \neq 0$, and we express this by saying r is significant).

$$\text{Percentage explained variation} = 100\ r^2 = 100(.5778)^2$$
$$= 33.39\%$$

This means the regression line reduces the sum of squared-distances (from points to line) by 33.39% versus using $\bar{y}$.* This is often expressed as 33.39% of the variation in freshmen GPA is explained by its association with high school predictor score.

A number of points should be made before we conclude.

1. *Causation is not proven.* A high correlation between two variables does not prove cause and effect. An unmeasured factor may be responsible for what only appears to be a co-movement (correlation) between x and y. Let's look at an example.

    A strong correlation exists between closeness to the Sun and cold temperatures in certain locations. The *closer* the Earth is to the Sun, the *colder* the temperature in University Park, Pennsylvania, for instance. In fact, this is also true for Annandale, Virginia, and most other college

---

*Because of rounding technique, this value of 33.39% may vary somewhat.

towns in the United States. In fact, it is true for almost every town in the United States. The closer the Earth is to the Sun on any particular day, in general, the colder the temperature. But just because a strong correlation exists in this case, please don't be tempted to conclude: closeness to the Sun "causes" cold temperatures. (Actually, the tilt of the Earth is the prime cause of temperature change. In winter in the Northern Hemisphere, the tilt is away from the Sun and temperatures drop. Ironically, the Earth is 2 million miles *closer* to the Sun at this point.)

The allure to assign causation to correlation results is great. Many initially exposed to the analysis fall victim. Resist it. It has lead many to rest their thinking into comfortable but false avenues of thought. Regression-correlation is an exploratory tool, which when used in conjunction with other evidence can help researchers in their often long, convoluted and, many times, frustrating quests to determine causation.*[26]

2. *Significance of r. r* must be significant before using the regression-line equation to predict population results. You must first prove at least some correlation exists in the population before logically using the sample regression line to make population predictions.

3. *Test of linearity.* Regression-correlation is a test of linearity, a test to see if a *straight-line* relationship exists in the data. It is not a test to see if *any* relationship exists. In other words, we could get a correlation coefficient of zero for our data (no linear correlation), however a very strong curvilinear relationship may indeed exist, as follows:

In each of these cases, $r \approx 0$ (little or no linear correlation exists) however there is a strong curvilinear relationship in both.

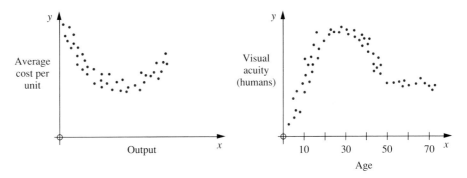

4. Some guidelines in using regression-correlation analysis.

a. Only *y*-values can be predicted from the regression equation we calculated, not *x*-values. A different regression-line equation must be calculated to predict *x*-values (although both will have the same correlation coefficient).

*Even if causation is known (or highly suspected), it is often difficult to determine whether *x* is causing *y* or the reverse, *y* is causing *x*. Refer to endnote 26 for references in education and further discussion.

b. Predictions should be made only within the framework of the subjects sampled. For instance, if you sampled SAT scores for men, don't use the results to predict SAT scores for women.

c. Similarly, prediction should be made only within the scope of the data. In other words, if oranges are tested at temperatures of 70° to 90° for spoilage, don't make predictions for oranges at 40° or at 110°.

d. Old data should not be assumed to give reliable current predictions. Population characteristics measured in 1986 may be quite different in 1996.

9.4 Using Samples to Estimate Population Characteristics

In sections 9.2–9.3, we presented techniques for describing bivariate data. If the only purpose of using these techniques is to describe a collection of data, the techniques need no qualification. They can be used without limitation. However, if the collection of data is to be used for the purpose of *estimating population characteristics,* then a number of assumptions must be met, as follows:

1. The researcher must achieve a valid random sample drawn from the population.*

2. The population has normality of arrays and homogeneity of variance, explained as follows:†

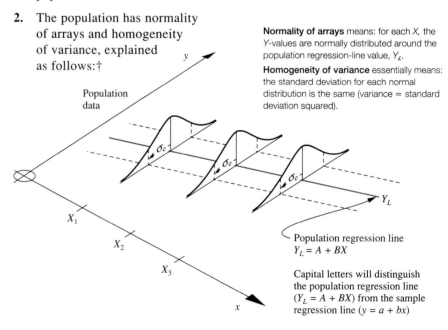

Normality of arrays means: for each X, the Y-values are normally distributed around the population regression-line value, Y_L.

Homogeneity of variance essentially means: the standard deviation for each normal distribution is the same (variance = standard deviation squared).

Population regression line
$Y_L = A + BX$

Capital letters will distinguish the population regression line ($Y_L = A + BX$) from the sample regression line ($y = a + bx$)

*In the case of fixed x's, a valid random sample of y's drawn from each population.

†Fortunately these limitations are robust, meaning slight to moderate violations can be tolerated without serious effect on results. The conditions have also been found to be satisfied in a broad variety of experiments in many diverse fields. However, serious violation can affect results, and techniques are available to test for conformity of these assumptions.

If we assume these population characteristics are satisfied, then we can proceed. Thus we know, for our data,

$y_L = a + bx \approx Y_L = A + BX$ the sample regression line is approximately equal to the population regression line.*

$S_e \approx \sigma_e$ the sample spread, S_e (the standard deviation of vertical distances from points to line) is approximately equal to its equivalent population spread measure, σ_e.†

In other words, the regression line and measure of spread for our sample will be approximately equal to the regression line and measure of spread for the population. Thus, for our data, these sample measures were calculated as follows:

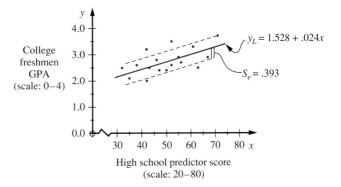

Wait a minute. I don't remember calculating S_e (the standard deviation of vertical distances from points to line). That's true, we didn't. Okay, I guess we better calculate it then. The calculation is as follows:

$$S_e = \sqrt{\frac{\Sigma(y - y_L)^2}{n - 2}}$$

Note: $(y - y_L)$ is the vertical distance from a point to the regression line, demonstrated below.

*In other words, given our population assumptions are met, a and b, the intercept and slope of the sample regression line, are unbiased estimators of A and B, the intercept and slope of the population regression line.

†In actuality, $S_e^2 \approx \sigma_e^2$, sometimes referenced as $S_{y/x}^2 \approx \sigma_{y/x}^2$ ($\sigma_{y/x}^2$ is the variance of the subpopulation of y-values for a particular value of x).

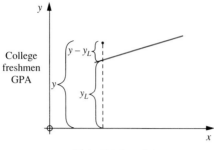

High school predictor score

If we subtract y_L (the y-distance up to the line) from y (the y-distance up to the point), we are left with the distance from the point to the line, denoted $(y - y_L)$. Thus,

$$(y - y_L) = \text{distance from point to line}$$

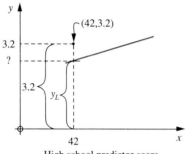

High school predictor score

Now let's use an example, say point (42,3.2), the first student from our original data collection.

To calculate the distance up to the line (y_L), we plug the x-value (42) into the regression equation, as follows:

$$y_L = 1.528 + .024x$$
$$= 1.528 + .024(42)$$
$$= 2.536$$

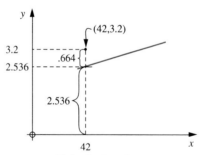

High school predictor score

Now we know $y = 3.2$ and $y_L = 2.536$, so,

$$y - y_L = \frac{\text{Vertical distance}}{\text{from point to line}}$$
$$3.2 - 2.536 =$$
$$.664 =$$

This .664 is the vertical distance from point to line.

To complete the formula, it is necessary to calculate the vertical distances for each point, square them, then add up all these squared values, as demonstrated in the following table.

HS predictor score x	College freshmen GPA y	y_L calculation $a + bx$	$=$	y_L	$(y - y_L)$	Squared $(y - y_L)^2$
42	3.2	$1.528 + .024(42)$	$=$	2.536	.664	.4409
55	2.9	$1.528 + .024(55)$		2.848	.052	.0027
61	3.3	$1.528 + .024(61)$		2.992	.308	.0949
32	2.5	$1.528 + .024(32)$		2.296	.204	.0416
67	2.9	$1.528 + .024(67)$		3.136	$-.236$	.0557
47	2.4	$1.528 + .024(47)$		2.656	$-.256$	.0655
52	3.5	$1.528 + .024(52)$		2.776	.724	.5242
71	3.7	$1.528 + .024(71)$		3.232	.468	.2190
35	2.1	$1.528 + .024(35)$		2.368	$-.268$	.0718
42	2.0	$1.528 + .024(42)$		2.536	$-.536$	.2873
38	2.6	$1.528 + .024(38)$		2.440	.160	.0256
43	2.5	$1.528 + .024(43)$		2.560	$-.060$	.0036
46	2.8	$1.528 + .024(46)$		2.632	.168	.0282
50	2.4	$1.528 + .024(50)$		2.728	$-.328$	.1076
52	2.6	$1.528 + .024(52)$		2.776	$-.176$	.0310
56	2.7	$1.528 + .024(56)$		2.872	$-.172$	.0296
63	2.5	$1.528 + .024(63)$		3.040	$-.540$	.2916

$$\Sigma(y - y_L)^2 = 2.3208$$

Note that rounding techniques will cause this value, 2.3208, to vary. Using more precise techniques, we obtained the value, 2.3189, which in section 9.3 was called *unexplained variation*.

Substituting 2.3208 into the formula

$$S_e = \sqrt{\frac{\Sigma(y - y_L)^2}{n - 2}} = \sqrt{\frac{2.3208}{17 - 2}} = .3933$$

$$S_e = .393$$

Much as any standard deviation can be viewed as a sort of average distance a data point is away from some central value, the standard deviation of vertical distances can be viewed similarly as a sort of average vertical distance a data point is away from the regression line, a central value. And, similarly, as we use s in single-variable data as an estimator of its equivalent population value, σ, we use S_e in bivariate data as an estimator of its equivalent population value, σ_e.

The above calculation is used for demonstration purposes only. In actual practice the following shortcut formula can be used:

$$S_e = \sqrt{\frac{\Sigma y^2 - a\Sigma y - b\Sigma xy}{n - 2}}$$ All these values were calculated in section 9.2

$$= \sqrt{\frac{(131.22) - (1.5281)(46.6) - (.0242)(2383.5)}{17 - 2}}$$

$$= .394$$

Note, the formula is very sensitive to rounding errors. Four decimal places are necessary for a and b to get near the same answer as above. Actually, even this is not totally accurate. It took seven decimal places for a and b to get the same result, .393.

Pictorially, we might represent S_e as shown at right.

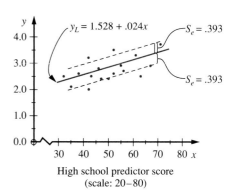

High school predictor score
(scale: 20–80)

Essentially, then, the sample regression line provides an estimator of the true population regression line, and the sample measure of spread, S_e, provides an estimator of the population measure of spread, σ_e.

Now suppose we wish to randomly select another individual from the Middletown College student population, can we guess this student's freshmen GPA? Well, if we knew that student's high school predictor score, we could use the sample regression line to give a reasonable estimate. Let's try an example.

Suppose the student's high school predictor score was known to be 55. We would then plug $x = 55$ into the regression-line equation to get 2.848, illustrated as follows:

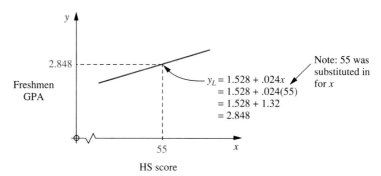

Notice we used the sample regression-line equation to calculate the best estimate GPA for that student, 2.848. But how accurate is this estimate? Certainly not every student with a high school predictor score of 55 will get a college freshmen GPA of 2.848. That's true. A GPA of 2.848 is our best estimate—that is, it estimates the true average freshmen GPA for a high school score of 55.*

But this raises two questions. How close is it to the true average freshmen GPA for high school score 55 and how much can GPAs actually vary for a particular student with high school score 55? To answer these questions, the following two intervals will be calculated.

1. The 95% confidence interval estimate of Y_L, the true *population* regression-line value at $x = 55$.

2. The 95% prediction interval for Y, a particular freshman's GPA who scored $x = 55$.

The following illustrates this.

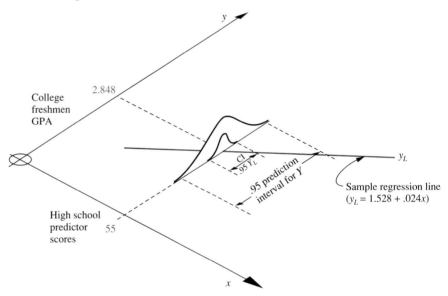

*Remember, to meaningfully use the sample regression line for population estimates, we must be reasonably assured at least some correlation exists in the population. Recall, we achieved for the data, a correlation coefficient of $r = +.58$, which was shown to be significant at .05. This means, at $\alpha = .05$, $\rho \neq 0$. In other words, some correlation exists in the population.

Although the sample regression line gives us a central estimate, around which to construct the two intervals, and S_e provides the sample measure of spread by which we may estimate the population measure of spread, σ_e, one more complicating factor arises. The spread of the intervals *also,* strangely enough, depends on the position of your x-value.

The following explains why this is so.

Suppose we were to sample $n = 17$ observations from the same population thousands of times and plotted the resulting regression lines on the same graph, as follows:

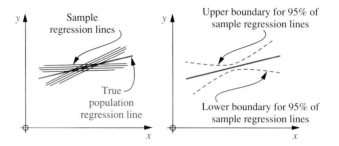

The sample regression lines will tend to pivot near some central x-value.* If we were to enclose the sample regression lines in a sort of middle-95%-of-the-lines zone, the boundary would be distinctly curved. In turn, this will make any confidence interval (CI) we may construct tight near the center of the x-values and spread out at the extremes. In fact, the interval is tightest at $\bar{x}$, the average x-value.

Because of this dependency on the position of x, we must factor in an adjustment that takes into consideration how far our particular x-value is away from $\bar{x}$. This adjustment is incorporated into the formula in the form of the square roots, shown as follows.

---

*This was discussed in a prior footnote, however, it's worth repeating. The sample regression lines all go through their central $\bar{x},\bar{y}$-value. And since sample averages approximate population averages, $\bar{x},\bar{y}$ will approximate μ_x,μ_y, the population center, giving the appearance that the sample regression lines are pivoting near some central value, which in fact they are.

95% CI of Y_L, the true population regression-line value

$$y_L \pm t\, S_e \sqrt{\frac{1}{n} + \frac{n(x - \bar{x})^2}{n\Sigma x^2 - (\Sigma x)^2}}$$

95% prediction interval for Y

$$y_L \pm t\, S_e \sqrt{1 + \frac{1}{n} + \frac{n(x - \bar{x})^2}{n\,\Sigma x^2 - (\Sigma x)^2}}$$

where

y_L = the y-value of the sample regression line for a given x

t = the t score for a 95% CI, df = $n - 2$ (in our case, df = $17 - 2 = 15$)

S_e = the sample measure of spread (the standard deviation of vertical distances from points to the regression line) and is used to estimate σ_e

n = the number of (x,y) observations

$$y_L \pm t\, S_e \sqrt{\frac{1}{n} + \frac{n(x - \bar{x})^2}{n\Sigma x^2 - (\Sigma x)^2}}$$

$$2.848 \pm 2.13(.393) \sqrt{\frac{1}{17} + \frac{17(55 - 50.1176)^2}{33,724}}$$

$$y_L \pm t\, S_e \sqrt{1 + \frac{1}{n} + \frac{n(x - \bar{x})^2}{n\Sigma x^2 - (\Sigma x)^2}}$$

$$2.848 \pm 2.13(.393) \sqrt{1 + \frac{1}{17} + \frac{17(55 - 50.1176)^2}{33,724}}$$

Note:

1. The value 33,724 under the square root was obtained from section 9.2. The value is the same as the b fraction denominator.

2. $\bar{x}$ = the average of all x's, $\Sigma x/n$ (in our case, $\Sigma x/n = 852/17 = 50.1176$, where the sum 852 was obtained from the calculations in section 9.2).

$$2.848 \pm 2.13(.393) \sqrt{.0588 + \frac{405.2426}{33,724}}$$

$2.848 \pm 2.13(.393) \sqrt{.0588 + .0120}$

$2.848 \pm 2.13(.393) \sqrt{.0708}$

$2.848 \pm 2.13(.393)(.2661)$

$2.848 \pm .2227$ (or approximately 2.63 to 3.07)

$$2.848 \pm 2.13(.393) \sqrt{1.0000 + .0588 + \frac{405.2426}{33,724}}$$

$2.848 \pm 2.13(.393) \sqrt{1.0000 + .0588 + .0120}$

$2.848 \pm 2.13(.393) \sqrt{1.0708}$

$2.848 \pm 2.13(.393)(1.0348)$

$2.848 \pm .8662$ (or approximately 1.98 to 3.71)

These calculations are sensitive to rounding technique so answers may vary. Thus,

$$2.63 \leq \quad \begin{matrix} \text{population} \\ \text{value} \\ Y_L \end{matrix} \quad \leq 3.07$$

$$1.98 \leq \quad \begin{matrix} \text{GPA for a} \\ \text{randomly} \\ \text{selected} \\ \text{student with } x = 55 \end{matrix} \quad \leq 3.71$$

We are 95% sure the population regression-line value, Y_L, is between 2.63 and 3.07, for a high school predictor score of 55.

If a particular student with HS score $x = 55$ is randomly selected, then we are 95% sure the student's GPA is in this interval.

(Actually, if we draw repeated samples of $n = 17$ and use the results of each to calculate a regression line and prediction interval, 95% of the intervals will include this student's GPA.)

Visually, this might be presented as follows:

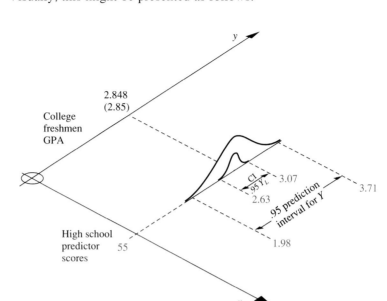

Summary

Origins of the Concept

What Gauss's discovery of the normal curve was to the early 1800s, Galton's discovery of regression-correlation was to the late 1800s; both precipitated a flurry of subsequent exploration, excitement, and useful application.

Bivariate data: a type of data where a linked *pair* of measurements is thought of as a single entity.

Although Galton provided the conceptual framework for regression-correlation, most notably with his parent-height–offspring height data expressed as a collection of bivariate entities, Karl Pearson's zeal was probably the pivotal factor that propelled regression-correlation analysis into international prominence (although mostly confined to the field of biological measurement). G. Udny Yule applied least-squares analysis to the calculation

of the regression line, which indelibly linked Galton's work to existing statistical theory and opened the door for all manner of advanced analysis (such as, multiple regression and ANOVA) and far broader application into the social sciences and many other fields.

Although regression-correlation has developed into an important tool in both the hard and soft sciences, keep in mind

A strong correlation between two variables does *not* prove cause and effect.

Review of Basic Techniques

Graphing and analyzing techniques were reviewed: points and lines plotted, locating a point given one coordinate, and calculating vertical distances (*y*-distances).

Organizing and Analyzing Bivariate Data

Several concepts concerning bivariate data were presented as follows.

A scatter diagram is often used to represent a collection of x,y (bivariate) observations (an example is shown as follows).

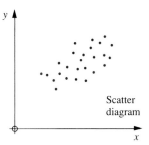

Scatter diagram

Yule's least-squares regression line is the preferred measure of central position for bivariate data. The constants a and b in the line equation represent the y-intercept and slope, respectively, calculated as follows:

$$a = \frac{\Sigma y \, \Sigma x^2 - \Sigma x \, \Sigma xy}{n \, \Sigma x^2 - (\Sigma x)^2}$$

$$b = \frac{n \, \Sigma xy - \Sigma x \, \Sigma y}{n \, \Sigma x^2 - (\Sigma x)^2}$$

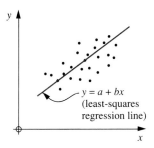

$y = a + bx$
(least-squares regression line)

S_e is the standard deviation of the vertical distances from points to regression line. Unfortunately, S_e does not provide us with an adequate measure of correlation.

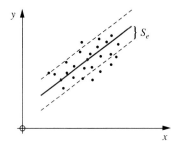

Correlation is defined as the ability of one variable to predict the value of another and r, the linear correlation coefficient, as a measure of this ability, calculated as follows:

$$r = \frac{n \, \Sigma xy - \Sigma x \, \Sigma y}{\sqrt{n \, \Sigma x^2 - (\Sigma x)^2} \, \sqrt{n \, \Sigma y^2 - (\Sigma y)^2}}$$

r varies as follows: $-1.00 \le r \le +1.00$

The closer r draws to -1.00 or $+1.00$, generally the closer data points cluster to a distinctly sloping regression line, the more confident we are the regression line will give reliable predictions of y and the stronger we say the correlation is.

A more visual interpretation of r, the one originally envisioned by Sir Francis Galton, defines r as the *standard slope*. In other words, r is the expected fractional standard deviation change in y for one standard deviation increase in x.

$$\text{Standard slope} = \frac{\text{standard deviation change in } y}{1 \text{ standard deviation increase in } x} = \frac{r}{1} = r$$

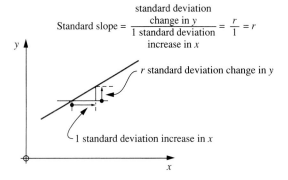

r standard deviation change in y

1 standard deviation increase in x

Test of Significance for r

Although $r \approx \rho$ (rho), that is, the sample correlation is approximately equal to the population correlation coefficient, we can very well achieve sizable sample correlations (r-values) when indeed no correlation exists in the population, that is, when $\rho = 0$. This is especially true the smaller the sample size. To test the hypothesis, $\rho = 0$, we compare our calculated r with a cutoff value obtained from the r table in the back of the text (df $= n - 2$) and accept $\rho = 0$ for any calculated r between ±cutoff value.

Explained Variation

The percentage explained variation can easily be found using $100 \; r^2$. Essentially explained variation is the extent sum of squared distances (from points to line) are reduced using the regression line, y_L, to represent the points rather than the average y-value, $\bar{y}$, to represent the points.

Confidence Interval Estimate of Y_L and Prediction Interval for Y

S_e and other values are used to estimate confidence and prediction intervals in the population.

Exercises

Note that full answers for exercises 1–5 and abbreviated answers for odd-numbered exercises thereafter are provided in the Answer Key.

9.1   Explain why the least-squares line is commonly referred to as a ''regression'' line.

9.2
a. Define co-relation (now, correlation).
b. Explain why S_e, the standard deviation of vertical distances from the x,y-observations to the regression line, is not a dependable measure of the strength of the correlation.

9.3   Over the past 100 years, there has been a high positive correlation between tonnage of tobacco sold in the United States and life expectancy of smokers. In other words, the more tobacco sold, the longer smokers live.

a. Does increased national consumption of tobacco cause increased life expectancy among smokers?
b. What unmeasured third variable might be causing both tobacco consumption and life expectancy to increase together?

9.4   Suppose Hoover College in Flat River, Missouri, a fictitious school, wished to test whether a particular high school predictor model (SAT scores combined with high school GPA) could be used to predict success at their school. The records of twelve students were randomly selected, yielding the following:

x (HS predictor score: 20–80 scale)	y (College GPA, freshmen year, for that student: 0–4 scale)
55	2.6
42	2.9
61	3.0
32	2.2
67	2.6
47	2.1
52	3.2
36	1.5
53	2.0
39	1.9
59	2.3
70	3.5

a. Construct a scatter diagram.
b. Calculate a and b. Write the least-squares equation in $y = a + bx$ form, then use the equation to predict freshmen GPA for a high school predictor score of: (i) 33 and (ii) 58.
c. Use the two x,y-points calculated in part b to plot the least-squares line.
d. Calculate the correlation coefficient, r, a measure of the strength of the relationship between x and y.

e. Is r significant at $\alpha = .05$? Explain how this decision was reached.

f. What percentage of the variation in freshmen GPA is "explained" by its relationship with high-school predictor score? Briefly discuss the meaning of this result.

g. Assuming the above is a valid random sample from the Hoover student population, what do the results indicate?

9.5 Use the results of example 9.4 to determine the following information for a student with a high school predictor score of 53.

a. Use the regression-line equation to find the best-estimate GPA for high school score 53.

b. Calculate S_e, the standard deviation of vertical distances from points to line.

c. Construct a 95% confidence interval estimate of the population value, Y_L.

d. Construct a 95% prediction interval of GPAs for a student in the population with high school score 53.

9.6 In a study on marriage satisfaction in women (as reported in *Glamour*, Dec. 1990, p. 58), researchers at the University of Michigan sought to explore the marketplace theory of dating, that is, to determine whether a women should spend time "shopping around" before settling down with the right man. Suppose we were to conduct a study of several randomly selected Ohio women, yielding the following data:

x (Length of time spent dating prior to marriage; including time spent with husband-to-be in years)	y (Level of marriage satisfaction ten years into the marriage; 0 to 100% scale)
10	47
4	43
6	58
1	72
.5	54
5	87
7	62

a. Construct a scatter diagram.

b. Calculate a and b. Write the least-squares equation in $y = a + bx$ form. Use this equation to predict level of marriage satisfaction for dating times of
(i) 5 years and (ii) 1 year.

c. Use the two x,y-points calculated in part b to plot the least-squares line.

d. Calculate r, the correlation coefficient. At $\alpha = .05$, is r significant?

e. What percentage of the variation in marriage satisfaction is explained by its relationship with length of time dating?

f. Assuming the above data is a valid random sample of what do the results indicate? Explain briefly.

9.7 Artificial skin (grown from human dermal fibroblast cells and cultivated in $4'' \times 8''$ sheets) can be kept alive for several weeks in a nutrient solution while being shipped and prepared for commercial use. The sheets are used in surgical operations by hospitals and by cosmetic companies, such as Estee Lauder and Helene Curtis, for testing skin products for irritation, toxicity, etc., and by others.

ArtTech, a leading supplier of the skin, wanted to determine if the *price* for the product has any influence on *demand* (reflected in number of sheets sold). Seven test markets yielded the following data.

x (Price: in hundreds of dollars per sheet)	y (Demand: in hundreds of thousands of sheets sold, scaled to national consumption)
$3	8
5	6
4	5
7	4
3	6
8	4
9	3

a. Construct a scatter diagram.

b. Calculate a and b. Write the least-squares equation in $y = a + bx$ form. Use the equation to predict demand for a price of (i) $8 and (ii) $3.50.

c. Use the two *x,y*-points calculated in part b to plot the least-squares line.
d. Calculate *r*, the correlation coefficient. At α = .05, is *r* significant?
e. What percentage of the variation in demand is explained by its relationship with price?
f. Assuming the above data is a valid random sample from the population (all national markets for the product), what do these results indicate?

9.8   Satisfaction in life has long been a topic of research at the University of California–Berkeley and at Wesleyan, Illinois, and Rutgers Universities. After more than a half century of research, a dominant theme seems to emerge, reflected in the following example.

Suppose nine randomly selected individuals, monitored over a thirty-year period, yielded the following data.

x (Satisfaction in life at age 16: 0 to 20 scale)	y (Satisfaction in life at age 46 for same person: 0 to 20 scale)
5	11
7	6
4	9
12	16
15	14
2	5
11	15
3	7
9	8

a. Construct a scatter diagram.
b. Calculate *a* and *b*. Write the least-squares equation in $y = a + bx$ form. Use this equation to predict satisfaction in life at age 46 for an individual whose satisfaction in life at age 16 is (i) 3 and (ii) 14.
c. Use the two *x,y*-points calculated in part b to plot the least-squares line.
d. Calculate *r*, the correlation coefficient. At α = .05, is *r* significant?
e. What percentage of the variation in satisfaction at 46 is explained by its relationship with satisfaction at 16?
f. Assuming the above data is a valid random sample from the general population, what do these results indicate? Briefly explain.

9.9   Cigarette smoking, the largest single preventable cause of premature death, costs this country in cigarette-related illnesses and lost productivity $2.17 for *each* pack of cigarettes smoked (according to 1989 studies by the Surgeon General). Smoking has been linked to a variety of health problems (heart attacks, fetal injury, respiratory problems) but perhaps none more dreaded than lung cancer.

Suppose regression-correlation analysis was used to assess the results of the following study, which monitored several thousand people over decades of cigarette usage and recorded the incidences of lung cancer, described below.

x (Cigarette usage: in packs smoked per day)	y (Risk multiple:* lung cancer for each group)
0	1
.5	5
1	12
1.5	18
2	23
2.5	27

*The risk multiple is best explained by example. For instance, for the second (*x,y*) pair given above (.5, 5), this means individuals who smoked .5 packs of cigarettes per day at the start of the study had 5 times the incidence of lung cancer during the course of the study as those who never smoked.

a. Construct a scatter diagram.
b. Calculate *a* and *b*. Write the least-squares equation in $y = a + bx$ form. Use this equation to predict the risk multiple for a person who smokes the following number of packs of cigarettes per day: (i) .7 and (ii) 2.2.
c. Use the two *x,y*-points calculated in part b to plot the least-squares line.
d. Calculate *r*, the correlation coefficient. At α = .01, is *r* significant?
e. What percentage of the variation in risk is explained by its relationship with number of packs smoked per day?
f. Assuming the above data is a valid random sample of the overall population, what do these results indicate? Briefly explain.

9.10 Girls emerge from adolescence feeling quite differently about themselves, according to several studies (at the University of Pittsburgh, Harvard University, and the Association of University Women). Suppose a number of randomly selected girls in the Ithaca-Binghamton area of New York State yielded the following data:

x (Girl's age in years old)	y (Self-esteem: 0 to 100 scale)
8	72
11	64
7	82
14	53
13	41
16	46
10	70
12	62

a. Construct a scatter diagram.

b. Calculate a and b. Write the least-squares equation in $y = a + bx$ form. Use this equation to predict self-esteem for a girl of age (i) 9 and (ii) 15.

c. Use the two x,y-points calculated in part b to plot the least-squares line.

d. Calculate r, the correlation coefficient. At $\alpha = .01$, is r significant?

e. What percentage of the variation in self-esteem is explained by its relationship with a girl's age?

f. Assuming the above data is a valid random sample of the total U.S. girl population, what do these results indicate? Briefly explain.

Endnotes

1. Sir Francis Galton studied medicine at Cambridge but never practiced. Independently wealthy, he explored Africa and studied and wrote about weather forecasting, anthropology, sociology, education, and finger printing. He witnessed in 1859 the publication of his cousin's famous book, the publication which to this day seems groundbreaking, Darwin's *Origin of Species* concerning evolution. Ten years later, Galton published *Hereditary Genius,* but the book was only a mild success. Galton was trying to show that talent runs in families (not exactly a surprising premise), and his presentation was mostly anecdotal, lacking a credible statistical base.

   Over the next twenty years, Galton sought to develop a statistical analysis compatible with this type of social science experiment. What resulted was an entirely new branch of mathematics called *regression-correlation analysis.* To this day, regression-correlation analysis along with subsequent breakthroughs, chi-square and ANOVA, are probably the most widely used statistical techniques in the social sciences. For further reading, refer to

   S. Stigler, *The History of Statistics,* (Cambridge: Belknap Press of Harvard University Press, 1986).

   H. Walker, *Studies in the History of Statistical Method,* (Baltimore: Williams & Wilkins Co., 1929).

   Sir F. Galton, "Regression Towards Mediocrity in Hereditary Stature," *Journal of the Anthropological Institute,* 15(1886):246–263.

2. Galton's actual table was presented with offspring height on the x-axis. The axes were switched for this demonstration to conform with the practice of presenting the intended independent variable, in this case, parent-height, on the x-axis.

3. Galton's rearrangement of the data seemed somewhat arbitrary, and since the method is not in contemporary use, no attempt is made to present the actual rearrangements in this demonstration. Actually, Galton claims to have ("smoothed") rearranged the entries "by writing at each intersection of a horizontal column and a vertical one, the sum of the entries in the four adjacent squares and using these to work upon" (Stigler, 1986). Walker (1929) claims Galton used the average of the four adjacent cells.

4. Galton first conducted experiments using peas (circa: 1874–1877). He used the weight of parent peas paired to offspring weight. He noted the same effect. Many separate normal distributions would form, each with a different mean (dependent on parent weight) and each with the *same* standard deviation. Refer to Francis Galton, "Typical Laws of Heredity," Nature 15(1877):492–495, 512–514, 532–533. For discussion, see Stigler (1986, pp. 281–283).

5. See endnote 1, last reference.

6. Actually the influence of improved nutrition on height in the 1900s has added a complicating factor. I suspect the referenced data might now read: parents 3″ taller than the average of their generation tend to produce offspring about 2″ taller than the average of their generation, although the true solution appears to be one of multiple, not simple, regression (not withstanding Mendelian laws of genetic inheritance, discussed in endnote 7).

7. H. Walker and J. Lev, *Elementary Statistical Methods* (Holt, Rinehart and Winston, Inc., 1969, p. 211).

8. It seems in 1900, with the rediscovery of Mendel's papers on crossing peas and the resulting formulation of the genetic basis of inheritance, these large-scale observations of Galton fell from favor and the Mendelian model persisted. In 1918, R. A. Fisher tried to weld the two approaches. For further reading on this: R. A. Fisher, ''The Correlation Between Relatives on the Supposition of Mendelian Inheritance,'' *Trans. Royal Society, Edin.* 52(1918):399–433. However, Galton's statistical insights (redefined by Yule, Pearson, and Fisher) still flourish today in regression-correlation analysis. And the concept of regression toward the mean is still of concern to modern researchers in the social sciences and other fields.

9. G. U. Yule, ''On the Theory of Correlation,'' *Journal of the Royal Statistical Society* 60(1897):812–854.

10. The concept of correlation was first discovered by Galton in 1888. Essentially, he used definition #2 below. However, today, there are a number of ways correlation is defined, as follows:

1. Pearson's product-moment definition, 1892.
2. As the slope of the regression line (when data is expressed in terms of their standard deviation from $\bar{x}$ and $\bar{y}$ (Galton, 1888, referenced below).
3. As explained, variation, r^2.
4. Brogden's method (as the expected proportional improvement in y from random selection to perfect selection). Refer to H. E. Brogden, ''On the Interpretation of the Correlation Coefficient as a Measure of Predictive Efficiency,'' *Journal of Educational Psychology,* 27(1946):65–76.

Refer to F. Galton, *Co-relations and Their Measurements, Chiefly from Anthropological Data,* ''Proceedings of the Royal Society of London'' 40(1888): 135–145. Also see ''Kinship and Correlation,'' *North America Review,* 150(1890):419–431.

11. Historical note: Initially, 1885–1892, not everyone understood nor appreciated Galton's work. There was one notable exception, F. Ysidro Edgeworth. Edgeworth, a classic literature scholar and self-trained mathematician, was the first to recognize the potential of Galton's statistical analysis. Edgeworth was trying to apply principles of statistics to the social sciences, mostly in economics, and was having the same difficulty Galton encountered, that of finding a sound statistical framework to describe the effect of one variable on another in a world of competing influences.

But Edgeworth was a shy and retiring man and it wasn't until after W. F. R. Welton's application of Galton's work to further studies in biology and the subsequent interest of Karl Pearson (then professor of applied mathematics and mechanics at University College London) that the work's impact imploded onto the scene. However, the emphasis at this early stage was clearly on biology. In fact it was the threesome of Pearson, Welton, and Galton who later in 1901 founded the prestigious journal *Biometrika*. Karl Pearson's research assistant, G. Undy Yule (1897–1899) bridged the gap with established statistical theory (via least-squares analysis) and with the continual attraction of productive students to Pearson's laboratory (Wm. Gossett [for more information on Wm. Gossett, see endnote 1 in chapter 7 and endnote 10 in chapter 10], Alice Lee, and others) the infant discipline was truly born. Pearson and Yule worked on multiple regression (late 1890s), whereas Pearson developed chi-square (1900).

12. Actually, in 1897, Yule had not realized the far-reaching implication of his own work. He initially chose to use the least-squares method ''solely for convenience of analysis'' (Yule, 1897). Calculus could be used to easily derive the equations (see endnote 17). It was not until 1899 and on that Yule further refined his work and realized the power of what he had initially stumbled upon ''for convenience.'' ''By the 1920s, Yule's approach came to predominate in applications in the social sciences, particularly in economics,'' (Stigler, 1986).

13. A. M. Legendre, 1805: *Nouvelles Methodes Pour la Determination des Orbits des Cometes,* (Paris: Courcier, 1805). Reissued with supplement, 1806. A second supplement was published in 1820. The method of least squares is found in the appendix of Legendre's work. Actually, Gauss claimed to have discovered the method of least squares ten years prior to Legendre's 1805 publication. This is one of the more famous on-going priority feuds in mathematics.

14. Likewise, Legendre found the center of gravity for several equal masses in space is the point that has the *least sum of squared distances* around it.

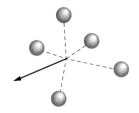

15. The five observations were 20° plus: 5', 7', 9', 9', and 10'. To solve for the position of *least* sum of squared distances, that is: $\Sigma (x - p)^2 = minimum,$ where $(x - p) =$ distance from observation, $x,$ to any position on number line, $p.$ Thus, $(5 - p)^2 + (7 - p)^2 + (9 - p)^2 + (9 - p)^2 + (10 - p)^2 = SS$ (sum of squared distances), which we must minimize. Multiply and collect terms, to get $336 - 80p + 5p^2 = SS.$ This parabola can be plotted on a graph to show the minimum (8',16) vertex value or calculus can be used. Take the derivative and set slope equal to zero, to get minimum at (8',16). This calculation gives the best estimate of central position, 20° 8', and the minimum sum of squared distances, 16.

16. Gauss actually tested several measures of spread: when he used absolute distances he called it $e_1,$ using squared distances, $e_2,$ cubed distances, $e_3,$ and on up to $e_6.$ He found e_2 (using squared distances) the most reliable measure. e_3 was second most reliable, and e_1 third (Walker, 1929, p. 72). In the early 1900s, Sir Ronald Fisher verified e_2 more efficient than e_1 (Bessel's function) (D. A. MacKenzie, *Statistics in Britain. 1865–1930,* [Edinburgh: Edinburgh Univ. Press, 1981, p. 207]).

17. In simplified form, Yule pursued the following basic line of reasoning: $\Sigma (y - y_{line})^2 =$ a minimum: where $(y - y_{line})$ is the vertical distance from the y-value of a data point to the y-value at the regression line. Then, $\Sigma [y - (a + bx)]^2 =$ a minimum. To solve for a and $b,$ take the partial derivative with respect to $a,$ set to zero, and rearrange to get an equation. Then take the partial derivative with

respect to *b,* set to zero, and rearrange to get a second equation. The two resulting equations are presented below, and these are to be solved as simultaneous equations for *a* and *b.*

$$na + b \Sigma x = \Sigma y$$
$$b \Sigma x^2 + a \Sigma x = \Sigma xy$$

18. In presenting Yule's results, some changes were instituted for clarity. First, Yule used the ratio or the number of out-paupers to in-paupers for the *y*-variable. This was transposed to the *x*-axis (as the intended independent variable) as percentage of welfare recipients receiving direct-money payment. Second, Yule's table on 580 poor-law unions was represented as a lesser number of dots on a scatter diagram. In Britain of the day, a poor-law union consisted of two or more parishes combined for administrative purposes. Thus, it was akin to an administrative district (Stigler, 1986), labeled welfare district in this presentation.

19. Actually, Yule conducted the analysis in response to data presented in a book by Charles Booth, *The Aged Poor.* The author had claimed, in effect, that the percentage of direct-money payment has no bearing on level of poverty in a district. Using the data presented in the book, Yule attempted to demonstrate that the data reflected the contrary. The magnitude of the correlation (.388, highly significant at *n* = 580) to Yule, hinted of some form of causality. However, it was pointed out to Yule by A. C. Pigou, professor of political economy at Cambridge, whose opinion was elicited by the Royal Commission in Britain which was meeting to consider reforms in the ''welfare'' system; Pigou submitted written testimony, part of which was a critique on Yule's work entitled, *The Limitations of Statistical Reasoning,* that an undetected, and as yet, unmeasured ''lurking variable'' was producing what appeared to be a causal relationship between the two measured variables presented by Yule, for instance, quality of the welfare managers in a district.

To this day, this still remains the dilemma of regression-correlation analysis. It does not prove cause and effect. There may very well be a third (or more) factor(s), as yet unknown or unmeasured, that is causing the two measured factors to appear to be in some way related. For further reading on the above case, refer to Stigler (1986).

20. Yule's study of 580 poor-law unions yielded a correlation coefficient of *r* = .388, highly significant for such a large sample size (note that this will have meaning only after completion of the material in this chapter).

In fact, in most of these early experiments, only very large samples were used (many hundreds of measurements). It wasn't until after Wm. Gossett's groundbreaking work with small-*n* sampling, in 1908 (Student's *t* distribution) that experiments with small-sample data gained credibility. Actually, it wasn't until Sir Ronald Fisher (1920s) endorsed Gossett's work by incorporating *t* values in ANOVA experiments that credibility was truly established.

21. Yule very quickly recognized this concept of a lurking or unmeasured variable. To some extent, his development of multiple regression reflected this concern. Unfortunately, no matter how many factors we include in the analysis, there is always the possibility that some vital factor affecting the data is missed.

22. In 1880, only one American college offered a course in statistics, Columbia, in the department of economics. Later, various universities, in the department of economics, offered statistics courses. In 1887, The University of Pennsylvania introduced a statistics course in the department of psychology. In 1889, Clark (anthropology); 1897, Harvard (biology); 1898, Illinois (mathematics); 1900, Columbia (education). By 1925, a majority of universities and colleges in the nation had courses in statistics, offered by various departments (mostly in the social sciences and mathematics, but also in education, business, public health, and agriculture). For further reading, see Walker (1929, pp. 151–163).

23. Historical endnote: As the military, industrial, and educational complexes grew markedly in the world in the late 1800s, so did the need to evaluate the intelligence of individuals. ''Clearly, teachers needed to be able to distinguish between different mental capacities in order to educate children suitably. Similar problems arose in military organizations and in industries in connection with attempts to fit individuals into appropriate positions'' (Tyler and Walsh, p. 44, referenced below).

In 1905, Binet-Simon published the first intelligence test for children based on observations of Binet's own daughters (Paris). Actually, Binet's tentative definition of intelligence might be of some interest: ''the tendency to take and maintain a definite direction; the capacity to make adaptations for the purpose of attaining a desired end; and the power of auto-criticism.'' Soon, the test was incorporated into other countries. The American version, the Stanford-Binet test, was first published in 1916 (revised 1937 and on), and unleashed an almost unprecedented era of testing. In 1939, the Wechsler-Bellevue Intelligence Scale was introduced and later, in 1955, the Wechsler Adult Intelligence Scale. Although ''there has been a decline in interest in pure intelligence tests since the 1920s, [to some extent because] it is difficult to ensure that test items are equally meaningful or difficult for members of different social groups, [there was] a corresponding increase in the number of mental tests that measure special aptitudes and personality factors'' (*New Columbia Encyclopedia,* 1975, p. 1348).

During the depression of the 1930s, thousands of people lost their jobs while many more thousands of students were graduating and seeking employment. Out of desperation, many were forced to enter new fields of work in which they had no prior experience or training. ''This situation led to the initiation of a large-scale program to develop vocational aptitude tests designed to help predict how successful a person would be at an occupation 'before' entering it. What grew out of this effort was the two and one-half hour General Aptitude Test Battery (GATB) which made it possible to evaluate an individual's suitability for hundreds of different jobs.'' Also, ''as more and more people have become interested in mental health and aware that psychological ills can be treated, demand for personality tests has grown'' along with psychiatric tests to identify illnesses, such as, schizophrenia (Tyler and Walsh, p. 3).

Suggested reading on this topic: L. Tyler and W. Walsh, *Tests and Measurement,* (Englewood Cliffs, N.J.: Prentice-Hall, Inc., 1979).

24. Actual validity studies normally include many hundreds of students and predictor factors are weighed using multiple-regression techniques.

25. For further information, contact the College Entrance Examination Board in Princeton, New Jersey.

26. For further discussion concerning experiments in education, refer to D. Campbell and J. Stanley, *Experiments in Quasi-Experimental Designs for Research,* (Boston: Houghton Mifflin Company, 1963, p. 65 and other pages). In one chapter, Campbell and Stanley pointed out, ''Almost inevitably we draw the implication that the educational levels of superintendents and stable leadership 'cause' better schools. [However, they contend] better schools . . . might well cause well-educated men to stay on, while poorer schools might lead the better-educated men to be tempted away into other jobs. Likewise, better schools might well cause superintendents to stay in office longer.''

This point of which is the cause and which is the effect was also discussed in a *Newsweek* article, ''Born or Bred'' (February 24, 1992, p. 49) concerning the origins of homosexuality. Male homosexuality has often been linked with father-son hostility, with the implication a hostile father in some way helps bring on the homosexuality in the son. It was contended, however, that homosexuals ''are extremely different when they're young and as a 'result' they can develop hostile relations with their fathers.'' In other words, is the father's hostility causing the homosexuality or is the homosexuality causing the father's hostility?

Most of us have heard about the link between cigarette smoking and lung cancer. In fact, the correlation is extremely high (see exercise 9.9 for actual data). However, we don't really know that it's not the lung cancer (or a lung-cancer gene) causing the cigarette smoking, perhaps by giving the person an early increased susceptibility to nicotine addiction. Cause and effect is a highly complex issue.

Chi-Square and ANOVA

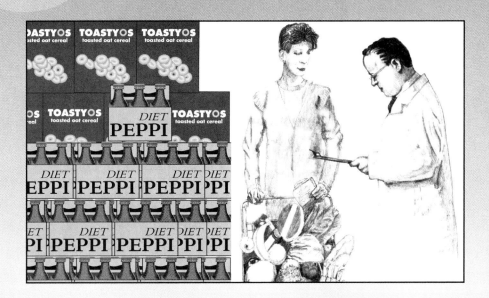

I n the 1920s, Sir Ronald Fisher became the focus of the statistical revolution,[1] verifying and redefining much of what had already been done and integrating new approaches, such as, analysis of variance, degrees of freedom, factorial and other experimental designs, into one unified entity. In a sense, 200 years of fragmented development crystalized under Fisher's guidance into much of the professional methods, theory, and techniques we use and study today.

Fisher realized early that estimating only a single population using very large-sample sizes (experiments at the time usually consisted of many hundreds of observations) was simply inadequate for the cost priorities and complex needs of modern industrial and scientific research. More efficient statistical tools were needed, ones capable of assessing the effects of multiple factors or treatments simultaneously while using only miminal sample sizes.

For such needs, two powerful statistical techniques have emerged and now dominate: chi*-square analysis (X^2) and analysis of variance (ANOVA). Although both techniques were introduced in two somewhat obscure papers (in 1900 and 1918, respectively)[2], the techniques have survived and grown to serve a staggering range of industrial and scientific uses. It would be impossible to demonstrate all the diverse applications and versatility of these two powerful techniques, however this chapter attempts to offer an introduction. ▼

10.1 Chi-Square X^2: Tests of Independence and Homogeneity

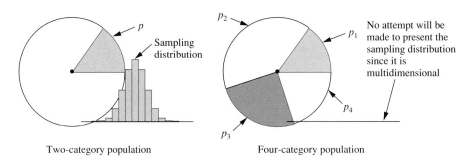

χ^2 Tests are based on sampling from two-or-more category populations such as those above.

The purpose of this and the next section is to demonstrate various uses of the $\mathbf{X^2}$ **test.** To a great extent, all X^2 tests are the same. They all record data in the form of frequencies or counts, that is, the number of subjects or objects falling into

All endnotes (with numbered references) are presented at the end of the chapter. It is not necessary to consult endnotes since they are mostly reference sources and historical fine points.

*Chi (X) is the twenty-second letter of the Greek alphabet, pronounced khi (rhymes with, sky).

distinct categories, with the test measuring z-score differences between the number of counts expected in a category (given some null hypothesis) and the number of counts actually observed in the experiment. The purpose of the test is most often to determine, much like regression-correlation, the extent one variable influences another.

Although the analysis and reasoning is much the same for all X^2 tests, the technical interpretation differs slightly. Some X^2 tests are said to assess *independence* between two population variables, others *homogeneity* (whether equal proportions exist in a series of populations), and others match sample results to some theoretical distribution to assess *goodness-of-fit*. We present these as three major groupings; keep in mind, however, that the mathematical analysis and interpretation is much the same for all.

One last introductory note: since the emphasis of this chapter will be in good part on marketing research, a brief introduction is presented as follows.

Although marketing research dates back to the late eighteen hundreds, these early research experiments were often little more than gathering testimonials from customers.[3] The actual courtship between statistics and marketing probably began in the 1930s when a far more scientific attitude on the part of consumer product firms sprung up.

To some extent, this new scientific attitude was due to advancements in statistical methodology, especially the new sampling procedures and methods developed by Sir Ronald Fisher, however, to a greater extent, it was probably more due to the harsh economic realities of the time.

In other words, this new scientific attitude was in a way forced on corporations by the tremendous upheaval of the Great Depression. Corporations had only recently automated production to meet a seemingly ever-surging demand that the boom of the 1920s had brought them to expect. Then suddenly, this enormous increase in capacity met with a series of cliff-like drops in demand (1930–1932), and the results were disastrous.*

Until 1925, consumer production firms were product oriented, as were most manufacturing firms at the time, with the philosophy that if a company produces a superior product that is more reliable and better built, people will beat down the door to buy it. Then when the Depression hit, with many corporations pushed to the edge of survival, executives desperately looked to the sales force to find ways to convince the consumer to buy. After a while, marketing departments sprung up to assess new markets and products. However, the enormous risks and outlays involved in entering new markets and introducing new products led

---

*It is difficult for us to imagine the intense suffering brought on by the Great Depression. No depression in the history of this nation—and we've had a few before this one (1873–1878, etc.)—had ever plummeted demand to such low levels, and for such an extended period of time, about 10 years. In its depth, unemployment exceeded 30%; today 10% is feared. People lived in constant terror of losing their jobs. It was not unusual to see families of five and six squatting on big city streets begging for food. With times like those, it is no wonder financially teetering corporations were willing to accept new approaches.

executives to the newly developed tools of statistical research. These tools promised cost efficiency and a reliable scientific base to assess actual consumer wants and needs.

Thus was born the marketing era. Corporations grew more sales oriented in the 1930s and 1940s, initially with the philosophy of trying to convince consumers to buy. Gradually from the 1950s on, corporations became more consumer oriented, with the philosophy of assessing consumer wants and needs, then satisfying them. Companies like Proctor & Gamble, General Foods, Colgate-Palmolive, and their advertising agency counterparts followed this wave into statistical testing of their products. By the late 1950s, statistical research formed an integral part of a great many new product introductions—from defining the early need for the product, through designing, packaging, and advertising strategies, eventually to testing the product in local markets, and gradually to expansion into national and worldwide distribution. This process often took (and still takes) several years with numerous statistical studies conducted along the way. One such study might proceed as follows.

X^2: Test of Independence

Example

ToastyOs, a new cereal, was introduced into a local test market last year to appeal to all adults. Because of its high nutrient content, management felt the more informed, perhaps more affluent, community would gravitate to the product. Therefore advertising and promotion was slanted toward this segment.

Now after a disappointing year on the shelves, there has been speculation based on focus group and panel research that taste preference for ToastyOs may be dependent on education level, that is, the more educated a person is, the more they might prefer ToastyOs. If indeed this is true, then Tabisco Foods, the maker of the product, felt they should capitalize on this market by focusing the major thrust of their marketing campaign on this segment. However, before committing to such a task, it was decided objective data should be gathered to verify these speculations. A research firm, Chi-Square Consultants, was called in. The consultants randomly sampled 500 individuals in the test market area who had tried the product, with the following results:

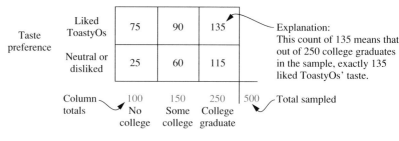

Explanation:
This count of 135 means that out of 250 college graduates in the sample, exactly 135 liked ToastyOs' taste.

	No college	Some college	College graduate	
Liked ToastyOs	75	90	135	
Neutral or disliked	25	60	115	
Column totals	100	150	250	500 Total sampled

Taste preference

Education level

a. Assuming a valid random sample, test at .05 whether the two variables in the population (education level and taste preference) are independent. Essentially, we are testing in this case whether differences in taste preference among the three education categories can be attributed to chance fluctuation or, indeed, to real differences in the population.

b. Briefly discuss the implications of these results for Tabisco.

c. Discuss how Tabisco might question the validity of such a study.

Overview

We will use a binomial example to explore the underlying rationale for all X^2 testing. Once the rationale is understood, a more convenient and more general formula will be introduced.

The previous chart is called a 2×3 contingency table, with two rows and three columns, and is comprised of six cells. Each cell identifies the number of respondents at the intersection of two classifications, one for education level, the other for taste preference. It is typical for such count data (also referred to as category, discrete, or proportion data) to be analyzed using X^2 techniques.

Specifically we will conduct a X^2 test of independence. The underlying purpose is to assess if one variable in the population is in any way influenced by or associated with another variable. Although tests of independence make no distinction as to which variable, x or y, might influence the other, in actual practice, we often suspect one particular variable (which we shall designate on the x-axis) as influencing the other. In other words, in this example, we might suspect that education level influences taste preference. But keep in mind that the x- and y-variables can be interchanged in a X^2 test of independence without any difference in result. Often in real-life experiments, we are not sure if the x-factors are influencing the y-factors or vice versa. Of course, in this particular experiment, it is doubtful whether the y-factors can influence the x-factors, but one never knows (and as with all statistical tests, cause and effect cannot be implied).

Solution

We start with the assumption that in the population, there is *no* influence or association, that the two variables are in no way related, and that indeed they are independent. So, for a test of independence, the initial conditions are as follows:

H_0: In the population, education level and taste preference are independent.	Essentially the test starts with the premise that any one variable in no way influences or is associated with the other.
H_1: Education level and taste preference are dependent.	Of course, if H_0 proves unlikely, then we assume the two variables are dependent. This implies some relationship between the two variables exists.
$\alpha = .05$	We accept the risk that when H_0 is true, 5% of the time we will reject it in error.

Any hypothesis test begins by assuming H_0 is true and, based on this assumption, constructing a set of expectations. If these expectations match closely to what we observe in the experiment, we merely accept H_0 as true. If however, the expectations do *not* match closely to what we observe, we suspect H_0 is not true (thus, we reject it).*

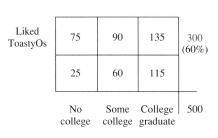

So, if H_0 is true, essentially taste preference and educational level are not related in any way, we can group all the top entries into one overall sum of 300 individuals who liked our product ($75 + 90 + 135 = 300$). Thus, out of a large random sample of 500, 300 liked ToastyOs.†
That's 60% (.60).

$$p_s = 60\%$$

$$\left(\tfrac{300}{500} \times 100 = 60\%\right)$$

In other words, the proportion of the overall sample of 500 who liked our product is 60%, and since large samples generally give good approximations of population characteristics, the population proportion should also be approxi-

---

*Although formally, the test of independence is based on the multiplication rule of probability (chapter 3, section 3.4) where two events are said to be independent if $P(A \text{ and } B) = P(A)P(B)$ and joint marginal probabilities used to estimate cell expected values (see upcoming footnote for a demonstration), we choose to model the analysis on a more practical footing, which will draw similarities to the test of homogeneity and allow an intuitive and less probabilistic interpretation; for those wishing the probabilistic interpretation, detailed footnotes follow.

†Formally in tests of independence, all row and column probabilities are calculated. This is shown below.

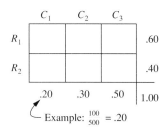

These are then used to estimate the expected cell probabilities, that is, the probability of two independent events occurring together (in one cell) is
$P(A \text{ and } B) = P(A)P(B)$
$P(R_1 \text{ and } C_1) = (.6)(.2) = .12$
$P(R_1 \text{ and } C_2) = (.6)(.3) = .18$
$P(R_1 \text{ and } C_3) = (.6)(.5) = .30$
etc., such that we get the following expected probabilities:

These probabilities are then used to determine the actual number we would expect in each cell by simply multiplying the probability by 500, the total in the study. Expected number = $P(A \text{ and } B) \times 500$ for each cell value.
$EV(R_1, C_1) = .12(500) = 60$
$EV(R_1, C_2) = .18(500) = 90$
$EV(R_1, C_3) = .30(500) = 150$
etc., such that we get the following expected values:

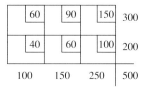

mately 60%. Thus, we can say that out of all the many thousands of people in the test market who have tried ToastyOs, approximately 60% would be expected to like our product, illustrated in the circle graph that will be presented shortly.

Now suppose we randomly select three valid samples from this population. How many in each sample would you expect to like ToastyOs? Answer: approximately 60%. Furthermore, if H_0 is indeed true, that education level and taste preference are truly independent, then each of the three education groups ($n = 100$, $n = 150$, and $n = 250$) can be viewed simply as three valid unbiased samples drawn from this population.* And you would expect each of these groups to prefer ToastyOs equally at 60%. To calculate the actual number we would expect to like ToastyOs in each group, we do the following.

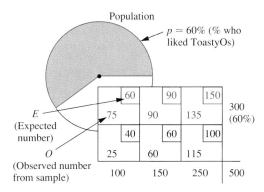

Expected values†
60% of 100 =   60 individuals
60% of 150 =   90 individuals
60% of 250 = 150 individuals

However, like any binomial problem, sometimes you get more or less than expected. To determine the likelihood of expecting one value and observing another, we calculate how many standard deviations (z scores) the observed value is away from the expected value. This is demonstrated on the following page.

*At this point, you might ask: didn't we bias the three samples by separating them by education level? And the answer is: not if H_0 is true, that is, not if education level has no influence on or association with taste preference. In other words, if education level in no way influences taste preference, then there is no bias and the three samples should be as valid and random as the original selection of 500.

†Formally in tests of independence, expected cell frequencies are calculated based on the marginal probabilities (as explained in an earlier footnote). Essentially for the upper left cell, our first calculation would be

$$\text{Expected value} = P(\text{Row}_1 \text{ and Column}_1) \times 500 = P(\text{Row}_1)P(\text{Column}_1) \times 500$$
$$= \tfrac{300}{500} \times \tfrac{100}{500} \times 500$$
$$= \tfrac{300}{500} \times 100$$

We can make two points from this: First, $300/500 \times 100$ is the same as the calculation above, denoted as 60% of 100 (essentially we obtained the 60% by the fraction 300/500). Second, we now have derived a shortcut formula for calculating expected values. Since $300 =$ Row total, $100 =$ Column total, and $500 =$ Grand total, we get

$$\text{Expected value} = \frac{(\text{Row total})(\text{Column total})}{\text{Grand total}}$$

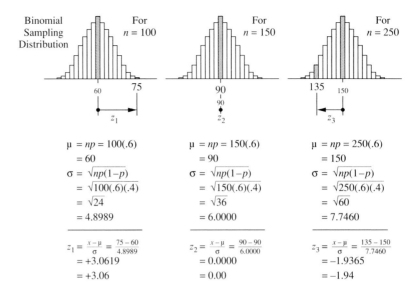

$\mu = np = 100(.6)$
$\quad = 60$
$\sigma = \sqrt{np(1-p)}$
$\quad = \sqrt{100(.6)(.4)}$
$\quad = \sqrt{24}$
$\quad = 4.8989$

$\mu = np = 150(.6)$
$\quad = 90$
$\sigma = \sqrt{np(1-p)}$
$\quad = \sqrt{150(.6)(.4)}$
$\quad = \sqrt{36}$
$\quad = 6.0000$

$\mu = np = 250(.6)$
$\quad = 150$
$\sigma = \sqrt{np(1-p)}$
$\quad = \sqrt{250(.6)(.4)}$
$\quad = \sqrt{60}$
$\quad = 7.7460$

$z_1 = \frac{x-\mu}{\sigma} = \frac{75-60}{4.8989}$
$\quad = +3.0619$
$\quad = +3.06$

$z_2 = \frac{x-\mu}{\sigma} = \frac{90-90}{6.0000}$
$\quad = 0.0000$
$\quad = 0.00$

$z_3 = \frac{x-\mu}{\sigma} = \frac{135-150}{7.7460}$
$\quad = -1.9365$
$\quad = -1.94$

If the z scores are too high, we suspect that H_0 is not true, that the observed values do not all originate from this $p = 60\%$ population. To assess this, the z scores are first squared (to eliminate the negatives) and then added. We call this sum, chi-square (X^2), which is calculated as follows:

$$X^2 \text{ (chi-square)} = z_1^2 + z_2^2 + z_3^2$$
$$= (3.0619)^2 + (0.000)^2 + (-1.9365)^2$$
$$= 9.3752 + 0.000 + 3.7500$$
$$= 13.1252 = 13.13$$

$$\boxed{X^2 = 13.13 \text{ (chi-square)}}$$

A high X^2 value indicates large z-score differences, bringing into question the truth of our initial assumption, H_0. Low X^2 values indicate small z-score differences, suggesting our initial assumption, H_0, may be correct.

Before we continue, note how long and tedious this process was. Fortunately, Karl Pearson* in 1900 algebraically rearranged the elements in the prior formula ($X^2 = \Sigma z^2$) and derived a more general formula that, first, shortens the calculations and, second, more importantly, can be used even when we sample from multicategory populations, such as, a three-category or four-category population. A truly remarkable achievement since calculating z score differences

---

*See endnote 2 for Pearson's original paper.

when sampling from multicategory populations can be quite a complex task indeed.* So, **for all X^2 problems, we will be using the general formula, as follows.**

The General Formula will be used for all X^2 tests	$$X^2(\Sigma z^2) = \Sigma \frac{(O - E)^2}{E}$$	where O = Observed value E = Expected value

To demonstrate its use, let's recalculate this X^2 value ($X^2 = 13.13$) as follows:

▼ The Chi-Square General Formula Applied to This Example

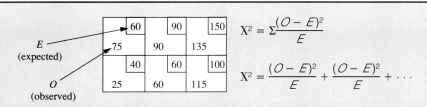

For each and every cell, the expected value is subtracted from the observed value. This quantity is squared and divided by E, as follows:

$$X^2 = \frac{(75 - 60)^2}{60} + \frac{(25 - 40)^2}{40}$$

$$+ \frac{(90 - 90)^2}{90} + \frac{(60 - 60)^2}{60} + \frac{(135 - 150)^2}{150} + \frac{(115 - 100)^2}{100}$$

$$= \frac{225}{60} + \frac{225}{40} + \frac{0}{90} + \frac{0}{60} + \frac{225}{150} + \frac{225}{100}$$

$$= \underbrace{3.750 + 5.625}_{9.375} + \underbrace{0.000 + 0.000}_{0.000} + \underbrace{1.500 + 2.250}_{3.750} = 13.125$$

$$\qquad\qquad z_1^2 \qquad\qquad\qquad z_2^2 \qquad\qquad\qquad z_3^2$$

Note each pair gives us the z^2 we calculated using the prior method (slight differences may occur due to rounding technique).

$$X^2 = 13.13$$

(This is the same answer as in prior calculation.)

Now, what exactly does this X^2 value of 13.13 mean? Well, it's a measure of the likelihood that the observed values could have originated from a $p = 60\%$ population. If the X^2 value is low (meaning, small z-scores differences), then we

*For three or more category populations, calculating actual z-score differences is a convoluted process because of the linear dependency of the variables. The sampling distributions are multidimensional and more sophisticated mathematics (matrix algebra) must be employed.

suspect H_0 is true. In other words, what you expected and what you observed were reasonably close. However, a *large* X^2 value (meaning, large z-scores differences) draws suspicion.

In fact, let's look at the actual z scores we obtained. Especially note the one that was $z_1 = 3.06$ (represented above as $z_1^2 = 9.375$). What is the probability of getting a z score three standard deviations away from the mean? Exceptionally rare indeed.

But this raises the question, at what point do we grow suspicious? In other words, how outrageously large do the z scores (and thus the X^2 value) have to be before we suspect H_0 is *un*true?

Well, we know from theory and many decades of experience that if H_0 were true, that is, education level and taste preference are truly independent, *and* if we were to perform this X^2 experiment hundreds of thousands of times and calculated the X^2 value for each experiment, that 95% of these X^2 values would fall at or below 5.99, depicted as follows:

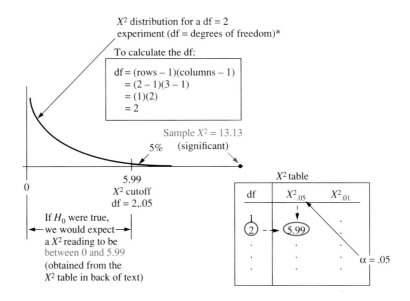

X^2 distribution for a df = 2 experiment (df = degrees of freedom)*

To calculate the df:

df = (rows − 1)(columns − 1)
  = (2 − 1)(3 − 1)
  = (1)(2)
  = 2

Sample X^2 = 13.13
(significant)

5%

0

5.99
X^2 cutoff
df = 2,.05

If H_0 were true,
←we would expect→
a X^2 reading to be
between 0 and 5.99
(obtained from the
X^2 table in back of text)

X^2 table

df	$X^2_{.05}$	$X^2_{.01}$
②	(5.99)	.
.	.	.
.	.	.
.	.	.

α = .05

*This is a X^2 experiment with two degrees of freedom (df = 2). X^2 distributions will be different, depending on the df value. Some examples are presented below. To calculate μ and σ for a particular X^2 distribution, we use the following formulas:

$\mu = df; \sigma = \sqrt{2(df)}$

For our example above:

$\mu = df = 2$.

So, referring to the above graph, if H_0 were true, you would expect a X^2 reading to be, on average, 2, and only 5% of the time to exceed 5.99.

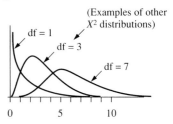

(Examples of other X^2 distributions)

df = 1
df = 3
df = 7

0   5   10

Note, our sample X^2 value was 13.13, indicating unusually large z-score differences, thus we conclude

H_0 is rejected

In other words, the differences between expected and observed values were too great to assume all educational groups preferred ToastyOs equally.

One last technical note before we present the conclusion to this problem.

Rule of Five
All "expected" cell frequencies in a X² experiment must be at least 5.*

All our expected cell frequencies greatly exceeded 5. Note:

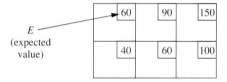

E
(expected value)

| 60 | 90 | 150 |
| 40 | 60 | 100 |

Answer

a. Because the X^2 value of the sample (13.13) exceeded the cutoff of 5.99, we reject H_0. This data supports the hypothesis H_1 that taste preference for ToastyOs and education level are dependent. Stated another way: the sample provides evidence that in the population, a relationship between education level and taste preference exists,† that the fluctuation in sample results was not merely "chance fluctuation," but fluctuation created by the interplay of the two variables in the population.

b. Implications for Tabisco: although the data provides evidence of dependence in the population, we have not as yet established the direction of the dependence. In other words, we might be concerned with which

---

*Cell frequencies must be at least 5 for reasons similar to why np and $n(1 - p)$ must be at least 5 in binomial experiments. Probabilities are based on sampling from a normal distribution, which occur in binomial (and multinomial) experiments when expected cell frequencies are greater than or equal to 5.

†Again, in X^2 experiments, as with regression-correlation, cause and effect cannot be implied. An *un*measured variable may be causing what only appears to be a relationship between the two measured variables.

education group might prefer our product more? To help us explore this, we translate the absolute numbers into percentage of group totals as follows.*

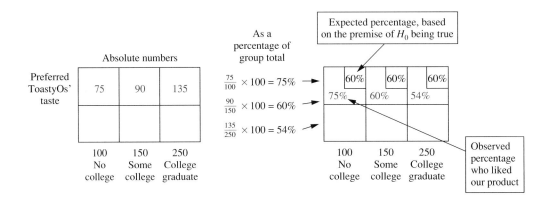

Note that 75% of the no college group prefers our product versus 54% of the college graduate group. If these results truly represent population tastes, the study clearly points to a surprising market for our product, the no college group. And thus the data suggests we might best consider this segment.

c. Questions of validity: At this point, we may be tempted to say, let's redirect our advertising and promotion to the no college group.† However, before such an important marketing decision can be made, Tabisco must address a number of questions concerning validity. In marketing (and other social) experiments, obtaining true valid data is often quite a difficult procedure.

One particular danger in such experiments is whether we achieved independence of response. This requirement dates back three centuries to the tossing of coin and dice experiments (review chapter 4, section 4.4). In fact, the entire mathematical foundation for binomial experiments is based on the premise that the outcome of one toss of a coin or die in no way influences the outcome of any other toss. In marketing experiments, this mostly translates to, the response of one participant in an experiment in no way influences the response of any other participant. Serious violation of this principle can render your data meaningless.

*Note that the absolute numbers do not reflect an accurate picture of which groups prefer our product more. This is because different group totals were involved.

†Actually, questions of marketing can grow quite complex. For instance, say the no college group prefers the upscale advertising of the product. In which case, a change in advertising may very well cause a drop in sales to this group.

For instance, if subjects are tested in the same room, one subject's response may very well influence another subject's response. In marketing textbooks, this is sometimes referred to as the *Asch phenomenon.* In the 1950s, psychologist S. E. Asch conducted experiments documenting the powerful influence groups and group norms have on an individual's behavior.[4] Dietz Leonhard[5] (1967) writes that people in group situations are often more motivated by fear of appearing ignorant of giving a ''wrong'' answer than by truth. Asch's work reenforces this, offering powerful experimental evidence that many people are more motivated by not being part of the group's majority than by their own better judgment.* For these reasons, data gathered from panel and focus groups can be misleading, and such data should be used with great caution.

Even if subjects were questioned (or tested) in isolation, human nature and interviewer/interviewee interaction are still at play. According to D. Leonhard, there is a tendency for people to be nice to the interviewer and also to be polite. Often the results of such questioning do not reflect the respondent's real feelings. Leonhard goes on to offer two well-known dictums of marketing research.

1. Man is least himself when he talks in his own person; [but] when he is given a mask he will tell the truth.

2. What people do not know, they substitute [in] with fantasy.

The point is that establishing validity in such experiments is no easy task and the marketing manager should be trained to be aware of the potential pitfalls of such research. In these experiments, where validity may be difficult to verify, statistical studies should be used in conjunction with other accumulated evidence before drawing any definitive conclusions.†   ▓

In the preceding example, X^2 testing was based on sampling from a two-category population, using the binomial sampling distribution. However, the X^2 test is quite powerful and broad in application. When situations arise involving multicategory populations (more than two categories), instead of using the binomial sampling distribution, we use the multinomial sampling distribution. Now, instead of two rows in the contingency tables, we have three or more rows (depending on the number of categories in the population).‡ However, we still

---

*This issue of group influence formed the thematic thrust of the 1950s motion picture, ''Twelve Angry Men,'' concerning jury deliberation, how two powerful opinions turned and twisted a jury's decision. This motion picture is well worth viewing for anyone considering a profession where group dynamics plays an important role.

†For further discussion and suggested readings in marketing research, see exercise 10.1, answer c.

‡For consistency from example to example, the contingency table is constructed such that each row contains data from only one distinct category in the population. Thus, a two-row contingency table would indicate we are sampling from a two-category population, and a three-row contingency table would indicate we are sampling from a three-category population, etc.

conduct the test in an identical manner using the general formula, which is presented again for your convenience.

This general formula is used for all X^2 tests regardless of the number of categories in the population(s).

$$X^2(\Sigma z^2) = \Sigma \frac{(O - E)^2}{E}$$

To recap, chi-square analysis is designed to evaluate frequency or count data (sometimes referred to as discrete, qualitative, category, or proportion data). In such experiments, we are sampling from a two *or more* category population, whose members are separated into distinct groupings according to some attribute or characteristic. When sampling from such populations, we are interested in the number we might expect from each category, based on a given hypothesis. These numbers are then compared to the numbers we actually observed. If the *z*-score differences are relatively small, reflected in low X^2 readings, we accept the given hypothesis, otherwise we reject it.

X^2: Test of Homogeneity

Very similar to tests of independence are tests of homogeneity. In these cases, certain marginal totals are fixed, such as in a medical experiment where a set number of patients are assigned to different experimental groups (thus, the column totals are fixed in advance).*

Although technically the experiment is viewed in a slightly different manner, the calculations are identical, so don't look for any differences in the mathematics. There are none. In fact, even the interpretation is nearly identical. Effectively we want to determine the extent to which one variable influences another. Let's look at how this works.

Example ——————— The *blood-brain barrier* is a unique layer of protective cells that block potentially harmful compounds (viruses, bacteria, and toxins) from entering the brain. However, it also blocks potentially *helpful* therapeutic drugs. This has been the source of much frustration in the search for successful treatments of neurological disorders, such as, psychiatric problems (depression, insomnia, schizophrenia), Alzheimer's and Parkinson's diseases, stroke, epilepsy, and certain types of dementia.[6]

---

*Note that in tests of independence, the marginal totals are unknown prior to sampling. It is only after sampling that the totals are established. In tests of homogeneity, certain marginal totals are fixed in advance of the experiment, which in effect makes each treatment a sampling from a different population.

We will stick to the convention of using the *x*-axis for the independent variable, thus we will fix only column totals and keep the row values random. Although, technically, we can reverse this, we will not choose to do so in this text.

For some time now, researchers have been experimenting with ways of breaching this protective layer of cells for the purpose of administering helpful drugs to the brain. It had been noted that caffeine, cocaine, and alcohol molecules penetrate the brain's barrier by partially dissolving in the fatty substance that constitutes the brain barrier wall and like ''greased piglets'' slip through. Substances that the brain needs to function, such as iron, amino acids, and immune cells, are actually escorted through by special ''receptor-escort'' cells. A third method of entry involves a compound that binds to the cell wall and produces a brief loosening, creating spaces between adjacent brain barrier cells through which small-molecule drugs might slither through—sort of ''temporary gaps.''

Suppose a team of researchers at the University of California assign 900 patients to four test groups for the administering of Alzaret, a drug used in the treatment of Alzheimer's disease (a fictitious drug), and obtained the following results:*

		Caffeine coated to dissolve in barrier cells (greased-piglet method)	Iron coated to fool cells into accepting it (receptor-escort method)	Immune-cell coat to fool cells into accepting it. (receptor-escort method)	Taken with loosening compound (temp-gap method)	
Level of patient improvement	Major improvement	50	55	50	25	
	Slight improvement	120	75	100	65	
	No improvement	80	70	150	60	
		250	200	300	150	Total: 900 patients

Same medication: administered by four methods

a. Assuming valid random selection, test at .05 whether differences in patient improvement among the four methods can be attributed to chance fluctuation or, indeed, to real differences among the methods. Essentially, are the four populations homogeneous, equally proportioned with respect to patient improvement, or not?

b. Briefly discuss the implications of this research for the medical team.

*Note, column totals are fixed, that is, a fixed number of patients were assigned to each group prior to treatment. This distinguishes the test of homogeneity from the test of independence, although the math and interpretation are nearly identical.

Solution

Initial conditions are established as follows:

H_0: The four methods are homogeneous with respect to patient improvement.
(In effect this means there is no difference among the four methods, that is, each will result in the same levels of patient improvement.)

H_1: The four groups are *not* homogeneous with respect to patient improvement.
(This means one or more methods is more effective than the others.)

In tests of homogeneity, it is assumed each group was sampled from a separate population, but each population has identical (homogeneous) characteristics. In other words, each population has the same proportion showing major improvement, slight improvement, and no improvement.

For this experiment

$$\alpha = .05$$

meaning, if H_0 is indeed true, there is a 5% chance we will reject it in error.

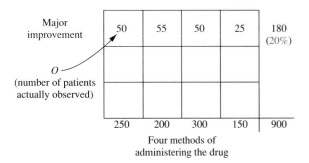

If H_0 is true, meaning we can expect the same percentage to show, for instance, major improvement in each treatment group, then we can get a close approximation of this common percentage by simply combining all those showing major improvement ($50 + 55 + 50 + 25 = 180$) and dividing by 900, the total in our sample:

$$\frac{180}{900} \times 100 = 20\%$$

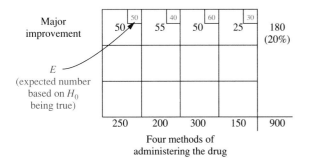

In other words, 20% of all those tested showed major improvement, and this should closely estimate this common percentage in the four populations. In other words, if H_0 were true (that is, there is no difference among the four treatments), then approximately 20% of the patients in each treatment group would be expected to show major improvement. Thus, we calculate 20% of each group to get the expected values.

Expected values for the first row
20% of 250 = 50
20% of 200 = 40
20% of 300 = 60
20% of 150 = 30

(Expected values are shown in upper right box of each cell.)

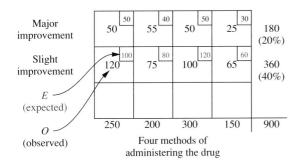

Major
improvement

Slight
improvement

E
(expected)

O
(observed)

Four methods of
administering the drug

Next, we repeat the process for the second row. The second row resulted in a total of 40% who showed slight improvement. Thus, we take 40% of each group total to get the expected values.

Expected values for the second row
40% of 250 = 100
40% of 200 = 80
40% of 300 = 120
40% of 150 = 60

Once we repeat the same process for the third row, the table is complete.

Major
improvement

	50		40		60		30		
50		55		50		25		180	
								(20%)	

Slight
improvement

	100		80		120		60		
120		75		100		65		360	
								(40%)	

No
improvement

	100		80		120		60		
80		70		150		60		360	
								(40%)	

| 250 | 200 | 300 | 150 | 900 |
| Caffeine coat | Iron coat | Immune coat | Loosening compound | |

Four methods of
administering the drug

Note (see
box at right)

For calculating the *expected values*, some find the following formula more convenient:

$$\text{Expected value} = \frac{(\text{Row total})(\text{Column total})}{\text{Grand total}}$$

As an example, to calculate the expected value, 30, noted at left, we do the following:

$$\text{Expected value} = \frac{(180)(150)}{900} = 30$$

To recap, the test begins by assuming H_0 is true, that is, by assuming all four treatment groups originated from four separate but equally proportioned populations. Then, based on this assumption, we calculate the number of patients we would *expect* in each cell. This expected value is then compared to the number of patients actually *observed,* illustrated as follows.

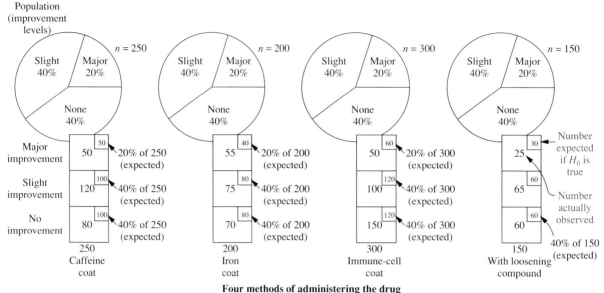

Four methods of administering the drug

Note that each cell has two values, the expected (in upper right box) and the observed. The expected value is based on H_0 being true, that indeed each population is equally proportioned. X^2 then measures the z-score differences from the expected to the observed values.* A high X^2 value means large differences, in which case, we would question the likelihood the observed values originated from the hypothesized populations, and conclude one or more treatments must offer different levels of patient improvement.

To calculate X^2, we use the general formula.†

$$X^2(\Sigma z^2) = \Sigma \frac{(O - E)^2}{E}$$

Note: For each cell, we subtract, $O - E$, square this value, then divide by E.

---

*Provided cell frequencies are at least 5, binomial (and multinomial) sampling distributions distribute normally.

†This has been footnoted before, but bears repeating. For a three-or-more category population, calculating actual z-score differences is quite complex because of the linear dependency of the variables. The sampling distribution is multidimensional and more sophisticated mathematics (matrix algebra) must be employed and the general formula is almost always used.

Remember, the calculation of $(O - E)^2/E$ must be performed for every cell, as follows:

$$X^2 = \frac{(50 - 50)^2}{50} + \frac{(120 - 100)^2}{100} + \frac{(80 - 100)^2}{100} + \frac{(55 - 40)^2}{40}$$
$$+ \frac{(75 - 80)^2}{80} + \frac{(70 - 80)^2}{80} + \frac{(50 - 60)^2}{60} + \frac{(100 - 120)^2}{120}$$
$$+ \frac{(150 - 120)^2}{120} + \frac{(25 - 30)^2}{30} + \frac{(65 - 60)^2}{60} + \frac{(60 - 60)^2}{60}$$

$$X^2 = 0.000 + 4.000 + 4.000 + 5.625 + .313 + 1.250$$
$$+ 1.667 + 3.333 + 7.500 + .833 + .417 + 0.000$$

$$\boxed{X^2 = 28.94}$$

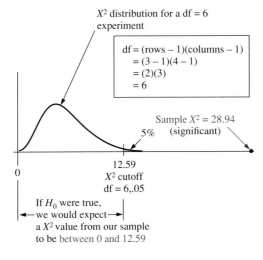

Now we know *if H_0 is true*, in other words, if the four populations are equally proportioned (homogeneous) *and if* we were to perform this X^2, df = 6, experiment hundreds of thousands of times, the differences between observed and expected values would be relatively modest, reflected in relatively modest X^2 readings (in fact, on average, if H_0 were true, X^2 would be* 6, with 95% of the X^2 values between 0 and 12.59).

Since our sample X^2 reading yielded 28.94, indicating relatively large *z*-score differences between observed and expected values, we conclude,

H_0 is rejected

*Recall from prior footnote, the average of a chi-square distribution equals its df. So, for the above,

$$\mu = df \text{ (degrees of freedom)}$$
$$= 6$$

Answer

a. Because the X^2 value of the sample (28.94) exceeded the cutoff value of 12.59, we reject H_0. This data supports the alternative hypothesis, H_1, that the populations are *not* all homogeneous (not equally proportioned) with respect to patient improvement. Stated another way, the samples provide evidence that in the four populations, levels of patient improvement are different, that the fluctuation in sample results is not merely chance fluctuation, but fluctuation due to real differences in patient improvement among the four treatment groups.

b. Implications: Although significance was achieved (H_0 rejected), we still do not know which cells or methods showed significant improvement. To help us better interpret the data, we translate the absolute numbers into percentages of column total, as follows:

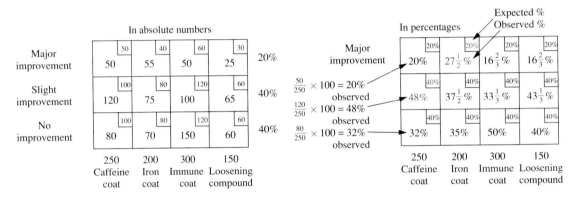

Post-evaluation testing for X^2 and ANOVA experiments usually involves sophisticated techniques reserved for more advanced courses, however, certainly a few simple observations can be made. Notice in the table, above right, that we had noticeably more patients than expected in the major improvement category using the iron coat technique ($27\frac{1}{2}\%$ versus 20% expected). Also more patients than expected were in the slight improvement category using the caffeine coat (48% versus 40% expected). It is probably safe to say, based on simple observation of the table, the caffeine and iron coat methods may indeed be worthy of further investigation.

10.2 Chi-Square: Test of Goodness-of-Fit

To a large extent, all X^2 testing is the same. Observed counts are compared to expected counts (which have been derived based on some given assumption, the null hypothesis). If the observed counts reasonably match expected counts, then you accept your initial assumption, the null hypothesis. The **goodness-of-fit test** is no different, only now, a single column of data* is presented and the degrees of freedom† (df) redefined as follows.

$$\boxed{df = r - 1}$$ number of rows (or cells) minus one

Other than these slight differences, we proceed in an almost identical manner. Let's use a contemporary marketing example to explain the details.

Example ——————— Seattle is known as one of the hottest diet soft drink markets in the nation. Suppose in Seattle, after a prolonged advertising war in diet soft drinks, one of the leading competitors, Sprinkle-Free, wanted to know if their market share had been maintained. (Note: Seattle and Milwaukee have the highest per capita diet soft drink consumption in the nation.)[7]

A market analyst was called in who randomly sampled 400 consumers from the metropolitan Seattle area, with the following results:

	Number of Consumers Selecting Brand	Prior Market Share
Peppi-Thin	130	30%
Cota-Light	140	35%
Sprinkle-Free	50	15%
Other Brands	80	20%
	(Data gathered after advertising war)	(Known distribution *prior* to advertising war)

a. Test the hypothesis that current market share has not changed from prior market share (at $\alpha = .05$).
b. Briefly discuss the implications for Sprinkle-Free.

---

*In keeping with the definition that each row in the table represents one category in our population, this necessitates presenting the table as a single column, such as that shown at right. In actual reports or even other textbooks, the same data may be presented as a single row, such as, | 130 | 140 | 50 | 80 |, however this is just a matter of convenience or style.

| 130 |
| 140 |
| 50 |
| 80 |

†The full degrees of freedom formula for goodness-of-fit tests is actually $df = r - 1 - m$, where m = the number of population parameters that must be estimated from the sample data. In the demonstration applications, no population parameters (characteristics) need be estimated, thus $m = 0$ for our examples and we ignore it. However, if we wish to use the goodness-of-fit test to determine if sample data had been drawn from a known population distribution, such as, a normal population (we will demonstrate this in exercise 10.10), and must use the sample data to estimate μ and σ, in which case, $m = 2$, then the formula is:
$df = r - 1 - m = r - 1 - 2 = r - 3$.

Solution

Establish initial conditions as follows:

H_0: In the population, market share has remained the same:
$$p_1 = 30\%$$
$$p_2 = 35\%$$
$$p_3 = 15\%$$
$$p_4 = 20\%$$

Essentially, the null hypothesis states that the market share has not changed, even after the advertising war.

H_1: Market share has changed.

If H_0 proves unlikely, then we assume H_1 is true.

$\alpha = .05$

We accept the risk that when H_0 is true, 5% of the time we will reject it in error.

Goodness-of-fit is probably the simplest of X^2 tests.* Essentially, we start by assuming our sample of $n = 400$ originated from the four-category population specified in the null hypothesis. We then calculate the expected value in each cell by taking 30%, 35%, 15%, and 20% of 400, respectively, as demonstrated below.

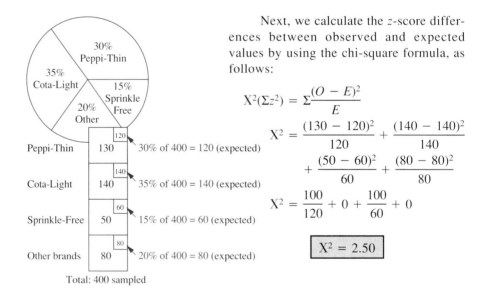

Next, we calculate the z-score differences between observed and expected values by using the chi-square formula, as follows:

$$X^2(\Sigma z^2) = \Sigma \frac{(O - E)^2}{E}$$

$$X^2 = \frac{(130 - 120)^2}{120} + \frac{(140 - 140)^2}{140}$$
$$+ \frac{(50 - 60)^2}{60} + \frac{(80 - 80)^2}{80}$$

$$X^2 = \frac{100}{120} + 0 + \frac{100}{60} + 0$$

$$\boxed{X^2 = 2.50}$$

Peppi-Thin   130   30% of 400 = 120 (expected)   [120]

Cota-Light   140   35% of 400 = 140 (expected)   [140]

Sprinkle-Free   50   15% of 400 = 60 (expected)   [60]

Other brands   80   20% of 400 = 80 (expected)   [80]

Total: 400 sampled

*Karl Pearson originally introduced X^2 as a goodness-of-fit test to determine if sample data could have originated from known population distributions, such as, the binomial or normal distributions.

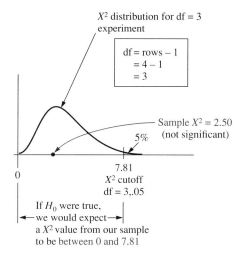

X^2 distribution for df = 3
experiment

df = rows – 1
= 4 – 1
= 3

Sample $X^2 = 2.50$
(not significant)

5%

0

7.81
X^2 cutoff
df = 3,.05

If H_0 were true,
we would expect
a X^2 value from our sample
to be between 0 and 7.81

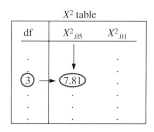

X^2 table

df	$X^2_{.05}$	$X^2_{.01}$
.	.	.
.	.	.
③ →	⑦.81	.
.	.	.
.	.	.
.	.	.

Now we know if H_0 were true (if the sample originated from a four-category population as specified in the null hypothesis), *and* if we were to perform this X^2, df = 3, experiment hundreds of thousands of times, the differences between observed and expected values would be relatively modest, reflected in relatively modest X^2 readings (in fact, on average, if H_0 were true, X^2 would be* 3, with 95% of the X^2 values between 0 and 7.81).

Since our sample X^2 reading yielded 2.50, indicating rather small z-score differences, we

Accept H_0
No change in market share

Answer

a. Because the X^2 value of the sample (2.50) was less than the cutoff value of 7.81, we accept H_0. In this case, we cannot prove the market share has changed.

b. Implications for Sprinkle-Free: Based on this data, we cannot assume any change in market share. Although the sample data actually showed a slight decrease in customers for Sprinkle-Free (50 versus 60 expected), this difference may very well be ''chance fluctuation.'' Remember, samples give approximations to population characteristics, and we would expect some fluctuation from sample to sample even though Sprinkle-Free's market share may indeed be constant at 15% in the population.   ◼

*For the above X^2 distribution, the average X^2 value would be 3. In other words, for the above X^2 distribution,

$$\mu = \text{df (degrees of freedom)}$$
$$= 3$$

We will continue with one more example from marketing, this time demonstrating goodness-of-fit for the uniform model. As background, the following is presented:

Although marketing research grew more sophisticated and prevalent in the 1930s, a number of research projects were conducted prior to this time. Some milestones were set when N. W. Ayer* conducted its first research project in 1879 and when Charles Parlin organized the nation's first commercial research department in 1911.

Supposedly, Parlin got his start rummaging through the garbage pails of Philadelphia counting soup cans.[8] Parlin was actually selling ad space for a magazine believed to appeal to working-class families. The Campbell Soup Co. at the time refused to place an ad in the magazine feeling that their soup appealed to higher-income families who could better afford the price† and, thus, advertising was targeted to this higher-income group. The Campbell Soup Co. reasoned that working class families made their own soup and therefore would not be interested in spending the extra money. Parlin, desiring to sell Campbell ad space in the magazine, took it on himself to count Campbell soup cans in garbage pails of various Philadelphia neighborhoods, or at least, so the story goes. Perhaps the experiment went something like this.

Example ——————— In an attempt to assess Campbell soup usage in the Philadelphia area, suppose C. C. Parlin randomly selected an equal number of garbage cans from various neighborhoods and counted the Campbell soup cans, with the following assumed results:

Socioeconomic Class of Neighborhood	Number of Campbell Soup Cans Found
Wealthy	42
Upper-middle class	60
Middle-working class	60
Lower-middle-working class	87
Lower-working class	51

a. Use X^2 analysis to test the hypothesis that Campbell soup usage is the same in all neighborhoods (at $\alpha = .01$).
b. Briefly discuss the implications of these results for the Campbell Soup Co.

*N. W. Ayer, an early venerated advertising agency.
†Campbell soup cost 10¢ at the time.

Solution

Notice in this experiment, no mention was made of prior Campbell soup usage. In other words, we have no documented evidence of how the proportions might break down by socioeconomic neighborhood, thus we start with the assumption,

H_0: In the population, usage is the same in all socioeconomic neighborhoods.

Essentially, H_0 states there is no difference in the percentage of usage among the various neighborhoods.

H_1: In the population, usage is *not* the same in all socioeconomic neighborhoods.

If H_0 proves unlikely, then we assume H_1 is true.

$\alpha = .01$

We accept the risk that when H_0 is true, 1% of the time we will reject it in error.

Essentially, we start by assuming our sample of $n = 300$ cans originated from a five-category population as specified in the null hypothesis, that is, where usage in all neighborhoods is equal. Since there are five types of neighborhoods, then each should yield $\frac{1}{5}$ or 20% of the Campbell soup cans found, demonstrated below left.

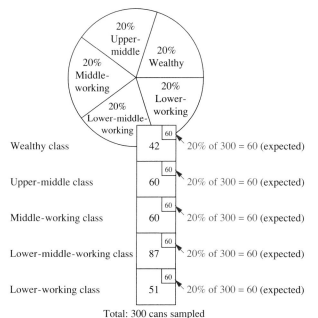

Wealthy class   42   20% of 300 = 60 (expected)

Upper-middle class   60   20% of 300 = 60 (expected)

Middle-working class   60   20% of 300 = 60 (expected)

Lower-middle-working class   87   20% of 300 = 60 (expected)

Lower-working class   51   20% of 300 = 60 (expected)

Total: 300 cans sampled

Next, we calculate the differences between observed and expected values by using the chi-square general formula, as follows:

$$X^2(\Sigma z^2) = \Sigma \frac{(O - E)^2}{E}$$

$$X^2 = \frac{(42 - 60)^2}{60} + \frac{(60 - 60)^2}{60} + \frac{(60 - 60)^2}{60}$$

$$+ \frac{(87 - 60)^2}{60} + \frac{(51 - 60)^2}{60}$$

$$X^2 = \frac{324}{60} + \frac{0}{60} + \frac{0}{60} + \frac{729}{60} + \frac{81}{60}$$

$$\boxed{X^2 = 18.90}$$

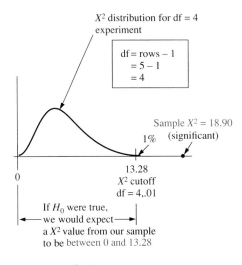

X^2 distribution for df = 4
experiment

df = rows − 1
= 5 − 1
= 4

Sample X^2 = 18.90
1%   (significant)

13.28
X^2 cutoff
df = 4,.01

If H_0 were true,
we would expect
a X^2 value from our sample
to be between 0 and 13.28

Now we know if H_0 were true, that is, indeed the sample originated from a five-category population with equal proportions, and if we were to perform hundreds of thousands of such experiments, the differences between observed and expected values would be relatively modest, reflected in relatively modest X^2 readings. (In fact, on average, if H_0 were true, X^2 will be* 4, with 99% of the X^2 readings falling between 0 and 13.28.)

Since our sample X^2 reading yielded 18.90 indicating relatively large z-score differences, we

Reject H_0

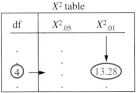

X^2 table

df	$X^2_{.05}$	$X^2_{.01}$
.	.	.
④	:	⟨13.28⟩
.	.	.

Answer

a. Because the X^2 value of the sample (18.90) exceeded the cutoff value of 13.28, we reject H_0. Thus the evidence supports the hypothesis, H_1, usage is *not* the same in all socioeconomic neighborhoods in Philadelphia.

b. Implications for Campbell: To help us better interpret the data, we translate the absolute numbers into percentage of total, as shown below.

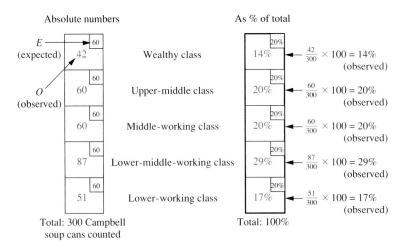

Absolute numbers

As % of total

E
(expected)   60   42   Wealthy class   20%   14%   ← $\frac{42}{300}$ × 100 = 14%
(observed)

O
(observed)   60   60   Upper-middle class   20%   20%   ← $\frac{60}{300}$ × 100 = 20%
(observed)

60   60   Middle-working class   20%   20%   ← $\frac{60}{300}$ × 100 = 20%
(observed)

60   87   Lower-middle-working class   20%   29%   ← $\frac{87}{300}$ × 100 = 29%
(observed)

60   51   Lower-working class   20%   17%   ← $\frac{51}{300}$ × 100 = 17%
(observed)

Total: 300 Campbell
soup cans counted

Total: 100%

*For the above X^2 distribution, the average X^2 value would be 4. In other words, for the above X^2 distribution,

$$\mu = \text{df (degrees of freedom)}$$
$$= 4$$

Again, post-evaluation testing for X^2 and ANOVA experiments usually involve sophisticated techniques reserved for more advanced courses, however certainly a few simple observations can be made. An examination of the previous table reveals a larger percentage of the Campbell business is coming from the lower-middle-working class (29% versus 20% expected) and relatively little from the wealthy class (14% versus 20% expected). Certainly if a valid random sample was achieved from the population, these results would indicate a stronger preference for the product in the lower-middle-working class than in the wealthy, the neighborhoods with the two extreme differences in the opposite direction.

C. C. Parlin's finding reflected similar results—that is, the working class used more canned soup than the wealthy. Parlin was quick to convince Campbell to become a client of the magazine, the *Saturday Evening Post.** ∎

10.3 Analysis of Variance (ANOVA) for Equal Sample Size

Sir Ronald Fisher[2] published a paper in 1918 that attempted to reconcile two theories of inheritance and, in the process, introduced a new statistical method, analysis of variance (ANOVA), which far outlived its original vehicle. The new technique gained fame in the 1920s when Fisher (then new at Rothamsted Experimental Station in Britain) applied the method to crop studies.[9] ANOVA has since expanded into many fields and today is probably the most widely used technique for industrial and scientific research, a testament to its underlying power and versatility.

We use the following example comparing the effects of advertising strategies on the sale of a consumer product. The example also demonstrates the difficulties in establishing validity for ''social'' experiments where precise experimental designs cannot readily be implemented.

Example

Twelve years ago, D'Oreal introduced a new lipstick into the U.S. called Entice. The lipstick, under its European name, had long been a favorite among young women in France and Switzerland, however in the original U.S. test market (12 years ago in Sacramento, Ca.) sales results were disappointing. The European strategy of targeting the product to avant-garde rebellious young women (ages 16 to 21) proved ineffective in Sacramento.

Finally, after two years of experimentation, a viable strategy emerged. Entice was targeted to single working women, 25 to 39, conservative and upwardly mobile. Sales thereafter were favorable and the lipstick, over a three-year period, was rolled out across the U.S. Since that period, sales have been relatively steady at approximately 1.3% market share. Management feels, however, that Entice is *under*performing and that more aggressive advertising would tap unexplored demand.

*As it turned out, the wealthy had servants to make the soup for them and thus had little need to purchase canned soup.

To test this strategy, TV ads were produced using two more aggressive experimental approaches. In addition, new ads were produced under the old soft-sell strategy. The three campaigns were run concurrently, each in a separate randomly selected test city, as follows:

Tulsa, Ok. received campaign A: Aggressive Approach

Springfield, Ill. received campaign B: Moderately Aggressive Approach

Lexington-Fayette, Ky. received campaign C: Soft-Sell (control group, prior strategy)

Advertising levels in each city were substantially increased (blitzed) to accelerate impact and, after several weeks, case shipment data was obtained from 9 randomly selected outlets in each city, with the following results:

	Campaign A	Campaign B	Campaign C
	9	16	13
	10	14	17
	12	19	18
Sales	8	21	15
Increase	8	15	10
(percentage increase above	13	17	12
normal levels of last year)	11	13	16
	10	20	16
	9	18	18
	(Tulsa)	(Springfield)	(Lex.-Fayette)
	Aggressive Ads	Moderately Aggressive Ads	Soft-Sell Ads (control group)

Three Different Advertising Strategies Were Tested

a. Calculate $\bar{x}$ and s^2 (average sales increase and variance) for each sample.
b. Assuming valid random samples, test at $\alpha = .05$ whether differences among the three sample $\bar{x}$'s can be attributed to chance fluctuation or, indeed, to actual differences in sales in the three cities.
c. Briefly discuss the implications of the results for D'Oreal.

Overview of ANOVA Test

Essentially, ANOVA uses sample differences as evidence to determine whether or not actual differences exist in the populations. Keep in mind, we only sampled 9 outlets out of the many thousands in each city. From these 9 outlets, we wish to determine if sample differences are substantial enough to warrant suspicion that differences actually exist in the populations.

In other words, one may very well take three samples (even from the *same* population) and get three different sample averages, three different $\bar{x}$'s, so differences in sample results in and of themselves will *not* prove differences exist in the populations. However, if the differences among the sample $\bar{x}$'s are significant, that is, of sufficient magnitude to offer evidence real differences exist, then we conclude at least one pair of populations is experiencing actual differences in sales.

If indeed all other relevant market influences hold reasonably constant during this test period (competition, market unemployment, and so on*), then we conclude the campaign strategies are associated with these differences.

One last word: Statistical studies do not prove cause and effect, and ANOVA is no exception. Even if we prove sale differences exist in the three test cities, we must always be open to the possibility that some, as yet unmeasured variable (and not the ad campaigns) may have caused the differences.† And the problem grows especially acute in social experiments where the experimental environment very often cannot be controlled or, sometimes, even adequately monitored. In an attempt to minimize the risk of unmeasured variables influencing results (and for other reasons), a number of preferred experimental designs have been developed and two are discussed shortly.

Population Assumptions

Whereas chi-square (X^2) tests for the most part are nonparametric, meaning they require no population assumptions, ANOVA is parametric, and requires samples be drawn from populations that are

---

a. Normally distributed,
b. With equal variances (essentially, equal standard deviations).

---

*In such experiments, it is important that we closely monitor market conditions to ensure any other factors that might influence results hold reasonably constant during the course of the experiment, such as competitive pricing, promotions, and other competitive influences. In addition, we should watch for any undue restrictions in channels of distribution to ensure a free flow of product to meet demand and monitor any unusual unemployment, strikes, or any potential local disruptive influences. And, of course, we must have some assurance that the three markets are demographically equivalent and somewhat representative of national consumer buying behavior and that the three ad campaigns are reasonably close in quality and attention-getting ability, so essentially, then, it's just the sales strategy that is being tested.

†The list of possible unmeasured variables for such experiments is theoretically endless and plagues all such research. In this case, some unmeasured variables might be corruption, payoffs, incompetence (not supplying stores with enough product, not unpacking deliveries, and so on), mistaken entries, demographic quirks in one or more test cities, local negative reaction to one of the actors playing in an ad, unethical behavior of a competitor (sabotaging displays, pricing, shelf position), and the list goes on.

We will discuss these population assumptions further as we proceed through the solution.* Actually, a further requirement of ANOVA testing is that samples be randomly selected and independent.†

Experimental Designs

Actually, the design of ANOVA experiments is quite important[10] to ensure population assumptions are met and to minimize unexplained factors influencing results. Many of the designs were developed by Sir Ronald Fisher in the 1920s in his work with agricultural experiments. Two popular designs are as follows.

The Completely Randomized Design: In this design, treatments are applied randomly to several homogeneous experimental units. In our case, this would require each of the three ad campaigns (the treatments) to be randomly assigned to several tiny homogeneous outlet zones throughout the country. In actual practice, it is not as yet feasible to isolate tiny outlet zones, perhaps a few thousand TV sets in a given shopping area. Of course, an entire city can be used as a unit, however this would be extremely expensive because several test cities must be used for each treatment (each ad campaign).

Instead, we usually direct one ad campaign to one city, one intact cluster, usually of several hundred or more outlet zones. However this leaves us open to unique influences peculiar to this one cluster (to this one city), which may not exist in another cluster, another city. Certainly great gains are taken to ensure test cities are as ''unpeculiar'' as possible (for further reading on this, refer to the article referenced in endnote 11).

Randomized Complete Block Design: In this design, all treatments are applied randomly to elements in several homogeneous blocks. In our case, we might use a city as a homogeneous block and divide the city into three typical zones and randomly apply a different ad campaign to each zone and do this for several homogeneous blocks (several cities). Unfortunately partitioning a city into three isolated media zones and delivering a different ad campaign to each zone is quite a task indeed. Although eventually, we might have the ability to design such experiments, at this point, they still remain somewhat beyond our grasp. So, in marketing experiments (and similar social experiments in education and the soft sciences) where we cannot readily conform to the experimental designs on which ANOVA is based, we must be careful to monitor all factors that may influence results. And even then, we are still not sure if some unknown, unexplained factors peculiar to one or more of the cities may not be causing what appears to be sale differences, which we think are differences due to the ad campaigns.

---

*Actually, the ANOVA test is quite robust, meaning, substantial departures from the requirement of normality can often be tolerated and, under certain conditions, even the equal standard deviation requirement can be relaxed without serious risk to final results, in the case where treatment groups are equal in sample size.

†In the ANOVA model, presented above, each group constitutes a random selection from its respective population and is independent of any other group. Threats to independence often grow acute in social experiments. For instance, the same subject must not be used in more than one group and one subject's response must in no way influence another subject's response.

Solution

a. The solution to the example presented at the beginning of this section is as follows. We first compute $\bar{x}$ and s^2 for each sample.

$$s^2 = \frac{\Sigma(x - \bar{x})^2}{n - 1}$$

Variance

$$s = \sqrt{\frac{\Sigma(x - \bar{x})^2}{n - 1}}$$

Note: the square root of the variance is the standard deviation.

x	$\bar{x}$	$x - \bar{x}$	$(x - \bar{x})^2$
9	10	-1	1
10	10	0	0
12	10	2	4
8	10	-2	4
8	10	-2	4
13	10	3	9
11	10	1	1
10	10	0	0
9	10	1	1
$90 = \Sigma x$			$24 = \Sigma(x - \bar{x})^2$

$$\bar{x} = \frac{\Sigma x}{n} = \frac{90}{9} = 10$$

$$s^2 = \frac{\Sigma(x - \bar{x})^2}{n - 1}$$

$$= \frac{24}{9 - 1} = 3$$

$n = 9$
$\bar{x} = 10$
$s^2 = 3$
For campaign A
(aggressive ads)

x	$\bar{x}$	$x - \bar{x}$	$(x - \bar{x})^2$
16	17	-1	1
14	17	-3	9
19	17	2	4
21	17	4	16
15	17	-2	4
17	17	0	0
13	17	-4	16
20	17	3	9
18	17	1	1
$153 = \Sigma x$			$60 = \Sigma(x - \bar{x})^2$

$$\bar{x} = \frac{\Sigma x}{n} = \frac{153}{9} = 17$$

$$s^2 = \frac{\Sigma(x - \bar{x})^2}{n - 1}$$

$$= \frac{60}{9 - 1} = 7.5$$

$n = 9$
$\bar{x} = 17$
$s^2 = 7.5$
For campaign B
(mod. aggressive ads)

x	$\bar{x}$	$x - \bar{x}$	$(x - \bar{x})^2$
13	15	-2	4
17	15	2	4
18	15	3	9
15	15	0	0
10	15	-5	25
12	15	-3	9
16	15	1	1
16	15	1	1
18	15	3	9
$135 = \Sigma x$			$62 = \Sigma(x - \bar{x})^2$

$$\bar{x} = \frac{\Sigma x}{n} = \frac{135}{9} = 15$$

$$s^2 = \frac{\Sigma(x - \bar{x})^2}{n - 1}$$

$$= \frac{62}{9 - 1} = 7.75$$

$n = 9$
$\bar{x} = 15$
$s^2 = 7.75$
For campaign C
(soft-sell ads)

For most advance work, ANOVA included, s^2 (and not s) is used as the measure of spread.

> s^2, the variance, will be used as our measure of spread, not the standard deviation.

Of course at this point you might ask, why add this complication when we were just getting used to s? If I didn't have to introduce it, I wouldn't. But unfortunately, for ANOVA testing, we combine measures of spread for several samples, and when we use the standard deviation, this often infuses unacceptable errors into our work, for two primary reasons. First, because of the larger number of calculations (square rooting) and increased rounding error and, second, on average, $s \neq \sigma$, whereas, on average, $s^2 = \sigma^2$.

In other words, if we were to draw a great many samples from the same population, on average, the s's would *not* give an unbiased estimate of σ (meaning, the s's will not on average equal the population standard deviation, σ). However, a great many s^2's (variances) drawn from the same population will indeed give us an unbiased estimate of σ^2 (meaning, the s^2's average *will* equal the population variance, σ^2).

Now, in preceding chapters we used s to estimate σ and tolerated this slight loss of accuracy because for large sample sizes, the bias is negligible. However, the bias is not so negligible for small-sample sizes, especially very small-sample sizes, and since ANOVA is conceived on the principles of small-sample work, this precision is essential,* thus we will use s^2 to estimate σ^2 (and not s to estimate σ).

---

*In one calculation in ANOVA, the number of groups, in effect, acts as a sample size. For instance, in our experiment with three treatment groups, it would mean we have a sample size = 3 (later, denoted as $k = 3$). And based on this very small-sample size of 3, we calculate a standard deviation that we use to estimate its population standard deviation (actually, variances are used; this is the calculation where the 3 $\bar{x}$'s are used to estimate $\sigma_{\bar{x}}^2$, the spread of the $\bar{x}$ distribution).

Actually, use of the variance, s^2, has other important advantages for advanced work, as we will learn later.*

To recap, because of increased calculation error and difficulty of use in advanced work and the fact that, on average, $s \neq \sigma$, with the bias especially pronounced in small-sample work, we must forgo use of the sample standard deviation as an estimator of the population standard deviation, and instead use the sample variance, s^2, as an estimator of the population variance, σ^2. Keep in mind, however, both s^2 and s essentially estimate the same phenomenon, that is, the spread of data in the population.

b. The initial conditions for ANOVA are as follows:

H_0: $\mu_A = \mu_B = \mu_C$ (no difference in average sales increase in the three test markets)	Essentially this says, the average sales increase will be the same in all cities. That is, if we computed the *average* sales increase for all the many thousands of outlets in Tulsa, this would equal the *average* sales increase for all the many thousands of outlets in Springfield, and this would equal the *average* sales increase for all the many thousands of outlets in Lexington-Fayette. In other words, no difference in average sales increase in the three test markets.
H_1: Not all μ's are the same	If H_0 proves unlikely, we accept H_1, that at least one μ is different.

$$\alpha = .05$$

This means, if H_0 is true, a 5% chance exists we will reject it in error.

Any hypothesis test begins by assuming H_0 is true, and based on this premise, constructing a set of expectations. So, indeed, if H_0 is true, that is, if the average sales increase in each city is identical (meaning, $\mu_A = \mu_B = \mu_C$) and we satisfy

---

*We will later learn that sample variances, such as, $\Sigma(x - \bar{x})^2/(n - 1)$, lend themselves to dissection, that is, the *sum-of-squared distances* component, $\Sigma(x - \bar{x})^2$, can be readily separated out and added to the sum-of-squared distances component of other samples, regardless of sample size. This greatly minimizes the calculations and adds tremendous versatility to the combining procedure, in that these sum-of-squared distances can be used as a gauge in assessment. In other words, the extent sum-of-squared distances are reduced (which is desirable in the error term) can be used to assess the value of one hypothesis over another, or to assess the power of a particular experimental design.

The advantage of using sum-of-squared distances in professional work cannot be overstated, and although it is often difficult to understand the concept, one must keep trying, because no true mastery of advanced work can be achieved without this understanding.

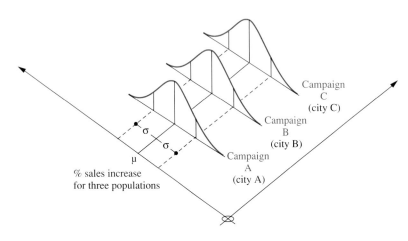

the initial population assumptions (normal distributions with equal standard deviations, $\sigma_A = \sigma_B = \sigma_C$) then all three populations, for sampling purposes, are identical, with common mean, μ, and common standard deviation, σ, shown at left.*

The diagram shows the percentage sale increases in all the many thousands of outlets in each test city. In other words, if we measured all the many thousands of outlets in city A and all the many thousands of outlets in city B and all the many thousands of outlets in city C, then plotted the results into three histograms, one for each city, the sales increases in each city would be normally distributed around a common average, μ, and the spread of values in each city, σ, would be identical.

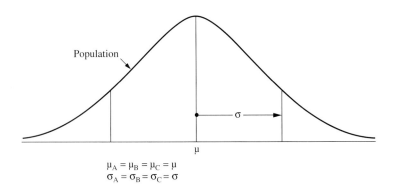

Now, whether we say we took three samples from three identical populations or three samples from the same population is immaterial. For simplicity, we will say we took three samples from the same population and describe this population by common mean, μ, and common standard deviation, σ, shown at left.

*In actuality, there may be many more outlets in one city than in the other, and the normal distributions may appear as shown at right. However, if the populations are quite large relative to the sample size (at least 20 times the sample size) and the distributions are normal with equal μ's and σ's, for sampling purposes they are considered statistically identical. In other words, if we draw samples from each, the theoretical $\bar{x}$ distributions and associated probabilities are identical.

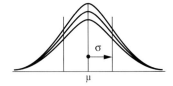

To recap, first we must have some assurance our population assumptions have been met (namely, equal standard deviations, $\sigma_A = \sigma_B = \sigma_C$, and that each population is normally distributed). The actual ANOVA test is then designed to determine if all the μ's are equal, or at least could be equal.* The test starts with the premise: *if* all the μ's were indeed equal, what can we expect? We then create a population identical to the three test populations and assume the three samples were drawn from this one population.

In the following paragraphs, we examine the logic of the test, then offer a method that greatly simplifies the calculations and is most often used in actual practice.

Logic of ANOVA

Essentially ANOVA compares two different estimates of the population variance, σ^2, one based on the spread of the three $\bar{x}$'s and the other based on the spread of data within each sample, essentially the average s^2, denoted as s^2_{pooled}.

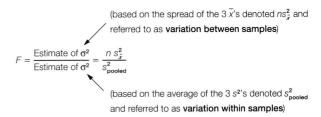

$$F = \frac{\text{Estimate of } \sigma^2}{\text{Estimate of } \sigma^2} = \frac{n\,s^2_{\bar{x}}}{s^2_{pooled}}$$

(based on the spread of the 3 $\bar{x}$'s denoted $ns^2_{\bar{x}}$ and referred to as **variation between samples**)

(based on the average of the 3 s^2's denoted s^2_{pooled} and referred to as **variation within samples**)

*Note: If H_0 is *not* true (not all μ's are the same), then the μ's can be different in several ways. One such arrangement of how the μ's can be different is shown below. Remember, there are many ways in which the μ's may be different—and this diagram is just one arbitrary example. Note that if H_0 proves false (not true), such as in this diagram, the population assumptions are still intact (namely, the populations are normally distributed with equal standard deviations) and that only the μ's are different. See endnote 10 for discussion on experimental design and references.

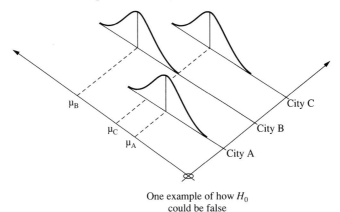

One example of how H_0
could be false

The rationale and calculations for the F fraction are as follows:

Top of F fraction: Essentially, the three $\bar{x}$'s are used to first estimate the spread of the $\bar{x}$ distribution and this value is then used to estimate σ^2, the population spread. The actual rationale and calculations are demonstrated in the large caption, which will be presented shortly; however, the calculation for the final result (117), the top of the F fraction, is shown as follows.

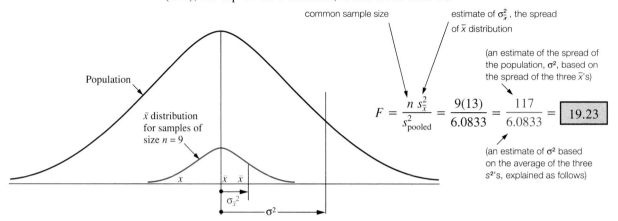

Bottom of F fraction: The three s^2's, are used to estimate σ^2, the population spread, by simply taking an average. This average is referred to as s^2_{pooled}.

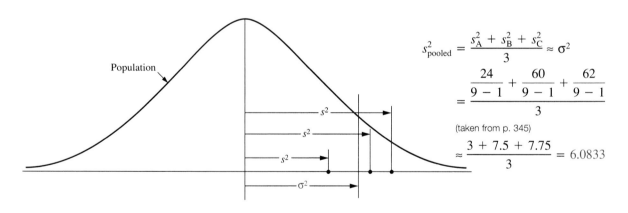

Essentially, if one sample s^2 provides a reasonable estimate of σ^2, then three combined (averaged) should offer an even better estimate.*

*Note the true σ^2 and s^2 distances would actually be much farther out than shown, however for demonstration purposes, they were put in the position of the standard deviation.

Rationale and Calculations for Top of F Fraction, $n\, s_{\bar{x}}^2$

The three $\bar{x}$'s are used to first estimate $\sigma_{\bar{x}}^2$, the spread of the $\bar{x}$ distribution and this value is then used to estimate σ^2, the spread of the population as follows.*

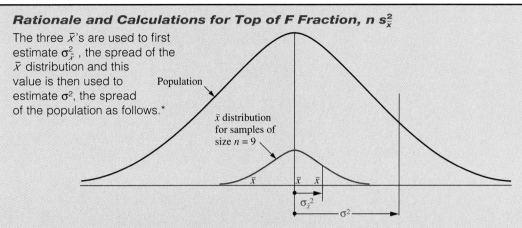

To estimate $\sigma_{\bar{x}}^2$, the spread of the $\bar{x}$ distribution, we use the 3 $\bar{x}$'s as if they were 3 randomly selected data points drawn from any normal population, in this case, drawn from the normal $\bar{x}$ population. Essentially, then, we have a sample of size $k = 3$ to calculate a sample average ($\bar{\bar{x}}$) and sample variance ($s_{\bar{x}}^2$) which is used to estimate the variance of the $\bar{x}$ distribution. In other words,

$\bar{\bar{x}} \approx \mu_{\bar{x}}$ the average of the 3 data points (the three $\bar{x}$'s) is approximately equal to† the average of the $\bar{x}$ distribution, $\mu_{\bar{x}}$.

$s_{\bar{x}}^2 \approx \sigma_{\bar{x}}^2$ the variance (spread) of the 3 data points is approximately equal to the variance (spread) of the $\bar{x}$ distribution, $\sigma_{\bar{x}}^2$. This calculation is shown at right.

Thus, $s_{\bar{x}}^2$ (13) is our estimate of the variance (spread) of the $\bar{x}$ distribution, $\sigma_{\bar{x}}^2$. Next, to estimate σ^2, the population variance, we use $s_{\bar{x}}^2$, the estimated spread of the $\bar{x}$ distribution. We know from preceding chapters, the formula for the standard deviation of the $\bar{x}$ distribution is as follows:

$$\sigma_{\bar{x}} = \frac{\sigma}{\sqrt{n}}$$

squaring both sides, we get

$$\sigma_{\bar{x}}^2 = \frac{\sigma^2}{n}$$

multiply both sides by n

$$n\,\sigma_{\bar{x}}^2 = \sigma^2$$

substitute in $s_{\bar{x}}^2$

$$\boxed{n\, s_{\bar{x}}^2 \approx \sigma^2}$$

$\bar{x}$	$\bar{\bar{x}}$	$(\bar{x} - \bar{\bar{x}})$	$(\bar{x} - \bar{\bar{x}})^2$
10	14	−4	16
17	14	3	9
15	14	1	1
42			26

$$\bar{\bar{x}} = \frac{\Sigma \bar{x}}{k} = \frac{42}{3} = 14$$

(k = number of $\bar{x}$'s)

$$\boxed{s_{\bar{x}}^2 = \frac{\Sigma(\bar{x} - \bar{\bar{x}})^2}{k - 1} = \frac{26}{3 - 1} = 13}$$

Note the spread of the 3 $\bar{x}$'s, $s_{\bar{x}}^2$, is used as an estimator of the spread of the $\bar{x}$ distribution, $\sigma_{\bar{x}}^2$. And the spread of the population is then estimated by multiplying this value, $s_{\bar{x}}^2$, by n. This is used in the above calculation for F.

---

*Note the true σ^2 and $\sigma_{\bar{x}}^2$ distances would actually be much farther out, however for demonstration purposes, they were put in the position of the standard deviation.

†Note, this is the standard deviation formula $s = \sqrt{\dfrac{\Sigma(x - \bar{x})^2}{n - 1}}$, studied in section 2.3, only now the formula has been squared and the notation slightly modified for this application.

Essentially, the F ratio, $F = n\, s_{\bar{x}}^2 / s_{pooled}^2$, compares two estimates of σ^2. One estimate of σ^2 is s_{pooled}^2, the bottom of the fraction, and this estimate is independent of whether H_0 is true or not. It is based solely on the pretest assumptions (normal population distributions with equal variances). It has nothing to do with the means, μ's, and therefore has nothing to do with whether H_0 is true or not. Essentially if the pretest population assumptions have been satisfied, s_{pooled}^2 is a good estimator of the common population variance, σ^2.

However, the second estimate of σ^2 is $n\, s_{\bar{x}}^2$, the top of the fraction, and is very dependent on whether H_0 ($\mu_A = \mu_B = \mu_C$) is true or not. If H_0 is true, the 3 $\bar{x}$'s will tend to cluster relatively close to each other (in accordance with the central limit theorem) and how spread apart they are should give us a reasonably decent estimate of $\sigma_{\bar{x}}^2$, the spread of the $\bar{x}$ distribution, which then should give us a reasonably decent estimate of σ^2, the population spread. In the case where H_0 is true, the F ratio would most probably be close to 1, that is, the top and bottom estimates of σ^2 would be near equal.

If H_0 is false, however (not all μ's are equal), the $\bar{x}$'s will scatter apart in an attempt to estimate their respective *unequal* μ's and, the more unequal the μ's, in general, the more the $\bar{x}$'s tend to scatter apart, resulting in estimates of $\sigma_{\bar{x}}^2$ and, thus, σ^2 generally much larger than the estimates of σ^2 based on s_{pooled}^2.* In this case, the top estimate of σ^2 would be much larger than the bottom estimate of σ^2 and the F ratio will tend to be far in excess of 1 (see footnote for further discussion).†

To recap, if H_0 is true, the F ratio should be close to one, and if H_0 is false, the F ratio tends to be much larger than one.

Now, with all this in mind, how do we conduct the ANOVA test? Pretty much the same way we conduct any hypothesis test.

We know from theory and experience that indeed if H_0 were true, that is, $\mu_A = \mu_B = \mu_C$ *and* if we were to conduct this df $= 2{,}24$ ANOVA experiment hundreds of thousands of times and calculate the F ratio for each experiment, that 95% of these F ratios (again, provided H_0 were true) would fall at or below

---

*The ANOVA test is based on the independence of s^2 and $\bar{x}$. Wm. Gossett showed that s^2 and $\bar{x}$ were uncorrelated and thus he assumed independent. Later, Fisher confirmed and experimentally verified these findings (see endnote 10 for further discussion).

†Recall that the bottom of the fraction (s_{pooled}^2) is independent of H_0, that is, independent of whether the three population μ's are equal or not. So the bottom of the fraction acts as a stable base, an anchor value we know is true regardless of whether H_0 is true or not.

However, the top of the fraction ($n\, s_{\bar{x}}^2$) is highly fluctuating, critically dependent on H_0 being true. If H_0 is true, the $\bar{x}$'s will cluster relatively close together and give a good estimate of σ^2, similar to the estimate obtained from s_{pooled}^2 and the F ratio will be close to 1 (one). However, if H_0 is false, the $\bar{x}$'s will scatter and give false (erroneous) estimates of σ^2, usually much larger than the true estimate obtained from s_{pooled}^2, and the F ratio will tend to be far in excess of 1 (one).

Technical note: If H_0 is true, the expected value of F is actually slightly higher than 1. Specifically for H_0 true,

$$E(F) = \frac{df_{error}}{df_{error}}$$

3.40 (of course with a great many F ratios clustered in the general vicinity of 1.00 [one]), as shown in the diagram below.

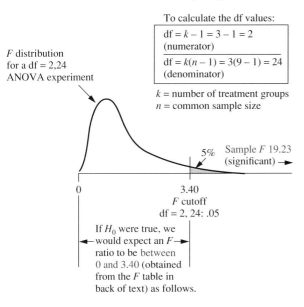

F distribution for a df = 2,24 ANOVA experiment

To calculate the df values:

df = $k - 1 = 3 - 1 = 2$
(numerator)

df = $k(n - 1) = 3(9 - 1) = 24$
(denominator)

k = number of treatment groups
n = common sample size

5%   Sample F 19.23 (significant) →

0

3.40
F cutoff
df = 2, 24: .05

If H_0 were true, we ← would expect an F → ratio to be between 0 and 3.40 (obtained from the F table in back of text) as follows.

df numerator

df denominator

$F_{.05}$ table

df | 1 ② 3 . . .

1

24 → 3.40

α = .05

Keep in mind, even if H_0 *is* true, the F ratio may very well exceed 1.00. This is because sample results, especially small-sample $\bar{x}$'s and s^2's, tend to fluctuate rather widely around the population values they estimate, which then causes the F ratio to fluctuate somewhat. However, we know it is quite unusual in a df = 2,24 ANOVA experiment for the F ratio (if H_0 is true) to exceed 3.40, in fact, this occurs only 5% of the time. Now, since our samples yielded an F ratio of 19.23, we

Reject H_0

In other words, the 3 $\bar{x}$'s were unreasonably spread out, producing an unusually large estimate of the population spread, σ^2 (top of fraction) far in excess of its true value, which had been properly estimated by s^2_{pooled}, the bottom of the fraction. Thus we,

Accept H_1

not all μ's are the same.

Answer

a. Calculate $\bar{x}$ and s^2 (the average sales increase and variance) in each sample.

City A (aggressive ads)	City B (mod. aggressive ads)	City C (soft-sell ads)
$\bar{x} = 10\%$	$\bar{x} = 17\%$	$\bar{x} = 15\%$
$s^2 = 3$	$s^2 = 7.5$	$s^2 = 7.75$

b. Assuming valid random samples, test at $\alpha = .05$ whether differences among the three sample $\bar{x}$'s can be attributed to chance fluctuation or, indeed, to actual differences in sales in one or more of the 3 cities.

For this, we conducted an analysis of variance test with the following result: $F = 19.23$. Because 19.23 exceeded the cutoff value of 3.40 (df $=$ 2,24, $\alpha = .05$), we accept H_1, the alternative hypothesis, that not all μ's are the same.

Generally when H_0 is rejected, we know *at least* one pair of treatments is significantly different, meaning, there were substantial enough differences in the samples such that we can conclude real differences exist in at least one pair of the populations. Certainly, the highest and lowest $\bar{x}$'s in cities B and A (17% versus 10%) would identify one pair. However, further analysis (beyond the scope of this text) is necessary to determine whether significant differences exist in the other pairs (that is, between cities B and C and cities A and C).

c. Briefly discuss the implications of the results for D'Oreal.

Given the test was properly conducted and we have reasonable assurances sample results are representative of U.S. purchase decisions we conclude that aggressive ads (tested in city A) should not be used for Entice lipstick since significantly better results were obtained using the moderately aggressive ads (tested in city B). However we still do not know if the moderately aggressive ads will result in better sales than our current soft-sell strategy (tested in city C). Posttest comparisons are needed to determine this and these techniques are covered in more advanced courses in statistics.* ■

Sum-of-Squared Distances

In actual practice, the F ratio is calculated using more efficient formulas that minimize rounding error and lend themselves to broader analysis. We will use the preceding example to help us understand the underlying rationale.

If you will recall, the calculation of the F fraction was obtained as shown in the following box. However, note how the calculation for $F = \dfrac{9(13)}{6.0833} = 19.23$ is now separated into sum of squared distances and calculations related to n, shown as follows.

---

*Actually, I conducted a posttest comparison of city B (moderately aggressive ads) and city C (soft-sell, our current approach) and found *no* significant difference. This means that although the sample $\bar{x}$'s were different (17% versus 15%) this was comfortably within the range of chance fluctuation, and we have *no* statistical evidence to show that sales in all the outlets in city B are actually any better or worse than sales in all the outlets of city C. From a marketing standpoint, since we cannot prove any increase in results in city B over city C, it might be safer at this point to stick with our current soft-sell approach.

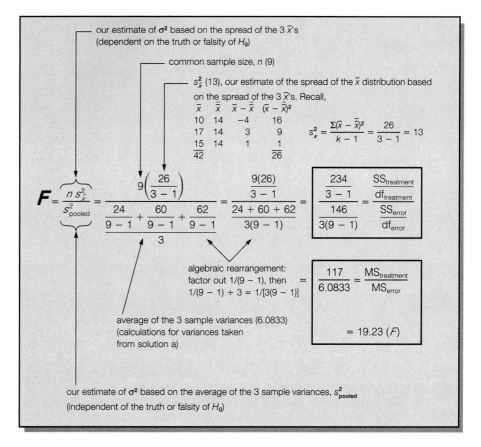

Notice the computations are partitioned into sum-of-squared distances related to the three $\bar{x}$'s ($SS_{treatment}$) and sum-of-squared distances related to the three s^2's (SS_{error}). Expressing the results in terms of sum-of-squared distances has many advantages in advanced work.* Generally in ANOVA experiments the above information is made available in the form of the following chart.

*Using sum-of-squared distances rather than variances has many advantages. First, sum-of-squared distances are additive, meaning, we can add the sum-of-squared distances from one sample to sum-of-squared distances of another sample, regardless of sample size (you can only add variances if the sample sizes are equal). Second, sum-of-squared distances lend themselves to integration into other techniques, such as, regression analysis, which also uses sum-of-squared distances. And third, the reduction of sum-of-squared distances in the error term can be used as a gauge to assess how much better one hypothesis explains results versus another hypothesis.

Summary Chart				
Source	**df**	**SS**	**MS**	**F**
Treatment	$k - 1 = 2$	234	117	19.23
Error	$k(n - 1) = 24$	146	6.0833	
Total	26	380		

Legend: $df_t = k - 1$
$\quad\quad\quad\; = $ Number of groups $- 1$
$\quad\quad\; df_e = k(n - 1)$
$\quad\quad\quad\; = $ Number of groups times one less than the common sample size
$\quad\quad\; SS_t = $ Sum-of-squared distances (treatment).
$\quad\quad\; SS_e = $ Sum-of-squared distances (error).
$\quad\quad\; MS_t = SS_t/df_t$ (Mean-square distance, treatment)
$\quad\quad\; MS_e = SS_e/df_e$ (Mean-square distance, error)

The ANOVA model is based on the fact we can prove SS_{total} (which is the sum-of-squared distances from $\bar{\bar{x}}$ to each of the data points, in our case, to 27 data points) is equal to $SS_{treatment}$ plus SS_{error}.

$$SS_{total} = SS_{treatment} + SS_{error}$$
$$380 = 234 + 146$$

Relatively easy to use formulas are available to calculate SS_{total} and $SS_{treatment}$ (and by subtraction, we can get SS_{error}). We demonstrate use of these formulas, as follows:

	City A		City B		City C	
	x	x^2	x	x^2	x	x^2
	9	81	16	256	13	169
	10	100	14	196	17	289
	12	144	19	361	18	324
	8	64	21	441	15	225
	8	64	15	225	10	100
	13	169	17	289	12	144
	11	121	13	169	16	256
	10	100	20	400	16	256
	9	81	18	324	18	324
	90	924	153	2661	135	2087

$$\boxed{SS_{total} = \Sigma x^2 - \frac{(\Sigma x)^2}{N}}$$

$$SS_{total} = \Sigma x^2 - \frac{(\Sigma x)^2}{N}$$
$$= 5672 - \frac{(378)^2}{27}$$
$$= 5672 - 5292$$
$$= 380$$
$$= 380$$

$$\boxed{SS_{total} = 380}$$

$$\boxed{SS_{treatment} = \frac{\Sigma T^2}{n} - \frac{(\Sigma x)^2}{N}}$$

$$SS_{treatment} = \frac{\Sigma T^2}{n} - \frac{(\Sigma x)^2}{N}$$
$$= \frac{49,734}{9} - \frac{(378)^2}{27}$$
$$= 5526 - 5292$$
$$= 234$$
$$= 234$$

$$\boxed{SS_{treatment} = 234}$$

$$\Sigma x^2 \text{ (sum of } x^2\text{'s)} = 924 + 2661 + 2087$$
$$= 5672$$
$$\Sigma x \text{ (sum of } x\text{'s)} = 90 + 153 + 135$$
$$= 378$$
$$\Sigma T^2 \text{ (sum of treatment totals squared)}$$
$$= 90^2 + 153^2 + 135^2$$
$$= 49{,}734$$

$$SS_{total} = SS_{treatment} + SS_{error}$$
$$380 = 234 + SS_{error}$$

$$\begin{cases} n = \text{common sample size} = 9 \\ N = \text{total number of values} = 27 \\ k = \text{number of groups} = 3 \\ T = \text{group totals (90, 153, 135)} \end{cases}$$

By subtraction,
$SS_{error} = 146$

$$F = \frac{SS_{treatment}/df_{treatment}}{SS_{error}/df_{error}} = \frac{SS_t/(k-1)}{SS_e/k(n-1)} = \frac{234/(3-1)}{146/[3(9-1)]} = \frac{117}{6.0833} = 19.23$$

10.4 Analysis of Variance (ANOVA) for Unequal Sample Size

Although use of unequal sample sizes for ANOVA experiments is not recommended,* situations do arise where this cannot be avoided (for instance, if subjects drop out of or fail to show up for an experiment or animals expire or are uncooperative and their results not counted, etc.) So, for situations where group sizes are unequal, the same formulas apply, however two formulas must be modified slightly to weigh the effects of the unequal group sizes.

$$SS_{treatment} = \Sigma \frac{T_i^2}{n_i} - \frac{(\Sigma x)^2}{N}$$

$$df_{error} = \Sigma(n_i - 1)$$

T_i: sum of values in i^{th} treatment group
n_i: sample size of ith treatment group
$N = \Sigma n_i$

(Note: when all n_i are equal, the formulas reduce to those presented earlier.)

*For unequal sample sizes we must be reasonably assured our population variances are equal. In general, the ANOVA test is quite robust, meaning, population assumptions can often be violated with minor effect on results, however unequal population variances and unequal sample sizes do not mix. As a general policy, always try to keep sample sizes equal in an ANOVA experiment.

There is no modification for SS_{total} or $df_{treatment}$. Let's use the same example to demonstrate, only this time, instead of sampling 9 outlets from each city, we will sample 5 from city A, 3 from city B, and 7 from city C, with the following results:

City A Treatment A* (aggressive ads)	City B Treatment B (moderately aggressive ads)	City C Treatment C (soft-sell ads)
9	16	13
10	14	17
12	$\underline{21}$	18
8	$T_2 = 51$	15
$\underline{11}$	$n_2 = 3$	14
$T_1 = 50$		12
$n_1 = 5$		$\underline{16}$
		$T_3 = 105$
		$n_3 = 7$

$$SS_{total} = \Sigma x^2 - \frac{(\Sigma x)^2}{N} = (9^2 + 10^2 + \cdots + 12^2 + 16^2) - \frac{(50 + 51 + 105)^2}{15}$$

$$= (3006) - \frac{42,436}{15} = 176.93$$

$$SS_{treatment} = \Sigma \frac{T_i^2}{n_i} - \frac{(\Sigma x)^2}{N} = \left(\frac{50^2}{5} + \frac{51^2}{3} + \frac{105^2}{7} \right) - \frac{(50 + 51 + 105)^2}{15}$$

$$= (2942) - \frac{42,436}{15} = 112.93$$

$$SS_{error} = SS_{total} - SS_{treatment}$$
$$= 176.93 - 112.93 = 64$$

The summary chart is as follows:

Note:
$df_t = k - 1$
$= 3 - 1$
$= 2$

$df_e = \Sigma(n_i - 1)$
$= (5 - 1) + (3 - 1) + (7 - 1)$
$= 4 + 2 + 6$
$= 12$

Summary Chart

Source	df	SS	MS	F
Treatment	2	112.93	56.465	10.59
Error	12	64	5.333	(significant)
Total	14	176.93		

*The word *treatment* originates from the early agricultural experiments where various soil treatments were tested for increased yield, such as the use of different fertilizer treatments that were tested on various $\frac{1}{40}$-acre test plots.

$$F \text{ Calculation}$$
$$F = \frac{SS_{treatment}/df_{treatment}}{SS_{error}/df_{error}} = \frac{112.93/2}{64/12} = \frac{56.465}{5.333} = 10.59$$

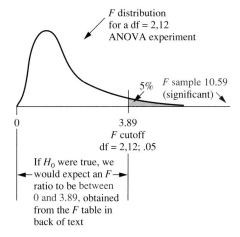

F distribution
for a df = 2,12
ANOVA experiment

5% F sample 10.59
(significant)

0   3.89
F cutoff
df = 2,12; .05

If H_0 were true, we
would expect an F
ratio to be between
0 and 3.89, obtained
from the F table in
back of text

In this case, $F_{cutoff} = 3.89$
$\quad\quad\quad\quad\quad 2,12; .05$

We have a df = 2,12 experiment.
Look across the top for a df of 2, then
proceed down for the df of 12.

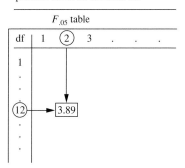

$F_{.05}$ table

Summary

Two powerful and versatile techniques were presented in this chapter: chi-square analysis (X^2) and analysis of variance (ANOVA). These techniques have served (and continue to serve) a staggering range of industrial and scientific uses.

Chi-square (X^2) analysis assumes we are sampling from a two-or-more category population. For this type of analysis, all data is recorded in the form of frequencies or counts, that is, the number of subjects or objects falling into distinct categories, with the test measuring z-score differences between the number of counts expected in a category (given some null hypothesis) and the number of counts actually observed in the experiment, and the purpose of the test is most often to determine the extent one variable influences another.

Three types of X^2 tests were presented, although the mathematics for all three is near identical. The three types are:

X^2 test of independence	to test if two variables are independent of each other
X^2 test of homogeneity	to test if several random samples could have originated from identical, homogeneous, category populations
X^2 goodness-of-fit	to test if observed counts could have originated from an expected distribution

All three tests are rather similar in execution and interpretation. Essentially each calculates z-score differences between observed and expected values. These z-score differences are then squared and summed up. This sum is X^2 (thus, $X^2 = \Sigma z^2$). However, this formula is often difficult to apply and the following algebraic rearrangement is almost always used in practice.

General Formula $X^2 = \Sigma \dfrac{(O - E)^2}{E}$

where

O = Observed count and E = Expected count

To establish the cutoff value, we refer to the X^2 tables, df = (rows − 1)(columns − 1) except in the case of the goodness-of-fit test, where df = rows − 1

   Analysis of variance (ANOVA) assumes we are sampling from several (two or more) measurement populations. Essentially sample differences are used to determine whether or not actual differences exist in the populations.

Population Assumptions

ANOVA requires independent samples be drawn from populations which are normally distributed with equal variances, essentially equal standard deviations.

Experimental Designs

The design of an experiment is quite important for ANOVA testing. Two population designs were discussed: the Completely Randomized Design and the Randomized Complete Block Design.

Logic of ANOVA

Essentially ANOVA compares two different estimates of the population variance, σ^2, one based on the spread of the three $\bar{x}$'s and the other based on the spread of data within each sample, essentially the average s^2, denoted s^2_{pooled}.

(estimate based on the spread of the $\bar{x}$'s and referred to as *variation between samples*)

$$F = \frac{\text{estimate of } \sigma^2}{\text{estimate of } \sigma^2} = \frac{n\, s^2_{\bar{x}}}{s^2_{pooled}}$$

(estimate based on the average s^2 denoted s^2_{pooled} and referred to as *variation within samples*)

s^2_{pooled}, the bottom of the fraction is independent of whether H_0 is true or not, however the second estimate of σ^2, $n\, s^2_{\bar{x}}$, the top of the fraction is very dependent on whether H_0 is true. If H_0 is true, the F ratio tends to be close to one; if H_0 is false, the F ratio tends to much larger than one.

   To establish the cutoff F from the tables,

For equal sample size $\begin{cases} df_{numerator} = k - 1 \\ df_{denominator} = k(n - 1) \end{cases}$

where $\begin{aligned} k &= \text{number of groups} \\ n &= \text{sample size in one group} \end{aligned}$

For unequal sample size $\begin{cases} df_{numerator} = k - 1 \\ df_{denominator} = \Sigma(n_i - 1) \end{cases}$

Sum-of-Squared Distances

In actual practice, the F ratio is calculated using more efficient formulas that minimize rounding error and lend themselves to broader analysis. Essentially,

$$F = \frac{SS_t / df_t}{SS_e / df_e}$$

where SS_t is the sum-of-squared distances related to the $\bar{x}$'s; SS_e is the sum-of-squared distances related to the s^2's; $df_t = df_{treatment}$, whereas $df_e = df_{error}$.

Exercises

Note that full answers for exercises 1–5 and abbreviated answers for odd-numbered exercises thereafter are provided in the Answer Key.

10.1 TOASTYOs, a new high-nutrient cereal was introduced into a local test market last year (in Springfield, Mo.) to appeal to the better informed, more affluent adult community, and advertising and promotion have been slanted to this segment. Sales have been disappointing, however, prompting the maker of the product, Tabisco Foods, to hire outside research consultants. The consultants gathered data on 600 randomly selected adults who had tried the product, with the following results of particular concern:

Taste preference	Preferred ToastyOs' taste	73	81	86
	Neutral or disliked	77	119	164
		150	200	250	600 total
		No college	Some college	College graduate

Education level

a. Assuming a valid random sample, test at .05 whether the two variables (education level and taste preference) are independent; essentially, in this case, whether differences in taste preference among the three education categories can be attributed to chance fluctuation or, indeed, to real differences in the population.

b. Briefly discuss the implications of these results for Tabisco.

c. Discuss how Tabisco might proceed with a decision.

10.2 The blood-brain barrier is a unique layer of protection cells that block potentially harmful compounds (viruses, bacteria, and toxins) from entering the brain. However, it also blocks potentially *helpful* therapeutic drugs. For some time now, researchers have been experimenting with ways of breaching this protective layer of cells for the purpose of administering helpful drugs to the brain. Suppose a team of researchers at a foundation for neurological studies in Cambridge randomly assigned 800 patients to four test groups for the administration of Penetril, a side effect-free drug for the treatment of Alzheimer's disease (a fictitious drug). Penetril was administered using four different methods, with the following results:

Major improvement	53	50	76	61
Slight improvement	42	28	45	45
No improvement	105	72	129	94
	200	150	250	200	800 total patients
	Caffeine coated to dissolve in barrier cells	Iron coated to fool cells into accepting it	Immune-cell coat to fool cells into accepting it	Taken with loosening compound

Same medication: administered by four methods

a. Assuming valid random selection, test at .05 whether differences in patient improvement can be attributed to chance fluctuation or, indeed, to real differences among the 4 methods; essentially are the four populations homogeneous, equally proportioned, with respect to patient improvement?

b. Briefly discuss the implications of this research.

10.3 Milwaukee is one of the hottest diet soft-drink markets in the nation. Suppose in Milwaukee after a prolonged advertising war, one of the leading diet soft-drink competitors, Peppi-Thin, wanted to know if their market share had been maintained. A study was commissioned where 300 diet soft-drink consumers from the metropolitan Milwaukee area were randomly sampled with the following results:

	Number of Consumers Selecting Brand	Prior Market Share
Peppi-Thin	74	30%
Cota-Light	72	25%
Sprinkle-Free	81	20%
Other Brands	73	25%

a. Test the hypothesis that current market share has not changed from prior market share (at $\alpha = .05$).
b. Briefly discuss the implications for Peppi-Thin.

10.4   At the turn of the century, in an attempt to assess Campbell soup usage in the Philadelphia area, C. C. Parlin randomly selected garbage pails from various neighborhoods and counted Campbell soup cans, suppose with the following results:

Socioeconomic Class of Neighborhood	Number of Campbell Soup Cans Found
Wealthy	71
Upper-Middle Class	86
Middle-Working Class	93
Lower-Middle-Working Class	82
Lower-Working Class	68

Use X^2 analysis to test the hypothesis that Campbell soup usage is the same in all neighborhoods (at $\alpha = .01$).

10.5   A new lipstick was introduced last year by D'Oreal called Entice. Entice uses a soft-sell advertising approach to promote itself, which management felt appropriate to its image; however national sales have been disappointing. Now the company wishes to test more aggressive advertising strategies. As such, TV ads were produced using two more aggressive experimental approaches. In addition, new ads were produced using the old soft-sell strategy.

The three campaigns were run concurrently, each in a separate test market, with advertising levels substantially increased (blitzed) to accelerate impact and, after several weeks, case shipment data was obtained from randomly selected outlets in each market, with the following results.

Study I

	Campaign A	Campaign B	Campaign C
Sales Increase (percentage increase above normal levels of last year)	15	16	13
	11	14	17
	18	19	18
	20	21	15
	16	15	10
	16	17	12
	12	13	16
	19	20	16
	17	18	18
	(Midland, Tx.) Aggressive Ads	(Indianapolis) Moderately Aggressive Ads	(Charleston, W.Va.) Soft-Sell Ads (control group)

Three Different Advertising Strategies Were Tested

a. Calculate $\bar{x}$ and s^2 (average sales increase and variance) for each sample.
b. Assuming valid random samples, test at $\alpha = .05$ whether differences among the three sample $\bar{x}$'s can be attributed to chance fluctuation or, indeed, to actual differences in sales in the three test cities.
c. Briefly discuss the implications of the results for D'Oreal.

Study II

Average Sales Increase above Normal Level for Each Campaign (in percentage increase above prior year, same period)	$n = 11$ $\bar{x} = 12.0$ $s^2 = 11.9$	$n = 11$ $\bar{x} = 16.0$ $s^2 = 9.5$	$n = 11$ $\bar{x} = 17.0$ $s^2 = 12.8$
	(Portland area) Aggressive Ads	(Ft. Wayne) Moderately Aggressive Ads	(Bloomington, Ill.) Soft-Sell Ads Control Group

Three Different Advertising Strategies Were Tested

d. Assuming valid random samples, test at $\alpha = .01$ whether differences among the three sample $\bar{x}$'s can be attributed to chance fluctuation or, indeed, to actual differences in sales in the three test cities?

e. Briefly discuss the implications of the results for D'Oreal.

Study III

	11	19	14
Sales Increase	9	12	18
above Normal	12	16	22
Level for Each	8	12	17
Outlet	4	24	13
(in percentage increase	10	11	15
above prior year, same		15	17
period)		18	
	(Wichita area)	(Dayton)	(Boise)
	Aggressive Ads	Moderately Aggressive Ads	Soft-Sell Ads Control Group

Three Different Advertising Strategies Were Tested

f. Calculate $SS_{treatment}$ and SS_{error} for the data.

g. Assuming valid random samples, test at $\alpha = .01$ whether differences among the three sample $\bar{x}$'s can be attributed to chance fluctuation or, indeed, to actual differences in sales in the three test cities.

10.6 Narcissism, a grossly inflated sense of self-importance with an unquenchable need for adoration, is increasingly being diagnosed by psychiatrists as the underlying problem in patients who complain of other problems, such as, inability to sustain relationships or severe depression after minor failure. Narcissism can produce exciting sports, entertainment, and political figures, who typically are charming and friendly to those who play into their needs for praise, superiority, or special treatment. However, narcissism can also produce petty office tyrants and obsessed workaholics. Typically, underneath, the narcissist feels life is empty and him- or herself worthless.

Treating such patients can be especially difficult, according to *New York Times* article, "Narcissism Looms Larger as Root of Personality Woes," (November 1, 1988, C1). Narcissists tend to be exploitative and insensitive to other's feelings and find it difficult to form the warm therapist-patient bond that other patients find natural. "Instead, they often become cold or even enraged when a therapist fails to play along with their inflated sense of themselves." This puts therapists in an untenable position because they must eventually deflate the narcissist's protective but self-deluding bubble. Narcissists confront disapproval or hurt (even minor disapproval or minor hurt) with sharp rage or deep depression. "A narcissist patient," according to one quoted therapist, "is likely at some point to attack or devalue the therapist. It's hard to sit with such people in your office."

Suppose the psychiatry department of a Massachusetts hospital conducted a 1-year study comparing various treatment methods with the following results:

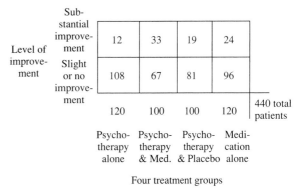

Four treatment groups

a. Assuming valid random samples, test at .05 whether differences in the level of improvement can be attributed to chance fluctuation or, indeed, to actual differences among the four treatments.

b. Briefly discuss the implications of these results.

10.7 Coronary bypass surgery, one of the most frequent in-hospital operations, consists of taking a vein or artery from somewhere in the body and using it to reroute blood around a clogged artery in the heart. However, choosing a particular surgeon to perform this operation may mean the difference between life and death, according to studies conducted in New England and Philadelphia hospitals (*Mass. Medical Society Journal Watch,* 8:5; *New York Times,* August 14, 1991, A11, also August 20, 1991, C3).

Suppose a Philadelphia hospital samples patient records on 1000 such surgeries, with the following outcomes:

	Patient not survived				60
Outcome of operations	18	9	12	21	60
	Patient survived				
	182	291	188	279	940
	Surgeon M 200	Surgeon K 300	Surgeon O 200	Surgeon S 300	1000 total patients

Four surgeons

a. Assuming a valid random sample, test at $\alpha = .05$ whether differences in the number of deaths can be attributed to chance fluctuation, or, indeed, to actual differences among the four surgeons.
b. Briefly discuss the implications of these results.
c. What other factors (besides the surgeon) might influence results?

10.8 Creativity and mental illness have long been thought related, although scant evidence exists. Studies conducted in the last decade at the University of Iowa-College of Medicine and at the University of Tennessee/Memphis may have identified a possible association. Suppose the National Institute of Mental Health commissions a study in which working professionals are randomly selected from three categories and classified as follows.

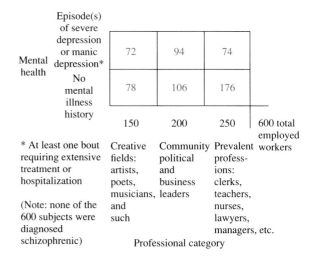

Mental health	Episode(s) of severe depression or manic depression*		
	72	94	74
	No mental illness history		
	78	106	176
	150	200	250	600 total employed workers

* At least one bout requiring extensive treatment or hospitalization

(Note: none of the 600 subjects were diagnosed schizophrenic)

Creative fields: artists, poets, musicians, and such

Community political and business leaders

Prevalent professions: clerks, teachers, nurses, lawyers, managers, etc.

Professional category

a. Assuming valid random samples, test at .05 whether differences in mental health can be attributed to chance fluctuation or, indeed, to actual differences among the three professional categories.
b. Briefly discuss the implications of these results.

10.9 Artificial insemination is chosen by over 80,000 women each year, resulting in 30,000 births annually. The procedure is administered primarily through private physicians, sperm banks, and fertility centers. However, a congressional survey, the most comprehensive to date, revealed a disturbing lack of donor-and-specimen testing for syphilis, hepatitis, genetic defects, and so on, by certain private practitioners.[12] Since the publication of the study (1988), some practitioners have ''improved significantly,'' according to one U.S. senator. Suppose, however, the American Fertility Society commissions a follow-up study investigating the records of 800 recently inseminated women, with the following results.

		Source of treatment			
Level of pretesting (of donor and donor specimen)	All important tests performed	152	198	130	
	Many important tests performed	89	95	56	
	Few important tests performed or tests were inadequately monitored	59	7	14	
		300 Private physician	300 Sperm bank	200 Fertility center	800 total women

Source of treatment

Assuming a valid random sample, test at .01 whether differences in pretesting can be attributed to chance fluctuation or indeed, to actual differences among the three sources of treatment.

10.10 Interest and Recall (I & R) testing is one technique used by large consumer-product firms to assess the raw ability of a TV ad to generate interest and be remembered. The test involves placing TV ads in isolated test markets with residents telephoned 24 hours later and questioned as to the extent they can recall the ad and the ad's key points. Generally, TV ads with over a 40 rating (40% of the people remembered substantial portions) are considered very strong.

Suppose Kurke Marketing Corporation in combination with *Advertising Age* magazine wanted to demonstrate that I & R scores are normally distributed and randomly selected 200 TV ads from Kurke's extensive quarter-century files and found the average I & R score to be $\bar{x} = 25$ with standard deviation, $s = 8$.

To test if I & R scores are normally distributed, each ad was classified by number of standard deviations (z scores) from the mean, as follows:

z Score from Mean	Number of Ads in Each Category	Normal Population Expected Percentage
Above 2	14	2.3%
+1 to +2	32	13.6%
0 to +1	48	34.1%
0 to −1	62	34.1%
−1 to −2	42	13.6%
Below −2	2	2.3%

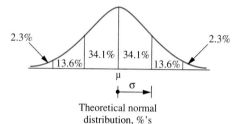

Theoretical normal distribution, %'s

> *Important*: For this experiment, df = $r - 1 - m$

where r = number of cells (or rows). Note that in experiments where we must estimate population characteristics with sample data, we subtract one df for each characteristic we estimate, denoted by m. In this case, we used $\bar{x}$ and s to estimate μ and σ, thus, $m = 2$, and

$$df = r - 1 - m$$
$$= 6 - 1 - 2$$
$$= 3$$

Assuming a valid random sample, test at .05 whether we can assume the data originated from a normal distribution.

10.11 Buttercups: Karl Pearson, in his original paper introducing X^2 (see endnote 2), applied his new technique to a species of plant of the genus, *Ranunculus* (native to Europe and common in North America), characterized by glossy yellow flowers and commonly referred to as crowfoot or buttercups.

The number of petals on the buttercup range from 5 to 11 and Pearson hypothesized the frequency of petals fit a skewed mathematical distribution that he had previously derived from other work. To test this hypothesis, he randomly selected 222 buttercups and counted the petals on each and classified the data as follows.

Category	Number of Buttercups in Each Category	Pearson's Theoretical Percentage
Five petals	133	61.667%
Six petals	55	21.847%
Seven petals	23	10.1281%
Eight, nine, ten, or eleven petals	11	6.307%

Assuming a valid random sample, test at .01 whether we can assume the sample data fits Pearson's distribution. (Note: round expected value to one decimal place.)

10.12 The number of ''ums,'' ''ahs,'' and ''ers'' in a speaker's presentation may not be a sign of nervousness, according to researchers at Columbia's[13] department of psychology, but rather an indication of the number of word or idea choices available to express the next utterance. In other words, the more options people have, the more likely they may be apt to say um, ah, or er.

Suppose a public speakers' organization tested the concept by randomly selecting an equal number of tape recorded minutes of several lecturers in various fields and recorded the number of interjections (ums, ahs, and ers), as follows:

Field	Number of Interjections
Art history	79
Chemistry	30
Literature	68
Mathematics	23

Assuming a valid random sample, test at .01 whether we can assume interjections (ums, ahs, and ers) are uniform across all fields.

10.13 Top-fashion houses, like most large businesses, seek new markets for their products and when countries make the transition from centrally planned economies to a free-market system, the top international fashion houses may wish to explore the market by introducing their line of women's sportswear (calf-length skirts, peplum jackets, cashmere shawls, and so on).

Suppose after several months of open competition in one such country, Odenfield of England wanted to determine if there were any early indications of customer loyalty and label identification. A comprehensive study involving major outlets monitored customers over randomly selected times making substantial purchases exclusively from a single designer, with the following results:

Fashion House	Number of Customers Purchasing Exclusively from One Designer
Yves St. Laurette (France)	70
Geoffrey Deane (U.S.A.)	78
Georgio Amati (Italy)	84
Odenfield (England)	92
Biesendorf (Germany)	76

Assuming a valid random sample, test at .01 whether we can assume brand loyalty is uniform among the five top international fashion houses.

10.14 Hi-tech concrete, a new generation of porous-free material, is slated to replace much of the infrastructure of this country (buildings, roads, sewers, and bridges) over the next few decades. The substance has such remarkable qualities that consideration is being given to consumer usage, such as, living room furniture, boats, automobile engine blocks, bottle caps, and soda cans. Not only is this new material vastly superior to ordinary concrete in strength and durability, but with additives such as plastic and steel fibers, it can be made to bend.

Study I

Suppose a team of researchers from Purdue University, experimenting with new plastic and steel fiber additives, produce five specimens for each experimental mixture and test as follows:

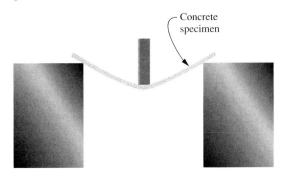

Concrete specimen

	Mixture I	Mixture II	Mixture III	Mixture IV
Bending Ability of Each Specimen (its elasticity; as compared to high-quality steel, which equals 10.0)	4.0	3.5	3.7	4.0
	3.9	3.4	3.2	3.8
	4.2	3.2	3.5	4.1
	3.8	3.3	3.3	4.4
	4.1	3.6	3.3	3.7

Four Different Mixtures Were Tested

a. Calculate $\bar{x}$ and s^2 (average bending ability and variance) for each group.

b. Assuming valid random samples, test at $\alpha = .05$ whether differences among the four sample $\bar{x}$'s can be attributed to chance fluctuation or, indeed, to actual differences in the mixtures.

c. Briefly discuss the implications of these results.

Study II

Suppose a team of researchers from Northwestern University, experimenting with new plastic and steel fiber additives, produce nine specimens for each experimental mixture and test as follows.

	Mixture V	Mixture VI	Mixture VII	Mixture VIII
Average Bending Ability for Each Group (its elasticity; as compared to high-quality steel, which equals 10.0)	$n = 9$ $\bar{x} = 5.00$ $s^2 = .10$	$n = 9$ $\bar{x} = 4.00$ $s^2 = .14$	$n = 9$ $\bar{x} = 4.20$ $s^2 = .10$	$n = 9$ $\bar{x} = 4.80$ $s^2 = .18$

Four Different Mixtures Were Tested

d. Assuming valid random samples, test at $\alpha = .05$ whether differences among the four sample $\bar{x}$'s can be attributed to chance fluctuation or, indeed, to actual differences in the mixtures.

e. Briefly discuss the implications of these results.

Study III

Suppose researchers at the National Institute of Standards and Technology, experimenting with new plastic and steel fiber additives, produce specimens for each experimental mixture and test as follows:

	Mixture A	Mixture B	Mixture C	Mixture D
Bending Ability for Each Specimen (its elasticity; as compared to high-quality steel, which equals 10.0)	4.2	5.1	3.7	4.1
	3.9	4.2	3.3	4.6
	4.4	4.8	3.9	3.7
	4.0	4.0	4.2	3.9
	4.7	3.9	3.3	
		4.5	3.6	
		4.1		

Four Different Mixtures Were Tested

f. Calculate $SS_{treatment}$ and SS_{error} for the data.

g. Assuming valid random samples, test at $\alpha = .05$ whether differences among the four sample $\bar{x}$'s can be attributed to chance fluctuation or, indeed, to actual differences in the mixtures.

10.15 Yellow stains on white clothing are stubborn discolorations believed to result from oxidated body oils that have reacted chemically with the clothing's cellulose fibers (as if it were a dye). In as little as one week, these reactions can permanently discolor white clothing with an ugly yellow stain and at this point the most potent laundry product will usually be of little value in removing them.

Study I

Suppose researchers at the Department of Textiles and Apparel at Cornell University, experimenting with new experimental formulations to remove these stains, washed six discolored strips in each of three experimental formulations, and recorded the results as follows:

Color chart

	Formulation		
	X	**Y**	**Z**
Color Rating	10	8	11
(after washing)	9	10	12
for Each Strip	10	12	14
0 = Dark Yellow, the	12	9	11
Initial Shade	8	10	10
20 = Pure White	11	8	14

All washed strips were matched to a color chart, a series of increasingly lighter shades, and rated from 0 to 20.

Three Different Formulations Were Tested

a. Calculate $\bar{x}$ and s^2 (average color rating and variance) for each group.
b. Assuming valid random samples, test at $\alpha = .05$ whether differences among the three sample $\bar{x}$'s can be attributed to chance fluctuation or, indeed, to actual differences in the stain-eliminating ability of the three formulations.
c. Briefly discuss the implications of these results.
d. Discuss in detail how one might design a valid experiment for the above test.

Study II

Suppose researchers at the Advanced Apparel Laboratory at the Fashion Institute of Technology, experimenting with new formulations to remove these stains, washed twelve discolored strips in each of three experimental formulations, and recorded the results as follows:

	Formulation		
	A	**B**	**C**
Average Color	$n = 11$	$n = 11$	$n = 11$
Rating for Each	$\bar{x} = 13.5$	$\bar{x} = 10.0$	$\bar{x} = 12.5$
Group (after	$s^2 = 9.6$	$s^2 = 7.3$	$s^2 = 6.8$
washing)			

0 = Dark Yellow, the Initial Shade

20 = Pure White

Three Different Formulations Were Tested

e. Assuming valid random samples, test at $\alpha = .05$ whether differences among the three sample $\bar{x}$'s can be attributed to chance fluctuation or, indeed, to actual differences in the stain-eliminating ability of the three formulations.
f. Briefly discuss the implications of these results.

Study III

Suppose researchers at the University of Georgia, experimenting with new formulations to remove these stains, wash discolored strips in each of three experimental formulations, and recorded the results as follows:

	Formulation		
	K	**M**	**N**
Color Rating for	8	12	9
Each Strip	7	10	11
(after washing)	9	13	9
0 = Dark Yellow, the	11	9	8
Initial Shade	8	11	12
20 = Pure White	7	15	
		11	

Three Different Formulations Were Tested

g. Calculate $SS_{treatment}$ and SS_{error} for the data.
h. Assuming valid random samples, test at $\alpha = .05$ whether differences among the three sample $\bar{x}$'s can be attributed to chance fluctuation or, indeed, to actual differences in the stain-eliminating abilities of the three formulations.

10.16 Office politics is serious business. It can make or break a career, according to experts (''Playing Office Politics,'' *Newsweek,* September 16, 1985). In fact, office politics is of such serious concern that many business schools (Harvard among them) offer courses exploring its virtues and pitfalls. In such courses, students learn strategies to elicit cooperation from fellow workers and to overcome obstacles, such as, turf protection and poor communications between departments, and (hopefully) are given some good old-fashion advice like never upstage your boss, cultivate necessary friendships, and beware of certain dangerous office types (such as, the manipulator or the spy, to name a few).

Suppose one university in New Jersey wished to offer such a course entitled, Power and Influence, modeled after a course taught at Harvard. However, administration felt their student body might be more receptive to alternative teaching methods. In fact, four teaching approaches were tested.

Study I

Suppose at semester's end, industry experts were brought in to evaluate achievement, with the following results.

	Lecture Method	Dramatic Enactments by Students	Seminar Case Discussion	Mix of First Three Methods
Achieve-ment Grade for Each Student (scale: 0–50)	30	28	32	32
	22	41	36	35
	27	33	31	27
	35	40	20	15
	21	24	41	32
	14	28	27	23
	26	37	30	25

Four Different Methods Taught by Same Instructor Using Similar Materials

a. Calculate $\bar{x}$ and s^2 (average grade and variance) for each group.
b. Assuming valid random samples, test at $\alpha = .01$ whether differences among the four sample $\bar{x}$'s can be attributed to chance fluctuation or, indeed, to actual differences in the teaching methods.
c. Briefly discuss the implications of these results.

d. Discuss in detail how one might design a valid experiment for the above test.

Study II

Suppose at semester's end, the instructor evaluated achievement, as follows:

	Lecture Method	Dramatic Enactments by Students	Seminar Case Discussion	Mix of First Three Methods
Average Student Achieve-ment Grade for Each Class (scale: 0–50)	$n = 7$ $\bar{x} = 31.0$ $s^2 = 39.7$	$n = 7$ $\bar{x} = 35.0$ $s^2 = 27.0$	$n = 7$ $\bar{x} = 34.0$ $s^2 = 37.2$	$n = 7$ $\bar{x} = 32.0$ $s^2 = 31.4$

Four Different Methods Taught by Same Instructor Using Similar Materials

e. Assuming valid random samples, test at $\alpha = .01$ whether differences among the four sample $\bar{x}$'s can be attributed to chance fluctuation or indeed, to actual differences in the teaching methods.
f. Briefly discuss the implications of these results.

Study III

Suppose the same study was repeated the following semester, with industry experts brought in to evaluate achievement, and the results were as follows:

	Lecture Method	Dramatic Enactments by Students	Seminar Case Discus-sion	Mix of First Three Methods
Achieve-ment Grade for Each Student (scale: 0–50)	28	38	27	19
	23	29	38	35
	25	36	23	32
	37	33	40	26
	25	37	36	29
	15	30		34
		35		

Four Different Methods Taught by Same Instructor Using Similar Materials

g. Calculate $SS_{treatment}$ and SS_{error} for the data.
h. Assuming valid random samples, test at $\alpha = .01$ whether differences in student achievement can be attributed to chance fluctuation or indeed, to actual differences in the teaching methods.

Endnotes

1. The prior focus was Professor Karl Pearson at University College, London, from about 1893 to 1914, known as the Biometric Period. Although the work concerned biological measurements and studies, the emphasis was clearly on theories concerning inherited characteristics, seemingly an attempt to follow Darwin's work. In fact, it was Pearson who founded (1901) and edited until his death (1936) the publication, *Biometrika*. Although Pearson is sometimes referred to as the founder of twentieth-century statistics, it was probably Sir Ronald Fisher who played the pivotal role in verifying, redefining, and unifying much of the cohesive entity we study today.

G. Udny Yule, initially a research assistant under Pearson, attempted to expand the analysis to social concerns, however Yule was somewhat an outsider. He later joined Fisher at Rothamsted. Even Wm. Gossett, who applied his small-sample techniques (1908) to sampling at the Guiness Brewery (he studied under Pearson for one year) was in a sense an isolated case in that his small-sample techniques were not broadly accepted into experimentation until Fisher in the 1920s adopted the *t* distributions in formulating the ANOVA tables.

For more reading on this, refer to D. A. MacKenzie, *Statistics in Britain: 1865–1930* (Edinburgh: Edinburgh University Press, 1981).

2. Karl Pearson, "On the Criterion That a Given System of Deviations from the Probable in the Case of a Correlated System of Variables is such that it can be Reasonably Supposed to have Arisen from Random Sampling," *The London, Edinburgh and Dublin Philosophical Magazine and Journal of Science*, Fifth Series, 50 (1900): 157–175. Reprinted in Karl Pearson's *Early Statistical Papers* (Cambridge University Press, 1948). Also, Sir Ronald Fisher, "The Correlation Between Relatives on the Supposition of Mendelian Inheritance," *Trans. Royal Society, Edinburgh,* 52 (1918): 399–433.

3. L. Boone and D. Kurtz, *Contemporary Marketing* (Chicago: The Dryden Press, Holt, Rinehart and Winston, 1989, pp. 7–13).

4. S. E. Asch, "Effects of Group Pressure Upon the Modification and Distortion of Judgments," in *Readings in Social Psychology,* ed. E. E. MacCoby et al. (New York: Holt, Rinehart and Winston, 1956, pp. 174–183).

5. D. Leonhard, *The Human Equation in Marketing Research* (New York: American Management Association, Inc., 1967, pp. 35–73).

6. *U.S. News & World Report,* "Blitzing the Defense," October 15, 1990, pp. 90–92. Also, *New York Times,* "Breaching the Brain's Wall to Deliver Drugs," September 4, 1991, D7.

7. C. Reiner, "Seattle-Area Pop Drinkers to Test Decaffeinated Pepsi," *Seattle Times,* August 3, 1982, E4.

8. L. Boone and D. Kurtz, *Contemporary Marketing* (Chicago: The Dryden Press, Holt, Rinehart and Winston, 1989, pp. 127–128).

9. R. A. Fisher and W. A. MacKenzie, "Studies in Crop Variation, II," *J. Agric. Sci.* 13 (1923): 311–320.

"Rothamsted Experimental Station [is] the world's oldest and England's most important agricultural experimental station . . . [with departments of] chemistry, soil microbiology, physics, botany, entomology, insecticides and fungicides, plant pathology, and statistics. . . . An important function of the station now is the training of post-graduate research workers" (*The New Columbia Encyclopedia,* New York: Columbia Press, 1975, p. 2363).

10. When ANOVA tests are properly designed, it is surprising the number of situations in which these population assumptions can be met (normal population distributions with equal standard deviations) and if the treatment groups truly have different effects, then essentially only the μ's change and usually not the normally-shaped distributions or common variance.

Although this phenomenon had been witnessed by Galton in sweet pea and human height experiments (1875–1889), which ultimately led to the formulation of regression analysis, it wasn't until Wm. Gossett showed that s^2 and $\bar{x}$ were uncorrelated and thus, he assumed, independent that the field was open to Fisher to make use of this in the development of ANOVA.

The design of the experiment is especially critical in ANOVA experiments to ensure the population assumptions are satisfied. For further reading on experimental design, refer to the following books by Sir Ronald Fisher:

The Design of Experiments (Edinburgh: Oliver & Boyd) *Statistical Methods for Research Workers* (New York: Hafner) several editions exist for each book.

11. "Forget Peoria, It's Now: 'Will It Play in Tulsa?' " *New York Times,* June 1, 1992, Al.

12. "A Scandal in Artificial Insemination," *New York Times Magazine,* "Good Health," Part 2, October 10, 1990, p. 23.

13. "Say Ah," *Columbia,* magazine of Columbia University, Summer 1991, p. 22.

Additional Topics

| n this chapter, we present additional topics of interest, categorized under the following broad headings.

Two-Category Population Sampling (Section 11.1)

Measurement Population Sampling (Section 11.2)

Nonparametric Tests (Section 11.3)

Examples are drawn from various fields, but the emphasis of the demonstration examples is on psychology and medicine, whereas homework exercises cover a broader spectrum. ▼

11.1 Two-Category Population Sampling

You might recall from sections 4.4 and 4.5, we discussed sampling from a two-category population. We continue this discussion here.

Small-*n* Binomial Tests

In chapter 4, recall we used the normal curve to approximate binomial probabilities when np and $n(1 - p)$ exceeded 5. However, when your sample size is small, specifically when np or $n(1 - p)$ drops below 5, the binomial sampling distribution grows increasingly more skewed or sloping (an example is shown in the sketch at right) and the normal curve *cannot* be depended on to give proper estimates. For these cases, we use the binomial formula, which is presented in the solution to the following example.*

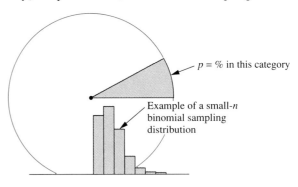

$p = \%$ in this category

Example of a small-*n* binomial sampling distribution

*At values very close to or at 5, the normal curve approximation can be used.

However, before presenting the example, a brief review of mathematical notation may be helpful, as follows.*

1. *The factorial symbol,* !: The factorial symbol, !, following a number, n, is defined as follows.

$$n! = n(n - 1)(n - 2) \ldots 1$$

Examples: $6! = 6(5)(4)(3)(2)(1) = 720$
$$4! = 4(3)(2)(1) \qquad = 24$$
Note: $1! = 1$
$$0! = 1 \text{ (by definition)}$$

2. *The combination symbol,* C_s^n: A combination may be used to count the number of ways we can achieve s successes in n selections and is defined mathematically as follows:

$$C_s^n = \frac{n!}{(n - s)!s!}$$

Examples: $C_3^5 = \frac{5!}{(5 - 2)!2!} = \frac{5!}{3! \, 2!} = \frac{120}{(6)(2)} = 10$

$C_1^7 = \frac{7!}{(7 - 1)!1!} = \frac{7!}{(6)!1!} = \frac{5040}{(720)(1)} = 7$

Example ———————— Viral epidemics in an exposed population generally strike a great many citizens, however, some people have the unique ability to produce resistant chemicals in infected cells which render the invading virus inactive. We refer to such people as having a natural immunity to the virus.

Suppose it is known that 10% of the U.S. population is naturally immune to HICNS, a deadly viral infection raging through the country (a fictitious infection). Now suppose we were to randomly select five individuals from the U.S. population, what is the probability that *exactly* 2 out of the 5 possess this natural immunity?

Solution This is a binomial experiment since we have a fixed number of selections (in this case, 5), each independent and each having the same probability a person with natural immunity will be chosen, which is 10%. Random selection from a large two-category population generally ensures satisfying binomial conditions.

*These topics were discussed in section 3.6.

Now, in this case, np is substantially less than 5 [Note: $p = .1$ and $n = 5$, such that $np = 5(.1) = .5$]. In these cases, the binomial sampling distribution is generally *not* normally distributed, and we employ the binomial formula to calculate probabilities, as follows:

▼ *Binomial Formula*

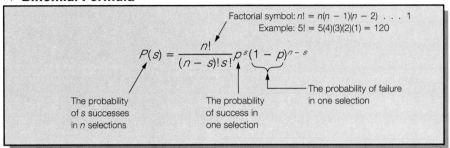

Factorial symbol: $n! = n(n - 1)(n - 2) \ldots 1$
Example: $5! = 5(4)(3)(2)(1) = 120$

$$P(s) = \frac{n!}{(n - s)!s!} p^s (1 - p)^{n - s}$$

The probability of s successes in n selections

The probability of success in one selection

The probability of failure in one selection

So, to find the probability of two successes in five selections, we substitute in $s = 2$ and $n = 5$ into the formula.*

$$P(s) = \frac{n!}{(n - s)!s!} p^s (1 - p)^{n - s}$$

Note: $p =$ the probability of success (natural immunity) in one selection, which is 10% or .1

$$P(2) = \frac{5!}{(5 - 2)!2!} (.1)^2 (.9)^{5 - 2}$$

$$= \frac{5!}{3! \, 2!} (.1)^2 (.9)^3$$

$$= \frac{120}{(6)(2)} (.01)(.729) = 10(.00729) = .0729$$

$$(7.29\%)†$$

Answer

The probability of obtaining *exactly* two successes (2 people having the natural immunity) out of five selections is 7.29%. In other words, if we were to randomly select five individuals from the population and do this several thousand times,

---

*Note the expression $n!/(n - s)!s!$ in the formula above is simply the combination C_S^n previously explained.

†Probabilities are generally expressed in decimal form, however we will use both decimal and percentage forms.

approximately 7.29% of these times the sample would contain exactly two with natural immunity, depicted as follows:

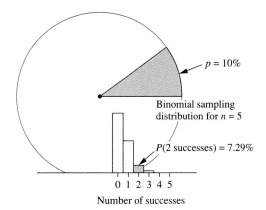

The rationale for the binomial formula is based on the multiplication rule for independent events studied in section 3.4, where the probability of several independent events occurring together was defined as the multiplication of each of their separate probabilities. For instance, the probability of 2 successes, followed by 3 failures would be as follows (S = success, X = failure):

$$\begin{aligned}
P(S, S, X, X, X) &= P(S)P(S)P(X)P(X)P(X) \\
&= (.1)(.1)(.9)(.9)(.9) \\
&= (.1)^2(.9)^3, \text{ which can be stated as} \\
&= (.1)^2(.9)^{5-2},
\end{aligned}$$

which is simply $p^s(1-p)^{n-s}$ in the binomial formula.

However, this is merely one arrangement by which we can obtain 2 successes in 5 selections. There are actually ten ways in which we can achieve this, listed below.

Ten Ways We Can Achieve 2 Successes in 5 Selections

S, S, X, X, X
S, X, S, X, X
S, X, X, S, X
S, X, X, X, S
X, S, S, X, X
X, S, X, S, X Note that each arrangement will have the same probability $(.1)^2(.9)^3$ which is $p^s(1-p)^{n-s}$
X, S, X, X, S
X, X, S, S, X
X, X, S, X, S
X, X, X, S, S

Counting the number of ways we can achieve s successes in n selections can be obtained more efficiently by using the formula* $n!/(n - s)!s!$ Note:

$$\frac{n!}{(n - s)!s!} = \frac{5!}{(5 - 2)!2!}$$
$$= \frac{120}{3!\,2!}$$
$$= \frac{120}{(6)(2)}$$
$$= 10 \text{ ways}$$

So, we can think of the binomial formula as merely the product of two components, as follows.

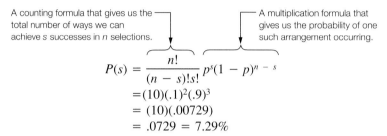

A counting formula that gives us the total number of ways we can achieve s successes in n selections.

A multiplication formula that gives us the probability of one such arrangement occurring.

$$P(s) = \frac{n!}{(n - s)!s!}\, p^s(1 - p)^{n - s}$$
$$= (10)(.1)^2(.9)^3$$
$$= (10)(.00729)$$
$$= .0729 = 7.29\%$$

Now let's practice.

PRACTICE 1: If 10% of a population has natural immunity to the HICNS virus, what is the probability out of 5 randomly selected individuals, exactly 1 will have his immunity?

ANSWER: 32.81%
Note:

$$P(1) = \frac{5!}{4!1!}\,(.1)^1(.9)^4$$
$$= 5(.06561)$$
$$= .32805$$
$$= 32.81\%$$

PRACTICE 2: If 10% of a population has natural immunity to the HICNS virus, what is the probability out of 5 randomly selected individuals, 1 *or less* has this immunity?

ANSWER: 91.86% (32.81% + 59.05%)
Note:

$P(1 \text{ or less}) = P(1) + P(0)$
$P(1) = 32.81\%$
(from Practice 1 example)

$$P(0) = \frac{5!}{5!\,0!\dagger}\,(.1)^0(.9)^5$$
$$= \frac{120}{(120)(1)}(1)(.59049)$$
$$= .59049 = 59.05\%$$

Thus,
$P(1) + P(0) = 32.81\% + 59.05\%$
$= 91.86\%$

*This formula $n!/(n - s)!s!$ can be denoted as C_s^n, $C_{n,s}$, $_nC_s$, or $\binom{n}{s}$ and referred to as a *combination*.

†Note: By definition, $0! = 1$ and $x^0 = 1$.

If we were to summarize the above results, it might appear as follows:

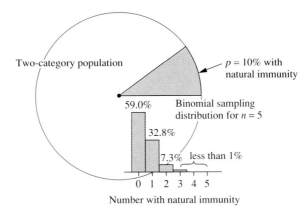

Note that the probability of three or more successes would be less than 1%, since the sum of the probabilities of 0, 1, or 2 successes = 99.1%, summarized as follows:
$P(0) = 59.0\%$ (rounded to one decimal place)
$P(1) = 32.8\%$
$P(2) = \underline{7.3\%}$
99.1% of the time we would expect 0, 1, or 2 people with this natural immunity in a random selection of 5 individuals.

Note the sloping shape of the sampling distribution. Skewed or sloping distributions are common when np or $n(1 - p)$ is less than 5. In these cases, the normal curve can *not* be depended on for proper estimates. Instead, each probability must be calculated separately, as demonstrated above. For your convenience, a table of common binomial probabilities is provided in the back of this text labeled, binomial tables.

Use of Binomial Tables

Many of the small-n binomial probabilities have been worked out in the binomial tables presented in the back of the text. For instance, for this example above, we look up $n = 5$, then locate the list of decimals under the column $p = .1$. This gives the various probabilities for 0 successes (.590), 1 success (.328), 2 successes (.073), and 3 successes (.008).

Note for 4 and 5 successes the spaces are blank. This means that the probability of obtaining 4 or 5 successes is negligible (less than .1%) and thus omitted.

Test of a Single Proportion, Two-Tailed (Large Sample, np and $n(1 - p)$ each ≥ 5)

In binomial experiments, a hypothesis test is normally carried out using proportions (a value divided by n). For example, 54 successes out of 1000 would be expressed, not as 54, but as the proportion 54/1000 or .054. Let's see how this works.

Suppose someone claims a certain attribute is present in 6% of the population. However, we do not know if this is true and decide to test using a random sample of 1000. If the sample yields 54 with the attribute, do we have evidence at $\alpha = .05$ to substantiate the claim of 6%?

The Logic of the Test.   Using methods already learned, we might solve the problem by first calculating the expected value (6% of 1000 = 60), then noting how many standard deviations 54 is away from this expected value. If 54 deviates from 60 by no more than 1.96 standard deviations (for α = .05), then we accept the claim, otherwise we reject it, as shown.

Using Proportions.   With proportions, we use an identical approach, only now the quantities are expressed as proportions (the values divided by n).* This is a common method for conducting a binomial test.

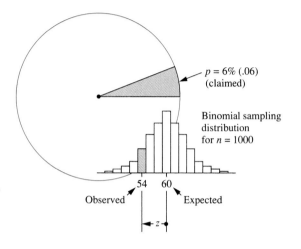

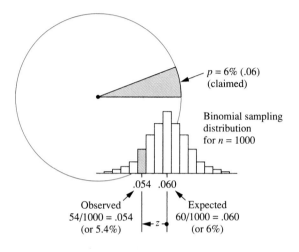

To calculate μ and σ for the sampling distribution:

The equivalent formula for calculating a z score using proportions is:

$\mu = np$
$= 1000(.06)$
$= 60$
$\sigma = \sqrt{np(1 - p)}$
$= \sqrt{1000(.06)(.94)}$
$= 7.51$

$z = \dfrac{x - \mu}{\sigma}$

$z = \dfrac{54 - 60}{7.51} = -.799$

$z = -.80$

$z = \dfrac{p_s - p}{\sigma_p}$

$z = \dfrac{.054 - .060}{.00751}$

$z = -.80$

where

$\sigma_p = \sqrt{\dfrac{p(1 - p)}{n}}$

$\sigma_p = \sqrt{\dfrac{.060(.940)}{1000}}$

$\sigma_p = .00751$

Since $z = -.80$ is within ±1.96, we accept claim of p = 6% (.06)

Since $z = -.80$ is within ±1.96, we accept claim of p = 6% (.06)

*Normally, proportions are expressed and calculated in decimal form, however for understanding purposes, especially in explanations, we will also use the percentage equivalent.

Note the two tests were identical, only the formulas on the right were slightly rearranged to accommodate proportions. The actual derivation of this rearrangement is shown below.

We start with the z formula for the binomial:

$$z = \frac{x - \mu}{\sigma}$$

Since proportions are values divided by n, we convert the values to proportions by simply multiplying the numerator and denominator by $1/n$.

$$z = \frac{(x - \mu)}{\sigma} \frac{1/n}{1/n}$$

$$z = \frac{x/n - \mu/n}{\sigma/n}$$

Note: $\frac{1/n}{1/n} = 1$, so the z formula has not changed in value. By having the values divided by n, each is now expressed as a proportion.

However, we can further simplify the formula by substituting in for μ and σ [Note: $\mu = np$ and $\sigma = \sqrt{np(1 - p)}$] and then reducing, as shown at right.

$$z = \frac{x/n - np/n}{\sqrt{np(1 - p)}/n} = \frac{x/n - p}{\sqrt{np(1 - p)/n^2}}$$

$$= \frac{x/n - p}{\sqrt{p(1 - p)/n}}$$

Substituting $\left\{ \begin{array}{l} p_s \text{ for } \dfrac{x}{n} \text{ and} \\[3mm] \sigma_p \text{ for } \sqrt{\dfrac{p(1 - p)}{n}}, \end{array} \right.$ we get the z formula for proportions, as follows.

▼ z Formula for Proportions

$$z = \frac{p_s - p}{\sigma_p}$$

where

$$p_s = \frac{x}{n} \text{ or } \frac{s}{n} \longleftarrow \text{(the number of successes over total selected)}$$

$$\sigma_p = \sqrt{\frac{p(1 - p)}{n}}$$

Keep in mind, both z formulas [$z = (x - \mu)/\sigma$ and $z = (p_s - p)/\sigma_p$] are identical. One is merely a rearrangement of the other. The former is employed when the binomial bars on the histogram represent whole-number values and the latter when the bars represent proportions (values divided by n).

Now let's see how this test might be conducted in an actual research project using proportions.

Example ———— Delusions, even bizarre delusions, are commonplace, according to *New York Times* article, "Delusion, Benign and Bizarre, Is Recognized As Common" (June 27, 1989, C3). Persecution was cited as the most common delusion (a person feels they are the focus of a plot, being spied on, followed, harassed, and so on). However, romantic fixation (feeling you are loved by someone of higher status, such as a celebrity), jealousy (groundless conviction your lover is unfaithful), grandiosity (conviction you have extraordinary power, insight, or heritage), and body fixation (belief something is wrong with your body despite contrary evidence) are other common delusions.

A belief is considered delusional, according to researchers, if a person holds to it no matter how bizarre and despite all evidence to the contrary. And although many people foster delusions, the problem starts when a person feels compelled to act on these deviant beliefs, such as, forever spying on or punishing a mate for groundless jealousies; incessantly writing, calling, or attempting to contact a person of higher status whom you believe is in love with you; or constantly visiting doctors for body problems, odors, skin infestations, and so on, which do not exist.

Although many people have isolated delusions that are harmless, for instance, a belief in flying saucers, others have severely deviant beliefs, which some researchers feel may be a forerunner to possible full-blown mental illness later in life.

Suppose it is claimed (based on research at the University of Wisconsin)* that 4% of college students foster severely delusional thinking and the National Institute of Mental Health tested the claim by randomly sampling 1000 college students nationally, yielding 32 with severely delusional thinking patterns. Does this provide evidence at $\alpha = .05$ that 4% of all college students foster severely delusional thinking patterns?

Solution

Use of proportions is a common method for conducting a single-sample binomial test. Initial conditions are set up as follows:

$H_0: p = 4\%$ (.04)
$H_1: p \neq 4\%$
$\alpha = .05$

Expected proportion of successes in sample:

$p = .04$ (4%)

Observed proportion of successes in sample:

$$p_s = \frac{s}{n} = \frac{32}{1000}$$
$$= .032 \ (3.2\%)$$

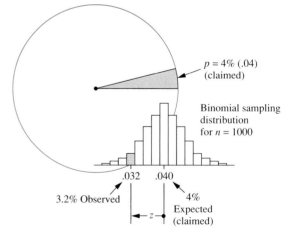

$p = 4\%$ (.04)
(claimed)

Binomial sampling distribution for $n = 1000$

.032   .040

3.2% Observed

4% Expected (claimed)

z

*For further reading, refer to "Delusion, Benign and Bizarre, Is Recognized as Common," *New York Times,* June 27, 1989, C3.

$$z = \frac{p_s - p}{\sigma_p} \qquad \text{where: } \sigma_p = \sqrt{\frac{p(1 - p)}{n}}$$

$$z = \frac{.032 - .040}{.0062} \qquad\qquad = \sqrt{\frac{.040(.960)}{1000}}$$

$$z = -1.29 \qquad\qquad\qquad = .0062$$

Accept H_0

Answer

Since our sample Z (-1.29) fell within our chance fluctuation range of ± 1.96 standard deviations of the expected value, we accept H_0 as possible, that is, 4% of all college students may indeed foster severely delusional thinking patterns.

Test of a Single Proportion, One-Tailed (Large Sample)

Had we worded the initial claim as 4% or more are believed to have severe delusional thinking patterns, the one-tailed hypothesis test format would have been employed, as follows:

H_0: $p \geq 4\%$
H_1: $p < 4\%$
$\alpha = .05$

Since in a one-tail test, all the α risk is placed in one tail, this affects the z-value at the cutoff, however, the remainder of the test is conducted in an identical manner. Note the sample z (-1.29) is still within chance fluctuation range and we again accept H_0.

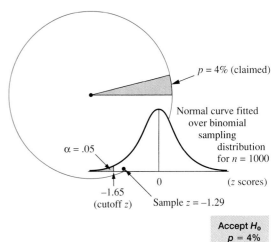

$p = 4\%$ (claimed)

Normal curve fitted over binomial sampling distribution for $n = 1000$

$\alpha = .05$

0   (z scores)

-1.65
(cutoff z)   Sample $z = -1.29$

Accept H_0
$p = 4\%$

Confidence Interval Estimate of p (Large Sample)

Had no initial claim been made and the National Institute of Mental Health merely wished to conduct the test on their own to determine an estimate of p, the proportion of college students who foster severely delusional thinking patterns, then a 95% confidence interval estimate of p would proceed as follows.*

*The rationale behind the construction of a confidence interval was presented in chapter 8 for measurement data. The rationale for proportion data is nearly identical. Essentially, the probability is high that p_s will be close to the value of p. Thus, knowledge of p_s (the sample proportion) gives us a point estimate of p, and the further away you are from p_s, the less probable it is that you will find p.

▼ *Formula for Confidence Interval Estimate of a Single Proportion*

$$p_s \pm z\,\sigma_p$$

In our case, it would be:

.032 ± 1.96(.0062)
.032 ± .012

$.020 \leq p \leq .044$

Thus, we are 95% confident that p the *true proportion* of college students nationwide with severely delusional thinking patterns is contained in the interval between .020 (2%) and .044 (4.4%).

Test of a Single Proportion (Small-*n*)

In small-*n* cases, where np or $n(1-p)$ is less than 5, the binomial tables in the back of the text can be used and the hypothesis test conducted intuitively. Let's see how this works.

Example ———— Suppose it is claimed that 10% of the population has a natural immunity to the HICNS virus. And we decide to test the claim by randomly selecting $n = 15$ individuals from the population and discover 6 with the natural immunity. Does this provide evidence at $\alpha = .05$ to support the claim?

Solution   Look up in the binomial probability tables in the back of the text the probabilities for $n = 15$, $p = .1$; these probabilities are represented below in the form of histogram bars.

H_0: $p = 10\%$ (.10)
H_1: $p \neq 10\%$
$\alpha = .05$

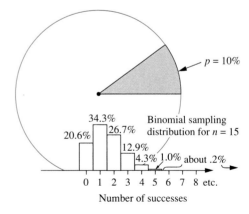

Note that the probability of 6 successes in 15 is about .2% or less. Thus we,

Reject H_0 $p \neq 10\%$

Look at the histogram bars. The most likely number of successes would be 0, 1, 2, or 3; 4 or 5 would be marginal and certainly 6 or more extremely rare.

Since we have statistical evidence to indicate $p \neq 10\%$ and the sample proportion ($\frac{6}{15}$) equals 40%, we would most likely conclude p is greater than the hypothesized 10%.

Difference Between Two Proportions

This test assumes *two* independent samples are drawn, one each from a separate binomial population. We will consider only the most common use of the test—that is, when both samples sizes are large and we wish to determine whether the two population proportions are equal, that is, if $p_1 = p_2$.

Example ————————

Suppose extremely deviant beliefs at college age are believed to be a possible forerunner to mental illness later in life. To explore this theory, suppose two groups of college students were tested in 1978 for aberrant thinking, then tested again in 1992 with the following result. Out of those with highly delusional thinking patterns in 1978, it was found 32 in 200 by 1992 had developed mental illness, and out of those nondelusional thinkers, 9 in 300 by 1992 had developed mental illness. At $\alpha = .05$, test $p_1 = p_2$.

Solution

The initial conditions are as follows:

$$H_0: p_1 = p_2$$
$$H_1: p_1 \neq p_2$$
$$\alpha = .05$$

Essentially the null hypothesis states that out of the two populations (those tested in 1978 as highly delusional and those tested in 1978 as nondelusional), the percentage who later develop mental illness will be equal.

Like any hypothesis test, we begin by assuming $H_0: p_1 = p_2$ is true, that is, mental illness will occur equally in both populations (in highly delusional thinkers and in nondelusional thinkers). We shall label p, the common population proportion (where $p = p_1 = p_2$). Now since we know,

$p_s \approx p$ a sample proportion is approximately equal to the population proportion, p,

we can use a sample proportion to estimate p. However (if H_0 is true, $p_1 = p_2$) we have two sample proportions, 32/200 and 9/300, to estimate this common population proportion, p. So we pool the results by dividing total successes by total sampled, as followed.

$$p \approx \frac{s_1 + s_2}{n_1 + n_2} = \frac{32 + 9}{200 + 300} = \frac{41}{500}$$
$$\approx .082 \quad = 8.2\%$$

In other words, out of 500 total, 41 had developed mental illness (or 8.2%).

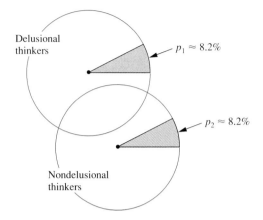

Thus, if indeed H_0: $p_1 = p_2$ is true (that is, if the proportion developing mental illness is equal in both populations), then our best estimate of this common proportion, p, is 8.2% (thus, $p = p_1 = p_2 \approx 8.2\%$), demonstrated at right.

Now suppose we randomly select two samples, one from each of these statistically identical populations, calculate the sample proportion for each, and subtract the two sample proportions from each other, $p_{s_1} - p_{s_2}$. Then we repeated this procedure thousands and thousands of times, each time recording the results of the subtraction into a histogram. Eventually we would discover the average subtraction would yield 0 (if H_0: $p_1 = p_2$ were true of course), and the distribution would be normally distributed with standard deviation, $\sigma_{p_{s_1} - p_{s_2}}$, shown as follows.

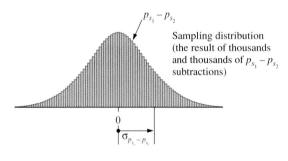

Sampling distribution (the result of thousands and thousands of $p_{s_1} - p_{s_2}$ subtractions)

$$\sigma_{p_{s_1} - p_{s_2}} = \sqrt{\frac{p(1-p)}{n_1} + \frac{p(1-p)}{n_2}}$$

In our case, since $p \approx .082$ (8.2%) $\sigma_{p_{s_1} - p_{s_2}}$ can be estimated as follows:

Thus,

$$\sigma_{p_{s_1} - p_{s_2}} \approx \sqrt{\frac{.082(.918)}{200} + \frac{.082(.918)}{300}}$$

$$\approx \sqrt{\frac{.075276}{200} + \frac{.075276}{300}}$$

$$\approx \sqrt{.0006272} = .02504$$

$$\sigma_{p_{s_1} - p_{s_2}} \approx .025$$

Because the distribution is normal with $\mu = 0$ and $\sigma_{p_{s_1} - p_{s_2}} \approx .025$, we would perform the hypothesis test as we would perform any hypothesis test, as follows:

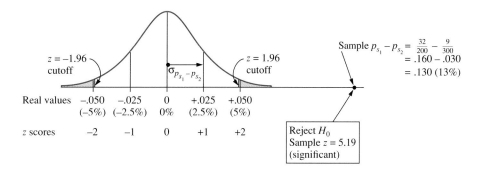

$$z = \frac{p_{s_1} - p_{s_2}}{\sigma_{p_{s_1} - p_{s_2}}}$$

$$z_{\text{sample}} = \frac{.160 - .030}{.02504}$$

$$z_{\text{sample}} = \frac{.130}{.02504} = 5.19$$

Answer

Since z_{sample} (5.19) exceeded the cutoff of 1.96, we reject H_0. The sample provides evidence that the two population proportions are not equal.

An examination of the data suggests mental illness later in life is more prevalent among college students who exhibit highly delusional thinking patterns than among those who test nondelusional. ■

Other Variations of This Test

First, this test can also be used when population p's are known to be different and we wish to test for this difference. In such cases, the analysis would be:

A. H_0: $p_1 - p_2 = k$ (a constant other than 0)

The sampling distribution would be centered around k, and the z formula would be

$$z = \frac{(p_{s_1} - p_{s_2}) - (p_1 - p_2)}{\sigma_{p_{s_1} - p_{s_2}}}$$

The formula for $\sigma_{p_{s_1} - p_{s_2}}$ would now contain p_1 over the n_1 denominator and p_2 over the n_2 denominator.

B. Note: This test can also be conducted as one tailed, with several possible null hypotheses:

$$p_1 - p_2 \geq 0 \text{ or } p_1 - p_2 \leq 0$$
$$p_1 - p_2 \geq k \text{ or } p_1 - p_2 \leq k$$

Second, this test of two proportions can also be performed using X^2 (chi-square); in which case, $z^2 = X^2$ (df = 1). Recall that we had calculated $\sigma_{p_{s_1} - p_{s_2}} = .02504$ from the prior example to get $z = 5.19$; when we square this value, the answer (26.94) will be near exact to the value obtained when we analyze the proportion data using chi-square techniques.

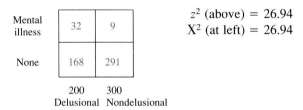

z^2 (above) = 26.94
X^2 (at left) = 26.94

Mental illness: 32, 9
None: 168, 291
200, 300
Delusional, Nondelusional

11.2 Measurement Population Sampling

All topics covered in this section involve sampling from two measurement populations, that is, populations described by μ and σ, such as the ones shown in the sketch below.

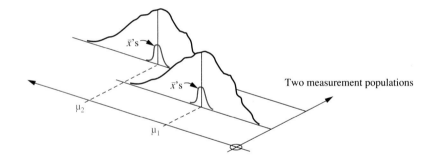

$\bar{x}$'s
$\bar{x}$'s
Two measurement populations
μ_2
μ_1

Difference of Two Population Means (Independent Case)

This test assumes two independent samples are drawn, one each from a different measurement population, and these samples are used to test whether the two population means are equal, that is, if $\mu_1 = \mu_2$. This is one of the most commonly used tests in psychology and other fields. Since different formulas are involved in different situations, we will start with the simplest case.

Large Samples
(Each n_1, $n_2 \geq 30$)
When both sample sizes are large (≥ 30), the formulas are relatively straightforward and *no* population preconditions are needed. Let's see how this works.

Example ——————— Nonverbal clues, such as hand gestures, head nods, etc., may not be as easy to interpret as many people think, according to experts. Although research indicates individuals can often gauge someone's feelings from their facial expression or assess obvious social traits, such as, gregariousness, more subtle traits like motivation, submissiveness, or whether someone is lying are often misjudged.

For instance, many employers mistake smiling, gesturing, and talking a lot for high motivation, when in fact these traits are not at all indicators of high motivation. "People think dominant people look you in the eye more . . . but it's actually the reverse." Submissive people look you in the eye more; also submissive people fiddle with objects more, gesture little, and slouch slightly while keeping their legs folded under the chair or table in interviews. Quickly and accurately assessing human traits is not only an important function for managers out to hire someone, but it is equally important for all of us in our daily lives in assessing those we wish to date, marry, or purchase a home or automobile from.

One very important assessment is whether a person is lying or not. Detecting a liar, however, may not be as easy as some think. Studies find that liars look you in the eye every bit as much as truthful people, and the fear of being disbelieved when telling the truth looks very much the same as the fear of being caught in a lie. However, many of us still feel we can tell when someone is lying.

Suppose a training institute wishing to compare the confidence people have in their ability to detect a liar versus their actual ability sets up an experiment in which two groups (one high in confidence and the other randomly selected citizens) were asked to view videotapes of several job interviews. Of the many applicants, all were eventually hired, however, it was later discovered 10 lied substantially during the interview and subsequently had to be fired. Everyone was asked to detect the 10 liars by watching and listening to the original videotapes. Each viewer was scored as to how many of the 10 liars were assessed correctly. Scores for each group were then averaged, with the following results:

Group 1	
High-Confidence Group	
(mostly court judges, police	**Group 2**
detectives, and psychiatrists)	**Average (control) group**
$n_1 = 50$ viewers	$n_2 = 55$ viewers
$\bar{x}_1 = 4.3$ liars assessed correctly	$\bar{x}_2 = 3.9$ liars assessed correctly
$s_1 = 1.5$	$s_2 = 1.8$

a. At $\alpha = .05$, test the claim $\mu_1 = \mu_2$.
b. Briefly discuss implications of the results.

Solution   The initial conditions are as follows:

$$H_0: \mu_1 = \mu_2$$
$$H_1: \mu_1 \neq \mu_2$$
$$\alpha = .05$$

Like any hypothesis test, we begin by assuming $H_0: \mu_1 = \mu_2$ is true, that is, the ability to detect liars is equal on average in both populations (in high-confidence judges, detectives, and psychiatrists and in average people), pictured below.

Study the populations for a moment. Note that individual scores can vary widely, however, $\bar{x}$'s (sample averages) tend to cluster quite close to μ and this is true even though the populations themselves may be irregularly shaped (refer to chapter 5, Central Limit Theorem, particularly, section 5.4: if $n \geq 30$, for almost any shaped population, the $\bar{x}$'s will distribute normally).

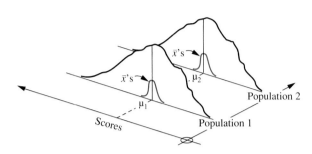

Sample 1	**Sample 2**
$n_1 = 50$	$n_2 = 55$
$\bar{x}_1 = 4.3$	$\bar{x}_2 = 3.9$
$s_1 = 1.5$	$s_2 = 1.8$

Now suppose we randomly select two samples, one from each of these two populations, calculate the sample $\bar{x}$'s, and subtract one from the other, $\bar{x}_1 - \bar{x}_2$. Then we repeated the procedure thousands and thousands of times, recording the results of each subtraction into a histogram. Eventually, we would discover the average subtraction would yield 0, and the sampling distribution would be normal in shape with standard deviation, $\sigma_{\bar{x}_1 - \bar{x}_2}$, calculated below.

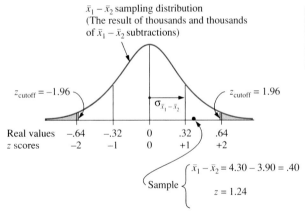

$\bar{x}_1 - \bar{x}_2$ sampling distribution
(The result of thousands and thousands of $\bar{x}_1 - \bar{x}_2$ subtractions)

$$\sigma_{\bar{x}_1 - \bar{x}_2} = \sqrt{\frac{\sigma_1^2}{n_1} + \frac{\sigma_2^2}{n_2}}$$

where s_1 and s_2 may be used to estimate σ_1 and σ_2.

In our case,

$$s_1 = 1.5, s_2 = 1.8$$
$$n_1 = 50 \quad n_2 = 55$$

$$\sigma_{\bar{x}_1 - \bar{x}_2} \approx \sqrt{\frac{(1.5)^2}{50} + \frac{(1.8)^2}{55}}$$

$$\approx \sqrt{\frac{2.25}{50} + \frac{3.24}{55}}$$

$$\approx \sqrt{.045 + .0589}$$

$$\approx \sqrt{.1039} = .3223$$

$$\boxed{\sigma_{\bar{x}_1 - \bar{x}_2} \approx .32}$$

The sample z score was calculated as follows:

$$z = \frac{\bar{x}_1 - \bar{x}_2}{\sigma_{\bar{x}_1 - \bar{x}_2}}$$

$$z = \frac{4.30 - 3.90}{.3223}$$

$$= \frac{.40}{.3223} = 1.24$$

Accept H_0
$z_{sample} = 1.24$
(not significant)

Answers

Since our sample z of 1.24 was within our cutoffs of ± 1.96, we accept H_0: $\mu_1 = \mu_2$. In other words, we conclude that the ability of those high in confidence to detect a liar is on average no better or worse than the average person.

According to *New York Times* article, "Non-Verbal Cues Are Easy to Misinterpret" (September 17, 1991, C1), professional judges, detectives, psychiatrists, etc. are "no better than anyone else in catching lies," and generally people's confidence in reading character very often exceeds their actual ability. Only one professional group was cited as exceeding chance in detecting liars: Government Secret Service Officers.

Well then, how do we tell a liar? According to one expert, mannerisms such as "looking away while talking, is hesitant or nervously fiddles with things" are not reliable indicators. In fact, "The more often a person has told a story, the less likely you'll be able to tell if it's a lie. . . . And if someone strikes you as charming and likable he's more likely to be seamless in any lies he tells you." However, there is hope. Although no guarantee of detecting a lie exists, you should probe further if someone exhibits rapidly shifting facial expressions (which indicates concealment of an emotion) or discrepancies between word content and manner of expression (voice, body, or facial gesture), also be wary of any sudden changes in several channels of expression. ■

Other Variations of This Test

Several variations of this test can be performed, such as, a one-tailed test. In fact, the test can also be used to test if $\mu_1 - \mu_2$ equals a constant. For instance, suppose we have strong reasons to believe $\mu_1 = 5.0$ and $\mu_2 = 3.5$, thus $\mu_1 - \mu_2 = 1.5$. In this case, the hypothesis test would be performed, as follows:

$\sigma_{\bar{x}_1 - \bar{x}_2}$ would be calculated in exactly the same manner (thus, $\sigma_{\bar{x}_1 - \bar{x}_2} = .32$), only now the center of the sampling distribution would be 1.5 (and not 0) and the test conducted as follows.

$\bar{x}_1 - \bar{x}_2$ sampling distribution
(thousands and thousands
of $\bar{x}_1 - \bar{x}_2$)

$z_{cutoff} = -1.96$ ←.32→ $z_{cutoff} = 1.96$

| .86 | 1.18 | 1.50 | 1.82 | 2.14 | Real values |
| -2 | -1 | 0 | +1 | +2 | z scores |

Sample $\begin{cases} \bar{x}_1 - \bar{x}_2 = 4.30 - 3.90 = .40 \\ \\ z = -3.41 \end{cases}$

Reject H_0
$z = -3.41$
(significant)

Initial conditions:

$H_0: \mu_1 - \mu_2 = 1.5$
$H_1: \mu_1 - \mu_2 \neq 1.5$
$\alpha = .05$

$$z = \frac{(\bar{x}_1 - \bar{x}_2) - (\mu_1 - \mu_2)}{\sigma_{\bar{x}_1 - \bar{x}_2}}$$

To calculate z_{sample},

$$z = \frac{(4.3 - 3.9) - (1.5)}{.3223}$$

$$= \frac{.4 - 1.5}{.3223} = -3.41$$

Small Samples
(Either n$_1$, n$_2$ < 30)

Whereas, in the large-sample case, no population preconditions are required, the opposite is true in the small-sample case. Because in the small-sample case (for either n_1 or $n_2 < 30$), the two populations must be normally or, at least, near normally distributed, and both population standard deviations must be equal, that is, $\sigma_1 = \sigma_2$, depicted as follows:

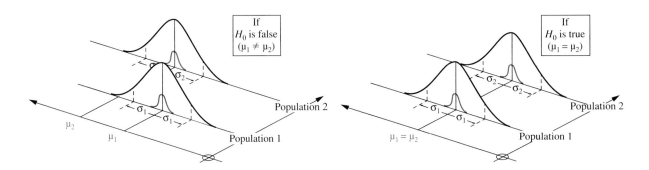

Notice whether H_0 is true or false, the preconditions are satisfied (normal populations with equal σ's). With proper design of experiments, these preconditions are often satisfied (see chapter 10, section 10.3, Experimental Designs for ANOVA, for further discussion*). To examine this small-sample test, we will use a slight variation of the prior example.

*ANOVA techniques (df = 1, $n_1 + n_2 - 2$) can also be used to analyze this data. In such cases, z^2 (or t^2) = F. The two-sample case, such as the one presented above, is often referred to as a two-sample t test.

Example ——————— Nonverbal clues, such as hand gestures, head nods, and so on may not be as easy to interpret as many people think. Suppose two groups (one high in confidence and the other average) were tested as to their ability to detect a liar. Each group viewed 25 job-interview videotapes where 10 of the applicants had substantially lied. Everyone was asked to detect the 10 liars by watching and listening to the videotapes.

<table>
<tr><td align="center">**Group 1**
High-Confidence Group
(mostly court judges, police
detectives, and psychiatrists)</td><td align="center">**Group 2**
Average Confidence
(control) Group</td></tr>
<tr><td>$n_1 = 18$</td><td>$n_2 = 24$</td></tr>
<tr><td>$\bar{x}_1 = 4.3$ liars correctly assessed</td><td>$\bar{x}_2 = 3.9$ liars correctly assessed</td></tr>
<tr><td>$s_1 = 1.5$</td><td>$s_2 = 1.8$</td></tr>
</table>

Test $\mu_1 = \mu_2$ at $\alpha = .05$ and briefly discuss results.

Solution First we must have some assurances our preconditions (normal populations with equal σ's) are satisfied.* Now, supposing preconditions are met, then the setup for the test would be as follows:

H_0: $\mu_1 = \mu_2$
H_1: $\mu_1 \neq \mu_2$
   $\alpha = .05$

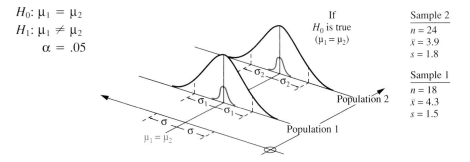

The test is performed in pretty much the same manner as any hypothesis test, only now we will introduce a formula change for $\sigma_{\bar{x}_1 - \bar{x}_2}$, to "weigh" the small-sample s's and use a t score because of the small-sample sizes. Recall a t score in effect compensates for the tendency of small-sample s's to underestimate σ, the population spread.

*Technical note: To test the precondition of normal populations, X^2 goodness-of-fit and other tests are available. To test $\sigma_1 = \sigma_2$, the F ratio is used; essentially, $F = s^2$(larger)/ s^2 (smaller), in our case, $F = (1.8)^2/(1.5)^2 = 1.44$. F cutoff, .05, df = (numerator $n - 1$), (denominator $n - 1$). So, for df = 23,17, F cutoff ≈ 2.20. Thus, we accept H_0: $\sigma_1 = \sigma_2$ (note: a two-tailed hypothesis test with $F_{.025}$ as upper bound would yield cutoff $F \approx 2.57$, whereas a one-tailed test, $F_{.05}$, would yield a more conservative 2.20).

First, let's introduce the weighting factor for the s's. Because larger-sample sizes generally provide better estimates, we weigh more heavily the value of the larger-sample s in estimating σ. To show how we do this, let's start with the formula for $\sigma_{\bar{x}_1 - \bar{x}_2}$ we originally used in the large-sample case, only now we will rearrange terms slightly and weigh the s's to provide a better estimate of σ, the common population spread, shown in the previous sketch.

$$\sigma_{\bar{x}_1 - \bar{x}^2} = \sqrt{\frac{\sigma_1^2}{n_1} + \frac{\sigma_2^2}{n_2}} \quad \text{**Large-sample formula**}$$
$$\text{**for** } \sigma_{\bar{x}_1 - \bar{x}_2}$$

However, since $\sigma_1 = \sigma_2$, we use the common symbol, σ (where $\sigma = \sigma_1 = \sigma_2$).

$$\sigma_{\bar{x}_1 - \bar{x}_2} = \sqrt{\frac{\sigma^2}{n_1} + \frac{\sigma^2}{n_2}}$$

Next, we factor out σ^2

$$\sigma_{\bar{x}_1 - \bar{x}_2} = \sqrt{\sigma^2 \left(\frac{1}{n_1} + \frac{1}{n_2} \right)}$$

To estimate this common variance, σ^2 (the population standard deviation squared), we use the sample estimates, s_1^2 and s_2^2, weighted, as follows:*

$$\text{Weighted Estimate of } \sigma^2 \approx \frac{(n_1 - 1)s_1^2 + (n_2 - 1)s_2^2}{n_1 + n_2 - 2}$$

▼ Small-Sample Formula for $\sigma_{\bar{x}_1 - \bar{x}_2}$

Substituting the weighted estimate of σ^2 into the formula, we get

$$\sigma_{\bar{x}_1 - \bar{x}_2} \approx \sqrt{\frac{(n_1 - 1)s_1^2 + (n_2 - 1)s_2^2}{n_1 + n_2 - 2} \left(\frac{1}{n_1} + \frac{1}{n_2} \right)}$$

Solving,

$$\sigma_{\bar{x}_1 - \bar{x}_2} \approx \sqrt{\frac{(18 - 1)(1.5)^2 + (24 - 1)(1.8)^2}{18 + 24 - 2} \left(\frac{1}{18} + \frac{1}{24} \right)}$$

$$\approx \sqrt{\frac{38.25 + 74.52}{40} \left(\frac{7}{72} \right)}$$

$$\approx \sqrt{(2.81925)(.09722)} = .5235$$

$$\sigma_{\bar{x}_1 - \bar{x}_2} \approx .52$$

Once the small-sample standard deviation is calculated, the hypothesis test is conducted in much the same manner as any hypothesis test. Essentially, we

*For equal n's, we merely use the average s^2 to estimate σ^2.

know from experience and theory, if we randomly select two samples repeatedly from two independent populations, and each time subtract the $\bar{x}$'s, that is, $\bar{x}_1 - \bar{x}_2$, then plot the result of these many thousands of subtractions into a histogram, the average subtraction would yield 0, with the sampling distribution normal in shape, with standard deviation $\sigma_{\bar{x}_1 - \bar{x}_2}$ (defined and calculated above). Only now, instead of a z score we use a t score, as follows.

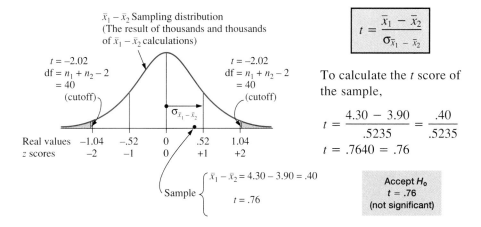

$$t = \frac{\bar{x}_1 - \bar{x}_2}{\sigma_{\bar{x}_1 - \bar{x}_2}}$$

To calculate the t score of the sample,

$$t = \frac{4.30 - 3.90}{.5235} = \frac{.40}{.5235}$$

$$t = .7640 = .76$$

Accept H_0
$t = .76$
(not significant)

Final notes:

1. As in large-sample testing, several variations of this small-sample test can be performed, such as, (a) one-tailed and (b) $\mu_1 - \mu_2 = a$ constant; the constant then is the center of the sampling distribution and now the t formula at right is used.

$$t = \frac{(\bar{x}_1 - \bar{x}_2) - (\mu_1 - \mu_2)}{\sigma_{\bar{x}_1 - \bar{x}_2}}$$

2. Techniques are also available when the precondition of equal σ's is violated. That is, if we conclude based on the F test that $\sigma_1 \neq \sigma_2$, we use the $\sigma_{\bar{x}_1 - \bar{x}_2}$ formula, at right, with the df approximately equal to the *smaller of* $n_1 - 1$ or $n_2 - 1$.

$$\sigma_{\bar{x}_1 - \bar{x}_2} = \sqrt{\frac{s_1^2}{n_1} + \frac{s_2^2}{n_2}}$$

Difference of Two Population Means (Dependent Case)

In the prior example, we assumed samples were drawn from two independent populations, that is, where individual values in the populations are in no way related. However, now we will explore experiments where each value in one population is related or linked to a specific value in the other population. Such populations are called **dependent.**

> **Dependent Populations**
> Each value in one population is related or linked to a specific value in another population.

"Before-and-after" studies typically involve dependent populations. For instance, a medical researcher might take a blood pressure reading on several individuals both before and after administering a new drug. The before blood pressure readings would be the first population, the after blood pressure readings the second population. However, each "before" reading in the first population is *linked* to a specific "after" reading in the second population, that is, both readings are on the same person. Thus the linkage is the *same* set of genes.

Another example might include assigning two pups from the same litter (and of the same sex) to two different treatment groups, and repeating this using several pairs of pups. Each reading of a pup in the first treatment group is related or *linked* to the reading of its sibling in the second treatment group, because of their *similar* genetic makeup.

The goal in these experiments is to make the pairs as similar as possible (perhaps matching several pairs of musicians, with each pair having as near as possible identical experience and competence or pairing students of the same age, sex, ethnic background, and GPA). In other words, the more similar the pair, the more the effects of the treatment can be assessed without concern about extraneous factors influencing results.*

Population assumptions: The preconditions for the test are that the two populations be dependent and each be normally distributed (or at least somewhat symmetrical).†

Example ——————— Personality: is it born or bred? Does our personality come programmed in our genes or is it molded from experience? As it turns out, certain traits may very well be fixed in our genes, according to researchers at University of Minnesota, Penn State, Indiana University, and the National Institute of Aging in Baltimore. For instance, research shows that three fundamental aspects of our personality change little throughout the course of our life, strongly suggesting genetic links

---

*Extraneous factors often mask true differences in an experiment or inject differences which do not exist. Therefore, controlling extraneous factors (factors other than those being measured) is always a goal in any experiment. And it has been shown, data properly paired generally reduces the effects of extraneous factors, which in turn reduces variability in the sampling distribution and increases our ability to identify true differences resulting from the treatment.

†Extreme departures from normality require use of other techniques; such techniques are discussed in section 11.3 (nonparametric tests).

(the three traits are: level of anxiety, friendliness, and openness to new experiences). Other research indicates genetics as influential as environment in determining motivation for achievement, leadership, extraversion, conscientiousness, and conservatism.*

Often such studies use identical twins (same egg) separated soon after birth to evaluate two different environmental upbringings on what is essentially the same set of genes. If a personality trait is strongly etched in those genes, then the twins later in life would be expected to score similarly in that trait, even though raised in different household environments and having never met each other.

Suppose a team of behavioral geneticists seeking to determine the origins of the personality trait of niceness (a combination of trust, sympathy, and cooperativeness), study 15 pairs of separated identical twins over several years, also evaluating their household environments. Each pair of twins is grouped by their household level of niceness, as follows.

Household Lower in Niceness (Score of Twin in That Household [0–20 scale])	Household Higher in Niceness (Score of Twin in That Household [0–20 scale])
8	13
6	6
7	12
11	10
6	9
8	12
9	14
6	12
3	9
12	11
8	13
9	7
10	15
9	15
8	7

If genetics plays a highly dominant role in this trait, the household environment would have little influence, and the twins should score similarly.

a. Test $\mu_1 = \mu_2$, at $\alpha = .05$.
b. Briefly discuss the implications of the results.

*The several references used for this example are presented at the end of the discussion.

Solution

Assuming our preconditions are satisfied (dependent, somewhat normal or symmetrical populations), then our initial assumptions proceed as follows:

$H_0: \mu_1 = \mu_2$
$H_1 \; \mu_1 \neq \mu_2$
$\alpha = .05$

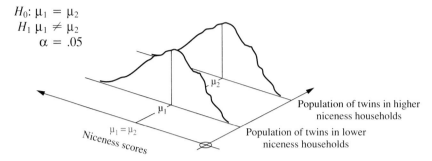

Although the samples drawn from these populations could be treated as independent (as they were in prior examples), it would be wasteful to lose the relationship between the twins. The pairs are strongly linked though their identical genes and, generally, linked-pair comparisons are more sensitive in identifying real population differences (since extraneous factors to a great extent are minimized). Thus, to treat these samples as anything but linked pairs would be to lose some sensitivity in the ability of the test to detect population differences.

But in either case, whether this test be independent or dependent, it starts with the same null hypothesis, $\mu_1 = \mu_2$, in this case, specifically that average niceness of twins in the lower niceness households equals average niceness of twins in higher niceness households. Effectively this states that household environment (parental influence, siblings, etc.) has little or no influence on this trait of niceness. Twins will score on average the same whether in low- or high-niceness households.

We will analyze these samples as dependent, as follows: The first step in any hypothesis test is to create the "expected" sampling distribution. In this way, we can compare what we would expect (based on H_0 being true) to what we actually obtained with our sample data. If what we actually obtained with our sample data is *near* to what we expected based on H_0 being true, we accept H_0—otherwise we reject it.

To create this expected sampling distribution we draw from many decades of experience and theory, which tell us, if we repeatedly sampled thousands and thousands of linked pairs from two dependent normal populations where $\mu_1 = \mu_2$, and calculated $x_2 - x_1$, the difference in scores for each of the thousands of linked pairs, and plotted the results of these subtractions (which we shall call d-values) into a histogram, the average subtraction (the average d-value) would be zero, with the distribution normal in shape, with standard deviation, σ_d, as follows:

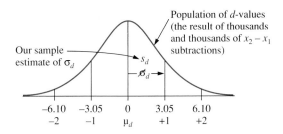

Population of d-values (the result of thousands and thousands of $x_2 - x_1$ subtractions)

Our sample estimate of σ_d

s_d

$\cancel{s}_d$

-6.10	-3.05	0	3.05	6.10
-2	-1	μ_d	+1	+2

Sample Standard Deviation

$$s_d = \sqrt{\frac{\Sigma(d - \bar{d})^2}{n - 1}}$$

where $d = x_2 - x_1$

$\bar{d}$ = Average d-value

n = Number of pairs

Although we know if H_0: $\mu_1 = \mu_2$ were true, minus differences and plus differences would be expected to be symmetrical around zero, and, indeed, experience has shown they are (in other words, the mean for the above distribution, μ_d, is 0), however we still do not know σ_d, the standard deviation. However we can estimate it using sample data.

$s_d \approx \sigma_d$ the standard deviation of the sample, s_d, is approximately equal to the standard deviation of the population, σ_d.

Thus s_d is calculated at right ($s_d = 3.05$), and the value is used to estimate σ_d, in the above distribution.

Sample Data		$x_2 - x_1$			
x_1	x_2	d	$\bar{d}$	$(d - \bar{d})$	$(d - \bar{d})^2$
8	13	5	3	2	4
6	6	0	3	-3	9
7	12	5	3	2	4
11	10	-1	3	-4	16
6	9	3	3	0	0
8	12	4	3	1	1
9	14	5	3	2	4
6	12	6	3	3	9
3	9	6	3	3	9
12	11	-1	3	-4	16
8	13	5	3	2	4
9	7	-2	3	-5	25
10	15	5	3	2	4
9	15	6	3	3	9
8	7	-1	3	-4	16
		45			130
		Σd			$\Sigma(d - \bar{d})^2$

To Calculate $\bar{d}$

$$\bar{d} = \frac{\Sigma d}{n}$$

$$\bar{d} = \frac{45}{15}$$

$$\boxed{\bar{d} = 3}$$

To Calculate s_d

$$s_d = \sqrt{\frac{\Sigma(d - \bar{d})^2}{n - 1}}$$

$$s_d = \sqrt{\frac{130}{15 - 1}} = 3.047$$

$$\boxed{s_d = 3.05 \approx \sigma_d}$$

However, the above is a distribution of *individual* pairs, and we are concerned with the *average* of 15 pairs. Recall, for the above data we have

$$n = 15$$
$$\bar{d} = 3$$

In other words, we know the *average* of 15 *d*-values. Thus we wish to know if this average of $d = 3$ is likely? Of course, one subtraction may very well yield 3 (just look at the above distribution; 3 is quite likely), however is it likely that a random sample of $n = 15$ subtractions would produce an *average* of 3, when the population average is 0?

In other words, the situation boils down to this: we randomly sample $n = 15$ from a population described by $\mu = 0$ and $\sigma = 3.05$. And we would expect the sample average to fall relatively close to the population average. How close? We simply use the central limit theorem formula to determine how close, as follows:

$$\sigma_{\bar{d}} = \frac{\sigma_d}{\sqrt{n}} \approx \frac{s_d}{\sqrt{n}}$$

To calculate $\sigma_{\bar{d}}$ for our example,

$$\sigma_{\bar{d}} \approx \frac{s_d}{\sqrt{n}} = \frac{3.05}{\sqrt{15}} = .7875$$

Thus, $\sigma_{\bar{d}} \approx .79$

Now once we know the spread of the $\bar{d}$ sampling distribution, the hypothesis test is conducted much like any other hypothesis test. Since this sample size is small, we use a *t* score adjustment at the cutoffs (df $= n - 1$). Recall, small-sample *s*'s tend to underestimate the population spread and essentially the *t* score compensates for this. Thus, we look up a cutoff *t* for df $= n - 1 = 15 - 1 = 14$ (.05, two-tailed test) and get ±2.14.

The hypothesis test is conducted as follows:

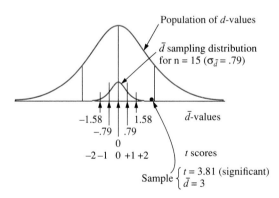

$$t = \frac{\bar{d} - \mu_d}{\sigma_{\bar{d}}}$$

To calculate the *t* score of the sample,

$$t = \frac{\bar{d} - \mu_d}{\sigma_{\bar{d}}} = \frac{3 - 0}{.7875}$$

$$= 3.81$$

Reject H_o
$t = 3.81$
(significant)

Answer

Since t_{sample} (3.81) is outside the cutoff values of ± 2.14, we reject H_0. Thus, we conclude $\mu_1 \neq \mu_2$. Essentially we conclude, twins do *not* exhibit the same level of niceness in lower versus higher niceness households. A glance at the data shows higher niceness households generally rearing higher niceness twins, suggesting environment plays a role in shaping this personality trait. ∎

This conforms to research by developmental psychologists at Pennsylvania State University, that niceness "is much more influenced by environment—mostly early environment—than by genes," and also conforms to the theories of Erik H. Erikson, famed Harvard professor, who theorized for example that a "sense of trust (an important element in 'niceness') . . . begins to develop from the infant's experience of a loving and supportive environment."

Erikson, with his wife, put forth a theory over 40 years ago theorizing (based on clinical data) that everyone passes through eight distinct stages of life (six occurring prior to adulthood), each with a conflict and resolution, where the person grapples "with a specific psychological struggle that contributes to a major aspect of personality." Many aspects relate to this trait of niceness (trust, sympathy, cooperativeness). For instance, basic trust, as stated above is learned in infancy, essentially through "consistency, continuity, and sameness of experience. . . . The first demonstration of . . . trust in the baby is the ease of his feeding, the depth of his sleep, the relaxation of his bowels."

In the second stage, just past infancy, the holding on and letting go stage, the danger is inflicting too much shame and doubt into the child; "this stage . . . becomes decisive for the ratio of love and hate, cooperation and willfulness, freedom of self-expression and its suppression." Certainly the personality trait of niceness includes some of these components. The third stage, the pre-school-play age, is the pleasure-in-attack-and-conquest stage, referred to as initiative versus guilt. In the boy, the emphasis is on intrusive modes of behavior; "in the girl it turns to modes of 'catching' in more aggressive forms of snatching or in the milder form of making oneself attractive and endearing." Empathy or care for others is gleaned from this and other stages (for further reading on the topic of personality traits, refer to Erik H. Erikson, *Childhood and Society,* [New York: W. W. Norton and Co., 1963, 2nd ed., pp. 247–274] and *New York Times* articles, "Major Traits Found Stable Through Life" [June 8, 1987, C1], "Erikson . . . View on Life" [June 14, 1988, C1], and "What a Child Is Given," [September 3, 1989, Magazine Section, p. 36]).

In summary, although this package of personality traits we labeled niceness seems to be learned, other aspects of personality, such as our anxiety level and openness for new experiences may indeed have strong genetic ties.

11.3 Nonparametric Tests

In the preceding section, we used an example concerning personality traits to demonstrate the test for the difference between two population means (dependent case). However, when population assumptions for the test cannot be met, namely, when the two populations are not normal nor symmetric,* we can use the *sign test* instead, since the sign test requires *no* prior population assumptions. Tests that require *no* prior assumptions about the shape of the population are called distribution-free and are part of a broader category called **nonparametric tests.**

> ### *Nonparametric Tests*
> A category of tests that allow a relaxation of some or all of the population requirements necessary in many of the tests we have studied.

However, nonparametric tests are not without disadvantages. If parametric assumptions can indeed be satisfied, then nonparametric tests are generally less sensitive than their parametric counterparts (that is, less sensitive in their ability to detect population shifts or differences), however this may be compensated for by using a larger-sample size. Also slightly different hypotheses may be involved. All this is discussed further as we work the examples.

Although there are several nonparametric tests, we present the following: the **sign test** (for two dependent samples), the **Wilcoxon rank-sum test** (for two independent samples), and the **Spearman rank-correlation test.**

Sign Test (For Two Dependent Samples)

In the case where we suspect severe violation of population assumptions in the test for the difference between two population means (dependent case) studied in section 11.2, we may elect to use a nonparametric test instead, which requires *no* assumptions concerning population characteristics, such as the sign test.†

To demonstrate this test, we will use the example studied in section 11.2 concerning personality traits. Recall that 15 pairs of identical twins, separated at birth and reared in different households, were later measured on the personality trait of niceness (along with their respective households), with the goal of assessing if the personality trait of niceness is genetically rooted.

---

*Generally, the parametric tests we have studied (those requiring population assumptions) are quite robust, meaning slight to moderate violation of these assumptions can often be tolerated without affecting the validity of the outcome.

†Actually the sign test can be used in several other situations, for instance, to test for a suspected median value of a population.

Niceness Score of Twin in Lower Niceness Household (0–20 scale) x_1	Niceness Score of Twin in Higher Niceness Household (0–20 scale) x_2	Sign of $x_2 - x_1$ Calculation
8	13	+
6	6	no difference
7	12	+
11	10	−
6	9	+
8	12	+
9	14	+
6	12	+
3	9	+
12	11	−
8	13	+
9	7	−
10	15	+
9	15	+
8	7	−

Note the column labeled, Sign of $x_2 - x_1$, above. Essentially this column indicates whether the score going from column 1 to column 2 had increased (+) or decreased (−). The sign test then converts the test to a binomial analysis (which is nonparametric) with the hypothesis that, in the population, if the treatment groups have the same or no effect on the measured variable, then the probability of selecting a pair of twins with a + difference should be about equal to the probability of selecting a pair of twins with a − difference. This is often expressed as $P(x_2 > x_1) = P(x_2 < x_1)$, however it can also be expressed simply as $P(+) = P(-)$. So the initial conditions are as follows:*

$$H_0: P(+) = P(-) = .5 \ (50\%)$$
$$H_1: P(+) \neq P(-) \neq .5 \ (50\%)$$

We shall use $\alpha = .05$.

*Technical note: The sign test essentially tests the null hypothesis, $H_0: \tau_d = 0$ (the median value of the $x_2 - x_1$ population $= 0$). In fact, if this $x_2 - x_1$ population proves to be symmetrical (not necessarily normal), we may employ the more powerful Wilcoxon test for paired data using the more explicit hypothesis, $H_0: \mu_1 = \mu_2$. This Wilcoxon test under the conditions that the parametric test can be used is only slightly less efficient than its parametric counterpart.

Thus, if H_0 is true (which we must assume to begin any hypothesis test), and if we randomly sampled several pairs of twins, we would expect approximately equal $+$ and $-$ differences. In other words, we are sampling from a two-category population consisting of 50% $+$'s, and when we randomly select $n = 14$, we conduct the hypothesis test as follows:

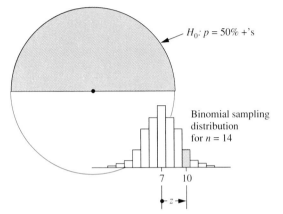

H_0: p = 50% +'s

Binomial sampling distribution for $n = 14$

7   10

z

| n = number of $+$'s and $-$'s (ties are not counted) |

Note that for $n = 14$, the binomial sampling distribution is approximately normally distributed. This is generally true if np and $n(1 - p)$ exceeds 5, which in this case it does.* Thus, we use the normal curve approximation with the following characteristics:

$$\mu = np$$
$$= 14(.5)$$

$$\boxed{\mu = 7}$$

$$\sigma = \sqrt{np(1 - p)}$$
$$= \sqrt{14(.5)(.5)}$$

$$\boxed{\sigma = 1.871}$$

We would expect about 7 $+$'s if H_0 were true ($\frac{1}{2}$ of $14 = 7$), however, we often get more or less than 7 when we sample $n = 14$. In our case, we obtained 10 $+$'s. This is $z = 1.60$ from our expected value of 7 (calculated at right). However, $z = 1.60$ is within the chance fluctuation range of ± 1.96 (for $\alpha = .05$). Thus, we must accept H_0.

To calculate sample z:

$$z = \frac{x - \mu}{\sigma}$$

$$z = \frac{10 - 7}{1.871} = 1.60$$

Accept H_0
z = 1.60
(not significant)

Essentially we accept $P(+) = P(-)$, which implies that the $+$ and $-$ differences were consistent with what we might expect when the central tendency of the two dependent populations are equal. Thus we might conclude, household niceness has little or no impact on twins' niceness, further implying the personality trait of niceness is primarily genetically determined. However, this conflicts with the result we obtained using the parametric test in section 11.2 with the same data (recall, in that test, $z = 3.81$, reject H_0, thus $\mu_1 \neq \mu_2$). That test implied a strong difference on average in twin scores.

Herein lies the problem with the sign test. It is not as sensitive as its parametric counterpart. In situations where its parametric counterpart can be used, the sign test would require an increase in sample size of approximately 60% (in this case, from $n = 15$ pairs to $n = 24$ pairs) to equate the two tests in their ability to detect population differences.

*For small-n, np or $n(1 - p) < 5$, we use the binomial tables in the back of the text to assess probabilities.

Wilcoxon Rank-Sum Test
(For Two Independent Samples, Each $n_1, n_2 \geq 10$)

Whereas the sign test just studied is a nonparametric test for dependent data, the *Wilcoxon rank-sum test* (equivalent to the Mann-Whitney U test) is a nonparametric test for independent data and can be used in place of the t test, studied in section 11.2, when the normality-of-population requirement of the t test cannot be satisfied. Let's look at an example.

Suppose in a study assessing American taste for exotic Chinese food, 25 randomly selected Americans were assigned to two treatment groups. Each group was served one dish, described below, which is served, according to the *Newsweek* article, "Pass a Snake, Hold the Rat" (July 29, 1991, p. 35) at China's "more exotic culinary establishments." Suppose participants were asked to rate the dish on a 20–80 scale, 50 being average, with the following results:

Group I (Braised Bear Paw, Boned and Simmered, Sliced Wafer-Thin, Served with Small Game Birds)		Group II (Snake Meat, Stir-Fried, Served with Civet Cat)	
	41		57
	21		50
	47		69
$n = 12$	43	$n = 13$	55
	64		48
$\bar{x} = 43.6$	28	$\bar{x} = 57.9$	61
	53		45
	24		31
	39		65
	52		56
	62		77
	49		63
			76

Suppose the normal curve population requirement for the independent-samples t test, studied in section 11.2, cannot be satisfied and we choose to use the Wilcoxon rank-sum test instead.

Population assumptions: The Wilcoxon test is not free of population assumptions, however the assumptions are less rigid than in the t test (studied in section 11.2). Essentially the difference is as follows: The t test assumes that populations are identical and normal in shape; μ's may be different however. Wilcoxon test assumes that populations are identical but any shape (normal, skewed, or otherwise); again μ's may be different, as shown in the sketch at left.

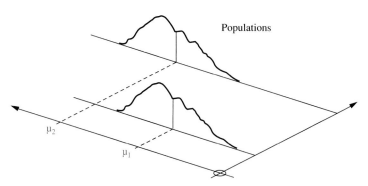

Populations

The purpose of the test is to see if these identically shaped populations have equal means, that is, if $\mu_1 = \mu_2$. Thus, our initial conditions are as follows.

H_0: $\mu_1 = \mu_2$ We assume the two distributions are identical in shape and we wish to test if the means are equal.

H_1: $\mu_1 \neq \mu_2$ If H_0 proves false, then we conclude $\mu_1 \neq \mu_2$.

We begin the test by ranking 1 to 25 all the values in the two groups from smallest to largest, circling values of the *smaller* group.

㉑	㉔	㉘	31	㊴	㊶	㊸	45	㊼	48	㊾	50	㊽	㊾	55	56	57	61	㊽	63	㊽	65	69	76	77
Rank 1	2	3	4	5	6	7	8	9	10	11	12	13	14	15	16	17	18	19	20	21	22	23	24	25

Next we add up the *ranks* of the smaller group, circled above, and denote this sum as R.

$$R = 1 + 2 + 3 + 5 + 6 + 7 + 9 + 11 + 13 + 14 + 19 + 21$$

$$\boxed{R = 111}$$

If H_0 is true, that is, $\mu_1 = \mu_2$, then we would expect about half the ranks in *each* group to be below the median rank of 13, and about half to be above it. In fact, for larger sample sizes, n_1 and n_2 each greater than 10, we know from theory and experience that the *sum of the ranks of the smaller group* (in the case of equal sample sizes, we take the smaller sum), would be normally distributed around μ_R with standard deviation, σ_R, defined and calculated as follows:

In our case,

$$\boxed{\mu_R = \frac{n_1(n_1 + n_2 + 1)}{2}}$$

$$\boxed{\sigma_R = \sqrt{\frac{n_1 n_2(n_1 + n_2 + 1)}{12}}}$$

$$\mu_R = \frac{12(12 + 13 + 1)}{2}$$

$$\sigma = \sqrt{\frac{12(13)(12 + 13 + 1)}{12}} = 18.3847$$

$$\boxed{\mu_R = 156}$$

$$\boxed{\sigma_R = 18.38}$$

The hypothesis test is performed in the same manner as any hypothesis test. *If H_0 were true*, repeated calculations of R would produce a distribution with an average μ_R of 156, with standard deviation, σ_R of 18.38 as follows:

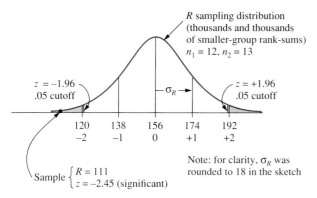

R sampling distribution
(thousands and thousands
of smaller-group rank-sums)
$n_1 = 12, n_2 = 13$

$z = -1.96$
.05 cutoff

σ_R

$z = +1.96$
.05 cutoff

120 138 156 174 192
-2 -1 0 $+1$ $+2$

Note: for clarity, σ_R was
rounded to 18 in the sketch

Sample $\begin{cases} R = 111 \\ z = -2.45 \text{ (significant)} \end{cases}$

To calculate the z score for the sample,

$$z = \frac{R - \mu_R}{\sigma_R}$$

In our case, sample $z = \dfrac{111 - 156}{18.3847} = -2.45$

Reject H_0
$z = -2.45$

Since the sample z (-2.45) was in the rejection area, we reject H_0 and conclude $\mu_1 \neq \mu_2$. Note that R was calculated for Group I (the bear-paw dish). Since this group had significantly *lower* rankings than expected, we conclude Americans are *less likely* to prefer the braised bear-paw dish and better prefer the stir-fried snake meat dish. According to the *Newsweek* article "Pass a Snake, Hold the Rat" (July 29, 1991, p. 35), the snake dish is supposed to be quite a spicy treat (in fact, snake soup was best preferred by the article's author over several exotic dishes, which also included scorpion and rat). Sounds yummy.

The test can also be performed as one tailed. For a small-sample test, n_1 or $n_2 < 10$, special methods and tables are required, not covered in this text.

Concerning the efficiency of this test versus the t test, if parametric conditions for the t test are satisfied, this Wilcoxon test would require an increase in sample size of only about 5% to equate the two in sensitivity, the ability to detect differences in the population means.

Spearman's Rank Correlation Coefficient, r_s

In chapter 9, we introduced r, the linear correlation coefficient, which was defined as a measure of prediction, a measure of the ability of one variable to predict the value of another. Here, we will introduce r_s, **Spearman's rank correlation coefficient,** which is defined in a similar way. r_s is a measure of the ability of one rank-variable to predict the value of another rank-variable. Similarly, as we used r to test if* $\rho = 0$, we will use r_s to also test if $\rho = 0$, that is, we use rank correlation to test if correlation exists in the population, however, now the test is nonparametric.

As with all nonparametric tests, population assumptions, mostly the requirement of normal populations, can be relaxed compared to its parametric equivalent, however, some sensitivity (the ability to detect correlation in the population) is sacrificed, although this can be compensated for by increasing the sample size. Other advantages include ranked data (data without precise proportional scaling) can also be tested and the Spearman test can also be used to detect

*Recall ρ (pronounced, rho) is the linear population correlation coefficient.

relationships other than linear ones, specifically any relationship that continually increases or continually decreases, not necessarily in a straight line. Let's see how this works.

According to the *Almanac of the American People*, "all couples have disagreements but certain problems are more likely to become so serious that couples seek therapy or a divorce." Extensive studies on the causes of divorce reported in the February 1985 issue of the *Journal of Marriage and the Family* revealed the following:

Cause of Divorce	Percentage Females Citing	Rank	Percentage Males Citing	Rank
Poor communication	70%	1	59%	1
Basic unhappiness	60	2	47	2
Incompatibility	56	3	45	3
Emotional abuse	55.5	4	25	6
Money problems	33	5	29	5
Sexual problems	32	6	30	4
Alcohol abuse of spouse	30	7	6	14
Infidelity of spouse	25	8	11	9
Physical abuse	22	9	3.6	16
In-laws	11	10	12	8
Children	9	11	4	15
Religious differences	8.6	12	6.5	12
Mental illness	5	13	7	11
Drug abuse of spouse	4	14	1.4	17
Infidelity (self)	3.9	15	6.2	13
Women's lib	3	16	15	7
Alcohol abuse (self)	1	17	9	10
Drug abuse (self)	.3	18	1	18

Note: Percentages were rounded to whole numbers except in cases that would produce a tie or zero.

Note the values in each column were ranked from 1 to 18 according to their relative size in that column.

Essentially r_s is a measure of the ability of one rank-variable to predict the value of another rank-variable and is calculated using the following formula.*

$$r_s = 1 - \frac{6\Sigma d^2}{n(n^2 - 1)}$$

Like r, r_s varies as follows:
$-1.00 \le r_s \le +1.00$

For each male-female pair, we subtract the rank of one from the other (column 2 rank minus column 1 rank) to obtain d, the difference in rank; d is then squared. The sum of all the d^2's, denoted Σd^2 is then used to calculate r_s using the above formula. For this experiment, $n = 18$ (18 ranks).

*Essentially the formula for r_s is Pearson's formula for r applied to ranks.

Cause of Divorce	Female (Rank)	Male (Rank)	d	d^2
Poor communication	1	1	0	0
Basic unhappiness	2	2	0	0
Incompatibility	3	3	0	0
Emotional abuse	4	6	2	4
Money problems	5	5	0	0
Sexual problems	6	4	−2	4
Alcohol abuse of spouse	7	14	7	49
Infidelity of spouse	8	9	1	1
Physical abuse	9	16	7	49
In-laws	10	8	−2	4
Children	11	15	4	16
Religious differences	12	12	0	0
Mental illness	13	11	−2	4
Drug abuse of spouse	14	17	3	9
Infidelity (self)	15	13	−2	4
Women's lib	16	7	−9	81
Alcohol abuse (self)	17	10	−7	49
Drug abuse (self)	18	18	0	0
			$\Sigma d^2 =$	274

$$r_s = 1 - \frac{6\Sigma d^2}{n(n^2 - 1)}$$

$$= 1 - \frac{6(274)}{18(324 - 1)}$$

$$= 1 - \frac{1644}{5814}$$

$$= 1 - .283$$

$$= .717$$

$$\boxed{r_s = .72}$$

Just as sample r is used in linear correlation to test is $\rho = 0$ (if the population correlation equals 0), we may use our sample r_s in rank correlation also to test if $\rho = 0$, if the population correlation equals 0, as follows:

H_0: $\rho = 0$
H_1: $\rho \neq 0$
$\alpha = .05$

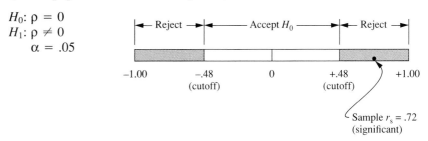

The cutoffs for this test (±.48) were obtained from the Spearman r_s table in back of the text. Look up $n = 18$, and then look down the column marked, two-tailed, .05 (for $\alpha = .05$).*

Reject H_0

*For $n > 30$, the sampling distribution is normally distributed with mean, $\mu_s = 0$ and standard deviation, $\sigma_s = 1/\sqrt{n-1}$; thus, for $n > 30$, the cutoffs can be calculated with the formula: r_s (cutoff) $= \pm z/\sqrt{n-1}$. For example, with $n = 50$ at $\alpha = .05$,
r_s(cutoff) $= \pm 1.96/\sqrt{50 - 1}$
$= \pm.28$.

Since the sample r_s of .72 exceeded the cutoff of .48, we conclude a significant positive correlation exists. In other words, there are strong similarities in the reasons why females and males divorce. Note the first six rankings were near identical for both sexes.

The *Almanac of the American People* goes on to present data on why couples seek therapy, with children problems being the second leading cause. However, this problem seems to be largely solvable with therapy since ''children'' was cited so infrequently in causes of divorce. On the other hand, few couples seek therapy for physical and emotional abuse, yet these were often cited as causes of divorce.

Some final notes: (1) In the case of tied ranks, we use the average rank for the tied cases. For instance, say ranks 16, 17, and 18 in column 1 represented the same tied percentages, we simply use 17, 17, 17 (the average of 16, 17, and 18) for all three. (2) Concerning the efficiency of this test versus the r test for significance: if parametric conditions for the r test are satisfied, the r_s test would require an increase in sample size of about 10% to equate the two in sensitivity, the ability to detect correlation in the population.*

Summary

Additional topics were presented in this chapter under three broad headings: two-category population sampling, measurement population sampling, and nonparametric tests.

Two-Category Population Sampling

All topics in this section concern sampling from one or more two-category populations.

Small-n Binomial Tests. When np or $n(1 - p)$ drops below 5, the normal curve cannot be depended on to give proper estimates. Instead, we may use the binomial formula to assess binomial probabilities.

$$P(s) = \frac{n!}{(n - s)!s!} \, p^s \, (1 - p)^{n - s}$$

where

$P(s)$ = the probability of s successes in n selections
p = the probability of success in one selection
$(1 - p)$ = the probability of failure in one selection
$n!$ = $n(n - 1)(n - 2) \ldots 1$

Many small-sample binomial probabilities have been worked out and are available in the binomial tables in back of the text.

Test of a Single Proportion. In binomial experiments, a hypothesis test is often carried out using proportions (values divided by n). The logic is identical to experiments demonstrated in section 4.5, however, now the formulas are rearranged to accommodate these values divided by n, and are presented as follows:

$$z = \frac{p_s - p}{\sigma_p}$$

where

$$\sigma_p = \sqrt{\frac{p(1 - p)}{n}}$$

p = population proportion to be tested
p_s = sample proportion obtained

For small-n test of proportions we use the binomial probabilities calculated with the binomial formula or obtained from the binomial tables in back of the text.

---

*This implies both r and r_s estimate ρ. Often in rank correlation the null hypothesis is stated as H_0: X and Y are mutually independent.

Confidence Interval Estimate of p. Using the sample proportion, we estimate the most probable location of *p* with the following formula:

$$p_s \pm z \; \sigma_p$$

For a 95% confidence interval, substitute in $z = 1.96$
Difference between Two Proportions (to test if
$p_1 = p_2$*).* This test assumes two independent samples are drawn, one each from a separate binomial population. In the large-sample case, the common proportion can be estimated pooling both sample results as follows:

$$p \approx \frac{s_1 + s_2}{n_1 + n_2}$$

and the $p_{s_1} - p_{s_2}$ sampling distribution is normally distributed around 0 with standard deviation,

$$\sigma_{p_{s_1} - p_{s_2}} = \sqrt{\frac{p(1 - p)}{n_1} + \frac{p(1 - p)}{n_2}}$$

Other variants of this test were also presented.

Measurement Population Sampling

All topics in this section involve sampling from two measurement populations, that is, populations described by μ and σ.
Difference of Two Population Means (independent case). For large samples (each $n_1, n_2 \geq 30$), to test if $\mu_1 = \mu_2$, the $\bar{x}_1 - \bar{x}_2$ sampling distribution is normally distributed around 0 with standard deviation,

$$\sigma_{\bar{x}_1 - \bar{x}_2} = \sqrt{\frac{\sigma^2}{n_1} + \frac{\sigma_2}{n_2}}$$

where σ_1 and σ_2 can be estimated with s_1 and s_2. No population assumptions are needed in this case.
For the small-sample case (either n_1 or $n_2 < 30$), to test if $\mu_1 = \mu_2$, the $\bar{x}_1 - \bar{x}_2$ sampling distribution is normally distributed around 0 with standard deviation,

$$\sigma_{\bar{x}_1 - \bar{x}_2} \approx \sqrt{\frac{(n_1 - 1)s_1^2 + (n_2 - 1)s_2^2}{n_1 + n_2 - 2} \left(\frac{1}{n_1} + \frac{1}{n_2} \right)}$$

Assumptions are populations normally distributed with equal variances (essentially, $\sigma_1 = \sigma_2$).

Difference between Two Population Means
(dependent case) to test $\mu_1 = \mu_2$*.* Dependent Populations: each value in one population is related or linked to a specific value in another population. The population of *d*-values (the result of thousands and thousands of linked-pair subtractions) is normally distributed around 0 with standard deviation, σ_d, estimated by the sample standard deviation of *d*-values, as follows:

$$\sigma_d \approx s_d = \sqrt{\frac{\Sigma(d - \bar{d})^2}{n - 1}}$$

The $\bar{d}$ sampling distribution is slightly different for each sample size however all distributions are normal or near normal (conforming to *z* or *t*) centered around 0 with standard deviation obtained using the central limit theorem, as follows:

$$\sigma_{\bar{d}} = \frac{\sigma_d}{\sqrt{n}} \approx \frac{s_d}{\sqrt{n}}$$

Assumptions are two dependent populations, each normally distributed or at least somewhat symmetrical.

Nonparametric Tests

When certain assumptions for a test cannot be satisfied, nonparametric tests may be used. Although several nonparametric tests exists, the following three were presented: the sign test, the Wilcoxon rank-sum test, and the Spearman rank correlation test. In general, when parametric assumptions can be satisfied, nonparametric tests require an increase in sample size to equate the two in their ability to detect population differences.
Sign Test (for dependent samples) to test $P(+)$
$= 50\%$*.* In the case where we suspect violation of population assumptions in the test for difference between two population means (dependent case), we may elect to use a nonparametric test instead that requires *no* assumptions concerning population characteristics, called the sign test, essentially described as follows.

If indeed two populations have the same central tendency, it is assumed the sign of the $x_2 - x_1$

calculation for several linked pairs would be approximately 50% +'s. Using this null hypothesis,

$$H_0: P(+) = 50\%$$

a binomial test is conducted comparing the number of + calculations actually achieved to this expected 50%.

Wilcoxon Rank-Sum Test (for independent samples, each $n_1, n_2 \geq 10$) to test $\mu_1 = \mu_2$. This Wilcoxon test is equivalent to the Mann-Whitney U test and is used when the requirement of normality of population cannot be satisfied in the difference between means test (independent case). However, the shape of the populations, which can be *non*normal, must be identical. In the test, all values are first ranked according to size and the ranks of the smaller group

are added. If $\mu_1 = \mu_2$ is true, this sum will have a sampling distribution centered around μ_R with standard deviation, σ_R, defined as follows:

$$\mu_R = \frac{n_1(n_1 + n_2 + 1)}{2} \qquad \sigma_R = \sqrt{\frac{n_1 n_2(n_1 + n_2 + 1)}{12}}$$

Spearman's Rank Correlation to test $\rho = 0$. r_s measures the ability of one ranked variable to predict the value of another ranked variable and is used to test if $\rho = 0$, with the statistic,

$$r_s = 1 - \frac{6\Sigma d^2}{n(n^2 - 1)}$$

Cutoff values are determined from Spearman r_s table in back of text.

Exercises

Note that full answers for exercises 1–5 and abbreviated answers for odd-numbered exercises thereafter appear in the Answer Key.

11.1
a. Suppose it is known that 20% of the population has a natural immunity to the HICNS virus. If we were to randomly select 7 individuals from the population, what is the probability that exactly 3 out of the 7 sampled possess this natural immunity?

b. Suppose it is *claimed* that 20% of the population have a natural immunity to the HICNS virus, and we decided to test the claim by randomly selecting $n = 12$ individuals from the population and discover 0 with the natural immunity. At $\alpha = .05$, test $p = 20\%$.

11.2
a. Suppose it is claimed that 3% of college students foster severely delusional thinking patterns and the National Institute of Mental Health tested the claim by randomly sampling 2000 college students nationally, yielding 72 with such delusional thinking. At $\alpha = .05$, use proportions to test $p = 3\%$.

b. Suppose the claim in part a had been worded as *3% or less* are believed to foster severely delusional thinking patterns, at $\alpha = .01$, test $p \leq 3\%$.

c. Suppose no claims were made and a random sample yielded 72 out of 2000 with severely delusional thinking patterns. Construct a 95% confidence interval estimate of p.

11.3 Two groups of college students were tested in 1978 for aberrant thinking, then tested again in 1992 with the following result: out of those with highly delusional thinking patterns in 1978, it was found by 1992 that 39 in 300 had developed mental illness, and out of the nondelusional thinkers, by 1992, 21 in 500 had developed mental illness. At $\alpha = .05$, test $p_1 = p_2$. Briefly discuss results.

11.4
a. Nonverbal clues, such as hand gestures, head nods, etc., may not be as easy to interpret as many people think. Suppose two groups (one high in confidence and the other average) were tested as to their ability to detect a liar. Each group viewed several job-interview videotapes where ten of the applicants had substantially lied. Everyone was asked to detect the ten liars by watching and listening to the videotapes.

High-Confidence Group	Average (Control) Group
$n = 38$	$n = 45$
$\bar{x} = 4.2$	$\bar{x} = 3.8$
$s = 1.8$	$s = 1.7$

At $\alpha = .05$, test the claim $\mu_1 = \mu_2$. Briefly discuss results.

b. Using the same setup in part a, suppose the high-confidence group consisted of 12, with $\bar{x} = 4.2$ and $s = 1.8$, and the average (control) group consisted of 14, with $\bar{x} = 3.8$ and $s = 1.7$. At $\alpha = .05$, test the claim $\mu_1 = \mu_2$. Briefly discuss results.

11.5 Suppose a team of behavioral geneticists seeking to determine the origins of the personality trait of niceness studied 17 pairs of identical twins, separated at birth and reared in different households. The twins were later measured on the personality trait of niceness (along with their respective households), with the goal of assessing if the personality trait of niceness is genetically rooted.

Each pair of twins were separated into one of two groups depending on their household level of niceness, with the following results.

Household Lower in Niceness: Score of Twin in that Household (0–20)	Household Higher in Niceness: Score of Twin in that Household (0–20)
9	14
5	7
8	8
12	11
7	9
9	14
10	12
5	11
4	8
13	11
8	12
9	8
11	16
10	14
9	8
14	18
7	11.5

At $\alpha = .05$, test $\mu_1 = \mu_2$ for these dependent samples. Briefly discuss results.

11.6 Inbreeding (reproduction between related individuals) often occurs when the offspring of an isolated group continually intermarry. In such cases, the gene pool is relatively small (sometimes starting with only a few couples; for example, the Old Order Amish in Lancaster, Pa., originated from only three couples) and over many generations certain characteristics, such as dwarfism or extra fingers, become more prevalent or exaggerated in the group. In biology, this is referred to as genetic drift (meaning, a drift away from the characteristics of the general population). This drift tends to grow more acute in populations where intermarriage is allowed among close kin, say for instance, siblings. In such cases, inherited defects (or, for that matter, excellences) tend to be transmitted in intense form. Contrary to some beliefs, inbreeding does not cause defects or deformities; however, inbreeding does increase their likelihood of showing up in the offspring.

a. Suppose because of inbreeding, the proportion of adolescents suffering a form of retardation in an isolated rural county is known to be 10%. If we were to randomly select 16 adolescents from this county, what is the probability *exactly* 3 out of the 16 sampled would have this retardation.

b. Suppose it is claimed that 20% of adolescents in another county suffer this retardation, and we decide to test the claim by randomly selecting 18 adolescents discovering 1 with this retardation. Does this provide evidence at $\alpha = .05$ to support the claim?

c. Suppose it is claimed that 5% of adolescents in some mountainous county suffer this retardation, and we decide to test the claim by randomly selecting 200 adolescents discovering 18 with the retardation. At $\alpha = .04$, use proportions to test the claim $p = 5\%$.

11.7 Memories are not concentrated in any one part of the brain, according to the late Harvard professor Karl Lashley after extensive neurological experiments with rats and other animals. Lashley concluded that memory in the brain is "nowhere and everywhere present." Lashley concluded this almost 70 years ago. And we've made scant progress since.

However now, with new brain-scan technology (PETs, SQUIDs, etc.) the door to the brain is slowly being unlocked.

According to current experts, the brain is a society of specialty zones where the memory of an experience may actually find its way into several parts of the brain. For instance, the first time you used a rubber ball may be recorded in a specialty zone governing its touch (soft, resilient), and in another governing shape (spherical); with the toss of the ball in another governing movement, and the look of the bounce in yet another governing spatial relations. And perhaps the cheer of parents into even yet another. In other words, this one short memory or experience may have been recorded into numerous specialty and subspecialty zones.

Essentially then, the brain has no memory bank per se. It seems to lay down a memory or knowledge into several zones pertaining to the particular actions or activities you are engaged in at the moment. And once consolidated into long-term memory, this long-term memory is rarely lost. However, age, lesions, disease, and other factors can slow down or disrupt the retrieval or linkage of various scattered elements of a memory, such as, when we attempt to link a name to a face.

a. Suppose because of brain lesions, it is known that a particular woman has only 5% recognition of famous faces (celebrities, U.S. presidents, etc.) and we presented her with 10 such randomly selected photos. What is the probability 0 of the 10 would be recognized?

b. Pharmaceutical houses have long been trying to develop a drug to improve memory with only very limited success. Suppose a new experimental memory drug was administered to a woman with $p = 5\%$ recognition and the woman was able to name 5 out of 14 random photos. At $\alpha = .01$, use proportions to test $p = 5\%$ recognition.

c. Suppose this memory drug was administered to a patient known to have a $p = 8\%$ recognition and the patient now was able to recognize 11 out of 72 photos. At $\alpha = .05$, use proportions to test $p = 8\%$ recognition.

11.8 Heart Disease is by far the number one killer in the nation, responsible for about half a million deaths annually. Although improved medical treatment may keep people alive longer, actual incidences of heart disease, attacks and angina, until the mid-1980s has remained stubbornly high. However, now, with the healthier life styles in recent years and known decreases in smoking rates and cholesterol and blood pressure levels, it was felt new studies were warranted.

Suppose very recent data (1990s) were presented to the American Heart Association that revealed 263 incidences of heart disease over the study period in one Massachusetts town out of 3,600 individuals monitored. Assume that 7% is the expected incidence of heart disease for this length study, age of population and other factors (based on identical studies conducted in the same Massachusetts town in the 1960s, 1970s and 1980s).

At $\alpha = .05$, test the hypothesis that heart disease has remained unchanged at $p = 7\%$. What conclusions can be drawn?

11.9 Writing fiction is easy; writing fiction that sells is hard. Ask any fiction writer. Although one can spend many semesters discussing what makes fiction great, one need only spend a few hours reading Scott Meredith's book, *Writing to Sell* (New York: Harper and Row, 1974, p. 58), to get a glimpse at what makes fiction sell. Meredith, a Manhattan literary agent, should know. He has represented a string of successful authors.

In one chapter, Meredith claims there is a plot skeleton or basic story structure ''underlying almost every piece of commercial fiction from fairy tales to classics to the stories in the magazine on the corner newsstand.'' This story structure runs something like this:

A sympathetic lead character finds himself in trouble of some kind and makes active efforts to get himself out of it. Each effort, however, merely gets him deeper into his trouble, and each new obstacle in his path is larger than the last. Finally, when things look blackest and it seems certain that the lead character is kaput, he manages to get out of his trouble through his own efforts, intelligence or ingenuity.

a. Suppose a university professor supports Meredith's claim and further asserts that a full 80% of all *great* literature follows this plot skeleton. Challenging this assertion, literature students randomly select 40 great novels, finding 26 that precisely follow this pattern. At $\alpha = .05$, use this evidence to test the professor's claim.

b. Suppose the professor had claimed *at least* 80% follow this pattern. At $\alpha = .03$, use the sample data in part a to test this claim.

c. Suppose no claim is made. Use the data, 26 out of 40 to construct a 95% confidence interval estimate of p, the population proportion.

11.10 Do jurors on a trial rearrange trial facts to reflect their own beliefs? According to experts, the answer is yes. In simulated trials, it was found that 45% of the references made by jurors after the trial was over referred to events that were *not* included in the courtroom testimony. The study also found that jurors' background led to crucial differences in their assumptions about human nature that played a powerful role in influencing their verdict. So important is the background of the jurors and the assumptions they bring with them that over 300 consulting firms now advise lawyers on jury selection.

In fact, for anywhere from $5000 to $100,000, lawyers can purchase actual simulated jury trials to test tactics, strategies, and arguments. It is believed these simulated trials provide insight into issues of importance and identify unsuccessful approaches. With such knowledge, consultants advise lawyers as to tactics for opening statements, strategies, and visual reenforcements.

Suppose two simulated jury trials were set up to test defense strategies for an upcoming highly publicized trial involving a famous U.S. family. The defense strategy was presented to a simulated jury of 36 individuals who later broke up into groups of 12 to deliberate and vote. An alternative defense strategy was presented to a different jury of 48 individuals who also broke up into groups of 12 to deliberate and vote. The results of the vote were as follows:

Strategy 1	**Strategy 2**
Out of 36 individuals, 19 voted not guilty.	Out of 48 individuals, 22 voted not guilty.

At $\alpha = .05$, test $p_1 = p_2$ and briefly discuss the implications of the results.

11.11 Hard to please females may be an important factor in the evolution of species, according to some experts. Seemingly meaningless male courtship rituals may actually be critical in selection of males by females. Whereas a male might go with just any pretty face, the females tend to need more time to pick and choose, which is believed to be highly influential in the development of certain male characteristics. For instance, such male show-off rituals as the spectacular acrobatics of male dolphins, the intricate chirping patterns of various species of male birds, and the many seemingly meaningless demonstrations, dances, and theatrics of males of various species are believed to give females time to assess a male's physical health and fitness as a father.

Although experts had originally thought distinctive male markings, such as a rooster's comb, a barn swallow's long tail, and exotic coloration of certain fish, may have been initially developed to elude predators or defend territory, new research seems to indicate these characteristics were developed for reasons no more than to curry favor with the gals. Strong markings are believed to signify good health (freedom from parasites, etc.) and therefore gain female preference. And through generations of this preferential mating, such markings become highly exaggerated in the species.

Suppose biologists wishing to explore rooster-hen mating preferences, randomly expose an equal number of hens to two types of roosters, with the following results:

Type I Roosters (long combs and bright wattles)	**Type II Roosters (average combs and wattles)**
Out of 90 roosters, hens mated with 72.	Out of 120 roosters, hens mated with 54.

At $\alpha = .05$, test $p_1 = p_2$ and briefly discuss the implications of the results.

11.12 A placebo (plasē′ bō), according to the *New Columbia Encyclopedia*, ''is an inert substance given instead of a potent drug. Placebo medications are sometimes prescribed when no drug is really needed because they make the patient feel well taken care of.

Placebos are also used as controls in scientific studies on the effectiveness of drugs. So-called double-blind experiments, where neither the doctor nor the patient knows whether the given medication is the experimental drug or the placebo, are often done to assure unbiased, statistically reliable results.''

a. Suppose in a double-blind study, 200 individuals were given an experimental drug for arthritis while 160 received a placebo. Patient improvement was rated on a 0–20 scale, with the following results:

Arthritis Drug	Placebo
$n = 200$	$n = 160$
$\bar{x} = 8.7$	$\bar{x} = 5.2$
$s = 2.4$	$s = 1.7$

At $\alpha = .05$, test $\mu_1 = \mu_2$. Briefly discuss implications of the results.

b. Suppose in another double-blind study for the drug, the following results were obtained:

Arthritis Drug	Placebo
$n = 11$	$n = 12$
$\bar{x} = 7.6$	$\bar{x} = 5.9$
$s = 2.3$	$s = 2.1$

At $\alpha = .05$, test $\mu_1 = \mu_2$. Briefly discuss the implications of the results.

11.13   Fear of the number 13, according to C. Panati's, *Extraordinary Origins of Everyday Things* (New York: Harper and Row, 1987, pp. 11–13), dates back at least to pre-Christian Scandinavian mythology. Twelve gods had been invited to a banquet. When a thirteenth god (the god of strife and evil) gate-crashed, Balder, the favorite of the gods, was killed in the struggle to evict this thirteenth god. The superstition was then reenforced by the tragic outcome of the Last Supper in which Christ and the apostles numbered thirteen and within a day, Christ was crucified. To this era, many still avoid the number 13, for instance, in assigning house addresses, airline seats (row 13 is often omitted), and apartment building floors.

Suppose a psychologist tested this superstition by randomly selecting the records of co-op apartment buildings recently built in several large cities, with the following results:

14th Floor was used to label the 13th Floor	13th Floor was used to label the 13th Floor
Percentage of floor space sold during first year	Percentage of floor space sold during first year

a. *Study I (large sample).*

$n = 80$ buildings	$n = 60$ buildings
$\bar{x} = 82\%$ of floor space sold on average	$\bar{x} = 62\%$ of floor space sold on average
$s = 8\%$	$s = 12\%$

At $\alpha = .05$, test $\mu_1 = \mu_2$. Briefly discuss implications.

b. *Study II (small sample).*

$n = 9$ buildings	$n = 11$ buildings
$\bar{x} = 78\%$ of floor space sold on average	$\bar{x} = 64\%$ of floor space sold on average
$s = 8\%$	$s = 12\%$

At $\alpha = .05$, test $\mu_1 = \mu_2$. Briefly discuss implications.

11.14   Dogs may not be man's best friend because of their usefulness, according to experts, but because of their submissive, groveling behavior in the face of human aggression. Such acts as cringing, looking away, and rolling over tend to deflect human anger and endear these creatures to our hearts.

The breeding of dogs, who are descendants of the wolf, is believed to have originated about 9000 years ago, and since that time they have been bred to suit the various needs and whims of man. For instance, the golden retriever was originally bred to hunt land and water birds, however now is probably more prized for its beauty and docility.

Suppose British Breeders, Inc., sensing a strong upper-class home market in America and Europe for a retriever with slightly exaggerated characteristics (golden and silver highlights, higher learning intelligence, and other improvements) have been secretly breeding over several generations this new breed of retriever.

As an experiment, pups of the older established breed and pups of this new breed were randomly placed in homes of several families requesting retriever pups. After a year of ownership, the families were asked to rate their dogs on several characteristics, which were summed into a 0 to 100 scale assessing overall desirability, with the following results:

a. *Study I: (large sample).*

Established Retriever	**New Retriever**
$n = 42$ family placements	$n = 35$ family placements
$\bar{x} = 84$ rating	$\bar{x} = 88$ rating
$s = 6$	$s = 4$

At $\alpha = .05$, test $\mu_1 = \mu_2$ and briefly discuss results.

b. *Study II: (small sample).*

Established Retriever	**New Retriever**
$n = 8$ family placements	$n = 9$ family placements
$\bar{x} = 86$ rating	$\bar{x} = 88$ rating
$s = 4$	$s = 3$

At $\alpha = .05$, test $\mu_1 = \mu_2$ and briefly discuss results.

11.15 The human brain until relatively recently has increased in size by several heaping tablespoons of gray matter every 100,000 years.* No other organ in the history of life is known to have grown faster. Human vulnerability coupled with an intense struggle for survival seems to have spurred on this vast growth in neural circuitry into what James Watson (of double-helix fame) calls ''the most complex thing we have yet discovered in our Universe.''

---

*''Starting about one million years ago, the fossil record shows an accelerated growth of the human brain. . . . Five hundred thousand years ago the rate of growth hit its peak. At that time the brain was expanding at a phenomenal rate of ten cubic inches every hundred thousand years (about 10 heaping tablespoons) . . . but after several hundred thousand years of very rapid growth the expansion of the brain slowed down, and in the last one hundred thousand years it has not changed in size at all'' R. Jastrow, *Until the Sun Dies,* W. W. Norton, Inc., 1977, p. 124.
     Other references: ''Mapping the Brain'' *Newsweek* (April 20, 1992, p. 60); ''Memory,'' *Newsweek* (September 29, 1986, p. 48); and ''Nerve Cell . . . ,'' *New York Times* (October 27, 1992, C1); ''Brain . . . ,'' *New York Times* (March 3, 1993, C1); and ''New Brain . . . ,'' *New York Times* (April 6, 1993, C1).

Now with modern scanners, we've been able to peek inside this vast tangle of 100 billion or so functioning cells (neurons), each with about 1000 protein branches covered with a chemical substance. A memory or learning experience is believed to be formed when this chemical substance is prodded by sensory input into reaching out and interconnecting with the branches and chemicals of other cells. Eventually, with all its memories, the brain becomes a vast network of chemical pathways and junctions interconnected throughout this huge inventory of brain cells.

Long-term learning is believed to be a permanent modification of these pathways and intelligence a function of how efficiently one prunes out unneeded pathways for a learning experience, with high intelligence associated with those who heavily prune, thus leaving more circuitry open for new learning. This implies the average person may think too much, that is, uses too many pathways to acquire, store, or retrieve bits of learning versus those supposedly higher in brain power.

Now, one way to measure thinking is energy consumption. To test this hypothesis, suppose energy consumption of the brain is measured on two groups, one randomly selected from a population reputed to be high in intelligence (not necessarily high in education, however), and the other, average in intelligence. Both groups contain individuals from diverse fields (musicians, artists, bankers, novelists, etc.). Each person was asked to perform a series of simple tasks (baking, threading a needle, playing a new video game, etc.) and after a short learning phase, the tasks were repeated and brain-energy consumption monitored (absorption of glucose), with the following results:

a. *Study I (large sample).*

Those Reputed as High in Intelligence	**Average in Intelligence**
$n = 32$	$n = 50$
$\bar{x} = 15.1$ units absorbed	$\bar{x} = 26.7$ units absorbed
$s = 7.3$	$s = 8.1$

b. *Study II (small sample).*

$n = 10$	$n = 16$
$\bar{x} = 14.5$ units absorbed	$\bar{x} = 24.5$ units absorbed
$s = 6.3$	$s = 5.8$

At $\alpha = .05$, test $\mu_1 = \mu_2$ and discuss the implications in each study.

11.16 Conducting reliable market research on movies is a difficult task indeed, for many reasons. First, in screenplay form, which is the easiest point at which to alter a story, the finished product is quite difficult to assess since the inputs of the actors, director, cinematographer, cutters, and so on, are usually so very vital in determining the desirability of the finished product. Of course, once the film is shot, we can run previews to various audiences, however, this is usually several months after completion and too late for major revision with crews and actors off shooting other movies.

In an attempt to overcome these obstacles, suppose Hybrid Films, Inc., planned to shoot a new movie on two separate shooting schedules. During the first schedule, Hybrid completes part of the movie including the end and enough key scenes to give a true sense of the film, but certainly not shooting all of it. These early rushes are roughly cut and assembled into a somewhat crude but viewable presentation for the primary purpose of assessing the film's potential profitability.

To assess this, the recently shot footage will be tested against a similar category film, also a shortened version, known to have broken even at the box office (no profit, no loss), which is used for comparison. The viewing group consists of 14 randomly selected young adults representative of the market for this type of movies. To ensure fairness, half viewed the comparison film first then viewed the experimental film, while the other half did the reverse, viewed the experimental film first then the comparison film.

None of the participants were shown titles, nor have they seen or heard of either film prior to viewing. The voting was on a 0 to 20 scale (10 being neutral), essentially assessing how strongly a person would recommend (greater than 10 vote) or not recommend (less than 10 vote) the movie to a friend. Each person voted for both films, with the following results.

Experimental Film Vote	Comparison Film Vote
12	14
10	10
13	15
10	6
10	13
5	9
14	16
8	12
12	15
14	16
10	9
10	13
8	12
11	15

At $\alpha = .05$, test $\mu_1 = \mu_2$ for these *dependent* samples. Briefly discuss the implication of the results.

11.17 Hope and optimism are emerging as keys to success in life (from academic achievement to bearing up under onerous jobs to surviving near-tragic illness) according to researchers at University of Kansas, Virginia Commonwealth University, University of Medicine and Dentistry in New Jersey, and other institutions.

One researcher found that people high in hope tend to (a) turn to friends for advice on how to achieve their goals; (b) tell themselves they can succeed at what they need to do; (c) even in tight spots, tell themselves things will get better as time goes on; (d) remain flexible enough to find different ways to get to their goals; (e) even if hope for one goal fades, aim for another (those low in hope tend to become fixated on one goal, persisting even when they find themselves blocked, staying at it and getting frustrated); and (f) show an ability to break a formidable task into specific, achievable chunks (people low in hope see only the large goal, and *not* the small steps to it along the way).

Suppose a known Western college wished to assess if a scale measuring hope and optimism could be used to predict academic success at their institution. Twelve pairs of incoming freshmen were used in the study with each pair matched closely as to high school GPA, ACT scores, family socioeconomic status, and ethnic background. Based on questionnaire scores, each in the matched pair was recorded in either the lower in hope or higher in hope category. After four years of college, their college GPAs were compared as follows:

Lower In Hope Group: College GPA (0–4 scale)	Higher In Hope Group: College GPA (0–4 scale)
2.2	2.1
1.8	2.9
3.0	3.8
3.1	3.0
3.4	3.3
2.7	3.5
2.0	3.9
3.7	3.5
3.0	3.2
2.9	3.6
3.3	3.4
1.9	2.8

At $\alpha = .05$, test $\mu_1 = \mu_2$ for these *dependent* samples. Briefly discuss the implications of the results.

11.18 Referring to exercise 11.16, assume the population assumptions for the dependent-samples test cannot be met, and we decide to evaluate the data using the sign test. At $\alpha = .05$, test $P(+) = P(-)$, and briefly discuss the results.

11.19 Referring to exercise 11.17, assume the population assumptions for the dependent-samples test cannot be met, and we decide to evaluate the data using the sign test. At $\alpha = .05$, test $P(+) = P(-)$, and briefly discuss the results.

11.20 Menopause now occupies more than $\frac{1}{3}$ of most women's lives, according to *New York Times,* "Can Drugs Treat Menopause . . ." (May 19, 1992, C1). Estrogen starts depleting in women at about the age of thirty-five and ends at about fifty in menopause. Except for hot flashes and some dizziness, the majority of women are not bothered by menopause, however, the danger lies in the prolonged absence of estrogen, which seems to cause brittle bones and may leave the women vulnerable to an assortment of ailments.

Suppose two groups of women in their late thirties, one given a hormonal drug and the other given a placebo, were monitored over several years and the ages at the onset of menopause recorded, as follows:

Hormonal Drug Group	Placebo Group
46.7	46.2
47.8	48.2
50.8	49.3
51.1	49.7
51.3	49.8
51.9	51.8
52.0	52.2
53.1	52.5
53.5	54.6
56.2	55.2
	55.6

At $\alpha = .05$, test $\mu_1 = \mu_2$ using the Wilcoxon rank-sum test for these two independent samples.

11.21 Body lubricant (hyaluranon or HA, a complex sugar) is found in all animals in the spaces between joints and cells. This lubricant helps keep tissue pliable, allowing skin to stretch, eyes to move, etc., and absorbs shock to the body from the impact of jumping and punching and, according to some, "is the most viscoelastic substance known to mankind."* Although injections into the skin of HA is sometimes used to repair wrinkles and scars, and in some countries for arthritis and here for treating the knees of racehorses, its cost when extracted naturally, often from the combs of roosters, is high, millions of dollars for one pound. However, laboratory-cultured HA is being developed that offers the promise of reduced costs and improved qualities.

Suppose in an experiment involving athletes with sports injuries, laboratory-cultured HA is compared to natural HA. Twenty-three athletes with severe joint pain were assigned to two groups, one receiving the laboratory HA and the other receiving the natural HA. After stress exercises, levels of pain

*Reference article: "Using the Body's Lubricant . . . ," *New York Times* (February 9, 1992, F7).

were monitored on a 0 to 100 scale (0 being no pain, 100 unbearable pain), with the following results:

Laboratory Cultured HA	Natural HA
26	28
30	29
31	33
37	34
42	38
48	40
55	41
58	45
64	47
66	59
74	68
	72

At $\alpha = .05$, test $\mu_1 = \mu_2$ using the Wilcoxon rank-sum test for these two independent samples.

11.22 In chapter 9, we had assessed if a correlation existed between high school predictor scores (high school GPA, SATs, and so on) and college freshmen GPA. Suppose for the following data, population assumptions for linear correlation cannot be met and we use the Spearman rank correlation coefficient to test if $\rho = 0$ at .05.

High-School Predictor Score	College Freshman GPA
36	1.5
39	1.9
42	2.9
47	3.1
53	2.0
55	3.6
59	2.3
61	3.0
67	2.8

11.23 Referring to chapter 9, exercise 9.11, use the Spearman rank correlation to test if $\rho = 0$ at .05. For convenience, the data is presented below in ascending ages.

Girl's Age	Level of Self-Esteem
7	82
8	72
10	70
11	64
12	62
13	41
14	53
16	46

Answer Key

Chapter One

1.1

This depends on which articles are chosen. A population may be thought of as all the possible values of a particular phenomenon that we wish to study, whereas a sample is part of that population.

1.2

a. The *ages* of all Chicago males divorced in 1988.
b. The symbol μ (m$\bar{u}$) is used to represent the average value of a population.
c. A random sample of 100 was used.
d. The symbol $\bar{x}$ is used to represent the average of a sample.
e. No. However, if a valid random sample was indeed achieved, then we can *approximate* the average age of the population as 35.2 years old.

1.3

First, we identify each member of the population with a number from 1 to 5000. Then we start anywhere in the random digit table, penciling off a section of four-digit numbers. Say for instance, we arbitrarily choose column 4, first number down to start. We circle numbers of 5000 or less in this section. If we wish 7 in our sample, then we circle only 7 numbers. Last, we locate the specific 7 Chicago males who were picked and determine their ages.

1.4

a. To calculate the average, $\bar{x}$, we sum up the numbers and divide by 7.

$$\bar{x} = \frac{26 + 61 + 39 + 43 + 31 + 22 + 37}{7} = 37 \text{ years old}$$

b. Use of small samples (under 30) come with certain restrictions, perhaps the most formidable being assurance that our population histogram is at least somewhat bell-shaped in appearance. Assuming this restriction is met and a valid random sample achieved, then we can state,

$$\bar{x} \approx \mu \qquad \text{the sample average ($\bar{x}$) is approximately equal to the population average (μ).}$$

Because $\bar{x} = 37$ years old from the calculation above in answer a, we can say that μ, the average age of *all* Chicago males divorced in 1988, is *approximately* 37 years old. One word of caution: it is generally preferable to keep your sample size as large as possible for closer approximations.

1.5

a. $p_s = \dfrac{18}{100}$

b. Percentage = Fraction $\times$ 100 = $\dfrac{18}{100} \times 100 = 18\%$

c. Because $p_s \approx p$ for valid random samples, the proportion of *all* Chicago males divorced in 1988 with blue eyes is *approximately* 18%.

1.7

a. If we had clocked *each and every* time over several months the nurse drew blood and recorded all these many hundreds of readings, this would be our population. Note the population would be all the *times* (in minutes) it took for a particular nurse-in-training to draw blood during this period.
b. We wish to determine the *average* length of time it took the nurse to draw blood over this several-month period. The symbol we use for population average is μ.

c. The sample is the six readings we actually clocked (10, 6, 5, 14, 6). We call this a sample of size $n = 6$ readings. To calculate $\bar{x}$, the average of the sample,

$$\bar{x} = \frac{10 + 6 + 5 + 14 + 6 + 13}{6} = 9.0 \text{ minutes}$$

d. A sample will be representative of a population if it is a valid random sample. For small samples (under 30) a further restriction is imposed. We must have assurance that the population histogram is at least somewhat bell-shaped in appearance.

i. Random selection implies that each population value (and every sample of the same size) has an equal chance of being selected. Although several approaches may be used, one approach might be as follows: Say the nurse performs this procedure approximately 20 times a day or 2000 times over several months. We could assign each $\frac{1}{2}$-hour interval the nurse is on duty over these several months a number from 1 to 2000, then use the random digit table to select values between 1 and 2000. Locate these particular $\frac{1}{2}$-hour work intervals and clock the nurse when she draws blood.

ii. Internal validity is our ability to obtain accurate, honest, and reliable readings of that which we set out to measure. In most real-life experiments, this determination can be quite complex. Certainly we should question the accuracy of our measuring instrument and whether factors beyond the control of the nurse are affecting the readings, that is, factors other than the pure abilities of the nurse, say for instance, interruption by a supervisor.

iii. A random sample may set the stage for achieving external validity, which is our ability to use the sample as representative of the population, however, random samples do not guarantee it. We must address the question: did our methods or presence in any way influence the results, thereby threatening external validity? Did the nurse know he or she was being monitored and was thereby affected? Did we monitor the nurse in an artificial setting, unlike normal surroundings and circumstances, thus possibly introducing an effect? All these questions must be addressed.

iv. When using small samples (under 30), for $\bar{x}$ to be approximately equal to μ we must have some assurance that the population histogram is at least somewhat bell-shaped in appearance. For this, we often draw on experience, that is, what has occurred in past similar experiments measuring the same phenomenon.

1.9

a. The heights of *all* Washington State University males.

b. The heights of 50 males sampled from the Washington State University campus.

c. Male I-D numbers are thrown into a rotating basket, thoroughly mixed, and 50 numbers are randomly drawn. These students are located and their heights measured. (Note: other answers are possible.)

Chapter Two

2.1

a.

Age category	Age category	Tally	Age category	Tally	Frequency or category total
15–19	15–19	/	15–19	卌	5
20–24	20–24		20–24	卌 卌	10
25–29	25–29		25–29	卌 卌 卌 ||	17
30–34	30–34		30–34	卌 卌 ||	12
35–39	35–39		35–39	||||	4
40–44	40–44		40–44	||	2
First, we must select suitable age categories or groups, which were given.	Second, we tally the data. For example, the first age on our list is 19.4 years old. We represent this by a slash in the 15–19 age category. Note: Had the age been 19.5, it would have been put in the 20–24 age category.		Third, we tally all 50 ages. When finished, the tally would appear as shown above. Also shown is the final count for each category called the frequency or category total.		

b.

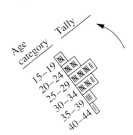
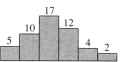

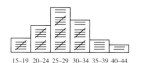

First, we encase the slashes in rectangular bars, and rotate the tally ¼ turn counter-clockwise.

The final position appears as above. For aesthetic reasons, the bars can be widened.

Histogram

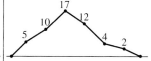

Last, we remove the tally slashes and, often, for ease of viewing, lightly shade or color the bars and put category totals atop each bar.

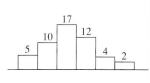

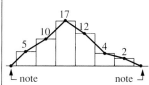

Frequency polygon

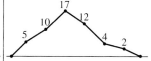

First, we start with a histogram.

Second, the midpoints of the histogram bars (represented as dots) are connected by straight lines, and anchored to the base line at each end (as noted in the diagram).

Third, the histogram bars are removed. To avoid drawing the histogram, often the frequency polygon is drawn directly from the tally data.

c. The population consists of all the ages of Countrygirl users nationwide.

d. The sample would be the ages of the 50 users randomly selected nationwide, thus the sample size is $n = 50$.

2.2

a. Since 21 out of 50 possess the attribute of brown hair, the proportion or fraction is $p_s = 21/50$.

b. To convert to a percentage, we multiply the fraction by 100.

$$\text{Percentage} = \text{Fraction} \times 100 = \frac{21}{50} \times 100 = 42\%$$

c. We can use either of two methods. We could convert 42% to degrees and measure with a protractor: 42% of 360° = .42(360) = 151.2° (see first diagram). Alternatively, we could estimate 42%. Since $\frac{1}{4}$ circle is 25% and $\frac{1}{2}$ is 50%, we proceed as shown in the second diagram. The completed solution might appear as shown in the last diagram.

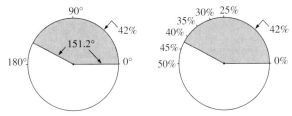

The completed solution might appear as follows:

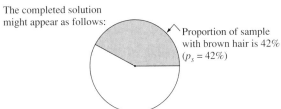

Proportion of sample with brown hair is 42% ($p_s = 42\%$)

2.3

a. Sample mean, $\bar{x}$, equals the sum of the values divided by n, the sample size. Because we had 8 values in our sample, the sample size is $n = 8$.

$$\bar{x} = \frac{\Sigma x}{n} = \frac{11 + 7 + 13 + 7 + 5 + 8 + 10 + 15}{8}$$

$$= 9\frac{1}{2} \text{ minutes}$$

b. Line up values according to size: 5, 7, 7, 8, 10, 11, 13, 15, and select the middle value. Since we have two middle values, we average the two, (8 + 10)/2 = 9 minutes.

c. The mode is the most frequently occurring value, which is 7 minutes.

d. Range = high value minus low value = 15 − 5 = 10 minutes. This can also be expressed as the data ranged from 5 to 15 minutes.

e. The standard deviation is calculated as follows:

x	$\bar{x}$	$x - \bar{x}$	$(x - \bar{x})^2$
5	9.5	−4.5	20.25
7	9.5	−2.5	6.25
7	9.5	−2.5	6.25
8	9.5	−1.5	2.25
10	9.5	.5	.25
11	9.5	1.5	2.25
13	9.5	3.5	12.25
15	9.5	5.5	30.25
			80.00

$$s = \sqrt{\frac{\Sigma(x - \bar{x})^2}{n - 1}}$$

$$= \sqrt{\frac{80.00}{8 - 1}}$$

$$= \sqrt{11.428}$$

$$= 3.38 \text{ minutes}$$

Note: sum of $(x - \bar{x})^2$ column = 80.00, which is used in calculation

f. The sample times must be *randomly* selected. Also, we must be assured that we achieved an honest, accurate, and reliable measure of these times (internal validity) and that our methods and presence in no way influenced the results (external validity). In addition, for small samples (under 30), we must be assured that our population histogram is at least somewhat bell-shaped in appearance.

2.4

Because the data is grouped, we use our grouped data formulas.

a. To calculate the mean, we sum our xf values and divide by 49.

Time Category	x	f	xf
3–5	4	7	28
6–8	7	15	105
9–11	10	16	160
12–14	13	8	104
15–17	16	3	48
		49	445

$$\bar{x} = \frac{\Sigma xf}{n}$$

$$= \frac{445}{49}$$

$$= 9.08 \text{ minutes}$$

sum of xf column (Σxf) = 445
$n = \Sigma f$ column = 49

b. First, we arrange the data in chart form, just as we did in the preceding example. Only this time, we include an additional x^2f column. The x^2f values are obtained by multiplying the x value by the xf value.

For example, in the first row, we obtained an x^2f reading of 112 by multiplying $4 \times 28 = 112$ (x times $xf = x^2f$).

Time Category	x	f	xf	x^2f
3–5	4	7	28	112
6–8	7	15	105	735
9–11	10	16	160	1600
12–14	13	8	104	1352
15–17	16	3	48	768
		49	445	4567

$$s = \sqrt{\frac{n(\Sigma x^2 f) - (\Sigma x f)^2}{n(n-1)}}$$

$$= \sqrt{\frac{49(4567) - (445)^2}{49(49-1)}}$$

$$= \sqrt{\frac{223{,}783 - 198{,}025}{49(48)}}$$

$$= \sqrt{10.951}$$

$$= 3.31 \text{ minutes}$$

$n = \Sigma f$ column $= 49$
sum of xf column $(\Sigma xf) = 445$
sum of x^2f column $(\Sigma x^2f) = 4567$

c. The modal class is the group that contains the most values. Since the group 9–11 minutes contains 16 observations, this is the modal class.

d. Histogram for 49 observed times from the nurse-in-training study.

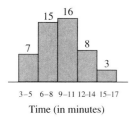

3–5 6–8 9–11 12–14 15–17
Time (in minutes)

e. Frequency polygon. Note: the midpoints of the tops of the histogram bars are connected by a straight line and anchored to the base line at each end.

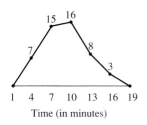

1   4   7   10   13   16   19
Time (in minutes)

2.5

a. To estimate the z score of a tree of 20 feet, we locate on the scale where a tree of 20 feet would be.

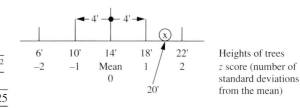

Heights of trees
z score (number of standard deviations from the mean)

A tree of 20 feet would be $1\frac{1}{2}$ standard deviations above the mean, therefore it has a z score of $+1\frac{1}{2}$ (or in decimal form, $z = +1.50$).

b. To estimate the z score of a tree with a height of 10 feet, we must first locate the position of 10 feet on the scale.

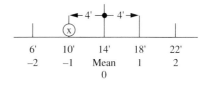

Heights of trees
z score (number of standard deviations from the mean)

Notice that a tree of 10 feet would be -1 standard deviation from the mean, therefore it has a z score of -1 (or in decimal form, $z = -1.00$).

c. To estimate the z score of a tree of 15 feet, we locate on the scale where a tree of 15 feet would be.

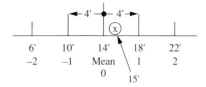

Heights of trees
z score (number of standard deviations from the mean)

A tree of 15 feet would be $\frac{1}{4}$ of a standard deviation above the mean, therefore it has a z score of $\frac{1}{4}$ (or in decimal form, $z = +.25$). In other words, since 15 feet is 1 foot above 14 feet and 1 foot is $\frac{1}{4}$ of your standard deviation of 4 feet, then $z = \frac{1}{4}$ or .25.

2.7

a. $\bar{x} = \dfrac{55}{5} = 11$

b. Median = 12

c. No mode

d. $14 - 7 = 7$

e. $s = \sqrt{\dfrac{34}{4}} = 2.92$

f. Assuming the population to be the puzzle-time scores of *all* second-graders in Westchester County, we must obtain a *valid random* sample from this population. In addition, for this small sample (under 30 observations), we must have some assurance that our population values are at least somewhat bell-shaped in appearance. Note: a threat to internal validity might include students' preknowledge of the puzzle, thus resulting in one or more unreliable time assessments.

2.9

a. $\bar{x} = \dfrac{2770}{70} = 39.57$

b. $s = \sqrt{\dfrac{70(137,500) - (2770)^2}{70(69)}} = 20.10$

c. Modal class is the 30–44 sales category.

d.   **e.**

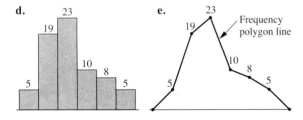

2.11

a. Because 7.6 ounces (oz) is $1\frac{1}{2}$ standard deviations above the mean (.4 oz + .2 oz above 7.0 oz), we say $z = 1\frac{1}{2}$ or 1.50.

b. Because 7.1 oz is $\frac{1}{4}$ standard deviation above the mean (.1 oz is $\frac{1}{4}$ of .4 oz), we say $z = \frac{1}{4}$ or .25.

c. Because 6.7 oz is $\frac{3}{4}$ standard deviation below the mean (6.7 is .3 below 7.0 and .3 is $\frac{3}{4}$ of .4), we say $z = -\frac{3}{4}$ or $-.75$.

2.13

a. 170 lbs is $1\frac{1}{4}$ standard deviations above the average of 155 lbs, thus we can say Jason's weight is $z = 1.25$. 138 is 2 standard deviations above the average of 118 lbs, thus we say Rebecca's weight is $z = 2.00$.

b. Since Rebecca is 2 standard deviations above her category average, versus $1\frac{1}{4}$ standard deviations for Jason, Rebecca would be considered more seriously overweight.

2.15

Bell-shaped or normal:

Heights of men, women, trees, giraffes

Machine output of various types: lengths, widths, diameters, volumes

Time a person needs to complete a repetitive task: brushing your teeth, changing a tire, running the $\frac{1}{4}$ mile

Scores on nationwide exams: SAT's, ACT's, IQ tests

Skewed (bulk of data at beginning values):

Salaries in a company, a town, a city, a state

Spread of epidemics (new cases per year)

Effect of medication in the body

Impact of a spurt of advertising on sales

Bimodal:

This often occurs when two populations are intermingled and compete for dominance, such as, if we measured the weights of students in a college classroom with both males and females or if we measured the reading scores of a group consisting of both 6th-graders and 8th-graders.

2.17

Values lined up in increasing order: 18 21 23 26 26 29 30 32 33 34 36 37 38 38 40 41 41 42 43 47.

a.

Score	Tally		Score	Pictogram
10–19	I		10–19	⭧
20–29	ⅢⅢ		20–29	⭧⭧⭧⭧⭧
30–39	ⅢⅢ III		30–39	⭧⭧⭧⭧⭧ ⭧⭧⭧
40–49	ⅢⅢ I		40–49	⭧⭧⭧⭧⭧ ⭧

b.

Stem	Leaves
1	8
2	1 3 6 6 9
3	0 2 3 4 6 7 8 8
4	0 1 1 2 3 7

c. Box-and-whisker plot

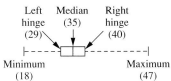

Left hinge (29)　Median (35)　Right hinge (40)

Minimum (18)　　　　Maximum (47)

d. $Q_1 = (26 + 29)/2 = 27.5$ (median of lower half, not including the overall median value of 35). Note left hinge in the box-and-whisker plot (answer c) includes the overall median value of 35 in its computation. Q_2 = overall median = 35. $Q_3 = (40 + 41)/2 = 40.5$.

Chapter Three

3.1

a. We cannot determine the true probability from this limited information. Consecutive events must be tallied under similar conditions *over the long run* before the cumulative probability fraction approaches the true probability.

b. Record the total number of tosses and bull's-eyes achieved under similar circumstances in the long run. Eventually this cumulative probability fraction will stick quite close to a single value. When this occurs, this value will be a good approximation of the true probability.

3.2

a. Because we have 15 chances for success (10 men and 5 women) out of 50 possibilities, P(curly black hair) = 15/50 = 30% (or .30).

b. Since 30 men are at the party, there are 30 chances for success out of 50 possibilities. Thus, P(male) = 30/50 = 60% (or .60).

c. Both conditions must be met for an *and* statement. Since only 10 meet these requirements of being male and having curly black hair, we have 10 chances for success out of 50, thus P(male *and* curly black hair) = 10/50 = 20% (or .20).

d. Satisfying either one or both conditions in an *or* statement will give us success. We have 30 men at the party plus an additional 5 women with curly black hair, thus there are 35 chances for success (male *or* curly black hair) out of 50, so P(male or curly black hair) = 35/50 = 70% (or .70). Note that the 10 men with curly black hair were already included in our initial count of 30 men.

e. Both conditions must be met. If 5 of 20 women have curly black hair, then 15 must *not* have curly black hair. Therefore, we have 15 chances for success out of 50 or P(female and not curly black hair) = 15/50 = 30% (or .30).

f. Because we know the person chosen is a male, we are now selecting from a pool of 30 possibilities (that is, 30 men). And out of this 30, there are 10 with curly black hair, in other words, 10 chances for success. Thus, P(curly black hair, given a male) = 10/30 = 33.3% (or in a decimal rounded to two decimal places, .33).

g. Since we already know the person is a male, there are 0 chances the person is a female. Thus, P(female, given a male) = 0/30 = 0% (or .00).

3.3

a.

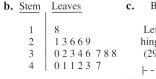

2	1	1	1	1
3	3	2	2	2
1　4	2　4	3　4	4　3	5　3
5	5	5	5	4

1,2	2,1	3,1	<u>4,1</u>	<u>5,1</u>
1,3	2,3	3,2	<u>4,2</u>	<u>5,2</u>
<u>1,4</u>	<u>2,4</u>	<u>3,4</u>	<u>4,3</u>	<u>5,3</u>
<u>1,5</u>	<u>2,5</u>	<u>3,5</u>	(4,5)	(5,4)

n = 20 equally likely outcomes

i. Since we have two chances for success (circled above) out of 20 equally likely outcomes, P(celebrity and celebrity) = 2/20 = 10%.

ii. For this, we have 12 chances for success (underlined above) out of 20, thus P(exactly one celebrity) = 12/20 = 60%.

iii. If we count the ways above (no markings), we get 6 chances for success, thus P(noncelebrity and noncelebrity) = 6/20 = 30%.

b.

1,1	2,1	3,1	<u>4,1</u>	<u>5,1</u>
1,2	2,2	3,2	<u>4,2</u>	<u>5,2</u>
1,3	2,3	3,3	<u>4,3</u>	<u>5,3</u>
<u>1,4</u>	<u>2,4</u>	<u>3,4</u>	(4,4)	(5,4)
<u>1,5</u>	<u>2,5</u>	<u>3,5</u>	(4,5)	(5,5)

n = 25 equally likely outcomes

i. Now we have 4 chances for success (circled above) out of 25 possibilities, thus P(celebrity and celebrity) = 4/25 = 16%.

ii. For this, we have 12 chances for success (underlined above), thus P(exactly one celebrity) = 12/25 = 48%.

iii. If we count the ways above (no markings), we get 9 chances for success: P(noncelebrity and noncelebrity) = 9/25 = 36%.

c. i. On the first pick, we have 2 chances for success out of 5 possibilities. However, if a celebrity was chosen and taken out of the party, we now have 1 chance of success out of 4 possibilities on the second pick.

P(celebrity and celebrity)
= P(celebrity first pick) P(celebrity second pick, given celebrity was chosen on first pick)
$= \dfrac{2}{5} \cdot \dfrac{1}{4} = \dfrac{2}{20} = 10\%$

Note: this is the same answer as exercise 3.3, part a(i).

ii. On the first pick, we have 2 chances for success out of 5 possibilities. However, since the first person was replaced, we still have 2 chances for success out of 5 possibilities on the second pick. Thus,

P(celebrity and celebrity)
$\quad = P$(celebrity first pick)
$\qquad P$(celebrity second pick)
$\quad = \dfrac{2}{5} \cdot \dfrac{2}{5}$
$\quad = \dfrac{4}{25}$ (or 16%)

Note: this is the same answer as exercise 3.3, part b(i).

d. i. On the first pick, we have 3 chances for success out of 5 possibilities. However, if a *non*celebrity was chosen, we now only have 2 chances for success (a *non*celebrity) out of 4 possibilities on the second pick. Thus,

P(noncelebrity and noncelebrity)
$\quad = P$(noncelebrity first pick)
$\qquad P$(noncelebrity second pick, given noncelebrity chosen on first pick)
$\dfrac{3}{5} \cdot \dfrac{2}{4} = \dfrac{6}{20}$ (or 30%)

Note: this is the same answer as exercise 3.3, part a(iii).

ii. On the first pick, we have 3 chances for success out of 5 possibilities. However, since the first person was replaced, we still have 3 chances for success out of 5 possibilities on the second pick. Thus,

P(noncelebrity and noncelebrity)
$\quad = P$(noncelebrity first pick)
$\qquad P$(noncelebrity second pick)
$\quad = \dfrac{3}{5} \cdot \dfrac{3}{5}$
$\quad = \dfrac{9}{25}$ (or 36%)

Note: this is the same answer as exercise 3.3, part b(iii).

e. i. There are two ways we can achieve success. First,

P(celebrity and noncelebrity) $= \dfrac{2}{5} \cdot \dfrac{3}{4}$
$\qquad\qquad = \dfrac{6}{20}$ (or 30%)

Second,

P(noncelebrity and celebrity) $= \dfrac{3}{5} \cdot \dfrac{2}{4}$
$\qquad\qquad = \dfrac{6}{20}$ (or 30%)

Thus, we have 12 chances for success out of 20 possibilities so P(exactly one celebrity) = 12/20 = 60% (same answer as exercise 3.3, part a[ii]).

ii. Again, there are two ways we can achieve success. First,

P(celebrity and noncelebrity) $= \dfrac{2}{5} \cdot \dfrac{3}{5}$
$\qquad\qquad = \dfrac{6}{25}$ (or 24%)

Second,

P(noncelebrity and celebrity) $= \dfrac{3}{5} \cdot \dfrac{2}{5}$
$\qquad\qquad = \dfrac{6}{25}$ (or 24%)

Thus, we have 12 chances for success out of 25 possibilities so P(exactly one celebrity) $= \frac{12}{25}$ = 48% (same answer as exercise 3.3, part b[ii]).

3.4

For this license plate, we have six selections.

____ ____ - ____ ____ ____ ____

For the first selection (the first letter) we have 26 choices (a, b, c, d, e, . . . , x, y, or z). We have the same 26 choices for the second letter. For the first digit, there are 9 choices (1, 2, 3, . . . , 9) but there are 10 choices for each of the remaining digits since 0 is included. Thus we have

$$\underline{26} \times \underline{26} \times \underline{9} \times \underline{10} \times \underline{10} \times \underline{10}$$
$$= 6{,}084{,}000 \text{ different license plates}$$

3.5

a. If we use a tree diagram to help us list the sample space of equally likely outcomes,

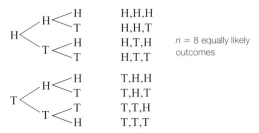

n = 8 equally likely outcomes

b. First, we count the number of times zero heads occurred in the eight equally likely outcomes, which was one time (T,T,T). Next, we count the number of times 1 head occurred, which was three times (H,T,T), (T,H,T), and (T,T,H). We record the number of times each of these occurred in the tally. Then continue for 2 heads and 3 heads. Rotating the tally $\frac{1}{4}$ turn produces the histogram

0 Heads	|
1 Head	|||
2 Heads	|||
3 Heads	|

Number of heads

c. When we toss a coin three times, there are eight equally likely outcomes, which are listed above. Out of these, we count how many will give us 2 heads. Since we have 3 chances for success (H,H,T), (H,T,H), and (T,H,H) out of 8 possibilities, the $P(2$ heads$) = \frac{3}{8}$ (or $37\frac{1}{2}\%$).

d. Since there is only one chance for success (T,T,T) out of eight possibilities, the $P(0$ heads$) = \frac{1}{8}$ (or 12.5%).

3.7

Empirical probability assumes the events have occurred under similar circumstances in the long run. Assuming this to be the case, our estimate would be 1455/8420 = 17.3%.

3.9

a. P(psych major) $= 12/36 = 33.3\%$.

b. P(business or education major) $= (11 + 8)/36$
$= 52.7\%$.

c. Both conditions must be met for an *and* statement. Since there are 17 (12 + 5) chances for success, the P(not a business and not an education major) $= \frac{17}{36}$
$= 47.2\%$.

3.11

a. Since it was stated the two side effects cannot occur together (known as mutually exclusive events), then the probability of finding them together in one person $= 0$.

b. $P(D) = .32$ (or 32 out of 100).

c. Since 32 out of 100 would be expected to experience dryness of the mouth, this implies 68 (100 minus 32 = 68) out of 100 would be expected to experience no dryness, thus P(no dryness) $= .68$.

d. Since 48 (32 plus 16) out of 100 experience either dryness or anxiety, then 52 (100 minus 48 = 52) out of 100 experience no dryness and no anxiety, thus P(no dryness and no anxiety) $= .52$.

e. This implies that 10 out of 100 experienced both side effects and were thus counted in the $P(D)$
$= .32$. So, if 32 out of 100 experience dryness and 10 of these 32 actually had both side effects, then 22 (32 minus 10 = 22) must have experienced *only* dryness, thus P(dryness only) $= .22$.

f. If 16 out of 100 experience anxiety and 10 of these 16 actually experience both side effects, then 6 (16 minus 10 = 6) must have experienced *only* anxiety, thus, P(anxiety only) $= .06$.

3.13

Without replacement after each pick, P(ace and ace and ace) $= \dfrac{4}{52} \cdot \dfrac{3}{51} \cdot \dfrac{2}{50} = \dfrac{24}{132{,}600}$ (or near 0%). With replacement after each pick, P(ace and ace and ace) $= \dfrac{4}{52} \cdot \dfrac{4}{52} \cdot \dfrac{4}{52} = \dfrac{64}{140{,}608}$ (near 0%).

3.15

Since all events are independent, P(H and H and H and H and H) $= P($H$)P($H$)P($H$)P($H$)P($H$) = \frac{1}{2} \cdot \frac{1}{2} \cdot \frac{1}{2} \cdot \frac{1}{2} \cdot \frac{1}{2}$
$= \frac{1}{32} = 3.1\%$.

3.17

Use the counting formula, $7 \times 10 \times 3 = 210$ different committees.

3.19

If we *randomly* sample, we know $p_s \approx p$ (the sample proportion is approximately equal to the population proportion): 25% of 60 is 15 ($.25 \times 60 = 15$). Therefore, we would expect *approximately* 15 women in our sample to possess the attribute of red hair. However, it is not unusual to get other values, say 14 women, 17 women, and so on.

3.21

a. Since the sum of the probabilities must add to 1.00, the missing $p(x) = .6$.

b.

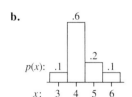

c. $\mu = \Sigma x\, p(x)$
$= 3(.1) + 4(.6) + 5(.2) + 6(.1)$
$= 4.3$
$\sigma = \sqrt{\Sigma(x - \mu)^2\, p(x)}$
$= \sqrt{(3 - 4.3)^2(.1) + \cdots}$
$= \sqrt{.169 + .054 + .098 + .289}$
$= \sqrt{.61} = .781$
$= .78$

3.23

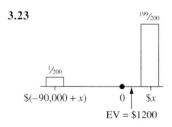

EV = $1200

$$EV = \mu = \Sigma x\, p(x)$$

$$EV = (-90,000 + x)\left(\frac{1}{200}\right) + x\left(\frac{199}{200}\right)$$

$$1200 = -450 + \frac{1}{200}x + \frac{199}{200}x$$

$$1200 + 450 = \frac{200}{200}x$$

$$\$1650 = x$$

3.25

a. Since task forces A,B and B,A are the same, we use a combination that counts the number of ways task forces of 2 can be chosen where order of selection is not important. Thus, $C_{6,2} = 6!/4!2! = 15$ ways.

b. In this case, order of selection is important, since task force A,B is now different from task force B,A. In this case, we can use a permutation, $P_{6,2} = 6!/4! = 30$ ways.

Chapter Four

4.1

a. This is not so. Although certain natural populations are normally distributed, much in nature has been shown to be skewed, bimodal, or take on a variety of distributive shapes. Where the normal distribution does seem to reoccur with amazing regularity is in sampling experiments (discussed in sections 4.4 and 4.5 and in chapter 5).

b. $n \to \infty$ The number of observations on the same phenomenon is enormously large.

$\Delta x \to 0$ The measurements are grouped into a histogram with exceedingly narrow intervals.

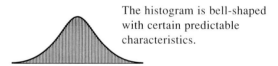

The histogram is bell-shaped with certain predictable characteristics.

c. Characteristics include: bell-shaped, fading at tails, symmetrical about μ, approximately 68% of the data lies within ±1 standard deviation, and approximately 95% of the data lies within ±2 standard deviations.

4.2

a.

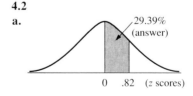

29.39% (answer)

0   .82   (z scores)

Normal curve table (on back cover)

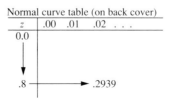

z	.00	.01	.02	. . .
0.0				

.8   → .2939

Since the table reads from $z = 0$ out to whatever z we look up, we simply look up $z = .82$. *Answer:* 29.39% of the data lies between $z = 0$ and $z = .82$.

b.

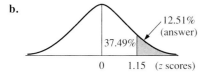

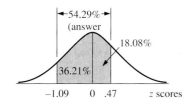

37.49% of the data lies between $z = 0$ and $z = 1.15$. However, we want the data *above* $z = 1.15$, so we subtract this from 50%.

$$50\% - 37.49\% = 12.51\%$$

Answer: 12.51% of the data lies above $z = 1.15$.

c. From $z = 0$ to $z = -1.09$ is 36.21% of the data. From $z = 0$ to $z = .47$ is 18.08% of the data. We add the two areas

$$36.21\% + 18.08\% = 54.29\%$$

Answer: 54.29% of the data lies between $z = -1.09$ and $z = .47$.

d. From $z = 0$ to $z = 2.78$ is 49.73% of the data, however we must subtract the data from $z = 0$ to $z = 1.53$ (which is 43.70%):

$$49.73\% - 43.70\% = 6.03\%$$

Answer: 6.03% of the data lies between $z = 1.53$ and $z = 2.78$.

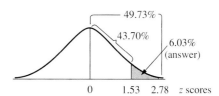

e. This is a working backward problem (given the percentage of data, find z). Since the normal curve reads from $z = 0$ out to $z = ?$, we merely look up 32%, or in decimal form, .3200. The closest value is .3212, which gives us $z = .92$.

Answer: 32% of the data can be found between $z = 0$ and $z = .92$ or between $z = 0$ and $z = -.92$ since the curve is symmetric.

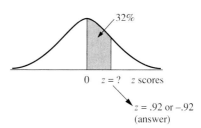

f. This is a working backward problem (given the percentage of data, find z). Since the normal curve table reads from $z = 0$ out to $z = ?$, we must look up 45%, or in decimal form, .4500. The closest value is

.4505, which gives us $z = -1.65$. (Note: since .4500 falls midway between .4495 and .4505, we round up to .4505.)

   Answer: The z score associated with the lower 5% of the data is $z = -1.65$. In other words, 5% of the data lies below $z = -1.65$.

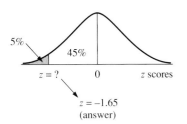

z = -1.65
(answer)

Normal curve table

z	.00	.01	.02	. . .	.05
1.6					.4505

g. This is a working backward problem (given the percentage of data, find z). Since the normal curve table reads from $z = 0$ out to $z = $?, we must look up 49% (half of 98%), or in decimal form .4900. The closest value is .4901, which gives us $z = 2.33$. Note we get $z = -2.33$ and $z = +2.33$.

   Answer: The z scores associated with the middle 98% of the data are $z = -2.33$ and $z = +2.33$.

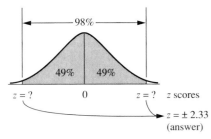

$z = \pm 2.33$
(answer)

Normal curve table

z	.00	.01	.02	.03	. . .
2.3				.4901	

4.3

a. We proceed by first using our formula to calculate the z score at the cutoff, then using our normal curve table to determine the percentage of data, as follows:

$$z = \frac{5'2'' - 5'5''}{2''} = \frac{-3''}{2''} = -1.50$$

The percentage of data from $z = 0$ to $z = -1.50$ is 43.32%, however we want the percentage of data below $z = -1.50$, so we subtract from 50% ($50\% - 43.32\% = 6.68\%$).

   Answer: 6.68% of the female students are under $5'2''$.

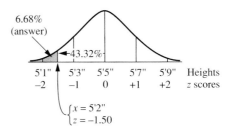

$\begin{cases} x = 5'2'' \\ z = -1.50 \end{cases}$

b. Calculate z scores at the cutoffs as follows:

$$\text{At } 5'2\tfrac{1}{2}'': z = \frac{5'2\tfrac{1}{2}'' - 5'5''}{2''} = \frac{-2\tfrac{1}{2}''}{2''} = -1.25$$

$$\text{At } 5'8'': z = \frac{5'8'' - 5'5''}{2''} = \frac{3''}{2''} = 1.50$$

The percentage of data from $z = 0$ to $z = -1.25$ is 39.44%; from $z = 0$ to $z = 1.50$ is 43.32%. Adding we get

$$39.44\% + 43.32\% = 82.76\%$$

Answer: 82.76% of the female students are between $5'2\tfrac{1}{2}''$ and $5'8''$.

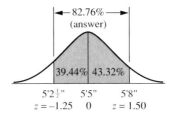

c. Calculate z scores at the cutoffs as follows:

At $5'8\frac{1}{2}''$: $z = \dfrac{5'8\frac{1}{2}'' - 5'5''}{2''} = \dfrac{3\frac{1}{2}''}{2''} = 1.75$

At $5'9\frac{1}{2}''$: $z = \dfrac{5'9\frac{1}{2}'' - 5'5''}{2''} = \dfrac{4\frac{1}{2}''}{2''} = 2.25$

The percentage of data from $z = 0$ to $z = 2.25$ is 48.78%. From this, we *subtract* the percentage of data from $z = 0$ to $z = 1.75$ (which is 45.99%).

$$48.78\% - 45.99\% = 2.79\%$$

Answer: 2.79% of the female students are $5'8\frac{1}{2}''$ to $5'9\frac{1}{2}''$.

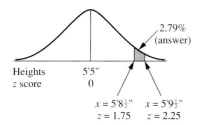

Heights   5'5"
z score   0

$x = 5'8\frac{1}{2}''$ $x = 5'9\frac{1}{2}''$
$z = 1.75$ $z = 2.25$

4.4

a. This is a working backward problem (given the percentage of data, find z). To find z at the cutoff, we look up 40%, or in decimal form, .4000. Remember: the table only reads from the center ($z = 0$) out. The closest value to .4000 is .3997, which gives $z = -1.28$.

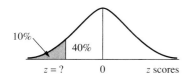

10%
40%

$z = ?$ 0 z scores

Normal curve table (on back cover)				
z	.00	.01	. . .	.08
0.0				
.				
.				
.				
1.2				.3997

Now we use $z = -1.28$ in our formula to solve for x at the cutoff. This requires some algebraic manipulation. First multiply both sides by 1.2; note this cancels the denominator of the fraction.

$$(1.2)(-1.28) = \dfrac{x - 21.0}{\cancel{1.2}} \cdot \dfrac{\cancel{1}}{\cancel{(1.2)}}$$

Now we get $(1.2)(-1.28) = x - 21.0$; refer to the calculations below for the remainder of the solution.

Answer: Below 19.46 minutes you would expect to find the fastest 10% of the memory times.

$$z = \dfrac{x - \mu}{\sigma}$$

$$-1.28 = \dfrac{x - 21.0}{1.2}$$

$$(1.2)(-1.28) = x - 21.0$$
$$-1.536 = x - 21.0$$
$$19.464 = x$$
$$\text{or } x = 19.46 \text{ minutes}$$

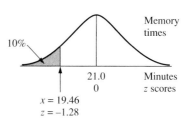

Memory times
10%

21.0   Minutes
0   z scores

$x = 19.46$
$z = -1.28$

b. This is a working backward problem (given the percentage of data, find z). To find z at the cutoff, we look up 25% (half of 50%), or in decimal form .2500. Remember: the table only reads from the center ($z = 0$) out.

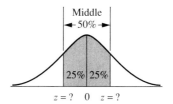

Middle
50%

25% 25%

$z = ?$ 0 $z = ?$

The closest value is .2486, which gives $z = .67$. Since it is symmetrical, we get $z = -.67$ and $z = +.67$.

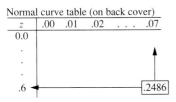

Normal curve table (on back cover)

z	.00	.01	.02	. . .	.07
0.0					
.					
.					
.					
.6					.2486

We substitute in $z = -.67$ and $z = +.67$ into our formula to solve for x at the cutoffs. (Note: to solve, multiply both sides by 1.2, then add 21.0 to both sides.)

$$z = \frac{x - \mu}{\sigma} \qquad z = \frac{x - \mu}{\sigma}$$

$$-.67 = \frac{x - 21.0}{1.2} \qquad +.67 = \frac{x - 21.0}{1.2}$$

$$x = 20.196 \qquad x = 21.804$$
$$x = 20.2 \text{ min} \qquad x = 21.8 \text{ min}$$

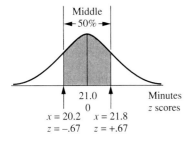

Answer: Between 20.2 minutes and 21.8 minutes you would expect to find the middle 50% of the memory times.

4.5

a. Here we are *randomly* selecting $n = 15$ members from a *large* two-category population. In practical terms, this defines binomial sampling. There is a fixed number of selections (15), each independent and each having the same probability a cured individual (a success) will be chosen, 60%.

In addition, the expected value (np) $= (15)(.60) = 9$, which is greater than 5, and $n(1 - p) = (15)(.40) = 6$, which is greater than 5. Thus, repeated samples of $n = 15$ will produce a sampling distribution that is approximately normally distributed and symmetrical around the expected value of 9, as follows:

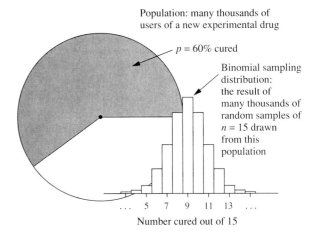

The mean and standard deviation of the histogram is

$$\mu = np \qquad\qquad \sigma = \sqrt{np(1 - p)}$$
$$= 15(.60) \qquad\quad = \sqrt{15(.60)(1 - .60)}$$
$$= 9 \qquad\qquad\quad = \sqrt{15(.60)(.40)}$$
$$\qquad\qquad\qquad\quad = 1.897$$
$$\qquad\qquad\qquad\quad = 1.90$$

Now a normal curve with these dimensions ($\mu = 9$, $\sigma = 1.90$) can be fitted over the histogram to estimate probabilities in any portion of the histogram.

To answer our question, what is the probability that out of $n = 15$ randomly selected individuals, we will find 7 or less cured, we proceed as follows:

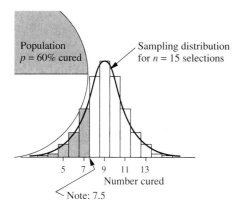

Note: 7.5

First, we shade the histogram bars representing 7 or less. Note the shading must extend to 7.5 to include the entire bar representing 7 cured. This half-unit adjustment is called your continuity correction factor. Now, we fit a normal curve over the histogram.

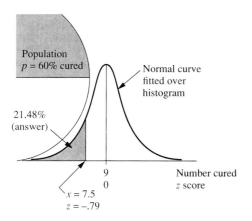

Second, we resketch the normal curve (for clarity) and shade the area 7.5 and below. Using $\mu = 9$ and $\sigma = 1.90$, we solve as we would any normal curve problem by first calculating the z score at the cutoff:

$$z = \frac{x - \mu}{\sigma} = \frac{7.5 - 9}{1.90} = -\frac{1.50}{1.90} = -.79$$

The percentage of data from $z = 0$ to $z = -.79$ is 28.52%. Subtract this from 50% to get 21.48%. Thus, 21.48% of the time we will achieve 7 or less cured when we randomly sample 15 from our population, and this can be visually represented as follows:

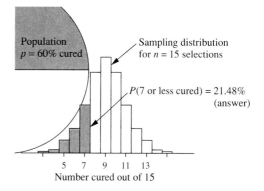

Number cured out of 15

b. This is a difficult question to answer. Generally, experiments in the hard sciences, such as in medicine, chemistry, physics, or biology, "past" results have been proven to be reasonably good predictors of "future" results, provided of course the conditions of the experiment are precisely duplicated. In the case of our medical experiment that would mean the drug was administered in an identical manner under identical conditions—and perhaps even to a similar set of subjects.

However, studies in other fields, such as in the soft or nonsciences, sociology, psychology, education, marketing, or communications, where we often measure attitudes or feelings or other human behavioral patterns, for instance, the desire to buy merchandise, experimental conditions are not so easy to control and thus results may or may not be duplicated from study to study, depending to a large extent on the researcher's ability to block out extraneous factors. For instance, in an educational study on the impact of some new instructional material, it would be difficult to eliminate the effects of a teacher's personality, a student's cultural bias, or parental cooperation from study to study, and these factors can impact the results significantly, regardless of the quality of the new instructional material. In other words, experiments in the soft and nonsciences, ''past'' results, in general, have not proven to be as good a predictor of ''future'' results (as in the hard sciences), mostly because the extraneous conditions of the experiment have not been able to be adequately controlled.

But wait. We don't have an experiment in the soft sciences here. Medicine is a hard science and therefore our past results should be a good predictor of future results, right? Even with experiments in the hard sciences, we cannot blindly accept past results without some follow-up assurances.

Say for instance in our medical study we wanted to administer the drug to a different cultural group, say the Negritos population of the Philippines, the native pygmy inhabitants of the Islands. Would you now expect a 60% cure rate? This is difficult to answer without some initial testing of the drug on the new patients and even then the answer may not be so cut and dried.

Say we randomly select $n = 15$ Negritos patients, administer the drug under identical conditions (as in past studies), only now with this group we get only 3 or 4 cured? What would you conclude? Notice the histogram from our prior sampling experiment, selecting $n = 15$ from a population of $p = 60\%$ cured.

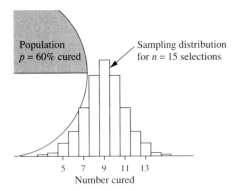

Population $p = 60\%$ cured

Sampling distribution for $n = 15$ selections

5   7   9   11   13
Number cured

Look at the histogram a moment. Even without doing any calculations, we can tell that the probability of getting 4 or less cured would be a rare event.

If you had to make a decision based on this one sample, what would you conclude? Most researchers would say: to get 4 or less cured out of 15 is such a rare event that we might suspect for this particular cultural population, for whatever reason, genetic differences, dietary or environmental factors or whatever, the cure rate for the new drug is *not* 60%, but probably less.

By the way, these are the types of decisions we will often be addressing in the remainder of the text.

Conclusion: applying past statistics to future studies takes a good deal of experience and judgment. In the hard sciences, experience has shown that results from well-conducted past studies are often good indicators of future results, of course provided experimental conditions are precisely duplicated. However, one should not blindly accept these past statistics, even in the hard sciences, without some further assurances. In the soft or nonsciences, past results must be applied with extreme caution since the multitude of variables that might affect results are often difficult to control, and thus results from these experiments may vary considerably from study to study.

4.7
a. Simply look up $z = .38$; 14.80%
b. $42.65\% + 50\%$; 92.65%
c. $50\% - 42.65\%$; 7.35%
d. $47.26\% - 27.94\%$; 19.32%
e. $9.87\% + 48.84\%$; 58.71%
f. $49.81\% - 44.84\%$; 4.97%
g. Look up 15% to get $z = .39$ (z could also equal $-.39$ if 15% was below μ); $z = .39$ or $-.39$
h. Look up 23.57% to get $z = -.63$. Thus, 73.57% of the data ($50\% + 23.57\%$) is above $z = -.63$
i. Middle 95% ($\pm47.5\%$ around μ); $z = \pm1.96$. Thus, middle 95% is between $z = -1.96$ and $z = +1.96$.

4.9

a. $z = -1.76$; $50\% - 46.08\%$; 3.92%

b. $z = .74$; $50\% + 27.04\%$; 77.04%

c. $z = -1.03$, $z = +.88$; $34.85\% + 31.06\%$; 65.91%

d. $z = 1.18$, $z = 1.76$; $46.08\% - 38.10\%$; 7.98%

e. Look up 35% to get $z = 1.04$; use z formula to solve for x: 10.91 cm

f. Middle 99% ($\pm 49\frac{1}{2}\%$ around μ); $z = \pm 2.58$; use z formula to solve for x's; 8.45 cm and 11.95 cm

4.11

a. $z = 1.45$; $50\% - 42.65\%$; 7.35%

b. $z = .27$, $z = 1.00$; $34.13\% - 10.64\%$; 23.49%

c. $z = .09$; $50\% + 3.59\%$; 53.59%

d. For the heaviest 65%, look up 15% (remember, the table reads from the center out); $z = -.39$; use z formula to solve for x: 6.47 oz

e. Middle 84% ($\pm 42\%$ around μ); $z = \pm 1.41$; use z formula to solve for x's; 5.35 oz and 8.45 oz

4.13

a. Since the population consists of 11% accepted, we would expect approximately 11% accepted in any random sample (11% of 60 = 6.6) or using the formula, $\mu = np = 60(.11) = 6.6$

b. $\mu = 6.6$; $\sigma = 2.424$; shade 8.5 and less; $z = .78$; $50\% + 28.23\%$; 78.23%

c. $\mu = 6.6$; $\sigma = 2.424$; shade 4.5 to 7.5; $z = -.87$, $z = .37$; $30.78\% + 14.43\%$; 45.21%

d. $\mu = 6.6$; $\sigma = 2.424$; shade 5.5 to 6.5; $z = -.45$, $z = -.04$; $17.36\% - 1.60\%$; 15.76%

4.15

a. Since the population consists of 75% with low wages, we would expect approximately 75% with low wages in any random sample (75% of 30 = 22.5) or, using the formula, $\mu = np = 30(.75) = 22.5$

b. $\mu = 22.5$; $\sigma = 2.372$; shade 19.5 and above; $z = -1.26$; $50\% + 39.62\%$; 89.62%

c. $\mu = 22.5$; $\sigma = 2.372$; shade 23.5 to 27.5; $z = .42$, $z = 2.11$; $48.26\% - 16.28\%$; 31.98%

4.17

a. Expected value, $\mu = np = 100(.23) = 23$; $\sigma = 4.208$; shade 19.5 to 25.5; $z = -.83$, $z = +.59$; $29.67\% + 22.24\%$; 51.91%

b. $\mu = 23$; $\sigma = 4.208$; shade 18.5 and less; $z = -1.07$; $50\% - 35.77\%$; 14.23%

c. Without knowledge of the details of the study (on which the $p = 23\%$ figure was based), it is difficult to answer this question. *If* the study was properly conducted, adhering to the conditions of internal and external validity and random selection of a sufficient size from the entire mouse population, we certainly would feel more comfortable accepting the $p = 23\%$ figure for all mice. However, often studies of this nature are rather limited. Caged mice are sometimes used, perhaps born in captivity and raised under artificial circumstances. Sometimes the conditions for the experiment can also be quite artificial. Generally, a number of independent studies, under varying conditions, are necessary before we can consider accepting research results of this nature for all mice.

Chapter Five

5.1

a. This is a typical "working backward (given the percentage of data, find z)" problem for the normal curve, only now we are dealing with the normal curve of the $\bar{x}$ distribution and must first calculate $\sigma_{\bar{x}}$, the standard deviation of the $\bar{x}$ distribution.

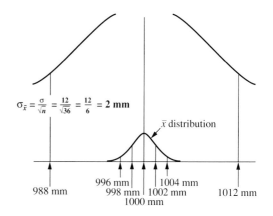

$$\sigma_{\bar{x}} = \frac{\sigma}{\sqrt{n}} = \frac{12}{\sqrt{36}} = \frac{12}{6} = 2 \text{ mm}$$

Since we know the percentage of data between the cutoffs is 95%, we merely look in the table of normal curve areas for the corresponding z scores. Remember: the table reads "half" the normal curve, so we must look up an area of 47.5% ($\frac{1}{2}$ of 95%), which in decimal form is .4750.

According to the table, the corresponding z scores are -1.96 and $+1.96$.

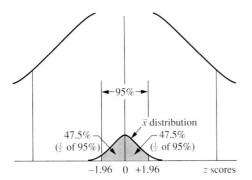

Substituting the z scores of -1.96 and $+1.96$ into our formula, we solve for the $\bar{x}$ at the cutoffs. Again, the z formula is coded with the symbol $\bar{x}$, which reminds us we are working in the $\bar{x}$ distribution.

$$z = \frac{\bar{x} - \mu}{\sigma_{\bar{x}}} \qquad z = \frac{\bar{x} - \mu}{\sigma_{\bar{x}}}$$

$$-1.96 = \frac{\bar{x} - 1000}{2} \qquad +1.96 = \frac{\bar{x} - 1000}{2}$$

Solving for $\bar{x}$:   Solving for $\bar{x}$:
$\quad \bar{x} = 996.08$ mm $\quad \bar{x} = 1003.92$ mm

The completed solution would appear graphically as follows:

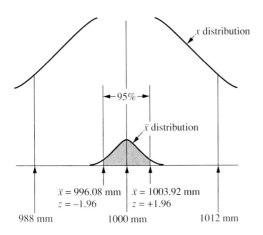

Answer: If we continually take random samples, with 36 pieces of material in each sample, and calculate the average length ($\bar{x}$) for each of these samples, then 95% of all the sample averages ($\bar{x}$'s) would be expected to fall between $\bar{x} = 996.08$ mm and $\bar{x} = 1003.92$ mm.

Since 95% is such an overwhelmingly high percentage of the $\bar{x}$'s, if one day we turned on the machine, took a random sample of 36 pieces and calculated a sample average ($\bar{x}$) that fell *outside* this 996.08 mm to 1003.92 mm range, would you suspect a malfunction in the machine? The answer is, yes. 95% of all the $\bar{x}$'s should fall between 996.08 mm and 1003.92 mm. Although with a properly functioning machine 5% of all the $\bar{x}$'s would fall *outside* this range, this 5% (1 out of 20) signifies quite a rare occurrence. Therefore, if you take a sample and obtain an $\bar{x}$ outside this range, the possibility of malfunction should be explored.

b. For $n = 144$, 95% of the $\bar{x}$'s would fall between $\bar{x} = 998.04$ and $\bar{x} = 1001.96$.

c. For $n = 576$, 95% of the $\bar{x}$'s would fall between $\bar{x} = 999.02$ and $\bar{x} = 1000.98$.

5.2

a. Formula calculations: $\sigma_{\bar{x}} = \dfrac{12}{\sqrt{36}} = 2$ mm;

$$\pm 2.58 = \frac{\bar{x} - 1000}{2}$$

Answer: If we continually take random samples, with 36 pieces of material in each sample, and calculate the average length ($\bar{x}$) for each of these samples, then 99% of all the sample averages ($\bar{x}$'s) would be expected to fall between $\bar{x} = 994.84$ mm and $\bar{x} = 1005.16$ mm.

Like the value 95%, industry and research often use this value of 99% to establish the critical cutoff lines to determine whether or not an operation may be malfunctioning.

b. For $n = 144$, 99% of the $\bar{x}$'s would fall between $\bar{x} = 997.42$ and $\bar{x} = 1002.58$.

c. For $n = 576$, 99% of the $\bar{x}$'s would fall between $\bar{x} = 998.71$ and $\bar{x} = 1001.29$.

5.3

Again, we do this problem as we would do any normal curve problem, however since we are dealing with the normal curve of the $\bar{x}$ distribution we must first calculate $\sigma_{\bar{x}}$, the standard deviation of the $\bar{x}$ distribution.

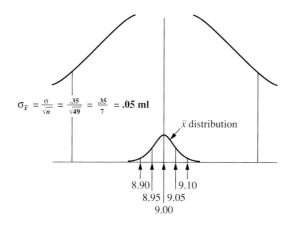

$$\sigma_{\bar{x}} = \frac{\sigma}{\sqrt{n}} = \frac{.35}{\sqrt{49}} = \frac{.35}{7} = .05 \text{ ml}$$

Calculating the z scores at the cutoffs of $\bar{x} = 8.92$ ml and $\bar{x} = 9.08$ ml:

$$z = \frac{\bar{x} - \mu}{\sigma_{\bar{x}}}$$
$$z = \frac{8.92 - 9.00}{.05}$$
$$z = -1.60$$

$$z = \frac{\bar{x} - \mu}{\sigma_{\bar{x}}}$$
$$z = \frac{9.08 - 9.00}{.05}$$
$$z = +1.60$$

Looking up the percentage of data between $z = 0$ and $z = -1.60$, we get 44.52%. To get the percentage of data *below* 8.92 ml, subtract this from 50%, which gives us 5.48%. Likewise, the data *above* 9.08 ml is 5.48%.

Graphically, the solution would appear as shown below:

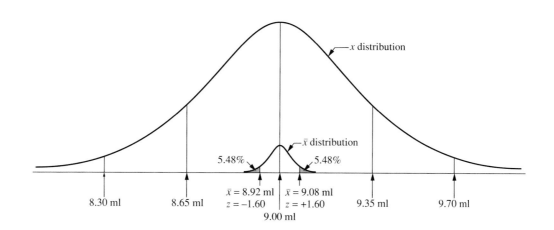

$$5.48\% + 5.48\% = 10.96\%$$

Answer: If we continually take random samples of 49 test tubes each, and calculate the average fill ($\bar{x}$) in each of these samples, then 10.96% (5.48% + 5.48%) of all the sample averages ($\bar{x}$'s) would be expected to fall either *below* $\bar{x}$ = 8.92 ml or *above* $\bar{x}$ = 9.08 ml.

If one day we set the process to fill at 9.00 ml, then randomly sample 49 test tubes and calculate an $\bar{x}$ either below 8.92 ml *or* above 9.08 ml, would this give us cause to suspect the process is malfunctioning? This is a difficult question to answer, since 10.96% or approximately 11 out of every 100 (11%) samples drawn from a properly operating process would be expected to fall either below 8.92 ml *or* above 9.08 ml. Generally, statisticians feel 11 out of 100 is too "common" an occurrence to suspect a process is malfunctioning, especially if the $\bar{x}$ is just barely below 8.92 ml or just barely above 9.08 ml. If the $\bar{x}$ is "close to" 8.92 ml or 9.08 ml, most likely the filling process would be allowed to continue. This begs the question: at what value of $\bar{x}$ do we begin to suspect a malfunction?

5.4

Again, we are dealing with the *average* market share in 64 stores and not the market share in an individual store. Thus, we use the $\bar{x}$ distribution. First, calculate $\sigma_{\bar{x}}$.

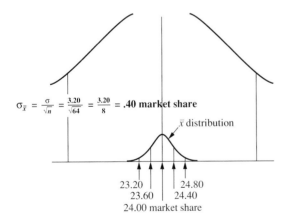

$$\sigma_{\bar{x}} = \frac{\sigma}{\sqrt{n}} = \frac{3.20}{\sqrt{64}} = \frac{3.20}{8} = .40 \text{ market share}$$

$\bar{x}$ distribution

23.20   24.80
23.60   24.40
24.00 market share

This is a typical "working backward (given the percentage of data, find z)" problem. We are given 1% and told this area is located at the *lowest* end. However, since our normal curve table reads from the center of the normal curve ($z = 0$) out to the z at the cutoff, we must stop and think. What is the percentage of data from the "center" to the cutoff? Since all the data left of center is 50%, the data to the cutoff must be 49% (50% minus 1%).

Looking up 49%, which in decimal form is .4900, we get a corresponding z score of -2.33.

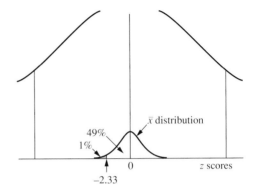

Substituting -2.33 into our formula and solving for $\bar{x}$ at the cutoff:

$$z = \frac{\bar{x} - \mu}{\sigma_{\bar{x}}}$$

$$-2.33 = \frac{\bar{x} - 24.00}{.40}$$

Solving for $\bar{x}$:

$$\bar{x} = 23.07 \text{ at the cutoff}$$

Graphically, the completed solution would appear as follows:

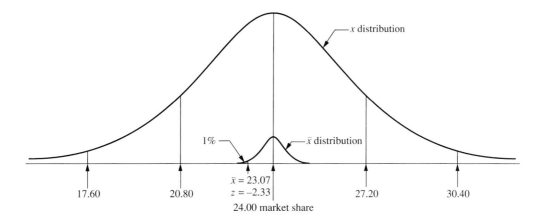

Answer: If we continually take random samples of 64 stores each, and calculate the average market share ($\bar{x}$) in each of these samples, then 1% of the $\bar{x}$'s would be expected to fall below a market share of 23.07.

5.5

a. Formula calculations: $\sigma_{\bar{x}} = \dfrac{4}{\sqrt{100}} = .4$;

$$z = \frac{31.5 - 32.0}{.4} = -1.25 \ (39.44\% \text{ Area})$$

$$z = \frac{32.5 - 32.0}{.4} = +1.25 \ (39.44\% \text{ Area})$$

Answer: If we continually take random samples, with 100 customers in each sample, and calculate the average age ($\bar{x}$) for each of these samples, then 78.88% of all the sample averages ($\bar{x}$'s) will fall between $\bar{x} = 31.5$ years old and $\bar{x} = 32.5$ years old.

b. Probability = 78.88%, the same as the answer in part a. If 78.88% of all the $\bar{x}$'s will fall between 31.5 and 32.5, then the probability of choosing one $\bar{x}$ and this $\bar{x}$ falling between these values is also 78.88%.

Remember: Probability = Area or Percentage of Data Under the Curve

c. Formula calculations: $\sigma_{\bar{x}} = \dfrac{4}{\sqrt{100}} = .4$;

$$-1.96 = \frac{\bar{x} - 32.0}{.4}$$

$$+1.96 = \frac{\bar{x} - 32.0}{.4}$$

Answer: If we continually take random samples, with 100 customers in each sample, and calculate the average age ($\bar{x}$) for each of these samples, then 95% of all the sample averages ($\bar{x}$'s) will fall between $\bar{x} = 31.22$ years old and $\bar{x} = 32.78$ years old.

5.7

a. For an "individual" account x, use $\sigma = 75.90$; $z = \pm.16$; 6.36% + 6.36%; 12.72%

b. For a sample "average" $\bar{x}$, use $\sigma_{\bar{x}} = 75.90/\sqrt{40}$ = 12; $z = \pm1.00$; 34.13% + 34.13%; 68.26%

c. $\sigma_{\bar{x}} = 13.02$; $z = \pm20/13.02 = \pm1.54$; 43.82% + 43.82%; 87.64%

5.9

a. Middle 90% ($\pm45\%$ around μ); $z = \pm1.65$. For an "individual" SAT score, use $\sigma = 96$ in z formula to solve for x: 90% of SAT verbal scores are between 271.6 and 588.4.

b. Middle 90% ($\pm45\%$ around μ); $z = \pm1.65$; for a sample "average," use $\sigma_{\bar{x}} = 96/\sqrt{144} = 8$ in z formula to solve for $\bar{x}$: 90% of $\bar{x}$'s will fall between 416.8 and 443.2.

c. $\sigma_{\bar{x}} = 14.813$; $z = \pm1.65$; use z formula to solve for $\bar{x}$: 90% of $\bar{x}$'s are expected to fall between 405.6 and 454.4.

5.11

a. For an "individual" doctor x, use $\sigma = 6.8$; $z = -.25$; 50% − 9.87%; 40.13%.

b. For a sample "average" $\bar{x}$, use $\sigma_{\bar{x}} = 6.8/\sqrt{55}$ = .917; $z = -1.85$; 50% − 46.78%; 3.22%.

c. $\sigma_{\bar{x}} = .917$; $z = -1.85$, $z = 1.42$; 46.78% + 42.22%; 89.00%.

Chapter Six

6.1

a. The probability of a Type I error is simply 2% (100% − 98%). Written in statistical terms, you would state α = .02. In other words, on a properly operating machine, 2% of the $\bar{x}$'s would fall *outside* your acceptance range for μ = 1000 mm, as shown:

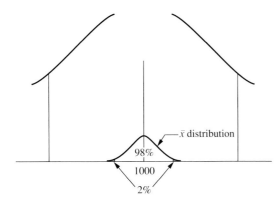

b. This is a typical "working backward (given the percentage of data, find z)" problem for the normal curve, only now we are dealing with the normal curve of the $\bar{x}$ distribution. Since we know the percentage of data between the cutoffs is 98% we merely look up the corresponding z scores. Doing this, we get $z = -2.33$ and $z = +2.33$. (Remember the table reads half the normal curve, so we must look up 49% ($\frac{1}{2}$ of 98%) or in decimal form .4900.)

Using $z = -2.33$ and $z = +2.33$, we calculate the values at the cutoffs with the resultant solution:

$$\sigma_{\bar{x}} = \frac{12}{\sqrt{36}} = 2$$

$$z = \frac{\bar{x} - \mu}{\sigma_{\bar{x}}} \qquad\qquad z = \frac{\bar{x} - \mu}{\sigma_{\bar{x}}}$$

$$-2.33 = \frac{\bar{x} - 1000}{2} \qquad +2.33 = \frac{\bar{x} - 1000}{2}$$

$$\bar{x} = 995.34 \text{ mm} \qquad \bar{x} = 1004.66 \text{ mm}$$

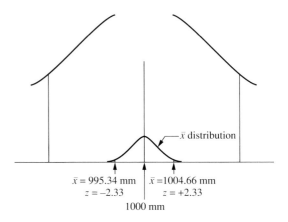

So, you would accept the machine as cutting properly, cutting at μ = 1000 mm, if you obtained a sample average between $\bar{x}$ = 995.34 mm and $\bar{x}$ = 1004.66 mm.

c. The following might be a Type I error scenario: You randomly sample 36 pieces and your sample average falls *outside* this 995.34 mm to 1004.66 mm range, say for instance you obtain a sample average of $\bar{x}$ = 994 mm. Based on this you reject the machine as operating properly and shut down production.

If indeed the machine is *not* operating properly, not cutting on average to μ = 1000 mm, then you have made no error. However, if the machine is okay, cutting properly at μ = 1000 mm, then you have made a Type I error.

6.2

a. If a machine is indeed cutting at μ = 995 mm, the sample averages ($\bar{x}$'s) now would cluster about μ = 995 mm, as follows:

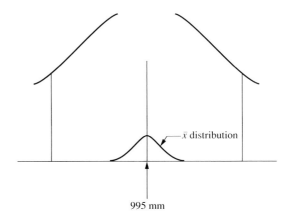

You would commit a Type II error if you accepted (in error) the machine cutting at $\mu = 1000$ mm. The only way you would accept a machine cutting at $\mu = 1000$ mm is if you took a random sample of 36 pieces and your sample average ($\bar{x}$) fell between 995.34 and 1004.66. So, to calculate β, the probability of a Type II error, you must calculate the percentage of $\bar{x}$'s that would fall between 995.34 and 1004.66 from a machine cutting at $\mu = 995$ mm.

Remember: the machine is cutting at $\mu = 995$ mm but you are unaware of this. The only information available to you is your one sample average, $\bar{x}$.

To find the probability of a Type II error in this problem, we calculate the percentage of $\bar{x}$'s that we would expect to fall between 995.34 and 1004.66 represented by the shaded region in the diagram.

$$\sigma_{\bar{x}} = \frac{\sigma}{\sqrt{n}} = \frac{12}{\sqrt{36}} = 2 \text{ mm}$$

$$z = \frac{\bar{x} - \mu}{\sigma_{\bar{x}}} \qquad z = \frac{\bar{x} - \mu}{\sigma_{\bar{x}}}$$

$$z = \frac{995.34 - 995}{2} \qquad z = \frac{1004.66 - 995}{2}$$

$$z = +.17 \qquad z = +4.83$$

(disregard since negligible data exists in this region)

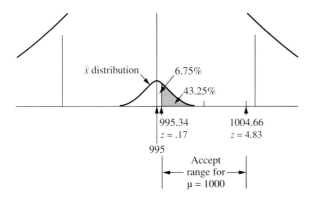

Looking up $z = +.17$, we get 6.75%, thus our Type II error risk is 43.25%. Since 43.25% (50.00% − 6.75%) of the $\bar{x}$'s coming from a machine cutting at $\mu = 995$ mm would be expected to fall between 995.34 mm and 1004.66 mm, this 43.25% is the risk you will obtain one of these $\bar{x}$'s and thus conclude (erroneously) the machine was cutting at $\mu = 1000$

mm. So, your risk of a Type II error when μ shifts to 995 mm is 43.25%. Written in decimal form (.4325) and rounded to two decimal places, you would say:

$\beta = .43$ which is the probability you will accept your initial assumption, $\mu = 1000$ mm, in error. This is your Type II error risk for a shift to $\mu = 995$ mm.

b. The following is a comparison of Type I and Type II error risks for basically the same cutting machine problem, but using different ranges for accepting $\mu = 1000$ mm.

From second demonstration example in section 6.2	From solutions 6.1a and 6.2a
Accept range for $\mu = 1000$: $\bar{x} = 997$ mm to 1003 mm	Accept range for $\mu = 1000$: $\bar{x} = 995.34$ mm to 1004.66 mm
Type I error risk: 13.36%	Type I error risk: 2%
Type II error risk: 15.87%	Type II error risk: 43.25%

Using a different acceptance range can change the risk for a Type I error. In this case, by "opening up" the accept range, we lowered our Type I risk from 13.36% to 2%. However, notice the Type II error risk increased from 15.87% to 43.25%. Imposing a lower α, merely increases β, which is the principle demonstrated.

6.3

a. A 10% Type I error risk implies you are using the middle 90% of the $\bar{x}$'s as your cutoffs for accepting $\mu = 1000$ mm, as follows:

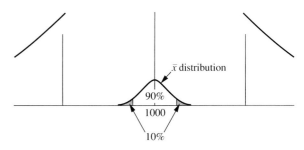

This is a typical "working backward (given the percentage of data, find z)" problem for the normal curve, only now we are dealing with the normal curve of the $\bar{x}$ distribution. Since we know the percentage of data between the cutoffs is 90%, we merely look up the corresponding z scores. Doing this, we get $z = -1.65$ and $z = +1.65$. (Remember the table reads half the normal curve, so we must look up 45% ($\frac{1}{2}$ of 90%) or in decimal form .4500.)

Using $z = -1.65$ and $z = +1.65$, we calculate the cutoff values as follows:

$$\sigma_{\bar{x}} = \frac{12}{\sqrt{36}} = 2$$

$$z = \frac{\bar{x} - \mu}{\sigma_{\bar{x}}} \qquad z = \frac{\bar{x} - \mu}{\sigma_{\bar{x}}}$$

$$-1.65 = \frac{\bar{x} - 1000}{2} \qquad +1.65 = \frac{\bar{x} - 1000}{2}$$

$$\bar{x} = 996.70 \text{ mm} \qquad \bar{x} = 1003.30 \text{ mm}$$

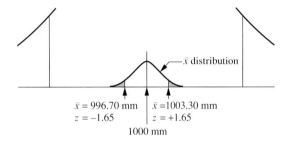

So, you would accept the machine as cutting properly, cutting at $\mu = 1000$ mm, if you obtained a sample average between $\bar{x} = 996.70$ mm and $\bar{x} = 1003.30$ mm.

b. If a machine is indeed cutting at $\mu = 1004$ mm, the sample averages ($\bar{x}$'s) now would cluster about $\mu = 1004$ mm, as follows:

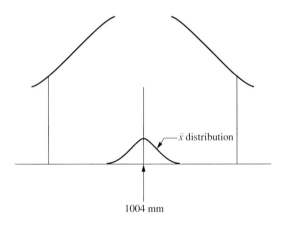

You would commit a Type II error if you accepted (in error) the machine cutting at $\mu = 1000$ mm. The only way you would accept a machine cutting at $\mu = 1000$ mm is if you took a random sample of 36 pieces and your sample average ($\bar{x}$) fell between 996.70 mm and 1003.30 mm. So, to calculate β, the probability of a Type II error, you must calculate the percentage of $\bar{x}$'s that would fall between 996.70 and 1003.30 from a machine cutting at $\mu = 1004$ mm.

Remember: the machine is cutting at $\mu = 1004$ mm but you are unaware of this. The only information available to you is your one sample average, $\bar{x}$.

To find the probability of a Type II error in this problem, we calculate the percentage of $\bar{x}$'s that we would expect to fall between 996.70 mm and 1003.30 mm represented by the shaded region in the diagram.

$$\sigma_{\bar{x}} = \frac{\sigma}{\sqrt{n}} = \frac{12}{\sqrt{36}} = 2 \text{ mm}$$

$$z = \frac{\bar{x} - \mu}{\sigma_{\bar{x}}} \qquad z = \frac{\bar{x} - \mu}{\sigma_{\bar{x}}}$$

$$z = \frac{996.70 - 1004}{2} \qquad z = \frac{1003.30 - 1004}{2}$$

$$z = -3.65 \qquad z = -.35$$

(disregard since negligible data exists in this region)

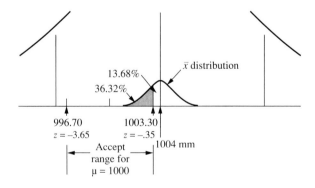

Looking up $z = -.35$, we get 13.68%, thus our Type II error risk is 36.32%. Since 36.32% (50.00% minus 13.68%) of the $\bar{x}$'s coming from a machine cutting at

$\mu = 1004$ mm would be expected to fall between 996.70 mm and 1003.30 mm, this is the risk you will obtain one of these $\bar{x}$'s and thus conclude (erroneously) the machine was cutting at $\mu = 1000$ mm. So, your risk of a Type II error when μ shifts to $\mu = 1004$ mm is 36.32%. Written in decimal form (.3632) and rounded to two decimal places, you would say:

$\beta = .36$ which is the probability you will accept your initial assumption, $\mu = 1000$ mm, in error. This is your Type II error risk for a shift to $\mu = 1004$ mm.

c. Power is the probability of making a correct decision by avoiding a Type II error. For this test, Power $= 63.68\%$ $(100\% - 36.32\%)$. Written in decimal form,

$$\text{Power} = .64 \qquad \text{Note: Power} = 1 - \beta$$
$$= 1.00 - .36$$
$$= .64$$

Explained another way, since 63.68% of the $\bar{x}$'s (50% + 13.68%) will be higher than 1003.30 mm as shown on the previous diagram, you have a 63.68% chance your sample average will be greater than 1003.30 mm. In that case, you would reject $\mu = 1000$ mm, which would be the correct decision. In other words, we have a 63.68% chance of making the correct decision (rejecting $\mu = 1000$ mm) and a 36.32% chance of making the wrong decision (accepting $\mu = 1000$ mm). Power is the probability of making the correct decision in this situation, and a Type II error is the probability of making the wrong decision.

6.4

a. The percentage of $\bar{x}$'s on a properly operating process that fall *outside* the 8.90 ml to 9.10 ml range is your Type I error risk. It is represented by the shaded regions in the following diagram. Calculating the percentage of data in the shaded regions we get:

$$\sigma_{\bar{x}} = \frac{\sigma}{\sqrt{n}} = \frac{.35}{\sqrt{49}} = .05$$

$$z = \frac{\bar{x} - \mu}{\sigma_{\bar{x}}} \qquad\qquad z = \frac{\bar{x} - \mu}{\sigma_{\bar{x}}}$$

$$z = \frac{8.90 - 9.00}{.05} \qquad z = \frac{9.10 - 9.00}{.05}$$

$$z = -2.00 \qquad\qquad z = +2.00$$

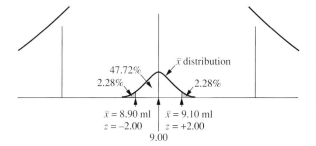

47.72%
2.28%
$\bar{x}$ distribution
2.28%

$\bar{x} = 8.90$ ml $\bar{x} = 9.10$ ml
$z = -2.00$ $z = +2.00$
9.00

The probability of a Type I error is 4.56% (2.28% + 2.28%). In other words, the percentage of $\bar{x}$'s you would expect to be *outside* the 8.90 ml to 9.10 ml range is 4.56%. Written in decimal form (.0456) and rounded to three decimal places you would say:

$$\alpha = .046$$

b. A Type I error is defined as rejecting your initial assumption in error, in this case, rejecting the process filling at $\mu = 9.00$ ml when the process is filling perfectly okay. This would occur if we randomly sampled 49 test tubes from a process filling at $\mu = 9.00$ ml, but we obtained a sample average ($\bar{x}$) *outside* the $\bar{x} = 8.90$ ml to $\bar{x} = 9.10$ ml acceptance range for $\mu = 9.00$ ml. When using this acceptance range, this will occur 4.56% of the time with a process that is filling properly.

The consequences would be to stop the filling process to search for a malfunction that doesn't exist. This is costly and time consuming and because we are dealing with polio and AIDS viruses, perhaps even dangerous.

c. The test tubes are now filling on average at $\mu = 9.14$ ml, so your sample averages ($\bar{x}$'s) would now cluster about 9.14 ml, as follows:

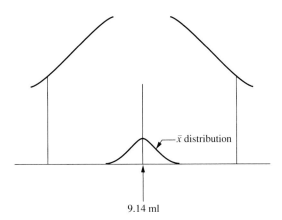

$\bar{x}$ distribution

9.14 ml

You would commit a Type II error if you accepted the process filling at $\mu = 9.00$ ml. The only way you would accept the process filling at $\mu = 9.00$ ml is if you took a random sample of 49 test tubes and your average ($\bar{x}$) fell between 8.90 ml and 9.10 ml. So, to calculate β, the probability of a Type II error, you must calculate the percentage of $\bar{x}$'s that would fall between 8.90 ml and 9.10 ml from a process filling at $\mu = 9.14$ ml.

Remember: the process is filling at $\mu = 9.14$ ml but you are unaware of this. The only information available to you is your one sample average, $\bar{x}$.

To find the probability of a Type II error in this problem, we calculate the percentage of $\bar{x}$'s that we would expect to fall between 8.90 ml and 9.10 ml, represented by the shaded region in the diagram that follows:

$$\sigma_{\bar{x}} = \frac{\sigma}{\sqrt{n}} = \frac{.35}{\sqrt{49}} = .05 \text{ ml}$$

$$z = \frac{\bar{x} - \mu}{\sigma_{\bar{x}}} \qquad z = \frac{\bar{x} - \mu}{\sigma_{\bar{x}}}$$

$$z = \frac{8.90 - 9.14}{.05} \qquad z = \frac{9.10 - 9.14}{.05}$$

$$z = -4.80 \qquad z = -.80$$

(disregard since negligible
data exists in this region)

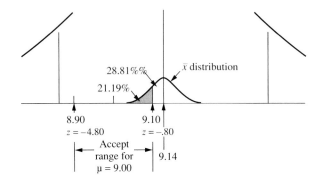

28.81%%

21.19%

$\bar{x}$ distribution

8.90
$z = -4.80$

9.10
$z = -.80$

Accept
range for
$\mu = 9.00$

9.14

Looking up $z = -.80$ we get 28.81%, thus our Type II error risk is 21.19%. Since 21.19% (50.00% minus 28.81%) of the $\bar{x}$'s coming from a process filling at $\mu = 9.14$ ml would be expected to fall between 8.90 ml and 9.10 ml, this 21.19% is the risk you will obtain one of these $\bar{x}$'s and thus conclude (erroneously) the process was filling at $\mu = 9.00$ ml. So, your risk of a Type II error when μ shifts to 9.14 ml is 21.19%. Written in decimal form (.2119) and rounded to two decimal places, you would say:

$\beta = .21$ which is the probability you will accept your initial assumption, $\mu = 9.00$ ml, in error; this is the Type II error risk for a shift to $\mu = 9.14$ ml.

d. A Type II error is defined as accepting your initial assumption in error, in this case, if we accept the process filling at $\mu = 9.00$ ml when the process is malfunctioning (not filling at $\mu = 9.00$ ml), we have committed a Type II error. This would occur if we randomly sample 49 test tubes from a malfunctioning process and this sample average ($\bar{x}$) falls *inside* the $\bar{x} = 8.90$ ml to 9.10 ml acceptance range for $\mu = 9.00$ ml. This will occur 21.19% of the time for a process malfunctioning, filling at $\mu = 9.14$ ml.

The consequences would be that we allow the process to continue filling at $\mu = 9.14$ ml. If this overfill will in some way cause subsequent problems, for instance, with the sealing of the test tubes or with spillage, then we would have consequences that could be quite serious.

6.5

a. This is simply a Type I error risk, rejecting $\mu = 9.00$ ml in error, which we calculated to be 4.56%.

b. This is a Type II error risk, accepting $\mu = 9.00$ ml in error. For a shift to $\mu = 9.14$ ml, we calculated this to be 21.19%.

6.7

a. For $n = 30$, $\sigma_{\bar{x}} = 12/\sqrt{30} = 2.191$; Type I error: using $\mu = 1000$, $z = \pm 1.37$; $50\% - 41.47\%$ $= 8.53\%$ in each tail; $8.53\% + 8.53\% = 17.06\%$. Type II error: using $\mu = 995$; $z = .91$ and $z = 3.65$ (disregard); $50\% - 31.86\% = 18.14\%$.

b. For $n = 100$, $\sigma_{\bar{x}} = 12/\sqrt{100} = 1.20$; Type I error: using $\mu = 1000$; $z = \pm 2.50$; $50\% - 49.38\% = .62\%$ in each tail; $.62\% + .62\% = 1.24\%$. Type II error: using $\mu = 995$, $z = 1.67$ and $z = 6.67$ (disregard); $50\% - 45.25\% = 4.75\%$.

c. If n increases from 30 to 100, the Type I error risk decreases from 17.06% to 1.24% and the Type II error risk decreases from 18.14% to 4.75%.

6.9

a. Opening the acceptance range *decreases* Type I error risk. When Type I error risk decreases, Type II error risk *increases*. In other words, as $\alpha \downarrow$, $\beta \uparrow$.

   Since power is essentially the probability of avoiding a Type II error, power will *decrease*. In other words as $\beta \uparrow$, power $\downarrow$ (remember, β + power $= 100\%$).

b. The preferred technique is to increase your sample size, n. This lowers both the Type I and Type II error risks for a fixed acceptance range. Since the Type II error risk decreases, power will increase. Remember: power is the ability of making the correct decision if μ shifts.

Chapter Seven

7.1

a. H_0: $\mu = .560$ mm   Calculation of $\sigma_{\bar{x}}$:
H_1: $\mu \neq .560$ mm
$\alpha = .01$ $\sigma_{\bar{x}} = \dfrac{\sigma}{\sqrt{n}} \approx \dfrac{s}{\sqrt{n}}$

$\sigma_{\bar{x}} \approx \dfrac{.030}{\sqrt{225}} \approx .002$

Assume H_0 true, use α to establish cutoffs. Our level of significance in this case is $\alpha = .01$ (1%), which in a two-tailed test implies we will accept the middle 99% of the $\bar{x}$'s as our boundary for accepting H_0 as true. We now look up the z scores corresponding to the middle 99% of the $\bar{x}$'s, which turn out to be $z = -2.58$ and $z = +2.58$. *Remember:* the normal curve table reads half the normal curve, starting from $z = 0$ out, so we must look up $\frac{1}{2}$ of 99% or $49\frac{1}{2}\%$, which in decimal form is .4950. Since .4950 falls

midway between $z = 2.57$ and $z = 2.58$, we round to the higher z score, 2.58.

   Substituting the z scores ± 2.58 into our z formula, we solve for the $\bar{x}$ at the cutoffs:

$$-2.58 = \frac{\bar{x} - .560}{.002} \qquad +2.58 = \frac{\bar{x} - .560}{.002}$$

cutoffs: $\bar{x} = .555$ mm $\bar{x} = .565$ mm

The completed solution would appear graphically as follows:

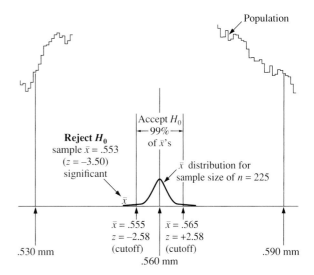

Note: the z score of the sample $\bar{x}$ (.553 mm) is calculated as follows:

$$z = \frac{.553 - .560}{.002}$$

$$z = -3.50 \text{ which is significant}$$

b. Yes, the results are significant. This sample ($\bar{x} = .553$ mm) provides evidence to indicate the average thickness of the fiber-optic thread in this shipment is significantly below specification. We conclude $\mu \neq .560$ mm for this shipment and, thus, reject the shipment.

7.2

a. H_0: $\mu \leq 140.0$ mg/day   Calculation of $\sigma_{\bar{x}}$:
H_1: $\mu > 140.0$ mg/day
$\alpha = .04$ $\sigma_{\bar{x}} = \dfrac{\sigma}{\sqrt{n}} \approx \dfrac{s}{\sqrt{n}}$

$\sigma_{\bar{x}} \approx \dfrac{48.2}{\sqrt{900}} \approx 1.607$

Assume H_0 true, use α to establish cutoffs. Assign the total α risk (.04 or 4%) to the *right* tail of the $\bar{x}$ distribution because we would reject H_0 only if our sample $\bar{x}$ falls significantly above 140.0 mg/day. To find the z score at the cutoff, we look up 46% in the normal curve table (50% $-$ 4%). Remember: the table reads from the center out. 46% in decimal form is .4600; the closest value is .4599, which is equivalent to $z = 1.75$.

Substituting the z score of $+1.75$ into our z formula, we solve for the $\bar{x}$ at the cutoff:

$$+1.75 = \frac{\bar{x} - 140}{1.607}$$

cutoff: $\bar{x} = 142.8$ mg/day

The completed solution might appear graphically as follows:

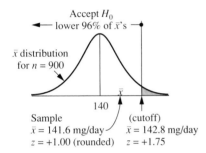

Accept H_0
lower 96% of $\bar{x}$'s

$\bar{x}$ distribution for $n = 900$

140

Sample
$\bar{x} = 141.6$ mg/day
$z = +1.00$ (rounded)

(cutoff)
$\bar{x} = 142.8$ mg/day
$z = +1.75$

Accept H_0
($\mu \leq 140.0$)

Note: the z score of the sample $\bar{x}$ (141.6) is calculated as follows:

$$z = \frac{141.6 - 140}{1.607}$$

$z = .996 = +1.00$ which is *not* significant

b. The results are *not* significant. This sample ($\bar{x} = 141.6$ mg/day) provides no evidence to refute the health organization's claim that the minimum effective dosage nationwide is on average 140.0 mg/day *or less;* therefore we accept H_0: $\mu \leq 140.0$ mg/day.

In reality, we also have the option to reserve judgment until more testing is done, but in either case, we cannot reject H_0.

7.3

a.

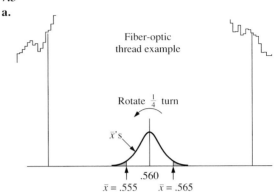

Fiber-optic thread example

Rotate $\frac{1}{4}$ turn

$\bar{x}$'s

.560

$\bar{x} = .555$ $\bar{x} = .565$

Cutoffs established, taken from exercise 7.1.
Rotate $\frac{1}{4}$ turn counterclockwise, extending cutoff lines to the right and shading rejection zone (as shown in the next diagram).

b.

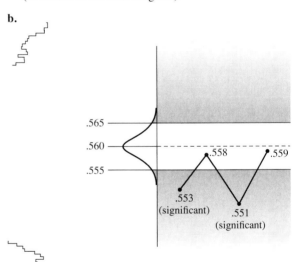

.565

.560   .558   .559

.555

.553
(significant)   .551
(significant)

$\bar{x} = .553$ mm $\bar{x} = .551$ mm
$\bar{x} = .558$ mm $\bar{x} = .559$ mm

Each sample $\bar{x}$ is plotted sequentially as the shipment comes in and is connected with a line (as shown above).

c.

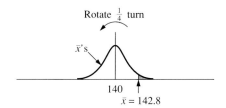

Elavil example

Rotate $\frac{1}{4}$ turn

$\bar{x}$'s

140

$\bar{x} = 142.8$

Cutoff established, taken from exercise 7.2.

Rotate $\frac{1}{4}$ turn counterclockwise, extending cutoff lines to the right, shading rejection zone (as shown in the following diagram).

d.

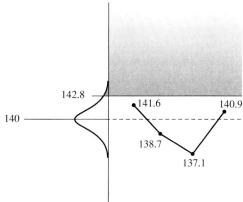

142.8

141.6   140.9

140

138.7

137.1

$\bar{x} = 141.6$ mg $\bar{x} = 137.1$ mg
$\bar{x} = 138.7$ mg $\bar{x} = 140.9$ mg

Each sample $\bar{x}$ is plotted sequentially as each result is obtained and connected with a line (as shown above).

7.4

a. H_0: $\mu = 100.0$ social IQ
H_1: $\mu \neq 100.0$ social IQ
$\alpha = .02$

Calculation of $\sigma_{\bar{x}}$:

$$\sigma_{\bar{x}} = \frac{\sigma}{\sqrt{n}} \approx \frac{s}{\sqrt{n}}$$

$$\sigma_{\bar{x}} \approx \frac{16.3}{\sqrt{12}} \approx 4.706$$

Small-sample conditions: mound-shaped population assured? (Yes); since s estimates σ, t scores must be used.

Assume H_0 true, use α to establish cutoffs. Our level of significance in this case is $\alpha = .02$ (2%), which in a two-tailed test implies we will accept the middle 98% of the $\bar{x}$'s as our boundary for accepting H_0 as true. We look up the t scores corresponding to a two-tailed test, $\alpha = .02$, which turns out to be $t = \pm 2.72$. (Note: for a large sample, the boundary would have been ± 2.33 standard deviations.) *Remember:* look down the df (degrees of freedom) column to 11, your sample size minus one (df $= n - 1 = 12 - 1 = 11$). $t = \pm 2.72$.

Substituting our t scores ± 2.72 into our (now t) formula, we solve for the $\bar{x}$ at the cutoffs:

$$t = \frac{\bar{x} - \mu}{\sigma_{\bar{x}}} \qquad t = \frac{\bar{x} - \mu}{\sigma_{\bar{x}}}$$

$$-2.72 = \frac{\bar{x} - 100}{4.706} \qquad +2.72 = \frac{\bar{x} - 100}{4.706}$$

cutoffs: $\bar{x} = 87.2$ $\bar{x} = 112.8$

The completed solution might appear graphically as follows:

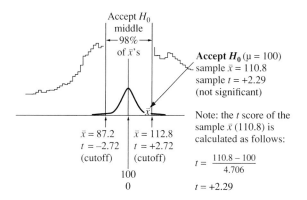

Accept H_0
middle
98%
of $\bar{x}$'s

Accept H_0 ($\mu = 100$)
sample $\bar{x} = 110.8$
sample $t = +2.29$
(not significant)

Note: the t score of the sample $\bar{x}$ (110.8) is calculated as follows:

$\bar{x} = 87.2$ $\bar{x} = 112.8$
$t = -2.72$ $t = +2.72$
(cutoff)   (cutoff)

100
0

$$t = \frac{110.8 - 100}{4.706}$$

$t = +2.29$

b. If the data constitutes a valid random sample of gifted students, the results indicate that gifted students on average score the same as ordinary students with social IQ, $\mu = 100$.

Validity: Intact groups, such as classes, violate one of the prime tenets of validity, that is, random selection. Results of such studies are often misleading.

Other aspects of validity are also quite difficult to guarantee in these types of experiments where the subjective judgment of an observer is critical. Before reliable data is obtained, often a great many similar studies must be conducted.

7.5

a. $n = 10$ Neanderthal skulls
$\bar{x} = 91.0$ cubic inch cranial capacity
$s = 4.714$ cubic inches

x	$\bar{x}$	$x - \bar{x}$	$(x - \bar{x})^2$	
87	91	−4	16	$\bar{x} = \dfrac{\Sigma x}{n}$
88	91	−3	9	
95	91	4	16	$= \dfrac{910}{10} = 91.0$
91	91	0	0	
89	91	−2	4	
98	91	7	49	$s = \sqrt{\dfrac{\Sigma(x - \bar{x})^2}{n - 1}}$
87	91	−4	16	
84	91	7	49	
95	91	4	16	$= \sqrt{\dfrac{200}{10 - 1}} = 4.714$
96	91	5	25	
910			200	
Σx			$\Sigma(x - \bar{x})^2$	

b. $H_0: \mu \leq 87.0$ cubic inches   Calculation of $\sigma_{\bar{x}}$:
$H_1: \mu \approx 87.0$ cubic inches
$\alpha = .05$

$$\sigma_{\bar{x}} = \frac{\sigma}{\sqrt{n}} = \frac{s}{\sqrt{n}}$$

$$\approx \frac{4.714}{\sqrt{10}} \approx 1.491$$

Small-sample conditions: mound-shaped population assured? (Yes); since s estimates σ, t scores must be used.

   Assume H_0 true, use α to establish cutoffs. Our level of significance in this case is $\alpha = .05$ (5%), which in a one-tailed test implies we will accept the lower 95% of the $\bar{x}$'s as our boundary for accepting H_0 as true. We look up the t score corresponding to a one-tailed test, $\alpha = .05$, which turns out to be $t = +1.83$ standard deviations. (Note: for a large sample, the boundary would have been $+1.65$ standard deviations.) *Remember:* look down the df, degrees of freedom, column to 9, your sample size minus one (df = $n - 1 = 10 - 1 = 9$), to get $t = +1.83$.

   Substituting our t score $+1.83$ into our (now t) formula, we solve for the $\bar{x}$ at the cutoff:

$$t = \frac{\bar{x} - \mu}{\sigma_{\bar{x}}}$$

$$+1.83 = \frac{\bar{x} - 87.0}{1.491}$$

cutoff: $\bar{x} = 89.73$

The complete solution might appear graphically as follows:

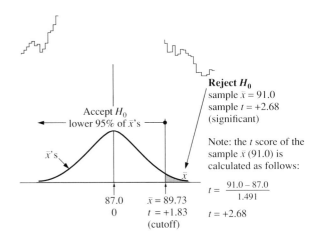

c. If the data constitutes a valid random sample of Neanderthal skulls, the results indicate Neanderthal man had a larger brain size than modern man. By the way, numerous skull findings over the past century and a half have confirmed this. Of course, this implies brain size is not a reliable indicator of intelligence. This should not come as a surprise, since (according to the *New Columbia Encyclopedia*) an elephant's brain weighs more than four times that of a human's, and a whale's brain, seven times; however, neither animal has the intelligence of an orangutan, whose brain weighs one-third as much as man's.

   Validity: Random selection may be in question here. Can we consider these ten skulls as a random selection from the Neanderthal population? Men's skulls are known to be larger than women's (and no mention was made of gender). The skulls could have all come from one family clan with similar genetic structure, and their particular skull traits may or may not be representative of the Neanderthal population.

   Other risks to validity seem minimal, barring researcher mistakes, shoddy technique, or questions of honesty.

7.7

a. *Reject H_0*: $\sigma_{\bar{x}} = .257$; cutoff $z = \pm 2.33$; cutoff $\bar{x} = 5.40$ to 6.60. We rejected H_0 because our sample $\bar{x}$ of 6.80 ($z = 3.11$) was in reject zone. Thus, we accept $H_1: \mu \neq 6.0$. The evidence indicates the average weight gain in men is greater than 6.0 lb.

b. *Accept H_0* ($\mu = 8.0$): $\sigma_{\bar{x}} = .369$; cutoff $z = \pm2.05$; cutoff $\bar{x} = 7.24$ to 8.76. We accepted H_0 because our sample $\bar{x}$ of 7.40 ($z = -1.63$) was in accept zone. There is no evidence to indicate a change in average weight gain for women.

c. Control chart for men:

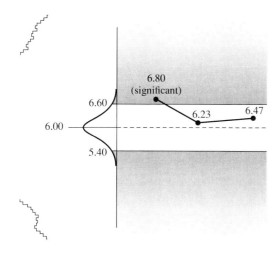

$\bar{x} = 6.80$ (significant)
$\bar{x} = 6.23$
$\bar{x} = 6.47$

d. Validity: Random selection is again the primary concern. Were those sampled volunteers? If they were, the results may be biased. Volunteers may have common predispositions, perhaps an intense fear of weight gain (or lack of concern). Even if our sample subjects were randomly selected from patients across the nation, we are not sure we have access to all those who have stopped smoking since many may not regularly attend a physician. Other risks to validity may include: Did the presence of the researcher influence results, as in the case where the patients know they are part of a study and ''perform'' (act differently than normal) for the researcher?

7.9

Reject H_0: $\sigma_{\bar{x}} = .136$; cutoff $z = \pm2.58$; cutoff $\bar{x} = -.35''$ to $+.35''$. We rejected H_0 because our sample $\bar{x}$ of $1.20''$ ($z = 8.82$) was in reject zone. Thus, we accept H_1: $\mu \neq 0''$ growth. The evidence indicates the average growth in short children is more than $0''$ (beyond expected).

7.11

a. *Accept H_0* ($\mu \leq 230$): A hypothesis test is not necessary in this example since the sample $\bar{x}$ (226 mg/dl) by inspection conforms to the null hypothesis ($\mu \leq 230$ mg/dl). In other words, there is no possibility of rejecting H_0 in this case. Thus, the evidence indicates the average heart attack victim's cholesterol level is 230 mg/dl or below.

b. *Reject H_0*: $\sigma_{\bar{x}} = 3.971$; cutoff $z = +1.65$; cutoff $\bar{x} = 236.55$ mg/dl. We rejected H_0 because our sample $\bar{x}$ of 241.1 ($z = 2.80$) was in the reject zone. Thus, we accept H_1: μ is greater than 230 mg/dl. The evidence indicates the average cholesterol level of a heart attack victim is greater than 230 mg/dl.

c. The results may differ because of violations to validity. Unless a random sample is ensured, results may vary considerably from study to study. Thus, results from nonrandom experiments are to be viewed with extreme caution and numerous such studies may have to be conducted before reliable predictions about μ can be obtained. This is often the case with medical studies where volunteers or patients from one hospital or region are used. Other violations to validity might include inability to accurately assess victim's cholesterol level immediately prior to attack or researcher error (accidental or deliberate).

7.13

a. *Accept H_0* ($\mu \leq \$11,809$): $\sigma_{\bar{x}} = \$593.41$; cutoff $z = 1.88$; cutoff $\bar{x} = \$12,924.61$. We accepted H_0 because our sample $\bar{x}$ of $\$12,053$ ($z = .41$) was in accept zone. The evidence gives us no grounds to reject the null hypothesis, thus we must accept the average household income for this district is $\$11,809$ *or below*. This district qualifies for poverty funds.

b. $\alpha = .01$: Lowering the α risk lowers the probability of falsely rejecting H_0 ($\mu \leq \$11,809$).

7.15

a. *Accept H_0* ($\mu \leq 11.7$ at $\alpha = .05$): $\sigma_{\bar{x}} = 1.306$; cutoff $t = +2.02$; cutoff $\bar{x} = 14.34$. We accepted H_0 because our sample $\bar{x}$ of 13.57 ($t = 1.43$) was in the accept zone. The evidence offers no grounds to reject Jessica's claim at $\alpha = .05$, thus we must accept it.

b. *Accept H_0* ($\mu \leq 11.7$ at $\alpha = .025$): $\sigma_{\bar{x}} = 1.306$; cutoff $t = 2.57$; cutoff $\bar{x} = 15.06$. We accepted H_0 because our sample $\bar{x}$ of 13.57 ($t = 1.43$) was in accept zone. The evidence offers no grounds to reject Jessica's claim at $\alpha = .025$, thus we must accept it.

7.17

a. *Reject H_0:* $n = 7$ children; $\bar{x} = \dfrac{-77}{7} = -11.0$;

$s = \sqrt{\dfrac{98}{7-1}} = 4.041$; $\sigma_{\bar{x}} = 1.527$; cutoff $t = \pm 3.14$;

cutoff $\bar{x} = -4.79$ to $+4.79$. We rejected H_0 because our sample $\bar{x}$ of -11.0 change in IQ ($t = -7.20$) was in reject zone. Thus, we must accept H_1: $\mu \neq 0$ change in IQ. The evidence indicates that abuse or neglect significantly lowers IQ.

b. If the data constitutes a valid random sample, these results are frightening. For the results would imply that abuse or neglect result in mentally handicapping a child. Of course, the study gave no indication whether the lowering of IQ is permanent or temporary (for further reading on this, refer to: ''Study Finds Severe Effect from Childhood Abuse,'' *New York Times,* February 18, 1991, p. 11).

Validity: For results to be representative of a population, random selection from that population must be assured. All too often, such studies are targeted to poverty areas where coincident factors may have a profound effect on results. For instance, the severe or grossly erratic behavior of parents on drugs causing depression or disorientation in the child. To generalize a conclusion that abuse or neglect lowers IQ, we must ensure that the sample is randomly selected from the general population; in addition, coincident factors must be isolated. This usually calls for advanced statistical techniques. At this point, let us say, experiments like this open up issues that are quite complex and usually treated in specialized statistics courses or in courses that concentrate on this particular subject matter.

Chapter Eight

8.1

a. Calculating a confidence interval proceeds in three steps, as follows:

First, use your sample average, $\bar{x} = 128.6$ lb as the center of your $\mu - \bar{x}$ (or μ) distribution. We know that in *any* normal distribution, 95% of the

data is between $z = -1.96$ and $z = +1.96$, so we shade this interval as follows:

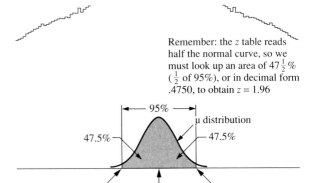

Remember: the z table reads half the normal curve, so we must look up an area of $47\frac{1}{2}\%$ ($\frac{1}{2}$ of 95%), or in decimal form .4750, to obtain $z = 1.96$

There is a 95% probability μ is contained within the shaded interval. Notice that the precise location of μ is unknown. However, we do know an *interval* where μ is most likely to be found.

Second, calculate the standard deviation of the μ distribution using the formula,

$$\sigma_{\bar{x}} = \frac{\sigma}{\sqrt{n}} = \frac{12.0}{\sqrt{900}} = \frac{12.0}{30} = .4 \text{ lb}$$

Third, we solve for the lower and upper limits of the interval using the z formula below. However, since $\bar{x}$ has replaced μ as the center of the distribution, to be technically correct, we must reverse $\bar{x} - \mu$ in the formula with $\mu - \bar{x}$. This gives us the correct minus and plus signs for our calculations. Frankly, even if we forget to do this technical switch, we would still obtain the correct answers, only in reverse order.

$$z = \frac{\mu - \bar{x}}{\sigma_{\bar{x}}} \qquad z = \frac{\mu - \bar{x}}{\sigma_{\bar{x}}}$$

$$-1.96 = \frac{\mu - 128.6}{.4} \qquad 1.96 = \frac{\mu - 128.6}{.4}$$

Solving for the lower and upper confidence limits of μ (and rounding to one decimal place), we get

$$\mu = 127.8 \text{ lb} \qquad \mu = 129.4 \text{ lb}$$

Graphically, the solution would appear as follows:

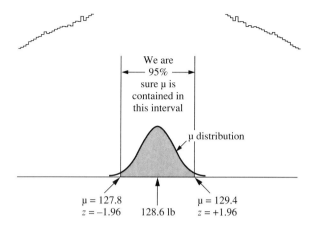

$\mu = 127.8$
$z = -1.96$ 128.6 lb $\mu = 129.4$
$z = +1.96$

Answer: Based on the sample taken, $\bar{x} = 128.6$ lb (and using $\sigma = 12.0$ lb), we are 95% sure the true average weight of European women (ages 18–34, heights 5′3″–5′7″) is contained in the interval between 127.8 lb and 129.4 lb.

Notice that μ, the true average weight of these European women, is really unknown. All we can state is that μ is between 127.8 lb and 129.4 lb. But even of this, we are not 100% sure. However, we are 95% sure. Could $\mu = 130$ lb? Yes. Although it is not likely. Note that sample results allow us to construct an interval where μ is most likely to be found, but the interval does not pinpoint μ's precise location. But this is still a rather valuable piece of information, since prior to the study we may have had little or no idea of the true average weight of these European women. At least now we can confidently state that the average weight of these European women is between 127.8 lb and 129.4 lb and we are 95% sure.

b. We solve using the same format as in part a, only now for a 99% confidence interval we use $z = \pm 2.58$.

Based on the sample taken, $\bar{x} = 128.6$ lb (and using $\sigma = 12.0$ lb), we are 99% sure the average weight of European women (ages 18–34, heights 5′3″–5′7″) is contained in the interval between 127.6 lb and 129.6 lb (rounded to one decimal place).

c. Yes. Based on *either* confidence interval (95% or 99%), we can state European women weigh "on average" less than equivalent American women. For European women, $\mu = 130.0$ lb is a highly unlikely situation based on this data.

d. A 99% confidence interval is less realistic in isolating the *true* location of μ. Most likely μ is close to $\bar{x}$, and far less likely to be at the extremes. Since a 99% confidence interval includes many more unlikely or extreme distances than a 95% confidence interval, a 99% confidence interval is often considered less realistic in defining μ's most probable location.

8.2

a. Again we proceed in three steps.

First, use your sample average $\bar{x} = 57.7$ hours as the center of the $\mu - \bar{x}$ (or μ) distribution. We know that in *any* normal distribution 90% of the data is between $z = -1.65$ and $z = +1.65$, so we shade this interval as follows:

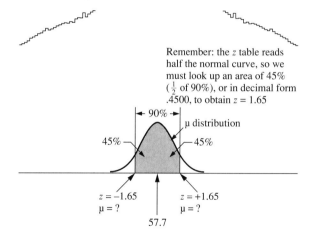

Remember: the z table reads half the normal curve, so we must look up an area of 45% ($\frac{1}{2}$ of 90%), or in decimal form .4500, to obtain $z = 1.65$

90%

μ distribution

45%   45%

$z = -1.65$ $z = +1.65$
$\mu = ?$ $\mu = ?$
57.7

There is a 90% probability μ will be contained in the shaded interval. Notice that the precise location of μ is unknown. However, we do have an interval where μ is most likely to be found.

Second, calculate the standard deviation of the μ distribution using the formula,

$$\sigma_{\bar{x}} = \frac{\sigma}{\sqrt{n}}$$

Often in real-life situations, σ, the standard deviation of the population, is either not supplied or unknown, as in this example. In this case, we use s, the sample standard deviation, in place of σ, since

$$s \approx \sigma$$

In other words, the individual measurements in one sample are most likely spread out in a manner similar to how the measurements in the entire population are spread out. If this is true, and a century of experience has confirmed this to be true, then we can use s, the standard deviation of the one sample, as a reasonably good estimator of σ, the standard deviation of the population. So now we calculate,

$$\sigma_{\bar{x}} = \frac{\sigma}{\sqrt{n}} \approx \frac{s}{\sqrt{n}} \approx \frac{6.0}{\sqrt{400}} \approx \frac{6.0}{20} \approx .3 \text{ hours}$$

Third, we solve for μ using the z formula. Remember: since $\bar{x}$ has replaced μ as the center of the distribution (to get the correct plus and minus signs), replace $\bar{x} - \mu$ with $\mu - \bar{x}$, as follows:

$$z = \frac{\mu - \bar{x}}{\sigma_{\bar{x}}} \qquad\qquad z = \frac{\mu - \bar{x}}{\sigma_{\bar{x}}}$$

$$-1.65 = \frac{\mu - 57.7}{.3} \qquad +1.65 = \frac{\mu - 57.7}{.3}$$

Solving for μ (and rounding to one decimal place), we get,

$$\mu = 57.2 \text{ hours} \qquad \mu = 58.2 \text{ hours}$$

Graphically, the solution would appear as follows:

Answer: Based on the sample taken, $\bar{x} = 57.7$ hours per week with $s = 6.0$ hr, we are 90% sure the true average number of hours Japanese teenagers put into class and home study per week is contained in the interval between 57.2 hours and 58.2 hours.

b. We don't know the precise average study time per week for Japanese students, however, we are 90% sure this average, whatever its value, is contained in the interval between 57.2 hours per week and 58.2 hours per week.

c. We solve using the same format as in part a, only now for a 95% confidence interval we use $z = \pm 1.96$. Based on the sample taken, $\bar{x} = 57.7$ hours with $s = 6.0$, we are 95% sure the average study time per week for Japanese students is contained in the interval between 57.1 and 58.3 hours.

d. We don't know the precise average study time of Japanese students, however, we are 95% sure the average is contained in the interval between 57.1 hours per week and 58.3 hours per week.

e. Essentially, the exercise is asking with what assurance do we know μ, the *true* average study time of Japanese students per week, is between *57.7 hr minus .2 hr* and *57.7 hr plus .2 hr.* In other words, what is the probability μ is contained between 57.5 hr and 57.9 hr?

This is a typical normal curve problem. We merely calculate the percentage of data between 57.5 hr and 57.9 hr. We proceed by using the same three steps we used in prior examples.

First, use your sample average, $\bar{x} = 57.7$ hr, as the center of the $\mu - \bar{x}$ (or μ) distribution. Now, we shade the interval between 57.5 hours and 57.9 hours, which we approximate as follows:

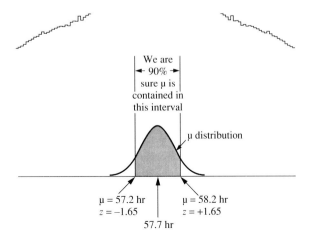

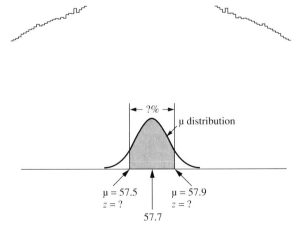

Second, calculate the standard deviation of the μ distribution using the formula,

$$\sigma_{\bar{x}} = \frac{\sigma}{\sqrt{n}}$$

Since σ is unknown, we use s, the standard deviation of the sample as an estimator, as follows:

$$\sigma_{\bar{x}} = \frac{\sigma}{\sqrt{n}} \approx \frac{s}{\sqrt{n}} \approx \frac{6.0}{\sqrt{400}} \approx \frac{6.0}{20} \approx .3 \text{ hours}$$

Third, we solve for the two z scores at the upper and lower limit. Remember to replace $\bar{x} - \mu$ with $\mu - \bar{x}$.

$$z = \frac{\mu - \bar{x}}{\sigma_{\bar{x}}} \qquad z = \frac{\mu - \bar{x}}{\sigma_{\bar{x}}}$$

$$z = \frac{57.5 - 57.7}{.3} \qquad z = \frac{57.9 - 57.7}{.3}$$

$$z = -.67 \qquad z = +.67$$

Looking up the area for $z = -.67$, we get 24.86%. Since the normal curve is symmetrical, a z of +.67 will yield an additional 24.86%. So, there is a 49.72% probability (24.86% + 24.86%) that μ will be contained in the interval from 57.5 hours to 57.9 hours. Rounded to one decimal place, 49.7%.

The completed solution would appear graphically as follows:

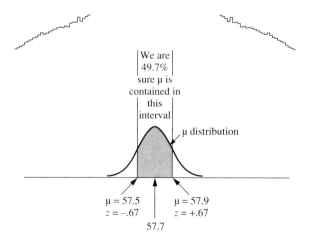

Answer: Based on the sample taken ($\bar{x} = 57.7$ hours with $s = 6.0$), we can assert with 49.7% probability that μ, the *true* average study time per week for Japanese students, is within .2 hours of $\bar{x}$, our sample average. By the way, .2 hours = 12 minutes.

8.3

To calculate n, the sample size necessary to meet these conditions, we proceed as follows:

The first step is basically the same. Use $\bar{x}$ as the center of the μ distribution. Since we know that in *any* normal distribution, 90% of the data falls between $z = -1.65$ and $z = +1.65$, we shade this interval as follows:

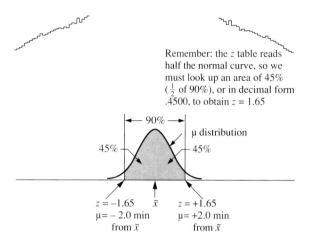

Remember: the z table reads half the normal curve, so we must look up an area of 45% ($\frac{1}{2}$ of 90%), or in decimal form .4500, to obtain z = 1.65

Notice that the cutoff at μ is exactly 2.0 minutes from $\bar{x}$. Another way of saying this is

$$\mu - \bar{x} = 2.0 \text{ minutes}^*$$

Also notice, we did not need to know the value of μ or the value of $\bar{x}$ to know $\mu - \bar{x}$.

The second step involves algebraically manipulating the two formulas, $z = (\mu - \bar{x})/\sigma_{\bar{x}}$ and $\sigma_{\bar{x}} = \sigma/\sqrt{n}$ into one combined formula for n (the derivation of which is given in section 8.3). Then we solve as follows:

$$n = \left(\frac{z\sigma}{\mu - \bar{x}}\right)^2 = \left[\frac{(1.65)(12.0)}{2.0}\right]^2 = (9.9)^2$$

$$n = 98.01, \quad \text{or}$$

$$n = 98 \text{ (rounded to nearest whole number)}$$

In the above formula, the quantity $\mu - \bar{x}$ is traditionally referred to as the ''maximum error'' in estimating μ. In our case, the financial department wanted the maximum error in estimating μ limited to 2.0 minutes (probably for an accurate cost assessment to properly bid on the job).

$^*\mu - \bar{x}$ could also equal -2.0 minutes.

Notice we had to know the value of σ, the standard deviation of the population, to continue. In real-life situations, it may be quite difficult to determine this, since we have not as yet taken the sample. Without a sample we have no s to estimate σ. However, we might guess past experience in assembling similar items may have allowed the toy designers to estimate σ at 12.0 minutes.

In fact, in section 8.3, second example, we used σ = 18.0 minutes and obtained quite a different answer. In practice, it is advised (when feasible) to actually take a prior sample ($n \geq 30$) and use the obtained s to estimate σ.

Answer: Thus, using a random selection of workers and conditions, we must assemble the new Walkie Doll 98 times to be 90% sure the *true* average time it takes to assemble the doll will be within 2.0 minutes of $\overline{x}$, the sample average.

8.4

a. No solution. Data unreliable.

Answer: Since the sample size was small ($n < 30$) and the medical journal gave no assurance the population was normally distributed, we must consider the data unreliable and not worthy of further analysis—this is especially true in view of the fact the sample size was exceedingly small, only $n = 6$. Had the sample size been $n = 25$ or thereabouts, maybe we might risk it.

Note: experience has shown that salary populations are often skewed. And, therefore, small samples drawn from such populations can very well give misleading results.

8.5

a. Notice the question states, "Assuming a normally distributed population." If this is a frivolous assumption, not based on fact, the sample data should be viewed with suspicion. However, we will assume the medical researchers have good reason to believe the population is normally distributed. In which case, we proceed with the same three steps we would use to calculate any confidence interval.

First, use your sample average, $\overline{x} = 38.3$ mg/dl as the center of your μ distribution. We know when using a large sample, 95% of the data would be expected to fall between $z = \pm 1.96$. However, now,

since we are using a *small sample*, we must adjust for the fact s most likely is causing $\sigma_{\overline{x}}$ to be underestimated in the formula $\sigma_{\overline{x}} = \sigma/\sqrt{n} \approx s/\sqrt{n}$. To compensate, we look up the t score adjustment, to obtain $t = \pm 2.31$ as follows:

Confidence Interval	90%	95%	98%	99%	Confidence Interval
6	1.94	2.39	3.14	3.71	6
7	1.89	2.36	3.00	3.50	7
8	1.86	2.31	2.90	3.36	8
9	1.83	2.26	2.82	3.25	9
10	1.81	2.23	2.76	3.17	10

$n - 1$

Remember: look down the 95% column to a df of 8 (your sample size minus one).

We now know 95% of the time μ would be expected to be contained in the interval within ±2.31 standard deviations of $\overline{x}$, that is, between $t = -2.31$ and $t = +2.31$, so we shade as follows:

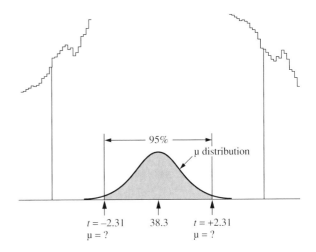

Second, we calculate $\sigma_{\overline{x}}$, as follows:

$$\sigma_{\overline{x}} = \frac{\sigma}{\sqrt{n}} \approx \frac{s}{\sqrt{n}} \approx \frac{7.83}{\sqrt{9}} \approx \frac{7.83}{3} \approx 2.61 \text{ mg/dl}$$

Third, we solve using the z formula, only now we use the letter t in place of z, noting that
i. s was used to estimate σ, and
ii. a small sample ($n < 30$) was probably used.

Technically, a t score is needed for all sample sizes when s is used to estimate σ. However, for large samples, the adjustment is considered negligible and for this reason we will ignore it for sample sizes $n \geq 30$.

$$z = \frac{\mu - \bar{x}}{\sigma_{\bar{x}}} \qquad\qquad z = \frac{\mu - \bar{x}}{\sigma_{\bar{x}}}$$

$$t = \frac{\mu - \bar{x}}{\sigma_{\bar{x}}} \qquad\qquad t = \frac{\mu - \bar{x}}{\sigma_{\bar{x}}}$$

$$-2.31 = \frac{\mu - 38.3}{2.61} \qquad +2.31 = \frac{\mu - 38.3}{2.61}$$

Solving for μ,

$$\mu = 32.27 \text{ mg/dl} \qquad \mu = 44.33 \text{ mg/dl}$$

Graphically, the solution would appear as follows:

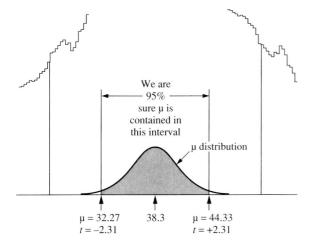

We are 95% sure μ is contained in this interval

μ distribution

$\mu = 32.27$ 38.3 $\mu = 44.33$
$t = -2.31$ $t = +2.31$

Answer: Based on the sample taken, $\bar{x} = 38.3$ mg/dl with $s = 7.83$, we are 95% confident the true average decrease in cholesterol levels among patients is contained in the interval between 32.27 mg/dl and 44.33 mg/dl.

b. Again, we first look for our assurance of a normally distributed population because our sample size is below 30 ($n < 30$). This is especially critical in this problem since our sample size is very small $n = 5$.

Since the problem states "Assuming a normally distributed population," we can proceed. To get started, we must calculate $\bar{x}$ and s for this data, as follows:

Decrease in Cholesterol (mg/dl blood)	$\bar{x}$	$x - \bar{x}$	$(x - \bar{x})^2$
33.0	32.0	1.0	1.00
38.2	32.0	6.2	38.44
28.0	32.0	−4.0	16.00
30.6	32.0	−1.4	1.96
30.2	32.0	−1.8	3.24
160.0			60.64
Σx			$\Sigma (x - \bar{x})^2$

$$\bar{x} = \frac{\Sigma x}{n} = \frac{160.0}{5}$$
$$= 32.0 \text{ mg/dl}$$

$$s = \sqrt{\frac{\Sigma (x - \bar{x})^2}{n - 1}}$$
$$= \sqrt{\frac{60.64}{5 - 1}}$$
$$= \sqrt{15.16}$$
$$= 3.894 \text{ mg/dl}$$

Knowing $\bar{x} = 32.0$ and $s = 3.894$, we can now proceed using our three-step procedure for confidence intervals.

First, use your sample average, $\bar{x} = 32.0$ as the center of your μ distribution. We know using a large sample, 95% of the data would be expected to fall between $z = \pm1.96$. However, now, since we are using a *small sample*, we must adjust for the uncertainty created when using s to estimate σ. To maintain a 95% confidence level, we look up the t score and obtain $t = \pm2.78$ as follows:

Confidence Interval	90%	95%	98%	99%	Confidence Interval
3	2.20	2.84	4.54	5.84	3
4	2.13	2.78	3.75	4.60	4
5	2.02	2.57	3.37	4.03	5

$n - 1$

Remember: look down the 95% column to a df of 4 (your sample size minus one)

We now know that 95% of the time μ would be expected to be contained in the interval within ± 2.78 standard deviations of $\bar{x}$, that is, between $t = -2.78$ and $t = +2.78$ as follows:

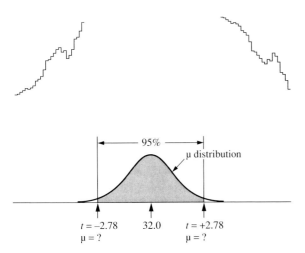

Second, we calculate $\sigma_{\bar{x}}$, as follows:

$$\sigma_{\bar{x}} = \frac{\sigma}{\sqrt{n}} \approx \frac{s}{\sqrt{n}} \approx \frac{3.894}{\sqrt{5}} \approx \frac{3.894}{2.236} \approx 1.742$$

Third, we solve using the z formula, only now we use the letter t in place of z, noting that
i. s was used to estimate σ, and
ii. a small sample ($n < 30$) was probably used.

Technically a t score is needed for all sample sizes when s is used to estimate σ. However, for large samples, the adjustment is negligible and for this reason we will ignore it for sample sizes $n \geq 30$.

$$z = \frac{\mu - \bar{x}}{\sigma_{\bar{x}}} \qquad z = \frac{\mu - \bar{x}}{\sigma_{\bar{x}}}$$

$$t = \frac{\mu - \bar{x}}{\sigma_{\bar{x}}} \qquad t = \frac{\mu - \bar{x}}{\sigma_{\bar{x}}}$$

$$-2.78 = \frac{\mu - 32.0}{1.742} \qquad +2.78 = \frac{\mu - 32.0}{1.742}$$

Solving for μ,

$$\mu = 27.16 \text{ mg/dl} \qquad \mu = 36.84 \text{ mg/dl}$$

Graphically, the solution would appear as follows:

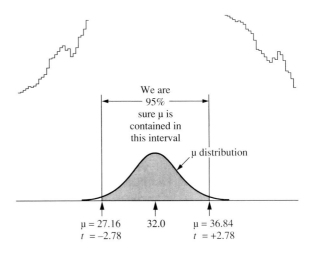

Answer: Based on the sample taken, $\bar{x} = 32.0$ mg/dl with $s = 3.894$, we are 95% confident the real average decrease in cholesterol levels among patients is somewhere between 27.16 mg/dl and 36.84 mg/dl.
c. *Study 1:* Although the precise average decrease in cholesterol among patients is unknown, this study showed the "average" is most likely contained between 32.27 mg/dl and 44.33 mg/dl and we are 95% sure of this.
   Study 2: This study showed we can expect to find the average decrease in cholesterol somewhere between 27.16 mg/dl and 36.84 mg/dl and we are 95% sure of this.
d. Let's take a look at the results visually.

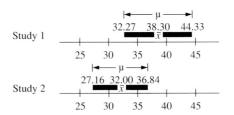

From the above, we might "guess" that μ is somewhere between 32.27 mg/dl and 36.84 mg/dl since both confidence intervals *overlap* in this region. Remember: confidence intervals give the most probable location of μ.

Another way to estimate μ is to average the two sample $\bar{x}$'s (provided the sample sizes are equal or near equal). For our example, we get $(32.0 + 38.3)/2 = 35.15$ mg/dl. Thus, μ, the *true* average decrease in cholesterol is most likely quite close to 35.15 mg/dl. For sizable differences in sample size, a weighted average can be taken.

Regarding σ, the population standard deviation, we know s, the sample standard deviation, is our best estimator. We obtained,

$$\text{for } n = 9, \quad s = 7.83$$
$$\text{for } n = 5, \quad s = 3.89 \text{ (rounded)}$$

Since a small sample s tends to *underestimate* σ, the smaller your sample, the more σ tends to be underestimated and the more we must be careful in our conclusions. A reasonable conclusion might be that σ is probably near 7.83 (the s we obtained from our larger sample) or maybe slightly more than 7.83 since s tends to underestimate σ for sample sizes $n < 30$. However, all of this is conjecture, and until larger samples are taken any statement should be made with great reservation.

e. Although the problem failed to mention the precise target population, it did say "a random selection of *patients*" was used. However, the word "patients" can refer to a number of categories, such as, hospital patients in for a variety of illnesses or hospital patients being treated for high cholesterol levels. Remember, patients hospitalized for cholesterol probably have extraordinarily high levels and may react to the drug differently than people with normal levels.

"Patients" could be hospital patients being treated for heart attacks, or perhaps patients being treated in a doctor's office. The target population from which the samples were drawn must be known before one can state meaningful conclusions.

For instance, if the word "patients" refers to patients who experienced heart attacks, then we might state that the drug will reduce cholesterol levels in *heart attack patients* on average approximately 35.15 mg/dl. However, in the general population, the drug may very well have a different effect. To determine the effect of the drug in the general population, we would have to sample from the general population.

8.7

a. $\sigma_{\bar{x}} = 2.24/\sqrt{64} = .28$; $z = \pm 1.28$; we are 80% sure μ is contained in the interval between 70.64″ and 71.36″.

b. $\sigma_{\bar{x}} = .28$; $z = \pm 1.96$; we are 95% sure μ is contained in the interval between 70.45″ and 71.55″.

c. $\sigma_{\bar{x}} = .28$; $z = \pm 2.58$; we are 99% sure μ is contained in the interval between 70.28″ and 71.72″.

d. $\sigma_{\bar{x}} = .28$; $\mu - \bar{x} = \pm .25$; $z = \pm .25/.28 = \pm .89$; $31.33\% + 31.33\% = 62.66\%$.

e. $n = [(1.65)(2.24)/.25]^2 = 218.57 = 219$ male students.

f. Yes. Our sample size was large enough that we obtained a reasonably precise estimation of μ. The 80%, 95%, and 99% confidence interval estimates of μ all showed μ to be *above* 70 inches. We can state with 99% assurance (let's use the most impressive statistic) that the average height of Michigan State men is above the national average.

8.9

a. $\sigma_{\bar{x}} = \$4.80/\sqrt{130} = \$.421$; $z = \pm 1.75$; we are 92% sure μ is contained in the interval between $23.76 and $25.24.

b. $\sigma_{\bar{x}} = \$.421$; $z = \pm 2.33$; we are 98% sure μ is contained in the interval between $23.52 and $25.48.

c. $\sigma_{\bar{x}} = \$.421$; $\mu - \bar{x} = \pm .50$; $z = \pm .50/.421 = \pm 1.19$; $38.30\% + 38.30\% = 76.60\%$.

d. $n = [(1.96)(4.80)/.5]^2 = 354$ shoppers.

8.11

a. $\sigma_{\bar{x}} = 6.20/\sqrt{10} = 1.961$; $\pm 1.83 = (\mu - 40.50)/1.961$; we are 90% sure μ is contained in the interval between $36.91 and $44.09.

b. $\sigma_{\bar{x}} = \$5.60/\sqrt{8} = 1.98$; $\pm 1.89 = (\mu - 34.75)/1.98$; we are 90% sure μ is contained in the interval between $31.01 and $38.49.

c. Small-sample sizes could be one reason. $\bar{x}$'s tend to spread out farther from μ as your sample size decreases. Another more practical reason is that tickets sold close to the concert date may command more money and if the surveys were conducted at different times, μ, the population average itself, could indeed be different.

d. Assuming a valid random sample was obtained in each study and the surveys conducted in the same time period (in other words, we are estimating the same μ), the confidence intervals *overlap* at approximately $37 to $38, so μ may very well be in or near this range.

Chapter Nine

9.1

Actually, the least-squares line is also commonly referred to as the best-fit or predictor line. The term "regression" stems from the original experiments of Sir Francis Galton. In one experiment, Galton compared parent-height versus offspring height and the line representing offspring height tended to "regress" toward the average of the population. In other words, tall parents tend to produce tall children but with heights somewhat closer (regressing) to the population average. For further reading, see section 9.0.

9.2

a. Correlation is defined as the ability of one variable to predict the value of another. For instance, using rainfall and crop output, if an increase in rainfall is accompanied by an increase (or decrease) in crop output, these two measures, inches of rainfall and bushels of output, may be said to correlate—i.e., knowledge of one allows us some form of prediction as to the value of the other.

b. The same S_e measure could accompany data with varying degrees of co-movement, varying degrees of correlation.

9.3

a. Apparently not, according to the Surgeon General's report, which stated cigarette smoking accounted for an estimated 31,600 deaths in 1985 (in women alone). Remember, correlation does not imply cause and effect. Correlation measures a co-movement between two measures that may or may not be actually related.

b. Time. In the last 100 years, the population has grown. This and other factors have resulted in sharp increases in tobacco consumption during the period. Also, in the last 100 years, medical advances and better nutrition have vastly improved life-expectancy rates for everyone, including smokers. So both have risen over the past 100 years, but *not* because of each other.

9.4

a.

x	y
55	2.6
42	2.9
61	3.0
32	2.2
67	2.6
47	2.1
52	3.2
36	1.5
53	2.0
39	1.9
59	2.3
70	3.5

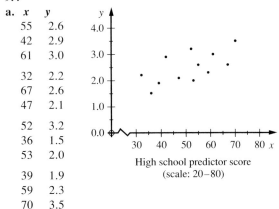

High school predictor score
(scale: 20–80)

b.

HS score x	College GPA y	x^2	xy	y^2
55	2.6	3025	143	6.76
42	2.9	1764	121.8	8.41
61	3.0	3721	183	9.00
32	2.2	1024	70.4	4.84
67	2.6	4489	174.2	6.76
47	2.1	2209	98.7	4.41
52	3.2	2704	166.4	10.24
36	1.5	1296	54	2.25
53	2.0	2809	106	4.00
39	1.9	1521	74.1	3.61
59	2.3	3481	135.7	5.29
70	3.5	4900	245	12.25
$\Sigma x = 613$	$\Sigma y = 29.8$	$\Sigma x^2 = 32{,}943$	$\Sigma xy = 1572.3$	$\Sigma y^2 = 77.82$

$$a = \frac{\Sigma y \, \Sigma x^2 - \Sigma x \, \Sigma xy}{n \, \Sigma x^2 - (\Sigma x)^2}$$

$$= \frac{(29.8)(32{,}943) - (613)(1572.3)}{12(32{,}943) - (613)^2}$$

$$= \frac{981{,}701.4 - 963{,}819.9}{395{,}316 - 375{,}769}$$

$$= \frac{17{,}881.5}{19{,}547}$$

$$= .9147951$$

$$\boxed{a = .915}$$

$$b = \frac{n\,\Sigma\,xy - \Sigma\,x\,\Sigma\,y}{n\,\Sigma\,x^2 - (\Sigma\,x)^2}$$

$$= \frac{12(1572.3) - (613)(29.8)}{19{,}547}$$

$$= \frac{18{,}867.6 - 18{,}267.4}{19{,}547}$$

$$= \frac{600.2}{19{,}547} = .0307054$$

$$\boxed{b = .031}$$

Substituting $a = .915$ and $b = .031$ into the line equation, $y = a + bx$, we get the following regression line equation for our data:

$$\boxed{\begin{array}{l} y_L = .915 + .031x \\ \text{GPA} = .915 + .031(\text{HS score}) \end{array}}$$

i. To predict freshmen GPA, y, for an x-value of 33 (HS score), we substitute 33 into the equation for x

$$y_L = .915 + .031(33)$$
$$= .915 + 1.023$$
$$= 1.938$$

(answers will vary depending on rounding technique)

$$\boxed{y_L = 1.9}$$

ii. To predict freshmen GPA, y, for an x-value of 58 (HS score), we substitute 58 into the equation for x

$$y_L = .915 + .031(58)$$
$$= .915 + 1.798 = 2.713$$

$$\boxed{y_L = 2.7}$$

c. In part b, we used the regression line equation to determine two x,y-points which were on the line, namely, $(33,1.9)$ and $(58,2.7)$. Since these points were on the regression line, we use these to locate the regression line's exact position, shown in the

first figure. In the second figure, the line is shown drawn through the original data points.

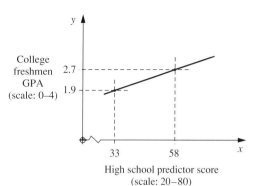

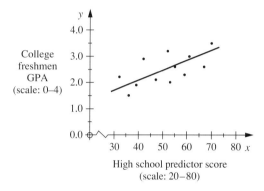

d. To calculate r, we use the following formula, which uses quantities also calculated in solution to part b:

$$\boxed{r = \frac{n\,\Sigma\,xy - \Sigma\,x\,\Sigma\,y}{\sqrt{n\,\Sigma\,x^2 - (\Sigma\,x)^2}\ \sqrt{n\,\Sigma\,y^2 - (\Sigma\,y)^2}}}$$

Formula to calculate correlation coefficient r

same as numerator in b calculation

$$= \frac{600.2}{\sqrt{19{,}547}\ \sqrt{12(77.82) - (29.8)^2}}$$

same as denominator in b calculation

$$= \frac{600.2}{139.8106\ \sqrt{933.84 - 888.04}}$$

$$= \frac{600.2}{(139.8106)(6.7676)}$$

$$= \frac{600.2}{946.182} = .63434$$

$$\boxed{\text{Thus, } r = .63}$$

r, the correlation coefficient, is a measure of the ability of one variable to predict the value of another, often referred to as a measure of the strength of the relationship between x and y. A second way of viewing r is as the *slope* of the regression line expressed in standard units (z-score changes). More specifically, r is the fractional standard deviation increase in y associated with one standard deviation increase in x. In other words, if you calculate the mean and standard deviation for all the x-values ($\bar{x}$ and s_x) and calculate the mean and standard deviation for all the y-values ($\bar{y}$ and s_y), then a .63 standard deviation increase in y is associated with one standard deviation increase in x. This can be expressed as a fraction, denoted as standard slope.

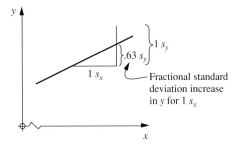

$$\text{Standard slope} = \frac{\text{Change in } y}{\text{Change in } x}$$

$$= \frac{rs_y}{s_x}$$

$$= \frac{.63\, s_y}{s_x}$$

$$= \frac{.63(1)}{(1)}$$

When $s_x = 1$ and $s_y = 1$, that is, when s_x and s_y are expressed in z score notation, then standard slope $= r = .63$.

$$= .63$$

Essentially, standard slope (r) offers us the slope of the line independent of scale or units a researcher might choose.

e. Test of significance for r. The problem arises that ρ, the correlation coefficient for the population, may very well equal zero (no correlation in the population) and we may be picking up a correlation in the sample by chance. In other words, we may

have randomly selected 12 students which produced an upward tilt in their x,y-measures by chance alone and indeed no real correlation exists in the population. To ensure against this, we conduct the following test:

$$H_0\colon \rho = 0$$
$$H_1\colon \rho \neq 0$$
$$\alpha = .05$$

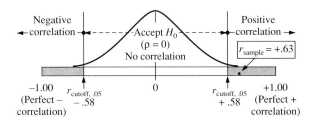

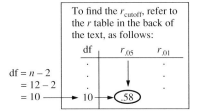

To find the r_{cutoff}, refer to the r table in the back of the text, as follows:

df	$r_{.05}$	$r_{.01}$
.		
.		
.		
10	.58	

$\text{df} = n - 2$
$\quad = 12 - 2$
$\quad = 10$

Significant at $\alpha = .05$
Reject H_0

Accept $H_1\colon \rho \neq 0$ (a correlation indeed exists in the population)

Since r_{sample} achieved significance, a correlation is assumed to exist in the population and, thus, we have some degree of confidence in using the sample regression line to predict y-values in the population.

f. Percentage explained variation attempts to assess "how-much-better" the regression-line fits the data points rather than using the $\bar{y}$-line. Specifically, it measures the extent the sum of squared distances (from points to line) are reduced when we use the regression line to represent the data points rather than using the $\bar{y}$-line.

Fortunately, we have a shortcut formula for calculating this value rather than go through the tedious process of computing the sum of squared distances around y_L (the regression line) and sum of squared distances around $\bar{y}$ (the average y), then calculating the percentage the squared distances are reduced. The shortcut formula for percentage explained variation involves the sample correlation coefficient, r, as follows:

Percentage
explained = $100r^2_{sample}$
variation

= $100(.63)^2$
= $100(.3969)$
= 39.69%

Using four decimal places for
r, the calculation becomes
$100(.6343)^2 = 40.23\%$

This result is often expressed as: 39.69% of the variation in college GPA is explained by its association with high-school predictor scores. In reality, this means, 39.69% of the sum of squared distances are reduced using the regression line to describe the points rather than the $\bar{y}$-line, indicating the regression line in some sense fits the data points better.

g. The data indicates that this particular HS predictor model (SAT scores plus high school GPA) does significantly correlate with college freshmen GPA at Hoover College. That is, a correlation in the population exists and the regression line may be used to estimate college freshmen GPAs. In actual validity studies, many hundreds of students are tested.

HS predictor models often show significant correlations in the range $r = .30$ to $.60$ when many hundreds of students at a college are tested. And indeed, the SAT predictor models do explain some of the variation in freshmen GPA for a good many college students across the nation. According to data provided by the College Entrance Examination Board (CEEB), $r = .50$ to $.60$ is typical for the predictor model presented above (although this varies from college to college), however even $r = .60$ is associated with only 36% of GPA variation [$100 r^2 = 100(.60)^2 = 36\%$]. Motivation, hard work, course load, a balance of social and extracurricular activities, and so on, would probably also explain a large percentage of the remaining GPA variation among students.

Interestingly enough, other high school information, such as, high school extracurricular and employment achievement, recommendations, interviews and interest, personality and creativity

measures tended to add little over the model we used above in predicting college GPA. One exception was a student's high school academic honors (*The College Board Technical Handbook*, T. Donlon, editor, N.Y.: CEEB, 1984, p. 168).

9.5

a. To calculate the best-estimate GPA (the regression-line estimate) for a student with a high school predictor score of 53, we simply plug in 53 for x in the regression line equation, as follows:

$y_L = a + bx$
= $.915 + .0307x$
= $.915 + .0307(53)$
= $.915 + 1.627 = 2.542$

$$\boxed{y_L = 2.54}$$

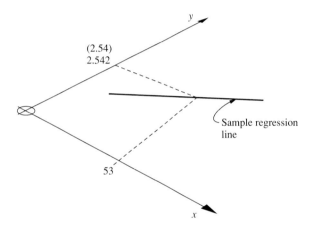

Note: additional decimal places are employed in answers 9.5 a–d because these estimates are highly sensitive to rounding technique.

b. To calculate S_e, the standard deviation of the vertical distances (from points to regression line), we can use the formula

$$S_e = \sqrt{\frac{\Sigma(y - y_L)^2}{n - 2}}$$

However, this requires us to calculate the distance from every point to the regression line. An arduous task indeed. Refer to the problem in section 9.2, solution d, for a demonstration. Instead, we shall use the shortcut formula (shown below). The formula is highly sensitive to rounding errors, so we must use several decimal places.

$$S_e = \sqrt{\frac{\Sigma y^2 - a\,\Sigma y - b\,\Sigma xy}{n-2}}$$ Note: all these values can be obtained from solution b of exercise 9.4.

$$= \sqrt{\frac{(77.82) - (.914795)(29.8) - (.030705)(1572.3)}{12 - 2}}$$

$$= \sqrt{\frac{(77.82) - (27.2609) - (48.2775)}{10}}$$

$$= \sqrt{\frac{2.2816}{10}} = .4776$$

$$\boxed{S_e = .478}$$

Pictorially, we might represent S_e as follows:

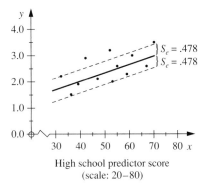

High school predictor score
(scale: 20–80)

c. To construct the 95% confidence interval for the equivalent population regression-line value, Y_L, we perform the calculations below.

$$y_L \pm tS_e \sqrt{\frac{1}{n} + \frac{n(x-\bar{x})^2}{n\,\Sigma x^2 - (\Sigma x)^2}}$$

where

y_L = sample regression-line estimate, 2.542
t = t score from table, 95% CI, df = $n - 2$
      = 12 − 2 = 10
S_e is obtained from prior exercise 9.4, solution d
n = number of (x,y) observations

$$2.542 \pm 2.23(.478) \sqrt{\frac{1}{12} + \frac{12(53 - 51.083)^2}{19,547}}$$

Note:
1) The value 19,547 under the square root was obtained from prior exercise 9.4, solution b. The value is the same as the denominator of the b fraction.
2) $\bar{x}$ = the average of all the x's, $\Sigma x/n$ (in our case, $\bar{x} = \Sigma x/n = 613/12 = 51.083$, where the sum 613 was obtained from the calculations in exercise 9.4, solution b).

$$2.542 \pm 2.23(.478) \sqrt{.0833 + \frac{12(3.6749)}{19,547}}$$

$$2.542 \pm 2.23(.478) \sqrt{.0833 + .00226}$$

$$2.542 \pm 2.23(.478)(.2925)$$

$$2.542 \pm .312 \begin{cases} 2.542 + .312 = 2.854 = \underline{2.85} \\ 2.542 - .312 = 2.230 = \underline{2.23} \end{cases}$$

$$\boxed{2.23 \leq \frac{\text{Population}}{Y_L} \leq 2.85}$$

Conclusion: We are 95% sure the population regression-line value, Y_L, is between 2.23 and 2.85 for a high school predictor score of 53, with our best estimate at 2.54.

d. To construct a 95% prediction interval of GPAs for a student with HS score of 53, we perform the following calculations. (Note: See prior solution for explanation of terms and values. The formula is identical, except for the addition of 1 under the square root.)

$$y_L \pm t\,S_e \sqrt{1 + \frac{1}{n} + \frac{n(x-\bar{x})^2}{n\,\Sigma x^2 - (\Sigma x)^2}}$$

$$2.542 \pm 2.23(.478) \sqrt{1 + \frac{1}{12} + \frac{12(53 - 51.083)^2}{19,547}}$$

$$2.542 \pm 2.23(.478) \sqrt{1.000 + .0833 + .00226}$$

$$2.542 \pm 2.23(.478)(1.0419)$$

$$2.542 \pm 1.1106$$

$$\boxed{\begin{array}{c} \text{Prediction} \\ 1.43 \leq \ \text{interval} \ = 3.65 \\ \text{of GPAs} \end{array}}$$

Conclusion: For a student with HS score 53, we are 95% sure the interval above includes the student's freshmen GPA.

Visually, the information in solutions a
through c might be presented as follows:

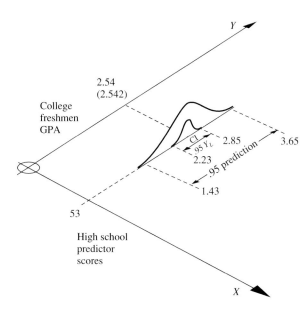

9.7
a.–c.

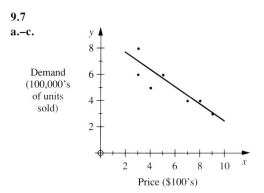

$$a = \frac{(36)(253) - (39)(179)}{7(253) - (39)^2} = 8.51$$

$$b = \frac{7(179) - (39)(36)}{250} = -.60$$

$$y = 8.51 - .60x$$

i. $y = 8.51 - .60(8) = 3.71$
ii. $y = 8.51 - .60(3.5) = 6.41$
Note that low demand (y)(fewer units sold) seems to
be associated with high price (x).

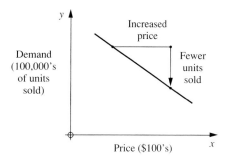

d. $r = \dfrac{-151}{\sqrt{250}\ \sqrt{118}} = -.88$ (significant at $\alpha = .05$, df $= n - 2 = 5$; cutoff $= \pm.75$)

e. Percentage explained variation $= 100(-.88)^2$
$= 77.44\%$. Using four decimal places for r, we get
77.30%.

f. The data indicates that as price increases, demand for
the product decreases. Assuming a valid random
sample, significant results would indicate a
correlation exists in the population, and this sample
regression line can be used as an estimator of the
relationship in the population (all national markets
for the product). However, because of the small-
sample size, great variation in the slope of the
regression line may occur. Remember, a sample only
estimates the characteristics of a population and
larger sample sizes tend to give better estimates.
    Although economic theory tells us that
Demand tends to decrease as Price increases, theory
cannot capture the many real-life factors that might
influence data. Essentially theory depends on other
factors being held constant, such as, competition,
promotion, channels of supply, and so on. In the real
world these factors often cannot be held constant. In
the case of health products, many more factors may
even affect the decision to buy, such as, prestige and
reputation of the supplier, perceptions that higher
priced products are safer to use, fear of being sued
by a patient, inconclusive statistical studies, or a
slight 'edge' brought to the public's attention, etc. In
other words, unless we test theory in combination
with actual market forces, we often cannot truly
assess the effect price has on demand for our
product.
    In this artificial-skin experiment above,
indeed, the results conform with established
economic theory. As Price increased, Demand
(reflected here in units sold) significantly decreased.

9.9

a.–c.

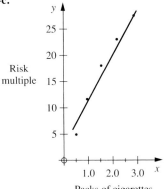

Packs of cigarettes
smoked per day

$$a = \frac{(86)(13.75) - (7.5)(155)}{6(13.75) - (7.5)^2} = .76$$

$$b = \frac{6(155) - (7.5)(86)}{26.25} = 10.86$$

$$y = .76 + 10.86x$$

i. $y = .76 + 10.86(.7) = 8.36$

ii. $y = .76 + 10.86(2.2) = 24.65$

Note that high incidences of lung cancer (y) seem to be associated with high smoking consumption (x).

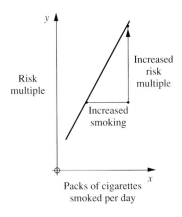

Packs of cigarettes
smoked per day

d. $r = \dfrac{285}{\sqrt{26.25}\ \sqrt{3116}} = .997$ (highly significant; $\alpha = .01$, df = 4; cutoff = ±.92)

e. % explained variation $= 100(.997)^2 = 99.40\%$.
Using four decimal places for r, we get 99.30%.

f. The data indicates the risk multiple for lung cancer is highly correlated to the number of packs of

cigarettes smoked. In other words, as the number of packs of cigarettes smoked increases, the risk of lung cancer increases markedly. Assuming the subjects of the experiment constitute a valid random sample from the overall population, the highly significant results would indicate the sample regression line is probably a close representation of the actual risk factors in the overall population.

The data for this study was based on information provided by the American Cancer Society, specifically the *Report of the Surgeon General, 1989 Executive Summary*, pp. 125–126. The study monitored female deaths from lung cancer in the United States during 1985. It was stated: "Cigarette smoking accounted for an estimated 82% of lung cancer deaths in women, or 31,600 deaths, in 1985."

Chapter Ten

10.1

a. H_0: In the population, education level and taste preference are independent.

Because 40% of the total sample liked ToastyOs (240/600) and large samples offer close approximations to population characteristics, we know approximately 40% of the population is expected to like ToastyOs.

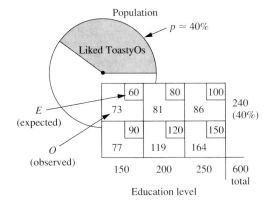

Now, *if* H_0 is true (education level and taste preference are independent) then the three education groups (150, 200, and 250) can be viewed as three valid unbiased samples drawn from this population (as valid and unbiased as the original 600). And we would expect each group to like ToastyOs in the same proportion 40%. This is calculated as follows:

Expected Values 40% of 150 = 60
40% of 200 = 80
40% of 250 = 100

Although the general formula is almost always used for χ^2 calculations we will demonstrate the underlying rationale (essentially, $\chi^2 = \Sigma z^2$), using this binomial experiment, as follows:

$$\chi^2 = z_1^2 + z_2^2 + z_3^2$$

Note that we had calculated the z-values at left, so substituting these into the previous equation, we get:

$$\chi^2 = z_1^2 + z_2^2 + z_3^2$$
$$= (2.1667)^2 + (.1443)^2 + (-1.8074)^2$$
$$= 4.6946 + .0208 + 3.2667$$
$$= 7.982$$

$$\boxed{\chi^2 \text{ sample} = 7.98}$$

In any χ^2 experiment, the observed values are compared to the expected values using z-score differences. χ^2 is then the sum of the z^2's.

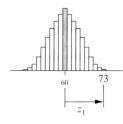

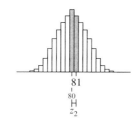

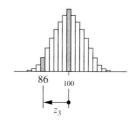

$\mu = np = 150(.4)$
$= 60$

$\sigma = \sqrt{np(1-p)}$
$= \sqrt{150(.4)(.6)}$
$= \sqrt{36}$
$= 6.0000$

$z_1 = \dfrac{x - \mu}{\sigma} = \dfrac{73 - 60}{6.000}$
$= 2.1667$

$\mu = np = 200(.4)$
$= 80$

$\sigma = \sqrt{np(1-p)}$
$= \sqrt{200(.4)(.6)}$
$= \sqrt{48}$
$= 6.9282$

$z_2 = \dfrac{x - \mu}{\sigma} = \dfrac{81 - 80}{6.928}$
$= .1443$

$\mu = np = 250(.4)$
$= 100$

$\sigma = \sqrt{np(1-p)}$
$= \sqrt{250(.4)(.6)}$
$= \sqrt{60}$
$= 7.7460$

$z_3 = \dfrac{x - \mu}{\sigma} = \dfrac{86 - 100}{7.7460}$
$= -1.8074$

General Formula

For all χ^2 testing, we will be using the general formula: $\chi^2 = \Sigma \dfrac{(O - E)^2}{E}$. For each cell, we subtract $O - E$, square this value, then divide by E. For the previous example,

$$\chi^2_{\text{sample}} = \frac{(73 - 60)^2}{60} + \frac{(77 - 90)^2}{90} + \frac{(81 - 80)^2}{80} + \frac{(119 - 120)^2}{120} + \frac{(86 - 100)^2}{100} + \frac{(164 - 150)^2}{150}$$

$$= \frac{169}{60} + \frac{169}{90} + \frac{1}{80} + \frac{1}{120} + \frac{196}{100} + \frac{196}{150}$$

$$= 2.8167 + 1.8778 + .0125 + .0083 + 1.9600 + 1.3067 = 7.982$$
$$= \boxed{7.98}$$

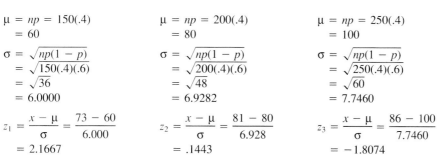

Note: each pair gives us the z^2 we calculated above (rounding techniques, however, may cause slight differences).

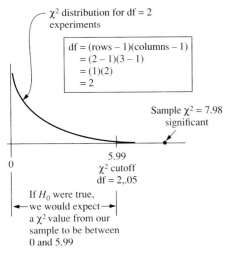

χ² distribution for df = 2
experiments

df = (rows – 1)(columns – 1)
= (2 – 1)(3 – 1)
= (1)(2)
= 2

Sample χ² = 7.98
significant

0

5.99
χ² cutoff
df = 2,.05

If H_0 were true,
← we would expect →
a χ² value from our
sample to be between
0 and 5.99

To obtain χ² cutoff from table

df	χ²$_{.05}$	χ²$_{.01}$
1		.
(2) →	(5.99)	.
.	.	.
.	.	.

If H_0 is true and if we were to perform this χ^2 df = 2 experiment many hundreds of thousands of times and plotted the resulting χ^2 values onto a histogram, its shape would resemble the curve above, with an average χ^2 value of 2 (for χ^2 experiments, the average χ^2 value equals its df) with 95% of the χ^2 values falling between 0 and 5.99. Since our sample χ^2 yielded 7.98,

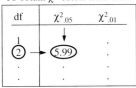

H_0 is rejected

Since the χ^2 value of the sample, 7.98, exceeded the cutoff value of 5.99, we reject H_0. This data supports the hypothesis, H_1, that a relationship between education level and taste preference exists in the population. In other words, we conclude that the fluctuation in sample results is not merely chance fluctuation, but fluctuation created by the interplay of the two variables in the population.

b. Although the data provides evidence of dependence in the population, we have not as yet established the direction of the dependence. In other words, we might be concerned with which education group might better prefer our product? To help us explore this, we translate each absolute number into a percentage-of-group total as follows:

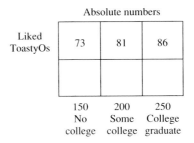

Absolute numbers

Liked
ToastyOs

| 73 | 81 | 86 |

150   200   250
No   Some   College
college college graduate

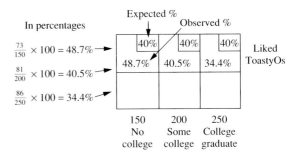

In percentages

Expected %   Observed %

$\frac{73}{150} \times 100 = 48.7\%$ →

$\frac{81}{200} \times 100 = 40.5\%$ →

$\frac{86}{250} \times 100 = 34.4\%$ →

| 40% | 40% | 40% | Liked
| 48.7% | 40.5% | 34.4% | ToastyOs

150   200   250
No   Some   College
college college graduate

Note that 48.7% of the no college group prefers ToastyOs versus only 40% expected. This suggests a market may exist in this segment that Tabisco may wish to explore with further research.

c. Although we might be tempted to suggest Tabisco Foods redirect their advertising and promotion to this group, the financial risks for such decisions are great. In marketing (and other social) experiments it is usually better advised that further evidence be gathered.

Because of the difficulty in establishing validity, often statistical experiments in marketing, education, and the soft sciences are treated as exploratory tools used in conjunction with other accumulated evidence. In this marketing example, perhaps sales records from the test market area could be examined, store clerks interviewed, or further consumer research conducted—perhaps in-depth interviews. In such interviews, a trained person is invited to the consumer's home to discuss various needs and uses for the product. Sometimes these chats last for hours, and in the hands of a highly experienced professional (using indirect questioning techniques and other methods of clinical psychology), consumers often reveal underlying motives and true feelings.*

After consideration of all the accumulated evidence, if we still have confidence in our findings, then perhaps a new test market, with redirected advertising, could be set up. Another advantage of in-depth interviewing is that if underlying motives for purchase can be identified, these may help us focus the advertising and promotion strategy. However, even if all this research were conducted properly, we are still not certain our efforts will

---

*In-depth interviews, according to Dietz Leonhard (Human Equation in Marketing Research, pp. 58–59, referenced below), ''makes small sampling possible and findings highly reliable.'' It also has the advantage of identifying underlying motives for purchase. For instance, certain types of automobiles might be purchased more to ''compensate for feelings of inadequacy,'' rather than for practical considerations. Extreme clothing styles may be ''compensation for feelings of being rejected or suppressed.'' Exotic or provocative perfumes may be purchased by young women because of the suppressed desire to be ''naughty.'' Businesspeople may purchase products because of a desire ''to be accepted by (other) business associates'' and to be praised by them. It is often the promise of companionship that motivates a teenage boy or girl to purchase a certain brand of soap, convinced the soap has the fragrance the opposite sex loves.

Knowledge of underlying motivation can often augment our effectiveness in marketing a product. For more readings on this, refer to publications by the American Management Association, New York, New York. Examples are

D. Leonhard, *The Human Equation in Marketing Research* (1967)

E. McKay, *Marketing Mystique* (1967)

J. Pope, *Practical Marketing Research*

---

result in actual sales at the grocery counter. For this, we put the product into a local test market, present our new advertising and promotion strategy, and wait for case-shipment data to confirm or negate our beliefs.

To recap, in social experiments, where validity may be difficult to verify, statistical studies should be used in conjunction with other accumulated evidence before drawing any definitive conclusions.

10.2

a. Initial conditions are established as follows:

H_0: The four methods are homogeneous with respect to patient improvement.

(In effect this means: there is no difference among the four methods, that is, each will result in the same levels of patient improvement.)

H_1: The four groups are *not* homogeneous with respect to patient improvement.

(Meaning one or more methods is more effective than the others.)

In tests of homogeneity, it is assumed each group was sampled from a separate population, but each population has identical (homogeneous) characteristics. In other words, each population has the same proportion showing major improvement, slight improvement, and no improvement. For this experiment $\alpha = .05$ meaning, if H_0 is indeed true, there is a 5% chance we will reject it in error.

Like any hypothesis test, we begin by assuming H_0 is true. That is, by assuming all four treatment groups were a sampling from four separate but equally proportioned (homogeneous) populations. We can estimate the common population proportions by calculating the percentage of the entire 800 patients who showed major improvement ($240/800 = 30\%$), who showed slight improvement ($160/800 = 20\%$), and no improvement ($400/800 = 50\%$).

	Caffeine coated to dissolve in barrier cells	Iron coated to fool cells into accepting it	Immune-cell coat to fool cells into accepting it	Taken with loosening compound	
Major improvement	53	50	76	61	240 (30%)
Slight improvement	42	28	45	45	160 (20%)
No improvement	105	72	129	94	400 (50%)
	200	150	250	200	Total: 800 patients

Level of patient improvement

Same medication: administered by four methods

These percentages (30%, 20%, and 50%) are next used to determine how many patients we would expect from each treatment group to show major, slight, and no improvement, respectively. This is calculated below.

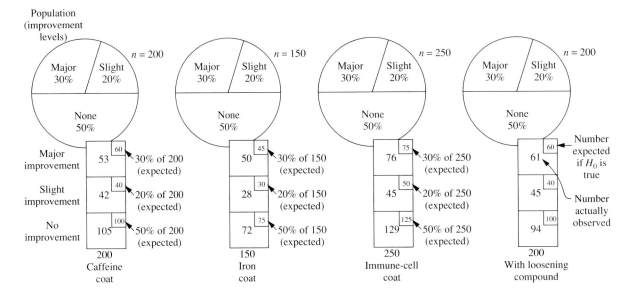

To calculate χ^2:

$$\chi^2 \; (\Sigma \, z^2) = \Sigma \frac{(O - E)^2}{E}$$

Note: For each and every cell, we subtract, $O - E$, square this value, then divide by E.

$$\chi^2 = \frac{(53-60)^2}{60} + \frac{(42-40)^2}{40} + \frac{(105-100)^2}{100} + \frac{(50-45)^2}{45} + \frac{(28-30)^2}{30} + \frac{(72-75)^2}{75}$$
$$+ \frac{(76-75)^2}{75} + \frac{(45-50)^2}{50} + \frac{(129-125)^2}{125} + \frac{(61-60)^2}{60} + \frac{(45-40)^2}{40} + \frac{(94-100)^2}{100}$$

$$\chi^2 = .817 + .100 + .250 + .556 + .133 + .120$$
$$+ .013 + .500 + .128 + .017 + .625 + .360 = 3.619$$

$$\boxed{\chi^2 = 3.62}$$

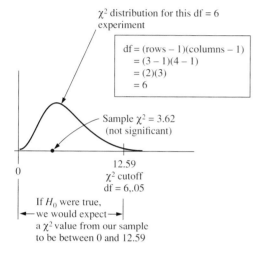

χ^2 distribution for this df = 6 experiment

$$df = (rows - 1)(columns - 1)$$
$$= (3 - 1)(4 - 1)$$
$$= (2)(3)$$
$$= 6$$

Sample χ^2 = 3.62
(not significant)

0

12.59
χ^2 cutoff
df = 6,.05

If H_0 were true,
←we would expect→
a χ^2 value from our sample
to be between 0 and 12.59

To obtain χ^2 cutoff from table

df	$\chi^2_{.05}$	$\chi^2_{.01}$
.	.	.
.	.	.
⑥ →	⟨12.59⟩	.
.	.	.
.	.	.

Now we know *if H_0 is true*, in other words, if the four populations are equally proportioned (homogeneous), *and if* we were to perform this χ^2 df = 6 experiment hundreds of thousands of times, the *z*-score differences between observed and expected values would be relatively modest, reflected in

relatively modest χ^2 readings (in fact, on average, if H_0 were true, χ^2 would be 6, with 95% of the χ^2 values between 0 and 12.59).

Since our sample χ^2 reading yielded 3.62, indicating relatively minor *z*-score differences between observed and expected values, we

Accept H_0

b. Since the χ^2 value of the sample (3.62) was less than the cutoff of 12.59, we accept H_0. The data supports the null hypothesis, that the four methods are indeed homogeneous with respect to patient improvement. In other words, we cannot statistically prove any difference in patient improvement exists among the four methods.

10.3

a. Establish initial conditions as follows:

H_0: In the population, market share has remained the same:
$p_1 = 30\%$
$p_2 = 25\%$
$p_3 = 20\%$
$p_4 = 25\%$

Essentially, the null hypothesis states the market share has not changed, even after the advertising war.

H_1: Market share has changed.

If H_0 proves unlikely, then we assume H_1 is true.

$\alpha = .05$

We accept the risk that when H_0 is true, 5% of the time we will reject it in error.

We start by assuming our sample of $n = 300$ originated from the four-category population specified in the null hypothesis. We then calculate the expected value in each cell by taking 30%, 25%, 20%, and 25% of 300, respectively, as demonstrated below.

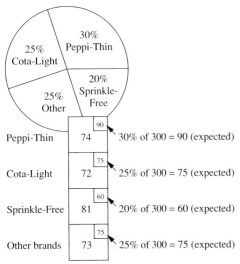

Peppi-Thin	74	30% of 300 = 90 (expected)
Cota-Light	72	25% of 300 = 75 (expected)
Sprinkle-Free	81	20% of 300 = 60 (expected)
Other brands	73	25% of 300 = 75 (expected)

Total: 300 sampled

Next, we calculate the differences between observed and expected values by using the chi-square formula, as follows:

$$\chi^2(\Sigma z^2) = \Sigma \frac{(O - E)^2}{E}$$

$$\chi^2 = \frac{(74 - 90)^2}{90} + \frac{(72 - 75)^2}{75}$$

$$+ \frac{(81 - 60)^2}{60} + \frac{(73 - 75)^2}{75}$$

$$\chi^2 = 2.844 + .120 + 7.350 + .053$$
$$= 10.367$$

$$\boxed{\chi^2 = 10.37}$$

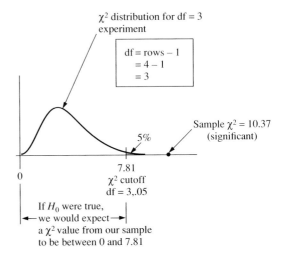

χ^2 distribution for df = 3 experiment

df = rows – 1
= 4 – 1
= 3

Sample $\chi^2 = 10.37$ (significant)

5%

7.81
χ^2 cutoff
df = 3, .05

If H_0 were true, we would expect a χ^2 value from our sample to be between 0 and 7.81

χ^2 table

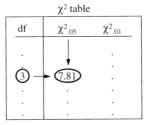

df	$\chi^2_{.05}$	$\chi^2_{.01}$
.		.
.		.
③	7.81	.
.		.
.		.

Now we know if H_0 were true (if the sample originated from a four-category population as specified in the null hypothesis), *and* if we were to perform this χ^2 df = 3 experiment hundreds of thousands of times, the differences between observed and expected values would be relatively modest, reflected in relatively modest χ^2 readings (in fact, on average, if H_0 were true, χ^2 would be 3, with 95% of the χ^2 values between 0 and 7.81).

Since our sample χ^2 reading yielded 10.37, indicating relatively large z-score differences between observed and expected values, we

Reject H_0

b. Since the χ^2 value of the sample (10.37 exceeded the cutoff value of 7.81), we reject H_0. Thus, the evidence supports H_1, the alternative hypothesis, that market share has changed.

To assess the implications for Peppi-Thin, we first translate the absolute numbers into observed and expected percentages, as follows:

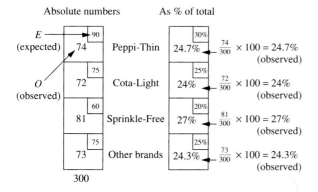

Absolute numbers   As % of total

E (expected) 90, 74 — Peppi-Thin — 30%, 24.7% ⟵ $\frac{74}{300} \times 100 = 24.7\%$ (observed)

O (observed) 75, 72 — Cota-Light — 25%, 24% ⟵ $\frac{72}{300} \times 100 = 24\%$ (observed)

60, 81 — Sprinkle-Free — 20%, 27% ⟵ $\frac{81}{300} \times 100 = 27\%$ (observed)

75, 73 — Other brands — 25%, 24.3% ⟵ $\frac{73}{300} \times 100 = 24.3\%$ (observed)

300

Although postevaluation testing for χ^2 and ANOVA experiments usually involve sophisticated techniques reserved for more advanced courses, certainly a few simple observations can be made concerning the most extreme differences, notably Peppi-Thin (30% expected, 24.7% observed) and Sprinkle-Free (20% expected, 27% observed). It would probably be safe to speculate that Sprinkle-Free gained real market share, and probably at the expense of Peppi-Thin.

10.4

H_0: In the population, usage is the same in all socioeconomic neighborhoods.

Essentially, H_0 states there is no difference in the percentages of usage among the various neighborhoods.

H_1: In the population, usage is *not* the same in all socioeconomic neighborhoods.

If H_0 proves unlikely, then we assume H_1 is true.

$\alpha = .01$

We accept the risk that when H_0 is true, 1% of the time we will reject it in error.

Essentially, we start by assuming our sample of $n = 400$ cans originated from a five-category population as specified in the null hypothesis, that is, where usage in all neighborhoods is equal. Since there are five types of neighborhoods, then each should yield $\frac{1}{5}$ or 20% of the Campbell Soup cans found, demonstrated below.

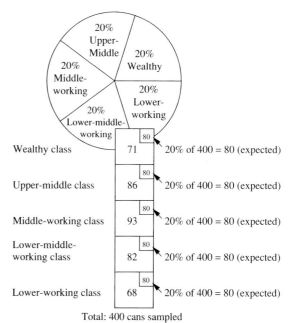

20% Upper-Middle / 20% Wealthy / 20% Middle-working / 20% Lower-working / 20% Lower-middle-working

Wealthy class — 71 — 20% of 400 = 80 (expected)

Upper-middle class — 86 — 20% of 400 = 80 (expected)

Middle-working class — 93 — 20% of 400 = 80 (expected)

Lower-middle-working class — 82 — 20% of 400 = 80 (expected)

Lower-working class — 68 — 20% of 400 = 80 (expected)

Total: 400 cans sampled

Next, we calculate the differences between observed and expected values by using the chi-square general formula, as follows:

$$\chi^2(\Sigma\, z^2) = \Sigma\, \frac{(O-E)^2}{E}$$

$$\chi^2 = \frac{(71-80)^2}{80} + \frac{(86-80)^2}{80} + \frac{(93-80)^2}{80}$$

$$+ \frac{(82-80)^2}{80} + \frac{(68-80)^2}{80}$$

$$\chi^2 = 1.013 + .450 + 2.113$$
$$+ .050 + 1.800 = 5.426$$

$$\boxed{\chi^2 = 5.43}$$

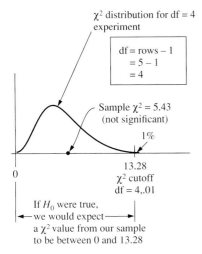

χ^2 distribution for df = 4 experiment

df = rows − 1
= 5 − 1
= 4

Sample χ^2 = 5.43
(not significant)

1%

13.28
χ^2 cutoff
df = 4,.01

0

If H_0 were true, we would expect a χ^2 value from our sample to be between 0 and 13.28

χ^2 table

df	$\chi^2_{.05}$	$\chi^2_{.01}$
.	.	.
.	.	.
④	.	13.28
.	.	.
.	.	.
.	.	.

Now we know if H_0 were true, that is, indeed the sample originated from a five-category population with equal proportions, and if we were to perform hundreds of thousands of such experiments, the differences between observed and expected values would be relatively modest, reflected in relatively modest χ^2 readings. (In fact, on average, if H_0 is true,* χ^2 will be 4, with 99% of the χ^2 readings falling between 0 and 13.28.)

Since our sample χ^2 reading yielded 5.43, indicating relatively small z-score differences between observed and expected values, we

Accept H_0

The results indicate no difference in Campbell soup usage among the different neighborhoods.

Actually, early in the century, C. C. Parlin used similar results to convince the Campbell Soup Company to advertise in the *Saturday Evening Post*, a magazine thought to appeal to working-class families. At the time, Campbell Company felt only the rich would pay the extra price for the luxury of canned soup.

10.5
a. We compute $\bar{x}$ and s^2 for each sample as follows:

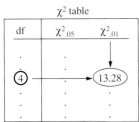

Variance

$$s^2 = \frac{\Sigma(x - \bar{x})^2}{n - 1}$$

$$s = \sqrt{\frac{\Sigma(x - \bar{x})^2}{n - 1}}$$

Note: the square root of the variance is the standard deviation.

x	$\bar{x}$	$x - \bar{x}$	$(x - \bar{x})^2$
15	16	−1	1
11	16	−5	25
18	16	2	4
20	16	4	16
16	16	0	0
16	16	0	0
12	16	−4	16
19	16	3	9
17	16	1	1
144			72
Σx			$\Sigma(x - \bar{x})^2$

$$\bar{x} = \frac{\Sigma x}{n} = \frac{144}{9} = 16$$

$$s^2 = \frac{\Sigma(x - \bar{x})^2}{n - 1}$$

$$= \frac{72}{9 - 1} = 9.00$$

$n = 9$
$\bar{x} = 16$
$s^2 = 9.00$
For campaign A
(aggressive ads)

*For the above χ^2 distribution, the average χ^2 value would be 4. In other words, for the above χ^2 distribution.

$$\mu = \text{df (degrees of freedom)}$$
$$= 4$$

x	$\bar{x}$	$x - \bar{x}$	$(x - \bar{x})^2$
16	17	-1	1
14	17	-3	9
19	17	2	4
21	17	4	16
15	17	-2	4
17	17	0	0
13	17	-4	16
20	17	3	9
18	17	1	1
153			60
Σx			$\Sigma(x - \bar{x})^2$

$$\bar{x} = \frac{\Sigma x}{n} = \frac{153}{9} = 17$$

$$s^2 = \frac{\Sigma(x - \bar{x})^2}{n - 1}$$

$$= \frac{60}{9 - 1} = 7.50$$

$n = 9$
$\bar{x} = 17$
$s^2 = 7.50$
For campaign B
(mod. aggressive ads)

x	$\bar{x}$	$x - \bar{x}$	$(x - \bar{x})^2$
13	15	-2	4
17	15	2	4
18	15	3	9
15	15	0	0
10	15	-5	25
12	15	-3	9
16	15	1	1
16	15	1	1
18	15	3	9
135			62
Σx			$\Sigma(x - \bar{x})^2$

$$\bar{x} = \frac{\Sigma x}{n} = \frac{135}{9} = 15$$

$$s^2 = \frac{\Sigma(x - \bar{x})^2}{n - 1}$$

$$= \frac{62}{9 - 1} = 7.75$$

$n = 9$
$\bar{x} = 15$
$s^2 = 7.75$
For campaign C
(soft-sell ads)

b. The initial conditions for ANOVA are as follows:

$H_0: \mu_A = \mu_B = \mu_C$
(no difference in average sales increase in the 3 test markets)

Essentially this says the average sales increase will be the same in all cities. That is, if we computed the *average* sales increase for all the many thousands of outlets in Midland, this would equal the *average* sales increase for all the many thousands of outlets in Indianapolis, and this would equal the *average* sales increase for all the many thousands of outlets in Charleston. In other words, no difference in average sales increase in the three test markets.

H_1: Not all μ's are the same.

If H_0 proves unlikely, we accept H_1, that at least one μ is different.
$\alpha = .05$
Meaning, if H_0 is true, a 5% chance exists we will reject it in error.

Any hypothesis test begins by assuming H_0 is true, and based on this premise, constructing a set of expectations.

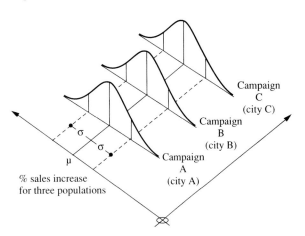

% sales increase
for three populations

So, indeed, if H_0 is true, that is, if the average sales increase in each city is identical (meaning, $\mu_A = \mu_B = \mu_C$) and we satisfy the initial population assumptions (normal distributions with equal standard deviations, $\sigma_A = \sigma_B = \sigma_C$, then all three populations, for sampling purposes, are identical, with common mean, μ, and common standard deviation, σ, shown below:

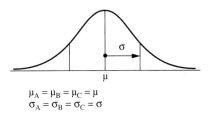

$$\mu_A = \mu_B = \mu_C = \mu$$
$$\sigma_A = \sigma_B = \sigma_C = \sigma$$

Essentially ANOVA compares two different estimates of the population variance, σ^2, one based on the spread of the 3 $\bar{x}$'s and the other based on the spread of data within each sample, essentially the average s^2, called s^2_{pooled}. Specifically,

(estimate based on the spread of the three $\bar{x}$'s and referred to as variation between samples)

$$F = \frac{\text{estimate of } \sigma^2}{\text{estimate of } \sigma^2} = \frac{n\, s^2_{\bar{x}}}{s^2_{pooled}} = \frac{(9)(1.00)}{8.0833}$$
$$= 1.11$$

(estimate based on the average of s^2's denoted s^2_{pooled} and referred to as variation within samples)

Top of Fraction Calculations

$\bar{x}$	$\bar{\bar{x}}$	$(\bar{x} - \bar{\bar{x}})$	$(\bar{x} - \bar{\bar{x}})^2$
16	16	0	0
17	16	1	1
15	16	−1	1
48			2

$$\bar{\bar{x}} = \frac{\Sigma \bar{x}}{k} = \frac{48}{3} = 16$$

(k = number of $\bar{x}$'s)

$$s^2_{\bar{x}} = \frac{\Sigma(\bar{x} - \bar{\bar{x}})^2}{k - 1} = \frac{2}{3 - 1} = 1.00$$

Substituting in common sample size, $n = 9$ and $s^2_{\bar{x}} = 1.00$ into the top of the F fraction ($ns^2_{\bar{x}}$), we get:
$(9)(1.00)$

Bottom of Fraction Calculations

$$s^2_{pooled} = \frac{s^2_A + s^2_B + s^2_C}{3}$$
$$= \frac{\dfrac{72}{9 - 1} + \dfrac{60}{9 - 1} + \dfrac{62}{9 - 1}}{3}$$
$$= \frac{9 + 7.5 + 7.75}{3} = 8.0833$$

Essentially, if one sample s^2 offers a good estimate of σ^2, then three combined (averaged) should offer an even better estimate.

In actual practice, the F ratio is calculated using more efficient formulas which minimize rounding error and lend themselves to broader analysis. To understand the underlying rationale of these formulas, the following is presented.

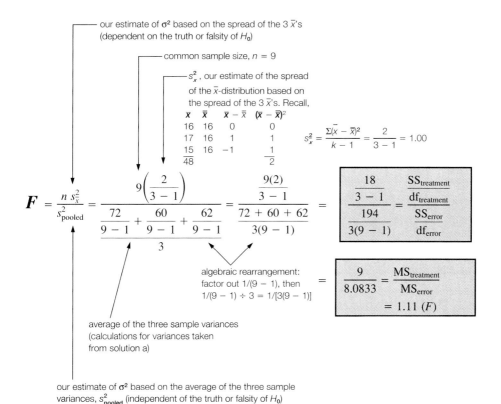

our estimate of σ² based on the spread of the 3 $\bar{x}$'s (dependent on the truth or falsity of H_0)

common sample size, $n = 9$

$s_{\bar{x}}^2$, our estimate of the spread of the $\bar{x}$-distribution based on the spread of the 3 $\bar{x}$'s. Recall,

x	$\bar{x}$	$\bar{x} - \bar{\bar{x}}$	$(\bar{x} - \bar{\bar{x}})^2$
16	16	0	0
17	16	1	1
15	16	−1	1
$\overline{48}$			$\overline{2}$

$$s_{\bar{x}}^2 = \frac{\Sigma(\bar{x} - \bar{\bar{x}})^2}{k - 1} = \frac{2}{3 - 1} = 1.00$$

$$F = \frac{n\, s_{\bar{x}}^2}{s_{pooled}^2} = \frac{9\left(\dfrac{2}{3-1}\right)}{\dfrac{72}{9-1} + \dfrac{60}{9-1} + \dfrac{62}{9-1}} = \frac{\dfrac{9(2)}{3-1}}{\dfrac{72 + 60 + 62}{3(9-1)}} = \begin{array}{c} \dfrac{18}{3-1} \\ \hline \dfrac{194}{3(9-1)} \end{array} = \dfrac{SS_{treatment}}{\dfrac{SS_{error}}{df_{error}}}$$

average of the three sample variances (calculations for variances taken from solution a)

algebraic rearrangement: factor out 1/(9 − 1), then 1/(9 − 1) ÷ 3 = 1/[3(9 − 1)]

$$= \dfrac{9}{8.0833} = \dfrac{MS_{treatment}}{MS_{error}}$$

$$= 1.11\ (F)$$

our estimate of σ² based on the average of the three sample variances, s_{pooled}^2 (independent of the truth or falsity of H_0)

Notice the computations are partitioned into sum-of-squared distances related to the three $\bar{x}$'s ($SS_{treatment}$) and sum-of-squared distances related to the three s^2's (SS_{error}). Expressing the results in terms of sum-of-squared distances has many advantages in advanced work. Generally in ANOVA experiments the above information is made available in the form of the following chart.

Source	df		SS	MS	F
Treatment	$k - 1 =$	2	18	9	1.11
Error	$k(n - 1) =$	24	194	8.0833	
Total		26	212		

Legend:
$df_t = k - 1$
  = Number of groups − 1
$df_e = k(n - 1)$
  = Number of groups times one less than the common sample size
SS_t = Sum-of-squared distances (treatment).
SS_e = Sum-of-squared distances (error).
$MS_t = SS_t/df_t$ (Mean-square distance, treatment)
$MS_e = SS_e/df_e$ (Mean-square distance, error)

Relatively easy to use formulas are available to calculate SS_{total} and $SS_{treatment}$ (and by subtraction, we can get SS_{error}). The following demonstrates use of these formulas.

City A		City B		City C	
x	x^2	x	x^2	x	x^2
15	225	16	256	13	169
11	121	14	196	17	289
18	324	19	361	18	324
20	400	21	441	15	225
16	256	15	225	10	100
16	256	17	289	12	144
12	144	13	169	16	256
19	361	20	400	16	256
17	289	18	324	18	324
144	2376	153	2661	135	2087

Σx^2(sum of all x^2's $= 2376 + 2661 + 2087$
$$= 7124$$
Σx(sum of all x's) $= 144 + 153 + 135$
$$= 432$$
ΣT^2(sum of treatment totals squared)
$$= 144^2 + 153^2 + 135^2$$
$$= 62{,}370$$

n = common sample size = 9
N = total number of values = 27
k = number of groups = 3
T = group total (144, 153, or 135)

$$SS_{total} = \Sigma x^2 - \frac{(\Sigma x)^2}{N}$$

$$SS_{treatment} = \frac{\Sigma T^2}{n} - \frac{(\Sigma x)^2}{N}$$

$$SS_{total} = \Sigma x^2 - \frac{(\Sigma x)^2}{N}$$
$$= 7124 - \frac{(432)^2}{27}$$
$$= 212$$

$$SS_{total} = 212$$

$$SS_{treatment} = \frac{\Sigma T^2}{n} - \frac{(\Sigma x)^2}{N}$$
$$= \frac{62{,}370}{9} - \frac{(432)^2}{27}$$
$$= 18$$

$$SS_{treatment} = 18$$

$$SS_{total} = SS_{treatment} + SS_{error}$$
$$212 = 18 + SS_{error}$$

By subtraction,
$$SS_{error} = 194$$

$$F = \frac{SS_{treatment}/df_{treatment}}{SS_{error}/df_{error}} = \frac{SS_t/k - 1}{SS_e/k(n - 1)} = \frac{18/3 - 1}{194/3(9 - 1)} = \frac{9}{8.0833} = 1.11$$

Now, with all this in mind, how do we conduct the ANOVA test? Pretty much the same way we conduct any hypothesis test.

We know from theory and experience that indeed if H_0 were true, that is, $\mu_A = \mu_B = \mu_C$ *and* if we were to conduct this df = 2,24 ANOVA experiment hundreds of thousands of times and calculated the F ratio for each experiment, that 95% of these F ratios would fall at or below 3.40 (with a great many F values clustered in the general vicinity of 1.00 [one]).

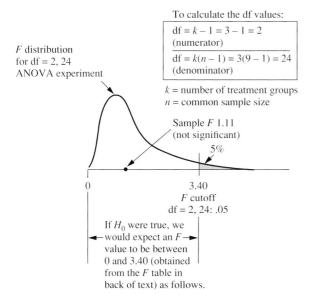

F distribution for df = 2, 24 ANOVA experiment

To calculate the df values:

df = $k - 1 = 3 - 1 = 2$ (numerator)

df = $k(n - 1) = 3(9 - 1) = 24$ (denominator)

k = number of treatment groups
n = common sample size

Sample F 1.11 (not significant)

5%

0

3.40
F cutoff
df = 2, 24: .05

If H_0 were true, we would expect an F value to be between 0 and 3.40 (obtained from the F table in back of text) as follows.

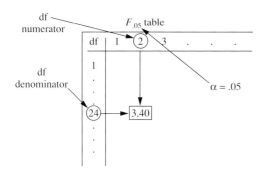

df numerator

$F_{.05}$ table

df | 1 | ② | 3 . . .

df denominator

$\alpha = .05$

1
.
.

②④ → ▸ 3.40

.
.
.

Keep in mind, even if H_0 is true, the F ratio may very well exceed 1.00. This is because sample results, especially small-sample $\bar{x}$'s and s^2's, tend to fluctuate rather widely around the population values they estimate, which then causes the F ratio to fluctuate somewhat. However, we know it is quite unusual in a df = 2,24 ANOVA experiment for the F ratio (if H_0 is true) to exceed 3.40, in fact, this occurs only 5% of the time.

Now, since our data yielded an F ratio of 1.11, we

Accept H_0

That is, we accept $\mu_A = \mu_B = \mu_C$. In other words, we cannot prove any differences in sales exist in the three test cities.

c. Since we could not prove any differences in sales in the three populations, we either reserve judgment until further evidence is available or assume the ad campaigns are equally as effective.

Unfortunately such studies are vulnerable to a multitude of possible unmeasured factors that may influence results. In the chapter (under the headings Overview of ANOVA and Experimental Designs) we offer a discussion of such influences and research designs which tend to minimize these effects.

d. H_0: $\mu_A = \mu_B = \mu_C$ (no difference in sale increases in the three cities)

H_1: at least one μ is different

$\alpha = .01$

$$F = \frac{n\, s_{\bar{x}}^2}{s_{pooled}^2} = \frac{(11)(7)}{11.4} = \boxed{6.75 \text{ significant}}$$

Important note: Since $\alpha = .01$, we must use the $F_{.01}$ table (not the $F_{.05}$ table) to determine the cutoff

df (numerator; top of table) = $k - 1$
$= 3 - 1$
$= 2$

df (denominator; side of table) = $k(n - 1)$
$= 3(11 - 1)$
$= 30$

$$F_{cutoff} = 5.39$$

Common Sample Size, $n = 11$

Calculation for $s_{\bar{x}}^2$:

$\bar{x}$	$\bar{\bar{x}}$	$\bar{x} - \bar{\bar{x}}$	$(\bar{x} - \bar{\bar{x}})^2$
12	15	-3	9
16	15	1	1
17	15	2	4
45			14

$$\bar{\bar{x}} = \frac{45}{3} = 15$$

$$s_{\bar{x}}^2 = \frac{14}{3 - 1} = 7$$

Calculation for s_{pooled}^2:

$$s_{pooled}^2 = \frac{11.9 + 9.5 + 12.8}{3}$$

$$= 11.4$$

Since our sample F (6.75) exceeded the cutoff of 5.39, *Reject H_0*.

We know at least one pair of treatments is significantly different, meaning there were substantial enough differences in the samples that we can conclude real differences exist in at least one pair of the populations. Certainly the highest $\bar{x}$ (city C, our current soft-sell strategy) and the lowest $\bar{x}$ (city A; aggressive ads) would identify one pair. However, further analysis (beyond the scope of this text) is necessary to determine whether differences exist in the other pairs, that is, between cities A and B and cities B and C).

e. Given the test was properly conducted and we have reasonable assurances that sample results are representative of U.S. purchase behavior, we conclude that our current soft-sell ads (tested in city C) would probably best be maintained. Clearly these outperformed aggressive ads (tested in city A), and there is certainly no evidence that moderately aggressive ads fare any better than our current soft-sell ads.

f. $H_0: \mu_A = \mu_B = \mu_C$ (no difference in sale increases in the three cities)

H_1: at least one μ is different

$\alpha = .01$

City A Treatment A (aggressive ads)	City B Treatment B (moderately aggressive ads)	City C Treatment C (soft-sell ads)
11	19	14
9	12	18
12	16	22
8	12	17
4	24	13
10	11	15
$T_1 = 54$	15	17
$n_1 = 6$	18	$T_3 = 116$
	$T_2 = 127$	$n_3 = 7$
	$n_2 = 8$	

$$SS_{total} = \Sigma x^2 - \frac{(\Sigma x)^2}{N} = (11^2 + 9^2 + 12^2 + \cdots + 15^2 + 17^2) - \frac{(54 + 127 + 116)^2}{21}$$

$$= (4653) - \frac{88,209}{21} = 452.57$$

$$SS_{treatment} = \Sigma \frac{T_i^2}{n_i} - \frac{(\Sigma x)^2}{N} = \left(\frac{54^2}{6} + \frac{127^2}{8} + \frac{116^2}{7}\right) - \frac{(54 + 127 + 116)^2}{21}$$

$$= (4424.41) - \frac{88,209}{21} = 223.98$$

$$SS_{error} = SS_{total} - SS_{treatment}$$

$$= 452.57 - 223.98 = 228.59$$

g.

Source	df	SS	MS	F
Treatment	2	223.98	111.99	8.82 (significant)
Error	18	228.59	12.70	
Total	20	452.57		

Note:
$df_t = k - 1$
$= 3 - 1$
$= 2$
$df_e = \Sigma(n_j - 1)$
$= (6 - 1) + (8 - 1) + (7 - 1)$
$= 18$

F Calculation:
$$F = \frac{SS_{treatment}/df_{treatment}}{SS_{error}/df_{error}} = \frac{223.98/2}{228.59/18} = \frac{111.99}{12.70} = 8.82$$

In this case, $F_{cutoff\ 2,\ 18;.01} = 6.01$. We have a df = 2, 18 experiment. Look across the top for a df of 2, then proceed down for the df of 18.

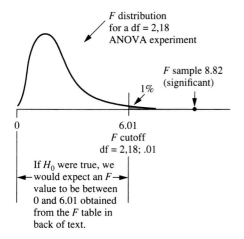

F distribution for a df = 2,18 ANOVA experiment

F sample 8.82 (significant)

1%

0

6.01
F cutoff
df = 2,18; .01

If H_0 were true, we would expect an F value to be between 0 and 6.01 obtained from the F table in back of text.

$F_{.01}$ table

df	1	②	3	. . .
1				
⑱		6.01		

Since our sample F (8.82) exceeded the cutoff of 6.01, *reject H_0*.

We know at least one pair of treatments is significantly different, meaning, there were substantial enough differences in the samples that we can conclude real differences exist in at least one pair of the populations.

10.7

a. H_0: $p_1 = p_2 = p_3 = p_4$ (no difference in patient survival rate among the four surgeons; common proportion, $p_s = 6\%$ for patients not surviving)

$$\chi^2 = 3 + .1915 + 4.5 + .2872 + 0 + 0 + .5 + .0319 = 8.51$$

χ^2 cutoff(df = 3; .05) is 7.81; since sample $\chi^2 >$ cutoff, reject H_0

b. Actual differences exist among the surgeons, with surgeon M exhibiting the highest proportion of patients not surviving (9% versus 6% expected) and surgeon K the lowest (3% versus 6% expected).

c. In a valid experiment, the researcher might randomly assign patients to the surgeons. However, this is rarely possible in real-life studies. Thus, other factors should be taken into account, such as, the age, sex, and physical condition of the patient. Advanced techniques (beyond the scope of this text) are used to adjust for such variations. In the article describing the actual studies, it was stated:

> In a multivariate analysis that adjusted for age, sex, comobid conditions, reoperation, ejection fraction, priority of surgery (emergent vs. scheduled), and other factors, researchers noted a . . . 4.2-fold difference among surgeons (2.2% to 9.3%).

To further substantiate validity (to rule out the effects of hospital environment, assisting staff, or equipment), it was noted:

> When surgeons moved from one hospital to another, their mortality rates remained about the same, even if other surgeons in the new hospital had quite different rates. (*Mass. Medical Society Journal Watch,* 8:5, p. 35)

10.9

H_0: For each source of treatment, the levels of pretesting are in the same proportion (common proportions: 60%, 30%, and 10%)

$\chi^2 = 4.36 + .01 + 28.03 + 1.8 + .28 + 17.63 + .83$
   $+ .27 + 1.8 = 55.01$

χ^2 cutoff (df = 4; .01) is 13.28; since sample $\chi^2 >$ cutoff; reject H_0
Actual differences exist among the three sources of treatment.

10.11

H_0: The sample buttercups conform to Pearson's distribution.

$$\chi^2 = .111 + .871 + .012 + .643 = 1.64$$

χ^2 cutoff (df = 3; .01) is 11.34; since sample $\chi^2 <$ cutoff, accept H_0
There is evidence to suggest buttercups fit Pearson's distribution.

10.13

H_0: Brand loyalty is uniform among the top fashion houses in this country.

$$\chi^2 = 1.25 + .05 + .20 + 1.80 + .20 = 3.50$$

χ^2 cutoff (df = 4; .01) is 13.28; since sample $\chi^2 <$ cutoff, accept H_0
There is evidence to suggest brand loyalty is uniform among the top fashion houses.

10.15

a. $n = 6$ $n = 6$ $n = 6$
   $\bar{x} = 10.00$ $\bar{x} = 9.50$ $\bar{x} = 12.00$
   $s^2 = 2.00$ $s^2 = 2.30$ $s^2 = 2.80$

b. H_0: $\mu_X = \mu_Y = \mu_Z$ (no difference in stain-eliminating ability of the three formulations)

$$F = \frac{ns_{\bar{x}}^2}{s_{pooled}^2} = \frac{(6)(1.75)}{2.3667} = 4.4366 = 4.44$$

F cutoff (df = 2, 15; .05) is 3.68

Note: $s_{pooled}^2 = \dfrac{2 + 2.3 + 2.8}{3} = 2.3667$

$s_{\bar{x}}^2 = 3.50/2 = 1.75$

Since sample F (4.44) > cutoff (3.68), reject H_0.

c. Generally, when using analysis of variance and H_0 is rejected, we state, at least one pair of treatments is significantly different (certainly the pair with the highest and lowest $\bar{x}$'s) and further analysis is necessary to determine whether significant differences exist with the other pairs. In this case, chemical formulation Z ($\bar{x} = 12.00$) has significantly better stain-eliminating abilities than chemical formulation Y ($\bar{x} = 9.50$). Further mathematical analysis is necessary to determine if chemical formulation Z is superior to chemical formulation X.

d. Designing statistically valid experiments is an art and a science best left to experts, however, some suggestions include: (i) cutting strips from the same stained cloth; (ii) then, randomly assigning these strips to the different formulations; (iii) using the same person to evaluate the stain-elimination ratings; and (iv) restricting the experiment to one material (for instance, a certain type and grade of cotton). Unfortunately, we are never 100% sure of the validity until many other researchers test the chemical formulations under varying circumstances and reach the same conclusions. Until then, it is best to be unmercifully self-critical.

e. H_0: $\mu_A = \mu_B = \mu_C$ (no difference in stain-eliminating ability of the three formulations)

$$F = \frac{n\,s_{\bar{x}}^2}{s_{pooled}^2} = \frac{(11)(3.25)}{7.9} = 4.5253 = 4.53$$

F cutoff (df = 2, 30; .05) is 3.32

Note: $s_{pooled}^2 = \dfrac{9.6 + 7.3 + 6.8}{3} = 7.9$

$s_{\bar{x}}^2 = 6.50/2 = 3.25$

Since sample $F >$ cutoff, reject H_0

f. Chemical formulation A ($\bar{x} = 13.5$, the highest) has significantly better stain-removing properties than formulation B ($\bar{x} = 10.0$, the lowest). Further mathematical analysis is necessary to determine a comparison to formulation C.

g. $SS_{treatment} = 34.1524$; $SS_{error} = 45.8476$; $SS_{total} = 80$

h. $F = \dfrac{SS_{treatment}/df_{treatment}}{SS_{error}/df_{error}} = \dfrac{34.1524/2}{45.8476/15} = 5.59$

F cutoff (df = 2, 15; .05) is 3.68
Since sample F (5.59) > cutoff (3.68), reject H_0
Actual differences exist in the formulations.

Chapter Eleven

11.1

a. This is a small-n binomial example since $np = 7(.2) = 1.4$, which is less than 5. To solve it, we can use either the binomial formula:

$$P(3) = \frac{7!}{(7-3)!3!} (.2)^3(1-.2)^{7-3}$$
$$= \frac{7 \cdot 6 \cdot 5 \cdot 4 \cdot 3 \cdot 2 \cdot 1}{(4 \cdot 3 \cdot 2 \cdot 1)3 \cdot 2 \cdot 1} (.2)^3(.8)^4$$
$$= (35)(.008)(.4096) = .114688 = .115 \ (11.5\%)$$

or simply look up the answer in the binomial tables in back of the text. In the binomial tables, look up $n = 7$, then down the column $p = .2$ (20%). This gives all the probabilities for 0, 1, 2, 3, etc. successes. $P(3 \text{ successes}) = .115 = 11.5\%$.

b. $H_0: p = 20\%$ (.2) for this hypothesis test. Since it is small-n, we can use either the binomial formula or the binomial tables in back of the text. If you use the tables, first look for $n = 12$, then down the column $p = .2$. This gives the probabilities for 0, 1, 2, etc. successes. $P(0 \text{ successes}) = .069 = 6.9\%$: $\alpha = .05$ implies reject for probabilities less than about $2\frac{1}{2}\%$ in either tail; since 6.9% exceeds this value, accept $H_0: p = 20\%$.

11.2

a. $H_0: p = 3\%(.030)$; expected proportion $p = .030$, with standard deviation, $\sigma_p = \sqrt{\dfrac{(.03)(.97)}{2000}}$
$$= \sqrt{.0000145} = .00381; \ p_s = 72/2000 = .036$$

$$\text{sample } z = \frac{p_s - p}{\sigma_p}$$
$$= \frac{.036 - .030}{.00381} = 1.57 \text{ (not significant)}$$

Since sample z (1.57) is within the z cutoffs of ± 1.96, thus we consider 72 successes out of 2000 selections chance fluctuation, and accept H_0; $p = 3\%$.

b. Placing the α risk (1%) in the right tail, we get a cutoff of $z = 2.33$ and reject above this value. Since our sample z (1.57) is less than 2.33, we accept H_0; $p \leq 3\%$.

c. To get a confidence interval, we use the formula: $p_s \pm z\sigma_p = .036 \pm 1.96(.00381) = .036 \pm .007$; thus we are 95% confident that the true population proportion is contained in the interval between .029 and .043 (2.9% and 4.3%).

11.3

$H_0: p_1 = p_2$; common proportion, $p = (39 + 21)/(300 + 500) = .075$ (7.5%). Expected value of the $p_{s_1} - p_{s_2}$ sampling distribution is 0 with standard deviation, $\sigma_{p_{s_1} - p_{s_2}} = .0192$ (calculated below); $p_{s_1} = 39/300 = .130$ and $p_{s_2} = 21/500 = .042$. Sample $z = 4.58$, as follows:

$$\sigma_{p_{s_1} - p_{s_2}} = \sqrt{\frac{(.075)(.925)}{300} + \frac{(.075)(.925)}{500}}$$
$$= \sqrt{.0002312 + .0001387}$$
$$= .0192327 = .0192$$

$$z = \frac{p_{s_1} - p_{s_2}}{\sigma_{p_{s_1} - p_{s_2}}}$$
$$= \frac{.130 - .042}{.0192}$$
$$= 4.58 \text{ (significant)}$$

Since sample $z = 4.58$ is outside the cutoffs of ± 1.96, we *reject* H_0. Thus we conclude $p_1 \neq p_2$. The samples provide evidence that the two population proportions are not equal. An examination of the data suggests mental illness later in life is more prevalent among those who exhibit highly delusional thinking patterns in college than among those who test nondelusional in college.

11.4

a. $H_0: \mu_1 = \mu_2$; the $\bar{x}_1 - \bar{x}_2$ sampling distribution has average value of 0 with standard deviation, $\sigma_{\bar{x}_1 - \bar{x}_2} = .3866$ (calculated below); we use $\bar{x}_1 = 4.2$ and $\bar{x}_2 = 3.8$ to get sample $z = 1.03$, as follows:

$$\sigma_{\bar{x}_1 - \bar{x}_2} = \sqrt{\frac{(1.8)^2}{38} + \frac{(1.7)^2}{45}}$$
$$= \sqrt{.08526 + .06422}$$
$$= .38662 = .3866$$

$$z = \frac{\bar{x}_1 - \bar{x}_2}{\sigma_{\bar{x}_1 - \bar{x}_2}}$$
$$z = \frac{4.2 - 3.8}{.3866}$$
$$z = 1.03 \text{ (not significant)}$$

Since our sample z of 1.03 is within our cutoffs of ± 1.96, we *accept* $H_0: \mu_1 = \mu_2$, no difference between the groups in the ability to detect a liar.

b. The setup is identical as in part a, only now we use our weighted $\sigma_{\bar{x}_1 - \bar{x}_2}$ formula for small samples and the t tables for the cutoff value (with df $= n_1 + n_2 - 2$). In this case, $\sigma_{\bar{x}_1 - \bar{x}_2} = .687$ and sample $t = .58$.

$$\sigma_{\bar{x}_1 - \bar{x}_2} = \sqrt{\frac{(12 - 1)(1.8)^2 + (14 - 1)(1.7)^2}{12 + 14 - 2}\left(\frac{1}{12} + \frac{1}{14}\right)}$$

$$= \sqrt{\frac{35.64 + 37.57}{24}(.08333 + .07143)}$$

$$= \sqrt{.4720824} = .68708 = .687$$

$$t = \frac{\bar{x}_1 - \bar{x}_2}{\sigma_{\bar{x}_1 - \bar{x}_2}}$$

$$= \frac{4.2 - 3.8}{.687}$$

$$= .58 \text{ (not significant)}$$

Since our sample t of .58 is within our cutoffs of ± 2.06 (df $= 12 + 14 - 2 = 24$), we accept H_0: $\mu_1 = \mu_2$, no difference between the groups in the ability to detect a liar.

11.5

H_0: $\mu_1 = \mu_2$. If the population assumptions (normal distributions with equal σ's) and H_0 were true, then we would know from theory and experience that thousands and thousands of $x_2 - x_1$ linked-pair subtractions would produce a population of results (called d-values, where $d = x_2 - x_1$) such that the average subtraction (or d-value) would yield 0 with standard deviation $\sigma_d \approx 2.60$ (s_d, the standard deviation of a sample of n) $= 17$ d-values was used to estimate its population standard deviation, σ_d, as shown below (calculations are given in the next column).

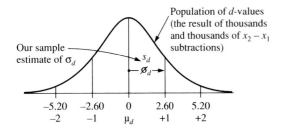

Population of d-values (the result of thousands and thousands of $x_2 - x_1$ subtractions)

Our sample estimate of σ_d

s_d

σ_d

−5.20	−2.60	0	2.60	5.20
−2	−1	μ_d	+1	+2

x_1	x_2	d	$\bar{d}$	$(d - \bar{d})$	$(d - \bar{d})^2$
9	14	5	2.5	2.5	6.25
5	7	2	2.5	−0.5	.25
8	8	0	2.5	−2.5	6.25
12	11	−1	2.5	−3.5	12.25
7	9	2	2.5	−0.5	.25
9	14	5	2.5	2.5	6.25
10	12	2	2.5	−0.5	.25
5	11	6	2.5	3.5	12.25
4	8	4	2.5	1.5	2.25
13	11	−2	2.5	−4.5	20.25
8	12	4	2.5	1.5	2.25
9	8	−1	2.5	−3.5	12.25
11	16	5	2.5	2.5	6.25
10	14	4	2.5	1.5	2.25
9	8	−1	2.5	−3.5	12.25
14	18	4	2.5	1.5	2.25
7	11.5	4.5	2.5	2.0	4.00
		42.5			108.00

To calculate $\bar{d}$,

$$\bar{d} = \frac{\Sigma d}{n}$$

$$= \frac{42.5}{17}$$

$$\boxed{\bar{d} = 2.5}$$

To calculate s_d,

$$s_d = \sqrt{\frac{\Sigma(d - \bar{d})^2}{n - 1}}$$

$$= \sqrt{\frac{108}{17 - 1}} = 2.598$$

$$\boxed{s_d = 2.60 \approx \sigma_d}$$

However, the above is a distribution of "individual" d-values. To estimate where "averages," where $\bar{d}$'s, might fall, we use the central limit theorem to calculate $\sigma_{\bar{d}}$, and conduct the hypothesis test as follows:

$$\boxed{\sigma_{\bar{d}} = \frac{\sigma_d}{\sqrt{n}} \approx \frac{s_d}{\sqrt{n}}}$$

To calculate $\sigma_{\bar{d}}$ for our example,

$$\sigma_{\bar{d}} \approx \frac{s_d}{\sqrt{n}} = \frac{2.60}{\sqrt{17}} = .63$$

$$\boxed{\sigma_{\bar{d}} \approx .63}$$

To calculate the t score of the sample,

$$\boxed{t = \frac{\bar{d} - \mu_d}{\sigma_{\bar{d}}}}$$

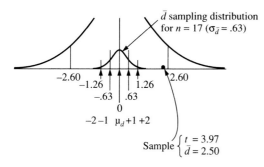

$\bar{d}$ sampling distribution for $n = 17$ ($\sigma_{\bar{d}} = .63$)

Sample $\begin{cases} t = 3.97 \\ \bar{d} = 2.50 \end{cases}$

$$\text{sample } t = \frac{2.5 - 0}{.63}$$

$$= 3.968 = 3.97$$

(which is outside the cutoff range of $t = \pm 2.12$; df = $17 - 1$, .05)

Reject H_0
$t = 3.97$
(significant)

We conclude $\mu_1 \neq \mu_2$, essentially the twins do *not* exhibit the same level of niceness in lower versus higher niceness households. A glance at the data shows higher niceness households generally rearing higher niceness twins, suggesting environment plays a significant role in shaping this personality trait.

11.7

a. Looking up $n = 10$ under the column $p = .05$ in the binomial tables, $P(0$ successes$) = .599 = 59.9\%$

b. H_0: $p = 5\%$. Looking up $n = 14$ under the column $p = .05$ in the binomial tables, $P(5$ successes$)$ = blank (less than .1%). Recall, $\alpha = .01$ for a two-tailed test implies we reject for probabilities less than approximately .5% in each tail. Since .1% < .5%, *reject H_0*. Thus we conclude $p \neq 5\%$. The data suggests recognition is greater than the hypothesized 5%, indicating the drug may have some effect on improving memory in these cases.

c. H_0: $p = 8\%$. For np, $n(1 - p) \geq 5$, such as in this example, we can use the normal curve to approximate binomial probabilities. $p = 8\%$ (.08) and $p_s = 11/72 = .153$; $\sigma_p = \sqrt{(.08)(.92)/72} = .032$; sample $z = (.153 - .080)/.032 = 2.28$ (significant). Since sample z (2.28) is outside the z cutoffs of ± 1.96 (for $\alpha = .05$), *reject H_0*. Thus we conclude $p \neq 8\%$. The data suggests recognition is greater than the hypothesized 8%.

11.9

a. H_0: $p = 80\%$. For np and $n(1 - p) \geq 5$, such as in this example, we can use the normal curve to approximate binomial probabilities. Facts: $p = 80\%$ (.80) and $p_s = 26/40 = .65$; $\sigma_p = \sqrt{(.80)(.20)/(40)}$ = .0632; sample $z = (.65 - .80)/.0632 = -2.37$ (significant).

Since sample z (-2.37) is outside the z cutoffs of ± 1.96 (for $\alpha = .05$), *reject H_0*. Thus we conclude $p \neq 80\%$. The data suggests a somewhat lower proportion than the hypothesized 80%.

b. H_0: $p \geq 80\%$. Placing the α risk in the left tail, we get a cutoff of $z = -1.88$ and reject below this value. Since our sample z (-2.37) was below this value, again *reject H_0*. Thus we conclude $p < 80\%$. The data suggests a somewhat lower proportion than 80%.

c. To get a confidence interval, we use the formula: $p_s \pm z \sigma_p = .65 \pm 1.96(.0632) = .650 \pm .124$; thus we are 95% confident that the true population proportion is contained in the interval between .526 and .774 (52.6% and 77.4%).

11.11

H_0: $p_1 = p_2$. Common population proportion, $p = (72 + 54)/(90 + 120) = .60$ (60%); $\sigma_{p_{s_1} - p_{s_2}} = \sqrt{(.60)(.40)/90 + (.60)(.40)/120} = .0683$; $p_{s_1} = 72/90 = .80$ (80%) and $p_{s_2} = .45$ (45%); sample $z = (.80 - .45)/.0683 = 5.124 = 5.12$ (significant).

Since sample z (5.12) is outside the z cutoffs of ± 1.96 (for $\alpha = .05$), reject H_0. Thus we conclude $p_1 \neq p_2$. The data suggests significant differences in the mating preferences of the hens. Roosters with long combs and bright wattles were greatly preferred, which conforms to studies. For further reading, see *New York Times* articles, "Hard-to-Please Females May Be a Neglected Evolutionary Force" (May 8, 1990, C1); "Dolphin Courtship . . . ." (February 18, 1992, C1); and "Female Adders Play Around . . . ." (February 4, 1992, C4).

11.13

a. H_0: $\mu_1 = \mu_2$. The $\bar{x}_1 - \bar{x}_2$ sampling distribution has average of 0 with $\sigma_{\bar{x}_1 - \bar{x}_2} = \sqrt{(8)^2/80 + (12)^2/(60)}$ = 1.789; sample $z = (82 - 62)/1.789 = 11.18$ (significant). Since sample z (11.18) is outside the

cutoffs of ± 1.96, reject H_0. Thus we conclude μ_1 $\neq \mu_2$, implying one group (the higher $\bar{x}$) sold significantly more units than the other.
b. H_0: $\mu_1 = \mu_2$. The $\bar{x}_1 - \bar{x}_2$ sampling distribution has average of 0 with $\sigma_{\bar{x}_1 - \bar{x}_2}$

$$= \sqrt{\frac{(9-1)(8)^2 + (11-1)(12)^2}{9 + 11 - 2} \left(\frac{1}{9} + \frac{1}{11} \right)}$$

$= 4.681$. Sample $t = (78 - 64)/4.681 = 2.99$ (significant). Since sample t (2.99) is outside the cutoffs of ± 2.10 (df $= 9 + 11 - 2 = 18$), reject H_0. Thus we conclude $\mu_1 \neq \mu_2$, implying one group (the higher $\bar{x}$) sold significantly more units than the other.

11.15

a. H_0: $\mu_1 = \mu_2$. The $\bar{x}_1 - \bar{x}_2$ sampling distribution has average of 0 with $\sigma_{\bar{x}_1 - \bar{x}_2} = \sqrt{(7.3)^2/32 + (8.1)^2/50}$ $= 1.726$. Sample $z = (15.1 - 26.7)/1.726 = -6.72$ (significant). Since sample z (-6.72) is outside the cutoffs of ± 1.96, reject H_0. Thus we conclude $\mu_1 \neq \mu_2$, implying one group (the higher $\bar{x}$) consumes significantly more brain energy (glucose) than the other group in performing the same set of thinking tasks. This conforms to the results obtained in studies (see ''Mapping the Brain,'' *Newsweek* [April 20, 1992]).

b. H_0: $\mu_1 = \mu_2$. The $\bar{x}_1 - \bar{x}_2$ sampling distribution has average of 0 with $\sigma_{\bar{x}_1 - \bar{x}_2}$

$$= \sqrt{\frac{(10-1)(6.3)^2 + (16-1)(5.8)^2}{10 + 16 - 2} \left(\frac{1}{10} + \frac{1}{16} \right)}$$

$= 2.416$. Sample $t = (14.5 - 24.5)/2.416 = -4.14$ (significant). Since sample t (-4.14) is outside the cutoffs of ± 2.06 (df $= 10 + 16 - 2 = 24$), reject H_0. Thus we conclude $\mu_1 \neq \mu_2$, implying one group (with the higher $\bar{x}$) consumes significantly more brain energy (glucose) than the other group in performing the same set of thinking tasks. See part a for reference.

11.17

H_0: $\mu_1 = \mu_2$. The $x_2 - x_1$ sampling distribution has average of 0 with $\sigma_d \approx s_d \approx .641$ (calculated as follows: $n = 12$; $\bar{d} = 6/12 = .5$; $s_d = \sqrt{4.52/(12-1)} = .641$). $\sigma_{\bar{d}} = .641/\sqrt{12} = .185$; sample $t = (.5 - 0)/.185$

$= 2.70$ (significant). Since sample t (2.70) is outside the cutoffs of ± 2.20 (df $= 12 - 1 = 11$), reject H_0. Thus we conclude $\mu_1 \neq \mu_2$, implying one group (the higher $\bar{x}$) received significantly higher GPAs than the other. An inspection of the data reveals the higher in hope group received the higher $\bar{x}$ and we might conclude that students high in hope tend to achieve better than those lower in hope, even though they entered college with nearly identical backgrounds (reference: ''Hope Emerges as Key to Success in Life,'' *New York Times* [December 24, 1991, C1]).

11.19

H_0: $P(+) = P(-) = 50\%$ (.5); $n = 12$ (ties are not counted), $\mu = 12(.5) = 6$; $\sigma = \sqrt{12(.5)(.5)} = 1.732$. Thus if H_0 were true, we would expect 6 +'s, however, we achieved 8 +'s. Sample $z = (8 - 6)/1.732 = 1.15$ (not significant). Since sample z (1.15) is inside the cutoffs of ± 1.96, we accept H_0, implying no provable difference in college GPAs between the two groups. Note this disagrees with the results obtained using the parametric test (exercise 11.17). The sign test is not as sensitive as its parametric counterpart in detecting population differences and generally larger sample sizes must be employed to compensate.

11.21

H_0: $\mu_1 = \mu_2$; $R = 1 + 4 + 5 + 8 + 12 + 15 + 16$ $+ 17 + 19 + 20 + 23 = 140$; $\mu_R = 11(11 + 12 + 1)/2$ $= 132$; $\sigma_R = \sqrt{11(12)(11 + 12 + 1)/12} = 16.248$; sample $z = (140 - 132)/16.248 = .49$ (not significant). Since sample z (.49) is within the cutoffs of ± 1.96, we accept H_0, $\mu_1 = \mu_2$, implying we have no evidence to indicate any difference in the pain-relieving effect of laboratory cultured HA versus natural HA.

11.23

H_0: $\rho = 0$; $d^2 = 49 + 25 + 9 + 1 + 1 + 25 + 16$ $+ 36 = 162$; $r_s = 1 - 6(162)/8(64 - 1) = 1 - 1.93$ $= -.93$ (significant). Since sample r_s ($-.93$) is outside the cutoffs of $\pm.74$, reject H_0. Thus we conclude $\rho \neq 0$, implying a correlation in the population. The negative sign further implies an inverse correlation, meaning as age increases, self-esteem tends to decrease.

Statistical Tables

z	0.00	0.01	0.02	0.03	0.04	0.05	0.06	0.07	0.08	0.09
Normal										
0.0	.0000	.0040	.0080	.0120	.0160	.0199	.0239	.0279	.0319	.0359
0.1	.0398	.0438	.0478	.0517	.0557	.0596	.0636	.0675	.0714	.0753
0.2	.0793	.0832	.0871	.0910	.0948	.0987	.1026	.1064	.1103	.1141
0.3	.1179	.1217	.1255	.1293	.1331	.1368	.1406	.1443	.1480	.1517
0.4	.1554	.1591	.1628	.1664	.1700	.1736	.1772	.1808	.1844	.1879
0.5	.1915	.1950	.1985	.2019	.2054	.2088	.2123	.2157	.2190	.2224
0.6	.2257	.2291	.2324	.2357	.2389	.2422	.2454	.2486	.2517	.2549
0.7	.2580	.2611	.2642	.2673	.2704	.2734	.2764	.2794	.2823	.2852
0.8	.2881	.2910	.2939	.2967	.2995	.3023	.3051	.3078	.3106	.3133
0.9	.3159	.3186	.3212	.3238	.3264	.3289	.3315	.3340	.3365	.3389
1.0	.3413	.3438	.3461	.3485	.3508	.3531	.3554	.3577	.3599	.3621
1.1	.3643	.3665	.3686	.3708	.3729	.3749	.3770	.3790	.3810	.3830
1.2	.3849	.3869	.3888	.3907	.3925	.3944	.3962	.3980	.3997	.4015
1.3	.4032	.4049	.4066	.4082	.4099	.4115	.4131	.4147	.4162	.4177
1.4	.4192	.4207	.4222	.4236	.4251	.4265	.4279	.4292	.4306	.4319
1.5	.4332	.4345	.4357	.4370	.4382	.4394	.4406	.4418	.4429	.4441
1.6	.4452	.4463	.4474	.4484	.4495	.4505	.4515	.4525	.4535	.4545
1.7	.4554	.4564	.4573	.4582	.4591	.4599	.4608	.4616	.4625	.4633
1.8	.4641	.4649	.4656	.4664	.4671	.4678	.4686	.4693	.4699	.4706
1.9	.4713	.4719	.4726	.4732	.4738	.4744	.4750	.4756	.4761	.4767
2.0	.4772	.4778	.4783	.4788	.4793	.4798	.4803	.4808	.4812	.4817
2.1	.4821	.4826	.4830	.4834	.4838	.4842	.4846	.4850	.4854	.4857
2.2	.4861	.4864	.4868	.4871	.4875	.4878	.4881	.4884	.4887	.4890
2.3	.4893	.4896	.4898	.4901	.4904	.4906	.4909	.4911	.4913	.4916
2.4	.4918	.4920	.4922	.4925	.4927	.4929	.4931	.4932	.4934	.4936
2.5	.4938	.4940	.4941	.4943	.4945	.4946	.4948	.4949	.4951	.4952
2.6	.4953	.4955	.4956	.4957	.4959	.4960	.4961	.4962	.4963	.4964
2.7	.4965	.4966	.4967	.4968	.4969	.4970	.4971	.4972	.4973	.4974
2.8	.4974	.4975	.4976	.4977	.4977	.4978	.4979	.4979	.4980	.4981
2.9	.4981	.4982	.4982	.4983	.4984	.4984	.4985	.4985	.4986	.4986
3.0	.4987	.4987	.4987	.4988	.4988	.4989	.4989	.4989	.4990	.4990
3.1	.4990	.4991	.4991	.4991	.4992	.4992	.4992	.4992	.4993	.4993
3.2	.4993	.4993	.4994	.4994	.4994	.4994	.4994	.4995	.4995	.4995
3.3	.4995	.4995	.4996	.4996	.4996	.4996	.4996	.4996	.4996	.4997
3.4	.4997	.4997	.4997	.4997	.4997	.4997	.4997	.4997	.4998	.4998
3.5	.4998	.4998	.4998	.4998	.4998	.4998	.4998	.4998	.4998	.4998

The normal curve table gives only the percentage of data starting from the middle ($z = 0$), out to whatever z score you look up. For instance, if you look up $z = 1.28$, you get .3997. This means .3997 or 39.97% of the data in the normal curve is found between $z = 0$ and $z = 1.28$.

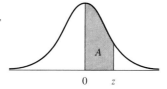

t				
Confidence Interval	90%	95%	98%	99%
Two-Tailed Hypothesis	.10	.05	.02	.01
One-Tailed Hypothesis	.05	.025	.01	.005
df				
1	6.31	12.71	31.82	63.66
2	2.92	4.30	6.97	9.92
3	2.35	3.18	4.54	5.84
4	2.13	2.78	3.75	4.60
5	2.02	2.57	3.37	4.03
6	1.94	2.45	3.14	3.71
7	1.89	2.36	3.00	3.50
8	1.86	2.31	2.90	3.36
9	1.83	2.26	2.82	3.25
10	1.81	2.23	2.76	3.17
11	1.80	2.20	2.72	3.11
12	1.78	2.18	2.68	3.05
13	1.77	2.16	2.65	3.01
14	1.76	2.14	2.62	2.98
15	1.75	2.13	2.60	2.95
16	1.75	2.12	2.58	2.92
17	1.74	2.11	2.57	2.90
18	1.73	2.10	2.55	2.88
19	1.73	2.09	2.54	2.86
20	1.72	2.09	2.53	2.85
21	1.72	2.08	2.52	2.83
22	1.72	2.07	2.51	2.82
23	1.71	2.07	2.50	2.81
24	1.71	2.06	2.49	2.80
25	1.71	2.06	2.49	2.79
26	1.71	2.06	2.48	2.78
27	1.70	2.05	2.47	2.77
28	1.70	2.05	2.47	2.76
29	1.70	2.05	2.46	2.76
30	1.70	2.04	2.46	2.75
40	1.68	2.02	2.42	2.70
50	1.68	2.01	2.40	2.68
100	1.66	1.98	2.36	2.63
∞	1.65	1.96	2.33	2.58

χ^2		
df	$\chi^2_{.05}$	$\chi^2_{.01}$
1	3.84	6.63
2	5.99	9.21
3	7.81	11.34
4	9.49	13.28
5	11.07	15.09
6	12.59	16.81
7	14.07	18.48
8	15.51	20.09
9	16.92	21.67
10	18.31	23.21
11	19.68	24.73
12	21.03	26.22
13	22.36	27.69
14	23.68	29.14
15	25.00	30.58
16	26.30	32.00
17	27.59	33.41
18	28.87	34.81
19	30.14	36.19
20	31.41	37.57
21	32.67	38.93
22	33.92	40.29
23	35.17	41.64
24	36.42	42.98
25	37.65	44.31
26	38.89	45.64
27	40.11	46.96
28	41.34	48.28
29	42.56	49.59
30	43.77	50.89

r		
df	$r_{.05}$	$r_{.01}$
1	.997	
2	.95	.99
3	.88	.96
4	.81	.92
5	.75	.87
6	.71	.83
7	.67	.80
8	.63	.76
9	.60	.73
10	.58	.71
11	.55	.68
12	.53	.66
13	.51	.64
14	.50	.62
15	.48	.61
16	.47	.59
17	.46	.58
18	.44	.56
19	.43	.55
20	.42	.54
25	.38	.49
30	.35	.45
35	.33	.42
40	.30	.39
45	.29	.37
50	.27	.35
60	.25	.33
70	.23	.30
80	.22	.28
90	.21	.27

Random Numbers

31871	60770	59235	41702	89372	28600	30013	18266	65044	61045
87134	32839	17850	37359	27221	92409	94778	17902	09467	86757
06728	16314	81076	42172	46446	09226	96262	77674	70205	98137
95646	67486	05167	07819	79918	83949	45605	18915	79458	54009
44085	87246	47378	98338	40368	02240	72593	52823	79002	88190
83967	84810	51612	81501	10440	48553	67919	73678	83149	47096
49990	02051	64575	70323	07863	59220	01746	94213	82977	42384
65332	16488	04433	37990	93517	18395	72848	97025	38894	46611
42309	04063	55291	72165	96921	53350	34173	39908	11634	87145
84715	41808	12085	72525	91171	09779	07223	75577	20934	92047
63919	83977	72416	55450	47642	01013	17560	54189	73523	33681
97595	78300	93502	25847	19520	16896	69282	16917	04194	25797
17116	42649	89252	61052	78332	15102	47707	28369	60400	15908
34037	84573	49914	59688	18584	53498	94905	14914	23261	58133
08813	14453	70437	49093	69880	99944	40482	04254	62842	68089
67115	41050	65453	04510	35518	88843	15801	86163	49913	46849
14596	62802	33009	74095	34549	76634	64270	67491	83941	55154
70258	26948	60863	47666	58512	91404	97357	85710	03414	56591
83369	81179	32429	34781	00006	65951	40254	71102	60416	43296
83811	49358	75171	34768	70070	76550	14252	97378	79500	97123
14924	71607	74638	01939	77044	18277	68229	09310	63258	85064
60102	56587	29842	12031	00794	90638	21862	72154	19880	80895
33393	30109	42005	47977	26453	15333	45390	89862	70351	36953
92592	78232	19328	29645	69836	91169	95180	15046	45679	94500
27421	73356	53897	26916	52015	26854	42833	64257	49423	39440
26528	22550	36692	25262	61419	53986	73898	80237	71387	32532
07664	10752	95021	17030	76784	86861	12780	44379	31261	18424
37954	72029	29624	09119	13444	22645	78345	79876	37582	75549
66495	11333	81101	69328	84838	76395	35997	07259	66254	47451
72506	28524	39595	49356	92733	42951	47774	75462	64409	69116
09713	70270	28077	15634	36525	91204	48443	50561	92394	60636
51852	70782	93498	44669	79647	06321	04020	00111	24737	05521
31460	22222	18801	00675	57562	97923	45974	75158	94918	40144
14328	05024	04333	04135	53143	79207	85863	04962	89549	63308
84002	98073	52998	05749	45538	26164	68672	97486	32341	99419
89541	28345	22887	79269	55620	68269	88765	72464	11586	52211
50502	39890	81465	00449	09931	12667	30278	63963	84192	25266
30862	61996	73216	12554	01200	63234	41277	20477	71899	05347
36735	58841	35287	51112	47322	81354	51080	72771	53653	42108
11561	81204	68175	93037	47967	74085	05905	86471	47671	18456

$F_{.05}$								
df	1	2	3	4	5	6	7	8	9
1	161.4	199.5	215.7	224.6	230.2	234.0	236.8	238.9	240.5
2	18.51	19.00	19.16	19.25	19.30	19.33	19.35	19.37	19.38
3	10.13	9.55	9.28	9.12	9.01	8.94	8.89	8.85	8.81
4	7.71	6.94	6.59	6.39	6.26	6.16	6.09	6.04	6.00
5	6.61	5.79	5.41	5.19	5.50	4.95	4.88	4.82	4.77
6	5.99	5.14	4.76	4.53	4.39	4.28	4.21	4.15	4.10
7	5.59	4.74	4.35	4.12	3.97	3.87	3.79	3.73	3.68
8	5.32	4.46	4.07	3.84	3.69	3.58	3.50	3.44	3.39
9	5.12	4.26	3.86	3.63	3.48	3.37	3.29	3.23	3.18
10	4.96	4.10	3.71	3.48	3.33	3.22	3.14	3.07	3.02
11	4.84	3.98	3.59	3.36	3.20	3.09	3.01	2.95	2.90
12	4.75	3.89	3.49	3.26	3.11	3.00	2.91	2.85	2.80
13	4.67	3.81	3.41	3.18	3.03	2.92	2.83	2.77	2.71
14	4.60	3.74	3.34	3.11	2.96	2.85	2.76	2.70	2.65
15	4.54	3.68	3.29	3.06	2.90	2.79	2.71	2.64	2.59
16	4.49	3.63	3.24	3.01	2.85	2.74	2.66	2.59	2.54
17	4.45	3.59	3.20	2.96	2.81	2.70	2.61	2.55	2.49
18	4.41	3.55	3.16	2.93	2.77	2.66	2.58	2.51	2.46
19	4.38	3.52	3.13	2.90	2.74	2.63	2.54	2.48	2.42
20	4.35	3.49	3.10	2.87	2.71	2.60	2.51	2.45	2.39
21	4.32	3.47	3.07	2.84	2.68	2.57	2.49	2.42	2.37
22	4.30	3.44	3.05	2.82	2.66	2.55	2.46	2.40	2.34
23	4.28	3.42	3.03	2.80	2.64	2.53	2.44	2.37	2.32
24	4.26	3.40	3.01	2.78	2.62	2.51	2.42	2.36	2.30
25	4.24	3.39	2.99	2.76	2.60	2.49	2.40	2.34	2.28
26	4.23	3.37	2.98	2.74	2.59	2.47	2.39	2.32	2.27
27	4.21	3.35	2.96	2.73	2.57	2.46	2.37	2.31	2.25
28	4.20	3.34	2.95	2.71	2.56	2.45	2.36	2.29	2.24
29	4.18	3.33	2.93	2.70	2.55	2.43	2.35	2.28	2.22
30	4.17	3.32	2.92	2.69	2.53	2.42	2.33	2.27	2.21
40	4.08	3.23	2.84	2.61	2.45	2.34	2.25	2.18	2.12
60	4.00	3.15	2.76	2.53	2.37	2.25	2.17	2.10	2.04
120	3.92	3.07	2.68	2.45	2.29	2.17	2.09	2.02	1.96
∞	3.84	3.00	2.60	2.37	2.21	2.10	2.01	1.94	1.88

$F_{.01}$

df	1	2	3	4	5	6	7	8	9
1	4052	4999.5	5403	5625	5764	5859	5928	5981	6022
2	98.50	99.00	99.17	99.25	99.30	99.33	99.36	99.37	99.39
3	34.12	30.82	29.46	28.71	28.24	27.91	27.67	27.49	27.35
4	21.20	18.00	16.69	15.98	15.52	15.21	14.98	14.80	14.66
5	16.26	13.27	12.06	11.39	10.97	10.67	10.46	10.29	10.16
6	13.75	10.92	9.78	9.15	8.75	8.47	8.26	8.10	7.98
7	12.25	9.55	8.45	7.85	7.46	7.19	6.99	6.84	6.72
8	11.26	8.65	7.59	7.01	6.63	6.37	6.18	6.03	5.91
9	10.56	8.02	6.99	6.42	6.06	5.80	5.61	5.47	5.35
10	10.04	7.56	6.55	5.99	5.64	5.39	5.20	5.06	4.94
11	9.65	7.21	6.22	5.67	5.32	5.07	4.89	4.74	4.63
12	9.33	6.93	5.95	5.41	5.06	4.82	4.64	4.50	4.39
13	9.07	6.70	5.74	5.21	4.86	4.62	4.44	4.30	4.19
14	8.86	6.51	5.56	5.04	4.69	4.46	4.28	4.14	4.03
15	8.68	6.36	5.42	4.89	4.56	4.32	4.14	4.00	3.89
16	8.53	6.23	5.29	4.77	4.44	4.20	4.03	3.89	3.78
17	8.40	6.11	5.18	4.67	4.34	4.10	3.93	3.79	3.68
18	8.29	6.01	5.09	4.58	4.25	4.01	3.84	3.71	3.60
19	8.19	5.93	5.01	4.50	4.17	3.94	3.77	3.63	3.52
20	8.10	5.85	4.94	4.43	4.10	3.87	3.70	3.56	3.46
21	8.02	5.78	4.87	4.37	4.04	3.81	3.64	3.51	3.40
22	7.95	5.72	4.82	4.31	3.99	3.76	3.59	3.45	3.35
23	7.88	5.66	4.76	4.26	3.94	3.71	3.54	3.41	3.30
24	7.82	5.61	4.72	4.22	3.90	3.67	3.50	3.36	3.26
25	7.77	5.57	4.68	4.18	3.85	3.63	3.46	3.32	3.22
26	7.72	5.53	4.64	4.14	3.82	3.59	3.42	3.29	3.18
27	7.68	5.49	4.60	4.11	3.78	3.56	3.39	3.26	3.15
28	7.64	5.45	4.57	4.07	3.75	3.53	3.36	3.23	3.12
29	7.60	5.42	4.54	4.04	3.73	3.50	3.33	3.20	3.09
30	7.56	5.39	4.51	4.02	3.70	3.47	3.30	3.17	3.07
40	7.31	5.18	4.31	3.83	3.51	3.29	3.12	2.99	2.89
60	7.08	4.98	4.13	3.65	3.34	3.12	2.95	2.82	2.72
120	6.85	4.79	3.95	3.48	3.17	2.96	2.79	2.66	2.56
∞	6.63	4.61	3.78	3.32	3.02	2.80	2.64	2.51	2.41

Spearman's r_s				
Two-Tailed Hypothesis	.10	.05	.02	.01
One-Tailed Hypothesis	.05	.025	.01	.005
n				
5	.90	—	—	—
6	.83	.89	.94	—
7	.71	.79	.89	—
8	.64	.74	.83	.88
9	.60	.68	.78	.83
10	.56	.65	.75	.79
11	.52	.62	.74	.82
12	.50	.59	.70	.78
13	.48	.57	.67	.75
14	.46	.55	.65	.72
15	.44	.53	.62	.69
16	.43	.51	.60	.67
17	.41	.49	.58	.65
18	.40	.48	.56	.63
19	.39	.46	.55	.61
20	.38	.45	.53	.59
21	.37	.44	.52	.58
22	.36	.43	.51	.56
23	.35	.42	.50	.55
24	.34	.41	.49	.54
25	.34	.40	.48	.53
26	.33	.39	.47	.52
27	.32	.39	.46	.51
28	.32	.38	.45	.50
29	.31	.37	.44	.49
30	.31	.36	.43	.48

For $n > 30$, the sampling distribution is normally distributed with mean, $\mu_s = 0$ and standard deviation, $\sigma_s = 1/\sqrt{n-1}$. Thus, for $n > 30$, r_s(cutoff) $= \pm z/\sqrt{n-1}$. For example, with $n = 50$ at $\alpha = .05$, r_s(cutoff) $= \pm 1.96/\sqrt{50-1} = \pm.28$.

Binomial Tables

n	s	.05	.1	.15	.2	.3	.4	.5
2	0	.902	.810	.723	.640	.490	.360	.250
	1	.095	.180	.255	.320	.420	.480	.500
	2	.002	.010	.023	.040	.090	.160	.250
3	0	.857	.729	.614	.512	.343	.216	.125
	1	.135	.243	.325	.384	.441	.432	.375
	2	.007	.027	.057	.096	.189	.288	.375
	3		.001	.003	.008	.027	.064	.125
4	0	.815	.656	.522	.410	.240	.130	.062
	1	.171	.292	.368	.410	.412	.346	.250
	2	.014	.049	.098	.154	.265	.346	.375
	3		.004	.011	.026	.076	.154	.250
	4			.001	.002	.008	.026	.062
5	0	.774	.590	.444	.328	.168	.078	.031
	1	.204	.328	.392	.410	.360	.259	.156
	2	.021	.073	.138	.205	.309	.346	.312
	3	.001	.008	.024	.051	.132	.230	.312
	4			.002	.006	.028	.077	.156
	5					.002	.010	.031
6	0	.735	.531	.377	.262	.118	.047	.016
	1	.232	.354	.399	.393	.303	.187	.094
	2	.031	.098	.176	.246	.324	.311	.234
	3	.002	.015	.042	.082	.185	.276	.312
	4		.001	.006	.015	.060	.138	.234
	5				.002	.010	.037	.094
	6					.001	.004	.016
7	0	.698	.478	.321	.210	.082	.028	.008
	1	.257	.372	.396	.367	.247	.131	.055
	2	.041	.124	.210	.275	.318	.261	.164
	3	.004	.023	.062	.115	.227	.290	.273
	4		.003	.011	.029	.097	.194	.273
	5			.001	.004	.025	.077	.164
	6					.004	.017	.055
	7						.002	.008
8	0	.663	.430	.272	.168	.058	.017	.004
	1	.279	.383	.385	.336	.198	.090	.031
	2	.051	.149	.238	.294	.296	.209	.109
	3	.005	.033	.084	.147	.254	.279	.219
	4		.005	.018	.046	.136	.232	.273
	5			.003	.009	.047	.124	.219
	6				.001	.010	.041	.109
	7					.001	.008	.031
	8						.001	.004

n	s	.05	.1	.15	.2	.3	.4	.5
9	0	.630	.387	.232	.134	.040	.010	.002
	1	.299	.387	.368	.302	.156	.060	.018
	2	.063	.172	.260	.302	.267	.161	.070
	3	.008	.045	.107	.176	.267	.251	.164
	4	.001	.007	.028	.066	.172	.251	.246
	5		.001	.005	.017	.074	.167	.246
	6			.001	.003	.021	.074	.164
	7					.004	.021	.070
	8						.004	.018
	9							.002
10	0	.599	.349	.197	.107	.028	.006	.001
	1	.315	.387	.347	.268	.121	.040	.010
	2	.075	.194	.276	.302	.233	.121	.044
	3	.010	.057	.130	.201	.267	.215	.117
	4	.001	.011	.040	.088	.200	.251	.205
	5		.001	.008	.026	.103	.201	.246
	6			.001	.006	.037	.111	.205
	7				.001	.009	.042	.117
	8					.001	.011	.044
	9						.002	.010
	10							.001
11	0	.569	.314	.167	.086	.020	.004	
	1	.329	.384	.325	.236	.093	.027	.005
	2	.087	.213	.287	.295	.200	.089	.027
	3	.014	.071	.152	.221	.257	.177	.081
	4	.001	.016	.054	.111	.220	.236	.161
	5		.002	.013	.039	.132	.221	.226
	6			.002	.010	.057	.147	.226
	7				.002	.017	.070	.161
	8					.004	.023	.081
	9					.001	.005	.027
	10						.001	.005
	11							
12	0	.540	.282	.142	.069	.014	.002	
	1	.341	.377	.301	.206	.071	.017	.003
	2	.099	.230	.292	.283	.168	.064	.016
	3	.017	.085	.172	.236	.240	.142	.054
	4	.002	.021	.068	.133	.231	.213	.121
	5		.004	.019	.053	.158	.227	.193
	6			.004	.016	.079	.177	.226
	7			.001	.003	.029	.101	.193
	8				.001	.008	.042	.121
	9					.001	.012	.054
	10						.002	.016
	11							.003
	12							

1. This table gives the probability of achieving s successes in n selections for various values of p (the probability of success on one selection).
2. Blank spaces indicate probabilities less than .001 (.1%).
3. To determine binomial probabilities for p = .6, .7, .8, .85, .9, and .95, we take advantage of certain similarities in the table. We look up the same n, then look in the (1 − p) column for (n − s) successes. For instance,

    For n = 10   We instead look up,   n = 10
        p = .8   p = .2 (1 − .8)
        s = 4   s = 6 (10 − 4)

Both have the same binomial probability of .006.

Binomial Tables (continued)

n	s	p (the probability of success on one selection)						
		.05	.1	.15	.2	.3	.4	.5
13	0	.513	.254	.121	.055	.010	.001	
	1	.351	.367	.277	.179	.054	.011	.002
	2	.111	.245	.294	.268	.139	.045	.010
	3	.021	.100	.190	.246	.218	.111	.035
	4	.003	.028	.084	.154	.234	.184	.087
	5		.006	.027	.069	.180	.221	.157
	6		.001	.006	.023	.103	.197	.209
	7			.001	.006	.044	.131	.209
	8				.001	.014	.066	.157
	9					.003	.024	.087
	10					.001	.006	.035
	11						.001	.010
	12							.002
	13							
14	0	.488	.229	.103	.044	.007	.001	
	1	.359	.356	.254	.154	.041	.007	.001
	2	.123	.257	.291	.250	.113	.032	.006
	3	.026	.114	.206	.250	.194	.085	.022
	4	.004	.035	.100	.172	.229	.155	.061
	5		.008	.035	.086	.196	.207	.122
	6		.001	.009	.032	.126	.207	.183
	7			.002	.009	.062	.157	.209
	8				.002	.023	.092	.183
	9					.007	.041	.122
	10					.001	.014	.061
	11						.003	.022
	12						.001	.006
	13							.001
	14							
15	0	.463	.206	.087	.035	.005		
	1	.366	.343	.231	.132	.031	.005	
	2	.135	.267	.286	.231	.092	.022	.003
	3	.031	.129	.218	.250	.170	.063	.014
	4	.005	.043	.116	.188	.219	.127	.042
	5	.001	.010	.045	.103	.206	.186	.092
	6		.002	.013	.043	.147	.207	.153
	7			.003	.014	.081	.177	.196
	8			.001	.003	.035	.118	.196
	9				.001	.012	.061	.153
	10					.003	.024	.092
	11					.001	.007	.042
	12						.002	.014
	13							.003
	14							
	15							

n	s	p (the probability of success on one selection)						
		.05	.1	.15	.2	.3	.4	.5
16	0	.440	.185	.074	.028	.003	.001	
	1	.371	.329	.210	.113	.023	.003	
	2	.146	.275	.277	.211	.073	.015	.002
	3	.036	.142	.229	.246	.146	.047	.009
	4	.006	.051	.131	.200	.204	.155	.028
	5	.001	.014	.056	.120	.210	.162	.067
	6		.003	.018	.055	.165	.198	.122
	7			.005	.020	.101	.189	.175
	8			.001	.006	.049	.142	.196
	9				.001	.019	.084	.175
	10					.006	.039	.122
	11					.001	.014	.067
	12						.004	.028
	13						.001	.009
	14							.002
	15							
	16							
18	0	.397	.150	.054	.018	.002		
	1	.376	.300	.170	.081	.013	.001	
	2	.168	.284	.256	.172	.046	.007	.001
	3	.047	.168	.241	.230	.105	.025	.003
	4	.009	.070	.159	.215	.168	.061	.012
	5	.001	.022	.079	.151	.202	.115	.033
	6		.005	.030	.082	.187	.166	.071
	7		.001	.009	.035	.138	.189	.121
	8			.002	.012	.081	.173	.167
	9				.003	.039	.128	.185
	10				.001	.015	.077	.167
	11					.005	.037	.121
	12					.001	.015	.071
	13						.004	.033
	14						.001	.012
	15							.003
	16							.001
	17							
	18							
20	0	.358	.122	.039	.012	.001		
	1	.377	.270	.137	.058	.007		
	2	.189	.285	.229	.137	.028	.003	
	3	.060	.190	.243	.205	.072	.012	.001
	4	.013	.090	.182	.218	.130	.035	.005
	5	.002	.032	.103	.175	.179	.075	.015
	6		.009	.045	.109	.192	.124	.037
	7		.002	.016	.055	.164	.166	.074
	8			.005	.022	.114	.180	.120
	9			.001	.007	.065	.160	.160
	10				.002	.031	.117	.176
	11					.012	.071	.160
	12					.004	.035	.120
	13					.001	.015	.074
	14						.005	.037
	15						.001	.015
	16							.005
	17							.001
	18–20							

Index